VOLUME ONE HUNDRED AND FIFTY SIX

Advances in
CANCER RESEARCH

Hepatobiliary Cancers: Translational Advances and Molecular Medicine

VOLUME ONE HUNDRED AND FIFTY SIX

ADVANCES IN
CANCER RESEARCH

Hepatobiliary Cancers: Translational Advances and Molecular Medicine

Edited by

ALPHONSE E. SIRICA
Emeritus Professor of Pathology and Distinguished Career Professor; Department of Pathology,
Virginia Commonwealth University School of Medicine,
Richmond, VA, United States

PAUL B. FISHER
Professor and Chairman,
Department of Human and Molecular Genetics,
VCU Institute of Molecular Medicine and VCU Massey Cancer School of Medicine Center, Virginia Commonwealth University, VA, United States

ACADEMIC PRESS
An imprint of Elsevier

Academic Press is an imprint of Elsevier
50 Hampshire Street, 5th Floor, Cambridge, MA 02139, United States
525 B Street, Suite 1650, San Diego, CA 92101, United States
The Boulevard, Langford Lane, Kidlington, Oxford OX5 1GB, United Kingdom
125 London Wall, London, EC2Y 5AS, United Kingdom

First edition 2022

Copyright © 2022 Elsevier Inc. All rights reserved.

No part of this publication may be reproduced or transmitted in any form or by any means, electronic or mechanical, including photocopying, recording, or any information storage and retrieval system, without permission in writing from the publisher. Details on how to seek permission, further information about the Publisher's permissions policies and our arrangements with organizations such as the Copyright Clearance Center and the Copyright Licensing Agency, can be found at our website: www.elsevier.com/permissions.

This book and the individual contributions contained in it are protected under copyright by the Publisher (other than as may be noted herein).

Notices
Knowledge and best practice in this field are constantly changing. As new research and experience broaden our understanding, changes in research methods, professional practices, or medical treatment may become necessary.

Practitioners and researchers must always rely on their own experience and knowledge in evaluating and using any information, methods, compounds, or experiments described herein. In using such information or methods they should be mindful of their own safety and the safety of others, including parties for whom they have a professional responsibility.

To the fullest extent of the law, neither the Publisher nor the authors, contributors, or editors, assume any liability for any injury and/or damage to persons or property as a matter of products liability, negligence or otherwise, or from any use or operation of any methods, products, instructions, or ideas contained in the material herein.

ISBN: 978-0-323-98392-1
ISSN: 0065-230X

> For information on all Academic Press publications
> visit our website at https://www.elsevier.com/books-and-journals

Publisher: Zoe Kruze
Developmental Editor: Naiza Ermin Mendoza
Production Project Manager: James Selvam
Cover Designer: Vicky Pearson Esser

Typeset by STRAIVE, India

Contents

Contributors   *xi*
Preface   *xv*

1. **Liver cancer risk-predictive molecular biomarkers specific to clinico-epidemiological contexts**   1
   Naoto Kubota, Naoto Fujiwara, and Yujin Hoshida

   1. Introduction   3
   2. Clinico-epidemiological contexts relevant to liver cancer risk   5
   3. Molecular indicators of high-risk liver   13
   4. Host and systemic factors associated with liver cancer risk   20
   5. Clinical implementation of molecular liver cancer risk assessment   23
   6. Conclusion   24
   Acknowledgments   25
   Conflict of interest   25
   References   25

2. **Inflammatory pathways and cholangiocarcinoma risk mechanisms and prevention**   39
   Massimiliano Cadamuro and Mario Strazzabosco

   1. Introduction   41
   2. Diseases prodromal to cholangiocarcinoma development   45
   3. Mechanisms of neoplastic transformation   53
   4. Role of inflammatory cells in modulating CCA malignant features   57
   5. Conclusions   62
   Grant support   62
   Conflict of interest statement   62
   References   62

3. **Causes and functional intricacies of inter- and intratumor heterogeneity of primary liver cancers**   75
   Subreen A. Khatib and Xin Wei Wang

   1. Introduction   77
   2. Liver cancer heterogeneity   83
   3. Spatial architecture and tumor heterogeneity   92

4. Concluding remarks and future perspectives	96
Acknowledgments	97
Author contributions	98
Declaration of interests	98
References	98

4. Implications of genetic heterogeneity in hepatocellular cancer — 103
Akanksha Suresh and Renumathy Dhanasekaran

1. Introduction	104
2. Approaches to studying heterogeneity in HCC	106
3. Molecular heterogeneity in HCC	110
4. Clinical implications of genetic heterogeneity in HCC	118
5. Therapeutic implications of genetic heterogeneity in HCC	122
6. Current perspectives and future directions	125
Acknowledgements	127
Grant support	127
Conflict of interest statement	128
References	128

5. Understanding the genetic basis for cholangiocarcinoma — 137
Mikayla A. Schmidt and Lewis R. Roberts

1. Introduction	138
2. Germline mutations in risk factors associated with cholangiocarcinoma	141
3. Overview of germline mutations and variants associated with cholangiocarcinoma	146
4. Key candidate gene studies of CCA risk variants	147
5. Pathogenic germline alteration (PGA) sequencing studies	154
6. Conclusion	160
Conflict of interest statement	160
Grant support	160
References	160

6. Novel insights into molecular and immune subtypes of biliary tract cancers — 167
Emily R Bramel and Daniela Sia

1. Introduction	170
2. The mutational landscape of biliary tract cancers	173
3. Multi-omics classifications of intrahepatic cholangicoarcinoma	179
4. Molecular classifications of other biliary tract cancers	185

5. Therapeutic implications	189
6. Conclusions and future perspectives	192
Acknowledgments	193
Grant support	193
Conflict of interest statement	193
References	193

7. Cancer-associated fibroblasts in intrahepatic cholangiocarcinoma progression and therapeutic resistance — 201
Aashreya Ravichandra, Sonakshi Bhattacharjee, and Silvia Affò

1. Introduction	202
2. Tumor microenvironment in iCCA	203
3. Cancer-associated fibroblasts in iCCA	204
4. Role of cancer-associated fibroblasts in iCCA	208
5. Therapeutic relevance of cancer associated fibroblasts in iCCA	216
6. Current and future perspectives	218
Acknowledgment	218
Grant support	218
Conflict of interest	218
References	219

8. Mechanisms and clinical significance of TGF-β in hepatocellular cancer progression — 227
Sobia Zaidi, Nancy R. Gough, and Lopa Mishra

1. Introduction	228
2. Overview of the TGF-β pathway	231
3. TGF-β signaling in liver homeostasis	233
4. Mouse models for TGF-β-Smad signaling in HCC development and progression	233
5. TGF-β pathway activity in HCC patients	236
Grant support	242
Conflict of interest statement	242
References	242

9. Matricellular proteins in intrahepatic cholangiocarcinoma — 249
Alphonse E. Sirica

1. Introduction	250
2. Overview of classes, structures, and complexities of matricellular proteins aberrantly expressed in iCCA	251

	3. Matricellular proteins as modulators of iCCA microenvironment and malignant progression	254
	4. Clinical relevance of matricellular proteins in iCCA	260
	5. Perspectives and conclusions	272
	Grant support	273
	Conflict of interest statement	273
	References	273

10. YAP1 activation and Hippo pathway signaling in the pathogenesis and treatment of intrahepatic cholangiocarcinoma — 283

Sungjin Ko, Minwook Kim, Laura Molina, Alphonse E. Sirica, and Satdarshan P. Monga

1. Introduction	286
2. Overview of Hippo-YAP1 signaling pathway	287
3. Biologic function and pathologic impact of Hippo-YAP1 pathway in ICCA	293
4. Clinical relevance and therapeutic potential of modulation of Hippo-YAP1 pathway in ICCA	302
5. Concluding remarks	308
Acknowledgments	308
Conflict of interest statement	308
References	308

11. Patient-derived functional organoids as a personalized approach for drug screening against hepatobiliary cancers — 319

Ling Li and Florin M. Selaru

1. Introduction	320
2. Hepatobiliary cancer treatment: Challenges	321
3. Tumor organoids for personalized drug screening	322
4. Conclusion and future perspectives	334
Grant support	335
Conflict of interest statement	273
References	335

12. Molecular therapeutic targets for cholangiocarcinoma: Present challenges and future possibilities — 343

Dan Høgdall, Colm J. O'Rourke, and Jesper B. Andersen

1. Introduction	345
2. Genomic alterations as therapeutic targets in CCA	348

3. Non-genomic alterations as therapeutic targets in CCA — 350
4. Present challenges and future possibilities for therapeutic targets in CCA — 355
5. Predictive Biomarkers & Treatment Resistance — 357
6. Current & future perspectives — 360
Acknowledgment — 360
Conflict of interest statement — 360
References — 361

13. Immunotherapies for hepatocellular carcinoma and intrahepatic cholangiocarcinoma: Current and developing strategies — 367
Josepmaria Argemi, Mariano Ponz-Sarvise, and Bruno Sangro

1. The immune biology of the liver — 369
2. The immunosuppressive environment of hepatocellular carcinoma — 370
3. Immunotherapy of HCC — 383
4. Immunotherapy of intrahepatic cholangiocarcinoma — 395
5. Concluding remarks — 398
Grant support — 398
Conflict of interest — 398
References — 398

14. Immunotherapy for hepatobiliary cancers: Emerging targets and translational advances — 415
Dan Li, Shaoli Lin, Jessica Hong, and Mitchell Ho

1. Introduction — 417
2. Emerging targets for liver cancer immunotherapy — 422
3. The role of GPC3 in HCC — 426
4. The role of mesothelin in iCCA — 433
5. Conclusion and future perspectives — 440
Acknowledgments — 440
Conflict of interest statement — 440
Grant support — 441
References — 441

Contributors

Silvia Affò
Institut d'Investigacions Biomèdiques August Pi i Sunyer, Barcelona, Spain

Jesper B. Andersen
Biotech Research and Innovation Centre (BRIC), Department of Health and Medical Sciences, University of Copenhagen, Copenhagen, Denmark

Josepmaria Argemi
HPB Oncology Area, Clinica Universidad de Navarra-CCUN; Hepatology Program, Centro de Investigacion Medica Aplicada (CIMA), Pamplona; Centro de Investigacion Biomedica en Red de Enfermedades Hepaticas y Digestivas (CIBERehd), Madrid, Spain

Sonakshi Bhattacharjee
Klinikum rechts der Isar, Technical University of Munich, Munich, Germany

Emily R Bramel
Division of Liver Diseases, Liver Cancer Program, Department of Medicine, Tisch Cancer Institute, Icahn School of Medicine at Mount Sinai, New York, NY, United States

Massimiliano Cadamuro
Department of Molecular Medicine (DMM), University of Padova, Padova, Italy

Renumathy Dhanasekaran
Division of Gastroenterology and Hepatology, Stanford University, Stanford, CA, United States

Naoto Fujiwara
Liver Tumor Translational Research Program, Simmons Comprehensive Cancer Center, Division of Digestive and Liver Diseases, Department of Internal Medicine, University of Texas Southwestern Medical Center, Dallas, TX, United States; Department of Gastroenterology, Graduate School of Medicine, The University of Tokyo, Tokyo, Japan

Nancy R. Gough
The Institute for Bioelectronic Medicine, Feinstein Institutes for Medical Research, Manhasset; Department of Medicine, Division of Gastroenterology and Hepatology, Northwell Health, New Hyde Park, NY, United States

Mitchell Ho
Laboratory of Molecular Biology, Center for Cancer Research, National Cancer Institute, Bethesda, MD, United States

Dan Høgdall
Biotech Research and Innovation Centre (BRIC), Department of Health and Medical Sciences, University of Copenhagen; Department of Oncology, Herlev and Gentofte Hospital, Herlev, Copenhagen University Hospital, Copenhagen, Denmark

Jessica Hong
Laboratory of Molecular Biology, Center for Cancer Research, National Cancer Institute, Bethesda, MD, United States

Yujin Hoshida
Liver Tumor Translational Research Program, Simmons Comprehensive Cancer Center, Division of Digestive and Liver Diseases, Department of Internal Medicine, University of Texas Southwestern Medical Center, Dallas, TX, United States

Subreen A. Khatib
Laboratory of Human Carcinogenesis, Center for Cancer Research, National Cancer Institute, Bethesda, MD; Department of Tumor Biology, Lombardi Comprehensive Cancer Center, Georgetown University Medical Center, Washington, DC, United States

Minwook Kim
Department of Developmental Biology, University of Pittsburgh School of Medicine, Pittsburgh, PA, United States

Sungjin Ko
Division of Experimental Pathology, Department of Pathology, University of Pittsburgh School of Medicine; Pittsburgh Liver Research Center, Pittsburgh, PA, United States

Naoto Kubota
Liver Tumor Translational Research Program, Simmons Comprehensive Cancer Center, Division of Digestive and Liver Diseases, Department of Internal Medicine, University of Texas Southwestern Medical Center, Dallas, TX, United States

Dan Li
Laboratory of Molecular Biology, Center for Cancer Research, National Cancer Institute, Bethesda, MD, United States

Ling Li
Division of Gastroenterology and Hepatology, School of Medicine, The Johns Hopkins University, Baltimore, MD, United States

Shaoli Lin
Laboratory of Molecular Biology, Center for Cancer Research, National Cancer Institute, Bethesda, MD, United States

Lopa Mishra
The Institute for Bioelectronic Medicine, Feinstein Institutes for Medical Research, Manhasset; Department of Medicine, Division of Gastroenterology and Hepatology, Northwell Health, New Hyde Park, NY; Center for Translational Medicine, Department of Surgery, The George Washington University, Washington, DC, United States

Laura Molina
Division of Experimental Pathology, Department of Pathology, University of Pittsburgh School of Medicine; Pittsburgh Liver Research Center, Pittsburgh, PA, United States

Satdarshan P. Monga
Division of Experimental Pathology, Department of Pathology, University of Pittsburgh School of Medicine; Pittsburgh Liver Research Center; Division of Gastroenterology, Hepatology, and Nutrition, University of Pittsburgh and UPMC, Pittsburgh, PA, United States

Colm J. O'Rourke
Biotech Research and Innovation Centre (BRIC), Department of Health and Medical Sciences, University of Copenhagen, Copenhagen, Denmark

Mariano Ponz-Sarvise
HPB Oncology Area, Clinica Universidad de Navarra-CCUN, Pamplona, Spain

Aashreya Ravichandra
Klinikum rechts der Isar, Technical University of Munich, Munich, Germany

Lewis R. Roberts
Division of Gastroenterology and Hepatology, Mayo Clinic College of Medicine and Science, Rochester, MN, United States

Bruno Sangro
HPB Oncology Area, Clinica Universidad de Navarra-CCUN; Hepatology Program, Centro de Investigacion Medica Aplicada (CIMA), Pamplona; Centro de Investigacion Biomedica en Red de Enfermedades Hepaticas y Digestivas (CIBERehd), Madrid, Spain

Mikayla A. Schmidt
Division of Gastroenterology and Hepatology, Mayo Clinic College of Medicine and Science, Rochester, MN, United States

Florin M. Selaru
Division of Gastroenterology and Hepatology, School of Medicine, The Johns Hopkins University, Baltimore, MD, United States

Daniela Sia
Division of Liver Diseases, Liver Cancer Program, Department of Medicine, Tisch Cancer Institute, Icahn School of Medicine at Mount Sinai, New York, NY, United States

Alphonse E. Sirica
Emeritus Professor of Pathology; Department of Pathology, Virginia Commonwealth University School of Medicine, Richmond, VA, United States

Mario Strazzabosco
Liver Center, Department of Internal Medicine, Yale University, New Haven, CT, United States

Akanksha Suresh
Division of Gastroenterology and Hepatology, Stanford University, Stanford, CA, United States

Xin Wei Wang
Laboratory of Human Carcinogenesis; Liver Cancer Program, Center for Cancer Research, National Cancer Institute, Bethesda, MD, United States

Sobia Zaidi
The Institute for Bioelectronic Medicine, Feinstein Institutes for Medical Research, Manhasset; Cold Spring Harbor Laboratory, Cold Spring Harbor; Department of Medicine, Division of Gastroenterology and Hepatology, Northwell Health, New Hyde Park, NY, United States

Preface

Primary liver cancers, which include hepatocellular carcinoma (HCC) and intrahepatic cholangiocarcinoma (iCCA), represent a heterogeneous class of highly lethal hepatobiliary cancers that notably over the last 10–20 years have been the focus of increasing concern, largely due to worldwide increases in their incidence and accompanying high mortality rates. Globally, HCC (∼75% of all cases) combined with iCCA (∼10%–15% of all cases) currently represents the sixth most commonly diagnosed cancer and the third leading cause of cancer deaths worldwide. Based on these considerations, there is a continuing and rapidly growing interest among clinicians, investigators, and patient advocates seeking greater mechanistic insights and novel biomarker-driven targeted approaches for more effectively personalizing the therapy and/or preventing these aggressive epithelial cancers.

The vast majority of hepatobiliary cancers are diagnosed at an advanced stage where treatment options are limited and patient survival outcomes are poor. Moreover, the biological and therapeutic challenges posed by HCC and iCCA as well as other biliary tract cancers are daunting, emphasizing an urgent need to review and assess ongoing and evolving basic, translational, and clinical research focused on addressing the critical obstacles that continue to limit progress toward achieving significant improvements in their clinical management and improving patient survival outcomes. Toward these goals, this thematic edition of *Advances in Cancer Research* is focused on providing comprehensive, authoritative, and timely reviews encompassing a range of basic, translational, and clinical research topics having both scientific significance and clinical relevance for HCC and iCCA, as well as additional biliary tract cancers.

The corresponding authors contributing to this volume are internationally recognized experts in the field of hepatobiliary cancers and their coauthors include a diverse group of established as well as highly promising early career investigators actively engaged in cutting-edge liver cancer research. The chapters herein provide a detailed, integrated, and balanced assessment of ongoing basic, translational, and clinical research, covering such important topics as HCC and iCCA risk mechanisms and clinical context-specific biomarkers [Chapter 1, Kubota, N. and Hoshida, Y.[a];

---

[a] Listed authors in the preface represent the first author followed by the corresponding author or senior author (Chapters 1–8 and 10–14) and sole author (Chapter 9).

Chapter 2, Cadamuro, M. and Strazzabosco, M.]; the complexity, intricacies, and challenges of inter- and intratumoral heterogeneity [Chapter 3, Khatib, S.A. and Wang, X.W.; Chapter 4, Suresh, A. and Dhanasekaran, R.]; inherited genetic variants [Chapter 5, Schmidt, M.A. and Roberts, L.R.]; molecular and immune subtypes and the critical role of the tumor microenvironment in promoting or modulating HCC and iCCA progression, therapeutic resistance, and immune evasion [Chapter 6, Bramel, E. and Sia, D.; Chapter 7, Ravichandra, A. and Affò, S.]; matricellular proteins, TGF-β, and Yap/Hippo pathway signaling in hepatobiliary cancer pathogenesis, malignant progression, and prognosis [Chapter 8, Zaidi, S. and Mishra, L.; Chapter 9, Sirica, A.E.; Chapter 10, Ko, S. and Monga, S.P.]; and critical driver mutations and other key oncogenic signaling pathways, hepatobiliary cancer immunobiology, novel cellular technologies for drug screening, current and emerging molecular therapeutic targets and immunotherapies [Chapter 11, Li, L. and Selaru, F.M.; Chapter 12, Høgdall, D. and Andersen, J.B.; Chapter 13, Argemi, J. and Sangro, B.; Chapter 14, Li, D. and Ho, M.]. Each chapter in this thematic volume also highlights the continuing vast challenges that still need to be overcome in order to achieve dramatic improvements in HCC and iCCA prevention and to enhance patient survival outcomes, as well as propose key areas of future research aimed at advancing these goals.

It is our hope that this thematic volume will serve as a valuable reference for basic, translational, and clinical researchers either established or new to the field. It is also hoped that this work will not only stimulate much-needed new and innovative research toward personalizing HCC and iCCA therapeutics and prevention strategies, but also will serve to encourage promising early career investigators to become engaged in the expanding and promising field of hepatobiliary cancer research.

ALPHONSE E. SIRICA, PhD, MS, AGAF, FAASLD
PAUL B. FISHER, MPh, PhD, FNAI

CHAPTER ONE

Liver cancer risk-predictive molecular biomarkers specific to clinico-epidemiological contexts

Naoto Kubota[a], Naoto Fujiwara[a,b], and Yujin Hoshida[a,*]

[a]Liver Tumor Translational Research Program, Simmons Comprehensive Cancer Center, Division of Digestive and Liver Diseases, Department of Internal Medicine, University of Texas Southwestern Medical Center, Dallas, TX, United States
[b]Department of Gastroenterology, Graduate School of Medicine, The University of Tokyo, Tokyo, Japan
*Corresponding author: e-mail address: yujin.hoshida@utsouthwestern.edu

Contents

1. Introduction	3
2. Clinico-epidemiological contexts relevant to liver cancer risk	5
2.1 Viral hepatitis	5
2.2 Metabolic disorders	10
2.3 Patient sex and race/ethnicity	11
2.4 Lifestyle factors and relatively rare contexts of hepatocarcinogenesis	12
3. Molecular indicators of high-risk liver	13
3.1 Dysregulations in hepatic transcriptome	13
3.2 Dysregulations in hepatic proteome and secretome	18
3.3 Metabolic dysregulations	18
3.4 Other liver-derived biomolecules	19
4. Host and systemic factors associated with liver cancer risk	20
4.1 Germline and somatic genetic variants/polymorphisms	20
4.2 Systemic inflammation and disorders	21
4.3 Extrahepatic microorganisms	22
5. Clinical implementation of molecular liver cancer risk assessment	23
5.1 Physicians' and patients' perspectives	23
5.2 Practical and regulatory issues for clinical implementation	24
6. Conclusion	24
Acknowledgments	25
Conflict of interest	25
References	25

Abstract

Hepatocellular carcinoma (HCC) risk prediction is increasingly important because of the low annual HCC incidence in patients with the rapidly emerging non-alcoholic fatty liver disease or cured HCV infection. To date, numerous clinical HCC risk biomarkers and scores have been reported in literature. However, heterogeneity in clinico-epidemiological context, e.g., liver disease etiology, patient race/ethnicity, regional environmental exposure, and lifestyle-related factors, obscure their real clinical utility and applicability. Proper characterization of these factors will help refine HCC risk prediction according to certain clinical context/scenarios and contribute to improved early HCC detection. Molecular factors underlying the clinical heterogeneity encompass various features in host genetics, hepatic and systemic molecular dysregulations, and cross-organ interactions, which may serve as clinical-context-specific biomarkers and/or therapeutic targets. Toward the goal to enable individual-risk-based HCC screening by incorporating the HCC risk biomarkers/scores, their assessment in patient with well-defined clinical context/scenario is critical to gauge their real value and to maximize benefit of the tailored patient management for substantial improvement of the poor HCC prognosis.

Abbreviations

AAV	adeno-associated virus
AFP	alpha-fetoprotein
AIH	autoimmune hepatitis
AUROC	area under the receiver operating characteristic
cPLS	cell-culture-derived prognostic liver signature
DAA	direct-acting antivirals
ELS	ectopic lymphoid structure
EPIC	European Prospective Investigation into Cancer and Nutrition
FDA	U.S. Food and Drug Administration
HBV	hepatitis B virus
HCC	hepatocellular carcinoma
HCV	hepatitis C virus
HH	hereditary hemochromatosis
HIR	hepatic Injury and Regeneration
LDT	laboratory developed test
M2BPGi	Mac-2-binding protein glycosylation isomer
MAFLD	metabolic dysregulation-associated fatty liver disease
MS	mass spectrometry
NA	nucleos(*t*)ide analogues
NAFLD	non-alcoholic fatty liver disease
NASH	non-alcoholic steatohepatitis
NMR	nuclear magnetic resonance spectroscopy
PBC	primary biliary cholangitis
PLS	prognostic liver signature
PLSec	prognostic liver secretome signature
SNP	single nucleotide polymorphisms
SVR	sustained virologic response
TexSEC	translation of tissue gene expression to SECretome
TLR	Toll-like receptor
VES	viral exposure signature

1. Introduction

Liver cancer is the sixth most common cancer type worldwide with more than 900,000 new cases annually, and the third leading cause of cancer-related death with more than 800,000 deaths each year (Sung et al., 2021). Hepatocellular carcinoma (HCC) is the dominant histological type of primary liver cancer that mostly develops in chronically diseased liver with cirrhosis, the terminal stage of progressive liver fibrosis (Fujiwara, Liu, Athuluri-Divakar, Zhu, & Hoshida, 2019). Chronic infection of hepatitis C virus (HCV) and hepatitis B virus (HBV) has been the major etiologies for HCC along with alcohol abuse, but with the recent remarkable progress in anti-viral drug development and the obesity epidemic, metabolic liver disease (i.e., non-alcoholic fatty liver disease [NAFLD] and newly proposed metabolic dysregulation-associated fatty liver disease [MAFLD]) is sharply emerging as the new predominant HCC etiology globally (Liu et al., 2021; Loomba, Friedman, & Shulman, 2021).

Because of the distinctly higher HCC incidence in patients with advanced fibrosis or cirrhosis compared to other cancer types in general population, practice guidelines from professional societies recommend semi-annual HCC screening with abdominal ultrasound with or without alpha-fetoprotein (AFP) in the chronic liver disease patients for detection of early-stage HCC amenable to curative therapies and improved survival (Marrero et al., 2018). However, with the rapidly growing target patient population as well as various provider- and patient-related barriers to implementation of the recommended protocol, the regular HCC screening is astoundingly underutilized in real-world clinical practice (utilization rate < 25%) (Wolf, Rich, Marrero, Parikh, & Singal, 2021). Thus, HCC risk prediction has been increasingly important to mitigate the burden on the HCC screening by identifying a subset of patients who need it most. A Markov-model-based simulation analysis showed that individual-risk-based personalized HCC screening, altering its intensity according to predicted HCC risk, is substantially more cost-effective compared to the current standard-care "one-size-fits-all" strategy uniformly applying the screening to all patients with cirrhosis (Goossens et al., 2017).

Clinical studies have shown that HCC risk, measured by annual incidence rate, hugely varies according to multiple clinical factors, particularly liver disease etiology often bound with geographic regions (e.g., HBV in Southeast Asia) (Fujiwara, Friedman, Goossens, & Hoshida, 2018).

While annual HCC incidence rate is 3%–7% in cirrhosis patients with active HCV infection, the rate in patients with histologically confirmed NAFLD cirrhosis is less than 1% (Sanyal et al., 2021; Simon et al., 2021). Given the recent dynamic and drastic shift of dominant liver disease etiology from viral hepatitis to metabolic disorders due to the widespread use of direct-acting antivirals and the global obesity epidemic, consideration about clinical context becomes more and more critical for cost-effective HCC risk prediction. To date, various clinical and/or molecular factors have been proposed as HCC risk indicators in diverse liver disease patient populations across the world, although none of them has been adopted in clinical practice to date (Kubota, Fujiwara, & Hoshida, 2020). In this chapter, we overview currently available HCC risk indicators in literature according to specific clinical contexts and discuss potential strategies for their clinical deployment and implementation toward the goal to refine management of the patients at risk of HCC development and improve patient prognosis (Fig. 1).

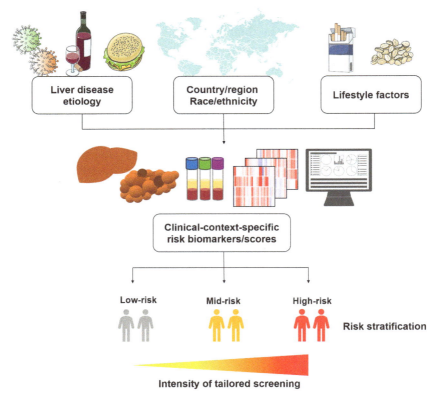

Fig. 1 HCC risk stratification considering clinico-epidemiological contexts.

2. Clinico-epidemiological contexts relevant to liver cancer risk

Besides liver disease etiology, HCC risk can be affected by various factors as depicted by the diverse HCC incidence across geographic regions despite shared etiology. For example, in Japanese cirrhosis patients with active HCV infection, 5-year cumulative HCC incidence reaches 30%, whereas the incidence is 17% in the Western countries; in the HBV-related cirrhosis patients in the endemic regions, 5-year cumulative HCC incidence is 15%, whereas the incidence is 10% in the Western countries (Fattovich, Stroffolini, Zagni, & Donato, 2004; Mancebo et al., 2013). These observations suggest the presence of multiple factors that influence HCC risk and confound each other, including patient race/ethnicity, exposure to environmental carcinogens, lifestyle-related factors such as diet and drinking, access and pattern of clinical care, among many others. Thus, to enable precise HCC risk assessment, it is ideal to derive and validate HCC risk prediction algorithms, controlling these influential factors. However, this is a challenging task that requires coverage of diverse global patient populations. To date, numerous clinical HCC risk scores have been defined and tested in Asian, European, and American patients (Table 1). Cross-regional validations have been conducted for some of the scores, although the scope of assessing the clinico-epidemiological confounding factors is still limited due to the highly biased distribution of the HCC risk-associated variables. Some scores such as ADRESS-HCC score (Flemming et al., 2014), aMAP score (Fan et al., 2020), and Toronto HCC Risk Index (THRI) (Sharma et al., 2017) were developed in multi-etiology cohorts, although variation in etiological composition may limit their general applicability. In contrast, some investigators have developed risk scores tailored for specific liver disease etiology (Ioannou, Green, Kerr, & Berry, 2019). In this section, we summarized recent findings about clinical factors associated with HCC risk according to the major liver disease etiology as the first-order clinico-pathological context that defines HCC risk.

2.1 Viral hepatitis

2.1.1 Hepatitis B virus

HBV has been the leading cause of chronic viral hepatitis and HCC, affecting 3.0%–5.5% of global population with a wide range of geographic variation; the prevalence of HBsAg positivity among general population is

Table 1 Clinical HCC risk scores validated in two or more independent cohorts.

Risk score	Variables	Major etiology	Major race/ethnicity	References
UM regression model	23 Clinical variables	HCV, cryptogenic, alcohol, other	Caucasian, Black, Hispanic	Singal et al. (2013)
aMAP risk score	Age, sex, albumin-bilirubin, platelets	HBV, HCV, HCV after SVR, non-viral	Asian, Caucasian	Fan et al. (2020)
ADRESS-HCC	Age, diabetes, race, etiology, sex, Child-Pugh score	HCV, alcohol, NASH, HBV, other	Non-Hispanic white, Hispanic/Latino, African American, Asian	Flemming, Yang, Vittinghoff, Kim, and Terrault (2014)
THRI	Age, sex, etiology, platelets	Viral, steatohepatitis, PBC, AIH	n.a.	Sharma et al. (2017)
Hughes et al	AFP	HCV, HBV	n.a.	Hughes et al. (2021)
CU-HCC	Age, albumin, bilirubin, HBV-DNA, cirrhosis	HBV	n.a.	Wong et al. (2010)
LSM-HCC	Liver stiffness, age, albumin, HBV-DNA	HBV	n.a.	Wong et al. (2014)
RWS-HCC	Sex, age, cirrhosis, AFP	HBV	Asian	Poh et al. (2016)
REACH-B	Sex, age, ALT, HBeAg, HBV-DNA	HBV	n.a.	Yang et al. (2011)
NGM1-HCC	Sex, age, family history of HCC, alcohol, ALT, HBeAg	HBV	n.a.	Yang et al. (2010)
NGM2-HCC	Sex, age, family history of HCC, alcohol, ALT, HBV-DNA	HBV	n.a.	Yang et al. (2010)

Model	Variables	Population	Ethnicity	Reference
GAG-HCC	Age, sex, HBV-DNA, core promoter mutations, cirrhosis	HBV	n.a.	Yuen et al. (2009)
Hung et al	Sex, age, ALT, previous liver disease, history of HCC, smoking, HBV/HCV infection	HBV	n.a.	Hung et al. (2015)
FIB-4	AST, ALT, platelets, age	HBV	n.a.	Suh et al. (2015)
D2AS risk score	HBV-DNA, sex, age	HBV	Asian	Sinn et al. (2017)
HCC-RESCUE	Age, sex, cirrhosis	HBV	Asian	Sohn et al. (2017)
GBM-based model	Cirrhosis, age, platelets, ETV or TDF, sex, ALT, HBV-DNA, albumin, bilirubin, HBeAg	HBV	Asian, Caucasian	Kim et al. (2021)
Modified REACH-B	Liver stiffness, sex, age, ALT, HBeAg	HBV treated with NA	Asian	Lee et al. (2014)
PAGE-B	Age, sex, platelets	HBV treated with NA	Caucasian	Papatheodoridis et al. (2016)
CAGE-B	cirrhosis, age	HBV treated with NA	Caucasian	Papatheodoridis et al. (2020)
SAGE-B	Liver stiffness, age	HBV treated with NA	Caucasian	Papatheodoridis et al. (2020)
Modified PAGE-B	Age, sex, platelets, albumin	HBV treated with NA	Asian	Kim et al. (2018)
APA-B	Age, platelets, AFP	HBV treated with NA	Asian	Chen et al. (2017)
CAMPAS model score	Cirrhosis, age, sex, platelets, albumin, liver stiffness	HBV treated with NA	Asian	Lee et al. (2020)
REAL-B	Sex, age, alcohol, diabetes, cirrhosis, platelets, AFP	HBV treated with NA	Asian	Yang et al. (2020)

Continued

Table 1 Clinical HCC risk scores validated in two or more independent cohorts.—cont'd

Risk score	Variables	Major etiology	Major race/ethnicity	References
AASL–HCC score	Age, albumin, sex, cirrhosis	HBV treated with NA	Asian	Yu et al. (2019)
CAMD score	Cirrhosis, age, sex, diabetes	HBV treated with NA	Asian	Hsu et al. (2018)
ALT Flare	ALT	HBV treated with NA	Asian, n.a.	Du et al. (2020)
Ganne-Carri et al	Age, alcohol, platelets, GGT, SVR	HCV	n.a.	Ganne-Carrié et al. (2016)
REVEAL-HCV	Age, ALT, AST/ALT ratio, HCV-RNA, cirrhosis, HCV genotype	HCV	n.a.	Lee et al. (2014)
ADRES score	SVR24, sex, FIB-4, AFP	HCV-SVR treated with DAA	Asian	Hiraoka et al. (2019)
HCC-SVR score	FIB-4, sex, AFP	HCV-SVR treated with DAA	Asian	Chun et al. (2020)
Sinn et al.	Age, sex, smoking, diabetes, total cholesterol, ALT	Non-HCV, HBV, Alcohol	Asian	Sinn et al. (2020)

AASL, age, albumin, sex, liver cirrhosis; ADRES, age, diabetes, race, etiology of cirrhosis, sex and severity of liver dysfunction; AFP, alpha-fetoprotein; AIH, autoimmune hepatitis; ALP, alkaline phosphatase; ALT, alanine aminotransferase; aMAP, age, male, albumin-bilirubin, platelets; APA; age, platelet, alpha-fetoprotein; AST, aspartate aminotransferase; CAGE, cirrhosis and age; CAMD, cirrhosis, age, male sex, and diabetes mellitus; CAMPAS, cirrhosis, age, male, platelets, albumin, liver stiffness; CU, Chinese University; DAA, direct-acting antivirals; ETV, entecavir; FIB-4, fibrosis-4; FILI, fibrosis improvement after lifestyle interventions; GAG, guide with age, gender, HBV-DNA, core promoter mutations and cirrhosis; GGT, gamma-glutamyltransferase; GMB, gradient-boosting machine; HbA1c, hemoglobin A1c; HBV, hepatitis B virus; HBeAg, hepatitis B e antigen; HCV, hepatitis C virus; HCC, hepatocellular carcinoma; LSM, liver stiffness measurement; n.a., not available/applicable; NA, nucleoside/nucleotide analogues; NASH, non-alcoholic steatohepatitis; NGM1 or 2-HCC, nomogram 1 or 2-HCC; PAGE, platelets, age, gender; PBC, primary biliary cholangitis; REAL, real-world effectiveness from the Asia Pacific Rim Liver Consortium; REVEAL, risk evaluation of viral load elevation and associated liver disease/cancer; RWS, Real-world risk score; REACH-B, risk estimation for HCC in chronic hepatitis B; SAGE, stiffness and age; SVR, sustained virologic response; TDF, tenofovir disoproxil fumarate; THRI, Toronto HCC risk index; UM, University of Michigan; VFMAP, virtual touch quantification, fast plasma glucose, sex, age, AFP.

7.5% in Africa, 5.9% in Western Pacific region, and 3.0% in Southeast Asia, whereas 1.5% in Europe and 0.5% in America (World Health Organization, 2021). National universal vaccination programs have substantially reduced incidence of HBV-related HCC. In Taiwan where the neonatal vaccination was pioneered, HCC incidence per 10^5 person-years was reduced from 0.92 to 0.23 with the vaccination (Chang et al., 2016). In the current care of HBV-infected individuals typically on treatment with viral-replication-suppressing oral nucleos(t)ide analogues (NAs) such as tenofovir and entecavir, annual HCC incidence rates are 0.0%–1.4% in noncirrhotic and 0.9%–5.4% in cirrhotic patients (Choi, Choi, & Lim, 2021). The regular HCC screening is particularly effective in patients with chronic hepatitis B to significantly reduce HCC-related mortality, while its clinical implementation needs improvement (Su et al., 2021; Yeo & Nguyen, 2021).

Clinically, co-infection of hepatitis delta virus and aflatoxin exposure are known to elevate risk of HBV-related HCC (Fujiwara et al., 2018). High blood HBV DNA level is an indicator of active viral replication, which is associated with increased incidence of HBV-related HCC and likely related to HBV DNA integration into host genome, enhancing carcinogenesis even in non-fibrotic/cirrhotic livers (Chen et al., 2006; Yang et al., 2011). Several clinical-variable-based HCC risk scores have been proposed mainly from the endemic regions, particularly East Asia and Europe to date, in HBV-infected individuals with or without NA treatment (Table 1). These scores consist of similar variables such as patient age and sex, AFP, and other biochemical tests, and their cross-region/population validations have shown generally good performance in predicting HCC risk with AUROCs of 0.70–0.80 (Kim et al., 2021; Voulgaris, Papatheodoridi, Lampertico, & Papatheodoridis, 2020). These scores may identify patients who need or do not need regular HCC screening and biomarker-based more refined HCC risk prediction.

2.1.2 Hepatitis C virus
HCV has been a major viral HCC etiology globally, especially in developed countries over the past decades, affecting more than 70 million individuals (1% of global population) (Roudot-Thoraval, 2021). The prevalence of HCV is relatively high in Eastern Mediterranean region (1.6%), Europe (1.3%), moderate in Africa (0.8%), and relatively low in Southeast Asia (0.5%), Western Pacific region (0.5%), and America (0.5%) (World Health Organization, 2021). Recent development of highly effective and safe oral direct-acting antivirals (DAA) has revolutionized the care of

patients with chronic hepatitis C, although majority of the patients are still undiagnosed and/or untreated (Yousafzai et al., 2021). HCC risk proportionally increases as liver fibrosis progresses, and approximately 80% of HCC patients have underlying cirrhosis (Fujiwara et al., 2018). HCV viral genetic diversity such as genotypes and quasispecies are known to influence the disease progression toward HCC development (Martinez & Franco, 2020). With the shift of anti-HCV therapies from the interferon-based regimens to DAAs, clinically relevant "high-risk" variants are also shifting (e.g., genotype 1b to 3) according to the therapeutic response. Pharmacological HCV clearance, namely sustained virologic response (SVR), substantially reduces future HCC incidence, although the HCC risk is not eliminated over a decade even after achieving an SVR (Baumert, Juhling, Ono, & Hoshida, 2017; Carrat et al., 2019). Clinical-variable-based HCC risk scores have been developed and/or tested in SVR patients, although their external validation is still limited compared to the HBV HCC risk scores (Table 1). These studies also suggest variation in their performance according to patient race/ethnicity, which should be elucidated in future studies.

2.2 Metabolic disorders
2.2.1 Alcohol abuse
Alcohol-related liver cirrhosis is one of the well-known risks of developing HCC, and the proportion of HCC attributed to alcoholic liver disease is estimated to be consistently 20%–25% over the past years (Massarweh & El-Serag, 2017). The HCC incidence rate in patients with alcoholic cirrhosis is reported to be 1.3%–3% annually (Morgan, Mandayam, & Jamal, 2004), and alcohol abuse shows synergistic effects with other etiologies including viral hepatitis on the progression of liver fibrosis and hepatocarcinogenesis (Palmer & Patel, 2012). In fact, several clinical HCC risk scores established in viral hepatitis patients such as NGM-HCC and REAL-B include alcohol abuse as one of their model variables (Ganne-Carrié et al., 2016; Yang et al., 2010, 2020).

2.2.2 Non-alcoholic fatty liver disease
The global prevalence of non-alcoholic fatty liver disease (NAFLD) is estimated to be 25% with an increasing trend (Huang, El-Serag, & Loomba, 2021; Younossi et al., 2019). With the newly proposed metabolic dysfunction-associated fatty liver disease (MAFLD), the prevalence is assumed to be even higher (50%) (Liu, Ayada, et al., 2021). Both prevalence and incidence of NAFLD-related HCC are predicted to significantly

increase over the next decade (Estes, Razavi, Loomba, Younossi, & Sanyal, 2018). The annual HCC incidence rates in histologically-confirmed NAFLD are less than 0.5% in patients with non-cirrhotic fibrosis and less than 1% even with cirrhosis (Sanyal et al., 2021; Simon et al., 2021). HCC development in non-cirrhotic liver is one notable feature of NAFLD, which is observed approximately in 30% of the patients (Mittal et al., 2016). These characteristics, i.e., sharply expanding at-risk population, low incidence rate, and carcinogenesis without involving cirrhosis, highlight urgent need for new strategies of HCC screening based on refined HCC risk prediction to enable cost-effective care of NAFLD patients (Singal & El-Serag, 2021). Several clinical HCC risk scores such as the Age, Diabetes, Race, Etiology of cirrhosis, Sex, and Severity (ADRESS)-HCC score and Toronto HCC risk index (THRI) have been tested in NAFLD patients in the context of multi-etiology patient cohorts (Table 1). There are also HCC risk scores specifically modeled in NAFLD patients (Ioannou et al., 2019).

2.3 Patient sex and race/ethnicity
2.3.1 Sex
Clinically, it is well known that male sex is associated with two- to threefold higher HCC incidence and mortality compared to females irrespective of geographic regions, which is interestingly not obvious in intrahepatic cholangiocarcinoma, the second common histological type of liver cancer (Massarweh & El-Serag, 2017). Liver cancer, dominantly HCC, is the third leading cause of cancer-related death in men, whereas it is the sixth leading cause in women (Sung et al., 2021). HCC prognosis is better in women compared to men in patients under 65 years old, while such difference disappears in older patients, suggesting involvement of sex hormones as an underlying biological factor (Rich et al., 2020). In addition to the biological difference, this sex disparity is likely attributable to complex interplays between multiple factors, including biological differences (e.g., sex hormone-related molecular pathways) as well as behavioral and environmental factors (drinking, smoking, dietary habit), and socioeconomic factors associated with patient race/ethnicity (Thylur et al., 2020), to be clarified in future studies. Particularly, there may be specific patient subgroups with unique (including opposite) association of sex with HCC risk. Given the unequivocal association with HCC risk, sex is indeed incorporated in most of the clinical HCC risk scores (Table 1).

2.3.2 Race/ethnicity

Patient race/ethnicity has been increasingly recognized as a confounding factor that influences HCC risk, which is often bound to specific HCC etiology prevalent in certain geographic regions, e.g., HBV infection in Southeast Asia and sub-Saharan Africa, where the patients reside or immigrate from (Fujiwara et al., 2018). Epidemiological studies have shown that the trend of HCC incidence is distinctly different between racial/ethnic groups, reflecting dynamic change in each regional population. For instance, HCC incidence has been decreasing in Asian and Pacific islanders that have been the leading HCC risk populations, whereas increasing in other racial/ethnic groups particularly Hispanics and Blacks compared to non-Hispanic whites over the past several decades in the United States (Franco, Fan, Jarosek, Bae, & Galbraith, 2018; Islami et al., 2017). HCC incidence rate in Hispanics is approximately two-times higher (21.2/100,000) compared to non-Hispanic Whites (9.3/100,000) (El-Serag, Sardell, Thrift, Kanwal, & Miller, 2021; Miller et al., 2021). Survival after HCC diagnosis is worse in Blacks likely due to diagnosis at late stages, whereas better in Asians and Hispanics, compared to Whites (Rich, Carr, Yopp, Marrero, & Singal, 2020). Race/ethnicity can be confounded with multiple factors, e.g., relative difference in prevalence of inheritable genetic variants associated with HCC risk, lifestyle (particularly dietary habits), and socioeconomic status influencing access to medical care, in a highly complex manner (Thylur et al., 2020). Given the complexity, no clinical HCC risk score incorporates race/ethnicity as its component. In addition, derivation and validation of clinical risk scores are substantially affected by biased distribution of race/ethnicity-related variables, and therefore generalization of the scores' performance is compromised. Nevertheless, several attempts to cross-validate some clinical risk scores across geographic regions (e.g., Asia vs. Europe), which may partially address the issue (Voulgaris et al., 2020). More studies at population level will be needed to clarify the influence of race/ethnicity on HCC risk under specific regional and clinical contexts.

2.4 Lifestyle factors and relatively rare contexts of hepatocarcinogenesis

Dietary factors such as Western diet and intake of anti-oxidative substances such as coffee have been associated with reduced HCC risk (Simon & Chan, 2020). Food contamination with carcinogen such as aflatoxin is a well-known HCC risk factor that leads to impaired tumor suppressor genes, e.g., *TP53*, and is associated with certain DNA mutational signatures (Letouze et al., 2017; McGlynn, Petrick, & El-Serag, 2021). Tobacco smoking is associated with 40%–70% increase of HCC risk (McGlynn et al., 2021). Use of generic drugs such as statins, aspirin, and metformin has been associated with lowered HCC risk, indicating their potential utility as chemopreventatives (Athuluri-Divakar & Hoshida, 2019). Other rare suspected habitual and/or dietary HCC risk factors include betel quid chewing, exposure to organic solvents, and heavy metals (e.g., toluene, trichloroethylene, cadmium, and lead) (Thylur et al., 2020). Physical activity is also associated with HCC risk as a potential intervention (Fujiwara et al., 2018). Among these factors, only smoking has been incorporated in clinical HCC risk scores (Table 1). Future studies should explore other factors, particularly modifiable by dietary, pharmacological, and/or physical interventions, to be incorporated in HCC risk scores that could guide therapeutic decision making.

3. Molecular indicators of high-risk liver

Various types of biomolecules such as nucleic acids, proteins, and metabolites, have been actively explored as indicators of elevated risk of HCC development as summarized in Table 2. It has been increasingly recognized that association of these factors with HCC risk is substantially affected by specific clinical contexts such as liver disease etiology as outlined in this section.

3.1 Dysregulations in hepatic transcriptome

Tissue transcriptome is one of the most widely studied omics data as a source of exploring HCC risk biomarkers (Kubota et al., 2020). One example is Prognostic Liver Signature (PLS) predictive of long-term HCC risk in patients with diverse clinical contexts such as HBV infection, HCV infection with or without pharmacological viral cure (i.e., SVR), alcohol abuse, and NAFLD/NASH, which is already implemented in an FDA-approved diagnostic assay platform as a Laboratory Developed Test (LDT) (Hoshida et al., 2008, 2013; King et al., 2015; Nakagawa et al., 2016; Ono et al., 2017). Of note, PLS is also implemented in a cell culture-based in vitro system, namely cell-culture-derived PLS (cPLS), for high-throughput HCC chemoprevention drug screening (Crouchet et al., 2021). A Hepatic Injury and Regeneration (HIR) signature was derived from liver tissues

Table 2 Molecular HCC risk indicators.

Type of biomarker	Biomarkers/scores	Variables	Major etiology	Major race/ethnicity	Reference
SNP	EGF SNP	EGF 61AG (rs4444903, A > G)	HBV, HCV	Asian, European, African	Jiang et al. (2015)
	IFNL3 SNP	IFNL3 (rs12979860: C > T, rs8099917: T > G)	HCV, HBV	n.a.	Qin et al. (2019)
	MICA SNP	MICA (rs2596542, C > T)	HCV, HBV	Asian, European	Luo, Wang, Shen, Deng, and Ye (2019)
	DEPDC5 SNP	DEPDC5 (rs1012068: T > G)	HCV	Asian	Liu et al. (2019)
	AURKA SNP	AURKA (rs1047972: G > A)	HCV	n.a.	Farid, Afify, Alshamoby, Abdelsameea, and Bedair (2021)
	TLL1 SNP	TLL1 (rs17047200: A > T)	HCV after SVR treated with IFN	Asian	Matsuura et al. (2017)
	CHI3L1 SNP Intergenic polymorphism	CHI3L1 (rs880633: T > C) intergenic polymorphism (rs597533: A > G)	HCV treated with DAA	n.a.	Mangoud, Ali, El Kassas, and Soror (2021)
	KIF1B or 1p36.22 SNP	KIF1B or 1p36.22 (rs17401966, A > G)	HBV	Asian	Luo, Zhang, Huang, and Hu (2019)
	STAT4 SNP	STAT4 (rs7574865, G > T)	HBV	Asian	Zhang, Xu, Liu, and Chen (2017)
	HLA-DQB1/HLA-DBA2 SNP	HLA-DQB1/HLA-DBA2 (rs9275319 A > G)	HBV	Asian	Jiang et al. (2013)

	lncRNA SNP	lnc-ACACA-1 rs9908998, lnc-RP11-150O12.3 rs2275959, rs1008547, rs11776545	HBV	Asian	Liu et al. (2021)
	PNPLA3 SNP	*PNPLA3* (rs738409: C>G)	NAFLD, alcohol, HCV	Caucasian	Singal et al. (2014)
	TM6SF2 SNP	*TM6SF2* (rs58542926: C>T)	Alcohol	Caucasian	Tang et al. (2019)
	HSD17B13 SNP	*HSD17B13* (rs72613567: TA)	Alcohol	Caucasian	Stickel et al. (2020)
	WNT3A-WNT9A SNP	*WNT3A-WNT9A* (rs708113: T>A)	Alcohol	n.a.	Trépo et al. (2021)
	MBOAT7 SNP	*MBOAT7* (rs641738: C>T)	NAFLD	Caucasian	Donati et al. (2017)
	CELSR2-PSRC1-SORT1 SNP	*CELSR2-PSRC1-SORT1* (rs599839; A>G)	NAFLD	n.a.	Meroni et al. (2021)
	IFNGR1 SNP	*IFNGR1* (rs1327474: G>A)	n.a.	n.a.	Aref et al. (2021)
Panel of SNPs	Genetic risk score	SNPs of *PNPLA3*, *TM6SF2*, *HSD17B13*	General population	Caucasian, n.a.	Gellert-Kristensen et al. (2020)
	Fat-genetic risk score (hepatic fat genetic risk score)	SNPs of *PNPLA3*, *TM6SF2*, *MBOAT7*, *GCKR*, and hepatic fat content	HCV treated with DAA	Italian, Egyptian	Degasperi et al. (2020)
	Polygenic risk scores	SNPs of *PNPLA3*, *TM6SF2*, *MBOAT7*, *GCKR*, and hepatic fat + HSD17B13	NAFLD, general population	n.a.	Bianco et al. (2021)
Tissue transcriptome	Prognostic liver signature (PLS)	186-Gene signature	HCV	Asian, Caucasian	Hoshida et al. (2013)
	HIR gene signature	233-Gene signature	HBV	Asian	Kim et al. (2014)

Continued

Table 2 Molecular HCC risk indicators.—cont'd

Type of biomarker	Biomarkers/scores	Variables	Major etiology	Major race/ethnicity	Reference
	Activated HSC gene signature	37-Gene signature	HBV	Asian	Ji et al. (2015)
	HSC signature	122-Gene signature	HCV, HBV	Caucasian, Asian	D. Y. Zhang et al. (2016)
	Ectopic lymphoid structure signature	12-Gene signature	HCV	Asian	Finkin et al. (2015)
	Immune mediated cancer field signature	172-Gene signature	HCV	Caucasian	Moeini et al. (2019)
Circulating biomolecule	cfDNA	Mutations of 4 genes, HBV integration	HBV	Asian	Qu et al. (2019)
	miRNA	7/8 miRNAs	HBV	Asian	Wang et al. (2016)
	miRNA	16 miRNAs	HBV, HCV	Asian	Huang et al. (2017)
	DNA methylation	TBX2 hypermethylation	HBV, HCV, alcohol	Asian	Wu, Yang, Wang, Chen, and Santella (2017)
	GlycoCirrhoTest/GlycoHCCRiskScore	Serum protein N-glycans	HCV	Caucasian	Verhelst et al. (2017)
	Serum glycan	M2BPGi	HCV	Asian	Yamasaki et al. (2014)
	Cytokine	IL-6	HCV	Asian	Nakagawa et al. (2009)
	Cytokine	IL-17	HBV, HCV	Asian	Liang et al. (2021)
	Cytokine	IL-27	HBV	Asian	Yuan et al. (2021)
	Cytokine	Myostatin	Alcohol	Asian	Kim et al. (2020)

Biomarker type	Biomarkers	Etiology	Ethnicity	Reference
Protein; PLSec/Secretome signature	High-risk; VCAM-1, IGFBP-7, gp130, matrilysin, IL-6, CCL-21 Low-risk; angiogenin, protein S	HCV-SVR (DAA), NAFLD/cryptogenic	Caucasian, Asian	Fujiwara et al. (2021)
Protein	IGF-1	HCV	Caucasian	Mazziotti et al. (2002)
Protein	OPN	Alcohol, HBV, HCV	n.a.	Duarte-Salles et al. (2016)
Protein; LCR1/LCR2	ApoA1, Hp/A2M	HCV, HBV, ALD, NAFLD	Caucasian, Sub-Saharan, North-Africa Middle East, Asian	Poynard et al. (2019)
Metabolites	14 Metabolites	HBV, HCV, alcohol	n.a.	Stepien et al. (2021)
Metabolites; HCC risk score R	2 Amino acids (Phe, Gln)	HBV, HCV	Asian	Liang et al. (2020)
Metabolites9	AST and 7 metabolites, including tyrosine, oleamide, lysoPC 16:1, lysoPC 20:3, 5-hydroxyhexanoic acid, androsterone sulfate, and TUDCA	HBV	Asian	Jee et al. (2018)
Metabolites	2 metabolites (phenylalanyl-tryptophan, glycocholate)	HBV	Asian	Luo et al. (2018)
Virome; viral exposure signature	61 Viral strains	HBV, HCV, HDV, aflatoxins, alcohol, NAFLD	Caucasian, Black, Asian	Liu et al. (2020)

AFP, alpha-fetoprotein; cfDNA, circulating free DNA; DAA, direct-acting antiviral agent; HBV, hepatitis B virus; HCV, hepatitis C virus; HDV, hepatitis D virus; HCC, hepatocellular carcinoma; HIR, hepatic injury and regeneration gene expression; HSC, hepatic stellate cell; IFN, interferon; M2BPGi, mac-2 binding protein glycosylation isomer; miRNA, microRNA; n.a., not available/applicable; NA, nucleoside/nucleotide analogues; NAFLD, non-alcoholic fatty liver disease; NASH, non-alcoholic steatohepatitis; SNP, single-nucleotide polymorphism; SVR, sustained virologic response.

in HBV-related HCC patients to predict de novo HCC recurrence (Kim et al., 2014). Transcriptomic signatures can be defined to detect pathogenic status of hepatic cell types such as hepatic stellate cells, the driver of liver fibrogenesis and carcinogenesis in viral hepatitis and NAFLD patients (Higashi, Friedman, & Hoshida, 2017; Ji et al., 2015; Zhang et al., 2016). Gene signatures of histological architecture can also be trained; a 12-chemokine-gene signature identified presence of HCC-risk-associated lymphocyte aggregate, ectopic lymphoid structure (ELS) in mainly HCV-infected patients (Finkin et al., 2015).

3.2 Dysregulations in hepatic proteome and secretome

Hepatic proteomic dysregulations that lead to alterations of secretome can have a potential as non-invasive biomarkers in circulation to monitor molecular pathogenic status of the liver. One classical example of such molecule is alpha-fetoprotein (AFP), which is an oncofetal protein and used as an HCC tumor marker. Interestingly, elevation of AFP can be observed years before clinically detectable HCC development, suggesting that it captures carcinogenesis-prone status of diseased liver tissue microenvironment (Hughes et al., 2021). Serum cytokines, including IL-6, IL-17, and IL-27, as well as other types of serum proteins such as laminin γ2 monomer and IGF-I, were reported to be associated with HCC risk (Kim, Kang, et al., 2020; Liang et al., 2021; Mazziotti et al., 2002; Nakagawa et al., 2009; Yamashita et al., 2021; Yuan et al., 2021) (Table 2). The hepatic-transcriptome-based PLS was translated into a serum-protein-based surrogate marker, Prognostic Liver Secretome signature (PLSec), by utilizing a computational pipeline translating tissue transcriptome to secretome (Translation of tissue gene expression to SECretome; TexSEC, www.texsec-app.org) (Fujiwara, Kobayashi, et al., 2021). PLSec in combination with AFP (PLSec-AFP) was validated for association with long-term HCC risk and hepatic decompensation in HCV-infected patients who achieved SVR and cirrhosis patients with mixed etiologies (Fujiwara, Fobar, et al., 2021; Fujiwara, Kobayashi, et al., 2021). The TexSEC pipeline also identified a secretome signature predictive of survival in severe alcoholic hepatitis patients (Fujiwara, Trépo, et al., 2021).

3.3 Metabolic dysregulations

Since liver is the major organ responsible for the essential metabolic functions in human body, measuring dysregulations of metabolites is a sensible approach to monitor molecular status of the liver at risk of carcinogenesis (Beyoğlu & Idle, 2020). Global metabolite profiling has been performed by using mass spectrometry (MS) and/or nuclear magnetic resonance spectroscopy (NMR) (Beyoğlu & Idle, 2013). Serum NMR analysis of European Prospective Investigation into Cancer and Nutrition (EPIC) cohort revealed that circulating metabolites were correlated with various lifestyles or environmental exposures associated with HCC risk (Assi et al., 2015). Unbiased liquid-chromatography-MS study of EPIC cohort identified 14 metabolites associated with long-term HCC risk, including 9 high-risk (N1-acetylspermidine, isatin, *p*-hydroxyphenyllactic acid, tyrosine, sphingosine, L,L-cyclo(leucylprolyl), glycochenodeoxycholic acid, glycocholic acid and 7-methylguanine) and 5 low-risk (retinol, dehydroepiandrosterone sulfate, glycerophosphocholine, γ-carboxyethyl hydroxychroman and creatine) metabolites (Stepien et al., 2021). In a Korean cohort, seven metabolites (tyrosine, oleamide, lysophosphatidylcholines 16:1, lysophosphatidylcholines 20:3, 5-hydroxyhexanoic acid, androsterone sulfate, and tauroursodeoxycholic acid) were identified by using non-targeted liquid-chromatography-MS (Jee et al., 2018). Plasma phenylalanine and glutamine levels were associated with HCC incidence in Asian patients with mainly viral hepatitis (Liang et al., 2020). Phenylalanyl-tryptophan and glycocholate were also identified as a serum metabolite biomarker panel in combination with AFP, which predicts presence of preclinical HCC before clinical diagnosis (Luo et al., 2018).

3.4 Other liver-derived biomolecules

Serum Mac-2-binding protein glycosylation isomer (M2BPGi) is a serum glycan biomarker that can predict the development of HCC in the patients with various etiologies including HCV and HBV treated with oral antiviral therapy (Hsu et al., 2018; Shinkai et al., 2018; Tseng et al., 2020; Yamasaki et al., 2014). GlycoCirrhoTest, a serum glycomics biomarker which could distinguish chronic liver disease patients with compensated cirrhosis from those with earlier stage of fibrosis, was reported to be associated with future HCC risk in cirrhotic patients (Verhelst et al., 2017). Circulating nucleic acids, e.g., cell-free (cf) DNA and RNA, are another class of non-invasive biomarker candidates. For example, *TBX2* hypermethylation in plasma was associated with increased HCC risk in an Asian cohort (Wu et al., 2017). Hepatocellular carcinoma screen (HCCscreen) score, which is composed of genetic alterations in cfDNAs (*TP53, CTNNB1,*

AXIN1, and *TERT*) as well as serum AFP and des-γ-carboxy prothrombin levels, was associated with HBV-related HCC occurrence within 6–8 months (positive predictive value of 0.17) (Qu et al., 2019). Circulating micro-RNA (miRNA) was also reported to be associated with HCC risk. An miRNA-based risk score consisting of 15 miRNAs with AFP was associated with 5-year HCC incidence in HBV-positive individuals with Korean ancestry (Wang et al., 2016). Another 16-miRNA-based risk score was also associated with HCC occurrence in HBV- or HCV-related cirrhotic patients (Huang et al., 2017). Nucleic acids and proteins loaded in extracellular vesicle (EV), a lipid bilayer-delimited particle secreted by liver cells, may serve as a new class of biomolecules to predict future HCC risk (Adeniji & Dhanasekaran, 2021).

4. Host and systemic factors associated with liver cancer risk

The above-mentioned clinical contexts are assumed to be accompanied with underlying various host genetic variants/aberrations and systemic molecular dysregulations that often represent cross-organ interactions to develop cancer-permissive hepatic tissue microenvironment. These factors in literature are overviewed in this section.

4.1 Germline and somatic genetic variants/polymorphisms

Inheritable genetic polymorphisms, particularly single nucleotide polymorphisms (SNPs), have been heavily studied given the easy access as biomarkers via readily available specimens such as buccal swab in daily clinical practice, and numerous HCC-risk-associated SNPs were reported (Kubota et al., 2020) (Table 2). Prevalence of risk alleles/genotypes is often biased across different racial/ethnic groups, suggesting their influence in determining race/ethnicity-specific susceptibility to HCC in response to clinical and environmental risk factors. Magnitude of association with HCC risk for these SNPs is generally modest, e.g., odds ratio of 1.5 or less, and multi-SNP "polygenic" scores have been explored (Bianco et al., 2021; Fujiwara et al., 2018). While germline DNA variants are easy to assay, they are unmodifiable by therapeutic intervention like sex.

Several SNPs were discovered and validated in patients with specific etiology. SNPs in *DEPDC5A* and *TLL1* genes were identified as HCC risk indicators in HCV-infected patients, whereas SNPs in *KIF1B* and *STAT4* genes were reported in HBV-infected patients (Liu et al., 2019; Luo, Zhang, et al., 2019; Matsuura et al., 2017; Zhang et al., 2017). SNPs in *MICA*, *IFNL3*, and *EGF* genes were associated with HCC risk among individuals with either HCV or HBV infection (Jiang et al., 2015; Qin et al., 2019). SNPs in *PNPLA3*, *TM6SF2*, and *HSD17B13* that were initially discovered for association with NASH, were also associated with HCC risk among NAFLD, HCV-SVR, and alcoholic hepatitis patients (Degasperi et al., 2020; Gellert-Kristensen et al., 2020; Kozlitina et al., 2014; Singal et al., 2014; Tang et al., 2019). Recently, a SNP in *WNT3A-WNT9A* has been identified as a risk factor of alcohol-related HCC (Trépo et al., 2021). Polygenic scores consisting of multiple SNPs (e.g., SNPs in *PNPLA3*, *MBOAT7*, *TM6SF2*, and *GCKR*) were currently actively explored in HCV-SVR and NAFLD patients (Bianco et al., 2021; Degasperi et al., 2020). Somatic DNA mutations in *PKD1*, *KMT2D*, and *ARID1A* genes in cirrhotic liver may have protective role from carcinogenesis (Zhu et al., 2019).

4.2 Systemic inflammation and disorders

Metabolic disorders, often associated with obesity and NAFLD, are well known drivers of systemic and hepatic inflammation mediated by various cytokines/chemokines such as interleukin-17 that lead to hepatocarcinogenesis, which could involve interaction between multiple organs/tissues such as adipose tissues, brain, and gut. Obesity, diabetes, and metabolic syndrome are associated with two- to threefold increase of HCC incidence (Fujiwara et al., 2018). Use of drugs to treat diabetes (e.g., metformin) and hyperlipidemia (e.g., lipophilic statins) is associated with reduced HCC risk, suggesting that biomarkers related to response to these drugs may have roles in HCC risk prediction and monitoring (Athuluri-Divakar & Hoshida, 2019).

Accumulation of ionic metal such as iron is known to increase oxidative stress and genetic damage that lead to carcinogenesis. Hereditary hemochromatosis (HH) is a genetic disorder with excessive iron absorption mostly caused by germline variants in *HFE* gene, and a known HCC risk condition, whereas iron overload can also elicit similar risk (Fujiwara et al., 2018). Autoimmune liver diseases such as autoimmune hepatitis (AIH) and primary biliary cholangitis (PBC) develop progressive liver fibrosis, and are at risk of developing HCC and/or cholangiocarcinoma once cirrhosis is established (Liang, Yang, & Zhong, 2012; Marrero et al., 2018). However, their incidence and prevalence are low (annual incidence of up to 1% or far less) because of the small patient population. In contrast, despite the similarly low incidence rates, prevalence is high in patients with NAFLD and HCV-SVR due to the vast size of patient population.

Alpha1-antitrypsin deficiency, caused by mutation in *SERPINA1* gene, is also associated with increased risk of developing both cirrhosis and HCC along with pulmonary emphysema (Strnad, McElvaney, & Lomas, 2020). Porphyrias, caused by dysfunctional heme biosynthesis, are associated with elevated risk of primary liver cancer (Wang, Rudnick, Cengia, & Bonkovsky, 2019). Other rare genetic diseases associated with HCC risk include glycogen storage disease type I and hereditary tyrosinemia type I (Dragani, 2010).

Genetic integration of adeno-associated virus (AAV), a virus belongs to the parvovirus group, into several cancer driver genes was identified in a small subset of HCCs, however, the potential risk of AAV on hepatocarcinogenesis is yet to be established (Nault et al., 2015; Schäffer et al., 2021).

4.3 Extrahepatic microorganisms
4.3.1 Gut microbiome

Intestinal bacterial flora (or gut microbiota) consists of more than 10^{14} microorganisms, representing more than 10^4 bacterial species, and changes in its composition (i.e., dysbiosis) are expected to serve as a source to identify HCC risk biomarkers as well as therapeutic targets to alter the risk (Schwabe & Greten, 2020; Zhou et al., 2020). Preclinical studies in rodent models have identified several cellular signaling such as toll-like receptor (TLR) pathway linking gut dysbiosis and HCC risk, which could be disrupted by gut sterilization, as well as physiological pathway such as bile acid metabolism that shape the gut-liver axis, affecting HCC risk (Thilakarathna, Rupasinghe, & Ridgway, 2021). Multiple studies in human HCC patients with various etiologies have shown that there are certain protective or harmful bacterial species with regard to HCC risk and biological/clinical consequence (Xu et al., 2021). Panels of gut bacteria (e.g., Enterococcus, Limnobacter, and Phyllobacterium) were shown to be associated with presence of HCC specific to or regardless of etiology (Hernandez et al., 2021; Ren et al., 2019; Zheng et al., 2020). Oral cyanobacteria was associated with HCC risk in patients with liver disease from mixed etiologies (Hernandez et al., 2021). Gut dysbiosis and its HCC risk association are likely influenced by specific dietary habits, host genetic factors, and geographic setting the patients reside, among many other factors, and therefore their characterization is expected to contribute to refining clinical-context-specific HCC risk prediction and chemopreventive intervention.

4.3.2 History of viral exposure

A Viral Exposure Signature (VES), composed of 61 viral strains determined using VirScan, a virome technology for detecting the exposure history to all known human viruses, was associated with future HCC development (Liu et al., 2020). Interestingly, this study revealed that some viruses such as human herpesvirus 5 were associated with future HCC development as previously unknown HCC risk factors (Fujiwara & Hoshida, 2020).

5. Clinical implementation of molecular liver cancer risk assessment

HCC risk assessment should ultimately be incorporated in the algorithm of regular HCC screening to improve early tumor detection. Even if research studies confirm clinical utility of new HCC risk biomarkers and/or scores according to specific clinical context, there are multi-fold obstacles to hamper their clinical translation as outlined in this section.

5.1 Physicians' and patients' perspectives

Even the guideline-recommended use of imaging- and blood-based biomarkers (i.e., ultrasound and AFP) is poorly utilized in real-world clinical practice (utilization rate < 25%) due to multiple provider- and patient-related factors (Wolf et al., 2021; Zhao & Nguyen, 2016). Obstacles on the physicians' side include limited knowledge about HCC screening particularly in primary care providers, logistical barriers such as time constraint in clinic, and the overwhelmingly large target patient population to regularly monitor for HCC development (Dalton-Fitzgerald et al., 2015; Dirchwolf et al., 2021; Fujiwara et al., 2018; Kubota et al., 2020; Simmons, Feng, Parikh, & Singal, 2019; Wolf et al., 2021). On the other hand, encouragingly, physicians are willing to tailor HCC screening tests if the probability of future HCC can be reliably estimated up-front (Kim et al., 2020). To facilitate use of new clinical HCC risk scores, web-based tools could be an option to consider: an example of such tool, HCC Risk Calculator (www.hccrisk.com), is available for representative HCC etiologies for clinical-context-specific risk prediction. On the patients' side, prohibitive factors include lack of knowledge about HCC screening, high costs for the tests, and physical access to the tests (Farvardin et al., 2017; Singal et al., 2021). Nevertheless, cirrhosis patients prioritize early HCC detection over potential screening-related harms or inconvenience (Woolen et al., 2021). Thus, education and outreach are critical in overcoming the issue for both physicians and patients to improve intake to the screening, considering clinical contexts that are often bound to racial/ethnic and socioeconomic disparities. Affordable and non-invasive tests will also lower the bar for clinical utilization of HCC screening.

5.2 Practical and regulatory issues for clinical implementation

A prior Markov-model-based simulation analysis showed that individual-risk-based personalization of HCC screening algorithm and modality is substantially cost-effective compared to the current "one-size-fits-all" strategy uniformly offering semi-annual screening with ultrasound and AFP (Goossens et al., 2017). Thus, once clinical utility of HCC risk markers and/or scores is confirmed, they may be considered for incorporation to develop such personalized HCC screening strategy (Fig. 1). However, assessment and determination of their clinical utility can be challenging due to heterogeneous performance of the biomarkers/scores across various clinical contexts/scenarios. For example, association of an HCC risk biomarker with magnitude of the risk (e.g., hazard ratio in Cox regression) hugely varies according to liver disease etiology (Fujiwara, Kobayashi, et al., 2021; Nakagawa et al., 2016). This raises a question: to what extent should HCC risk biomarkers/scores be tailored to each specific clinical context/scenario. If a risk biomarker/score is defined in more specific clinical context, its performance is expected to be improved, while its generalizability will be compromised. In addition, clinical validation of candidate HCC risk biomarkers may become more challenging due to the confined patient population by clinical context, especially given the requirement of rigor in the phased step-wise biomarker validation for regulatory approval of the assays implemented in clinical diagnostic platform (Fujiwara et al., 2019). Smaller target patient population to apply the biomarkers may also diminish commercial interest and incentive for clinical assay development.

6. Conclusion

In this chapter, we overviewed clinical and molecular indicators of HCC risk in patients with chronic liver diseases with specific consideration about clinical context that compromises development and application of new HCC risk biomarkers and scores across heterogeneous patient populations. Given the complexity of the heterogeneity, future studies should clarify performance and utility of candidate HCC risk biomarkers and/or scores with proper characterization of target patient population to gauge their real value under well-specified clinical context/scenario to ultimately improve the poor prognosis of HCC patients. Furthermore, clinical utility of the biomarkers/scores should be revisited and redefined along with the ever-evolving clinical demographics of liver disease patients with the progresses in development of new therapeutics such as direct-acting antivirals and chemopreventive agents.

Acknowledgments

This work was supported by Uehara Memorial Foundation funding to N.F. and US NIH (DK099558, CA233794, CA226052), European Commission (ERC-2014-AdG-671231, ERC-2020-ADG-101021417), and Cancer Prevention and Research Institute of Texas (RR180016) to Y.H.

Conflict of interest

The authors declare no conflict of interest relevant to the contents of this manuscript.

References

Adeniji, N., & Dhanasekaran, R. (2021). Current and emerging tools for hepatocellular carcinoma surveillance. *Hepatology Communications*. https://doi.org/10.1002/hep4.1823.

Aref, S., Zaki, A., El Mahdi, E. M., Adel, E., Bahgat, M., & Gouda, E. (2021). Predictive value of interferon γ receptor gene polymorphisms for hepatocellular carcinoma susceptibility. *Asian Pacific Journal of Cancer Prevention*, 22(6), 1821–1826. https://doi.org/10.31557/apjcp.2021.22.6.1821.

Assi, N., Fages, A., Vineis, P., Chadeau-Hyam, M., Stepien, M., Duarte-Salles, T., et al. (2015). A statistical framework to model the meeting-in-the-middle principle using metabolomic data: Application to hepatocellular carcinoma in the EPIC study. *Mutagenesis*, 30(6), 743–753. https://doi.org/10.1093/mutage/gev045.

Athuluri-Divakar, S. K., & Hoshida, Y. (2019). Generic chemoprevention of hepatocellular carcinoma. *Annals of the New York Academy of Sciences*, 1440(1), 23–35. https://doi.org/10.1111/nyas.13971.

Baumert, T. F., Juhling, F., Ono, A., & Hoshida, Y. (2017). Hepatitis C-related hepatocellular carcinoma in the era of new generation antivirals. *BMC Medicine*, 15(1), 52. https://doi.org/10.1186/s12916-017-0815-7.

Beyoğlu, D., & Idle, J. R. (2013). The metabolomic window into hepatobiliary disease. *Journal of Hepatology*, 59(4), 842–858. https://doi.org/10.1016/j.jhep.2013.05.030.

Beyoğlu, D., & Idle, J. R. (2020). Metabolomic and lipidomic biomarkers for premalignant liver disease diagnosis and therapy. *Metabolites*, 10(2). https://doi.org/10.3390/metabo10020050.

Bianco, C., Jamialahmadi, O., Pelusi, S., Baselli, G., Dongiovanni, P., Zanoni, I., et al. (2021). Non-invasive stratification of hepatocellular carcinoma risk in non-alcoholic fatty liver using polygenic risk scores. *Journal of Hepatology*, *74*(4), 775–782. https://doi.org/10.1016/j.jhep.2020.11.024.

Carrat, F., Fontaine, H., Dorival, C., Simony, M., Diallo, A., Hezode, C., et al. (2019). Clinical outcomes in patients with chronic hepatitis C after direct-acting antiviral treatment: A prospective cohort study. *Lancet*, *393*(10179), 1453–1464. https://doi.org/10.1016/s0140-6736(18)32111-1.

Chang, M. H., You, S. L., Chen, C. J., Liu, C. J., Lai, M. W., Wu, T. C., et al. (2016). Long-term effects of hepatitis B immunization of infants in preventing liver cancer. *Gastroenterology*, *151*(3), 472–480.e1. https://doi.org/10.1053/j.gastro.2016.05.048.

Chen, C. H., Lee, C. M., Lai, H. C., Hu, T. H., Su, W. P., Lu, S. N., et al. (2017). Prediction model of hepatocellular carcinoma risk in Asian patients with chronic hepatitis B treated with entecavir. *Oncotarget*, *8*(54), 92431–92441. https://doi.org/10.18632/oncotarget.21369.

Chen, C. J., Yang, H. I., Su, J., Jen, C. L., You, S. L., Lu, S. N., et al. (2006). Risk of hepatocellular carcinoma across a biological gradient of serum hepatitis B virus DNA level. *JAMA*, *295*(1), 65–73. https://doi.org/10.1001/jama.295.1.65.

Choi, W. M., Choi, J., & Lim, Y. S. (2021). Effects of tenofovir vs entecavir on risk of hepatocellular carcinoma in patients with chronic HBV infection: A systematic review and meta-analysis. *Clinical Gastroenterology and Hepatology*, *19*(2), 246–258.e249. https://doi.org/10.1016/j.cgh.2020.05.008.

Chun, H. S., Kim, B. K., Park, J. Y., Kim, D. Y., Ahn, S. H., Han, K. H., et al. (2020). Design and validation of risk prediction model for hepatocellular carcinoma development after sustained virological response in patients with chronic hepatitis C. *European Journal of Gastroenterology & Hepatology*, *32*(3), 378–385. https://doi.org/10.1097/meg.0000000000001512.

Crouchet, E., Bandiera, S., Fujiwara, N., Li, S., El Saghire, H., Fernández-Vaquero, M., et al. (2021). A human liver cell-based system modeling a clinical prognostic liver signature for therapeutic discovery. *Nature Communications*, *12*(1), 5525. https://doi.org/10.1038/s41467-021-25468-9.

Dalton-Fitzgerald, E., Tiro, J., Kandunoori, P., Halm, E. A., Yopp, A., & Singal, A. G. (2015). Practice patterns and attitudes of primary care providers and barriers to surveillance of hepatocellular carcinoma in patients with cirrhosis. *Clinical Gastroenterology and Hepatology*, *13*(4), 791–798.e791. https://doi.org/10.1016/j.cgh.2014.06.031.

Degasperi, E., Galmozzi, E., Pelusi, S., D'Ambrosio, R., Soffredini, R., Borghi, M., et al. (2020). Hepatic fat-genetic risk score predicts hepatocellular carcinoma in patients with cirrhotic HCV treated with DAAs. *Hepatology*, *72*(6), 1912–1923. https://doi.org/10.1002/hep.31500.

Dirchwolf, M., Marciano, S., Ruf, A. E., Singal, A. G., D'Ercole, V., Coisson, P., et al. (2021). Failure in all steps of hepatocellular carcinoma surveillance process is frequent in daily practice. *Annals of Hepatology*, *25*, 100344. https://doi.org/10.1016/j.aohep.2021.100344.

Donati, B., Dongiovanni, P., Romeo, S., Meroni, M., McCain, M., Miele, L., et al. (2017). MBOAT7 rs641738 variant and hepatocellular carcinoma in non-cirrhotic individuals. *Scientific Reports*, *7*(1), 4492. https://doi.org/10.1038/s41598-017-04991-0.

Dragani, T. A. (2010). Risk of HCC: Genetic heterogeneity and complex genetics. *Journal of Hepatology*, *52*(2), 252–257. https://doi.org/10.1016/j.jhep.2009.11.015.

Du, Y., Du, B., Fang, X., Shu, M., Zhang, Y., Chung, H., et al. (2020). ALT flare predicts hepatocellular carcinoma among antiviral treated patients with chronic hepatitis B: A cross-country cohort study. *Frontiers in Oncology*, *10*, 615203. https://doi.org/10.3389/fonc.2020.615203.

Duarte-Salles, T., Misra, S., Stepien, M., Plymoth, A., Muller, D., Overvad, K., et al. (2016). Circulating osteopontin and prediction of hepatocellular carcinoma development in a large European population. *Cancer Prevention Research (Philadelphia, Pa.), 9*(9), 758–765. https://doi.org/10.1158/1940-6207.CAPR-15-0434.

El-Serag, H. B., Sardell, R., Thrift, A. P., Kanwal, F., & Miller, P. (2021). Texas has the highest hepatocellular carcinoma incidence rates in the USA. *Digestive Diseases and Sciences, 66*(3), 912–916. https://doi.org/10.1007/s10620-020-06231-4.

Estes, C., Razavi, H., Loomba, R., Younossi, Z., & Sanyal, A. J. (2018). Modeling the epidemic of nonalcoholic fatty liver disease demonstrates an exponential increase in burden of disease. *Hepatology, 67*(1), 123–133. https://doi.org/10.1002/hep.29466.

Fan, R., Papatheodoridis, G., Sun, J., Innes, H., Toyoda, H., Xie, Q., et al. (2020). aMAP risk score predicts hepatocellular carcinoma development in patients with chronic hepatitis. *Journal of Hepatology, 73*(6), 1368–1378. https://doi.org/10.1016/j.jhep.2020.07.025.

Farid, A. A., Afify, N. A., Alsharnoby, A. A., Abdelsameea, E., & Bedair, H. M. (2021). Predictive role of AURKA rs 1047972 gene polymorphism and the risk of development of hepatocellular carcinoma. *Immunological Investigations, 1-11*. https://doi.org/10.1080/08820139.2021.1920609.

Farvardin, S., Patel, J., Khambaty, M., Yerokun, O. A., Mok, H., Tiro, J. A., et al. (2017). Patient-reported barriers are associated with lower hepatocellular carcinoma surveillance rates in patients with cirrhosis. *Hepatology, 65*(3), 875–884. https://doi.org/10.1002/hep.28770.

Fattovich, G., Stroffolini, T., Zagni, I., & Donato, F. (2004). Hepatocellular carcinoma in cirrhosis: Incidence and risk factors. *Gastroenterology, 127*(5 Suppl. 1), S35–S50. https://doi.org/10.1053/j.gastro.2004.09.014.

Finkin, S., Yuan, D., Stein, I., Taniguchi, K., Weber, A., Unger, K., et al. (2015). Ectopic lymphoid structures function as microniches for tumor progenitor cells in hepatocellular carcinoma. *Nature Immunology, 16*(12), 1235–1244. https://doi.org/10.1038/ni.3290.

Flemming, J. A., Yang, J. D., Vittinghoff, E., Kim, W. R., & Terrault, N. A. (2014). Risk prediction of hepatocellular carcinoma in patients with cirrhosis: The ADRESS-HCC risk model. *Cancer, 120*(22), 3485–3493. https://doi.org/10.1002/cncr.28832.

Franco, R. A., Fan, Y., Jarosek, S., Bae, S., & Galbraith, J. (2018). Racial and geographic disparities in hepatocellular carcinoma outcomes. *American Journal of Preventive Medicine, 55*(5 Suppl. 1), S40–s48. https://doi.org/10.1016/j.amepre.2018.05.030.

Fujiwara, N., Fobar, A. J., Raman, I., Li, Q. Z., Marrero, J. A., Parikh, N. D., et al. (2021). A blood-based prognostic liver secretome signature predicts long-term risk of hepatic decompensation in cirrhosis. *Clinical Gastroenterology and Hepatology*. https://doi.org/10.1016/j.cgh.2021.03.019.

Fujiwara, N., Friedman, S. L., Goossens, N., & Hoshida, Y. (2018). Risk factors and prevention of hepatocellular carcinoma in the era of precision medicine. *Journal of Hepatology, 68*(3), 526–549. https://doi.org/10.1016/j.jhep.2017.09.016.

Fujiwara, N., & Hoshida, Y. (2020). Viral exposure signature associated with liver cancer risk. *Trends in Molecular Medicine, 26*(8), 711–713. https://doi.org/10.1016/j.molmed.2020.06.008.

Fujiwara, N., Kobayashi, M., Fobar, A. J., Hoshida, A., Marquez, C. A., Koneru, B., et al. (2021). A blood-based prognostic liver secretome signature and long-term hepatocellular carcinoma risk in advanced liver fibrosis. *Med (New York, N.Y.), 2*(7), 836–850.e810. https://doi.org/10.1016/j.medj.2021.03.017.

Fujiwara, N., Liu, P. H., Athuluri-Divakar, S. K., Zhu, S., & Hoshida, Y. (2019). Risk factors of hepatocellular carcinoma for precision personalized care. In Y. Hoshida (Ed.), *Hepatocellular carcinoma: Translational precision medicine approaches* (pp. 3–25). Cham (CH): Humana Press. Springer Nature Switzerland AG.

Fujiwara, N., Trépo, E., Raman, I., Li, Q. Z., Degré, D., Gustot, T., et al. (2021). Plasma-signature-model for end-stage liver disease score to predict survival in severe alcoholic hepatitis. *Clinical Gastroenterology and Hepatology*. https://doi.org/10.1016/j.cgh.2021.02.041.

Ganne-Carrié, N., Layese, R., Bourcier, V., Cagnot, C., Marcellin, P., Guyader, D., et al. (2016). Nomogram for individualized prediction of hepatocellular carcinoma occurrence in hepatitis C virus cirrhosis (ANRS CO12 CirVir). *Hepatology, 64*(4), 1136–1147. https://doi.org/10.1002/hep.28702.

Gellert-Kristensen, H., Richardson, T. G., Davey Smith, G., Nordestgaard, B. G., Tybjaerg-Hansen, A., & Stender, S. (2020). Combined effect of PNPLA3, TM6SF2, and HSD17B13 variants on risk of cirrhosis and hepatocellular carcinoma in the general population. *Hepatology, 72*(3), 845–856. https://doi.org/10.1002/hep.31238.

Goossens, N., Singal, A. G., King, L. Y., Andersson, K. L., Fuchs, B. C., Besa, C., et al. (2017). Cost-effectiveness of risk score-stratified hepatocellular carcinoma screening in patients with cirrhosis. *Clinical and Translational Gastroenterology, 8*(6), e101. https://doi.org/10.1038/ctg.2017.26.

Hernandez, B. Y., Zhu, X., Risch, H. A., Lu, L., Ma, X., Irwin, M. L., et al. (2021). Oral cyanobacteria and hepatocellular carcinoma. *Cancer Epidemiology, Biomarkers & Prevention*. https://doi.org/10.1158/1055-9965.EPI-21-0804.

Higashi, T., Friedman, S. L., & Hoshida, Y. (2017). Hepatic stellate cells as key target in liver fibrosis. *Advanced Drug Delivery Reviews, 121*, 27–42. https://doi.org/10.1016/j.addr.2017.05.007.

Hiraoka, A., Kumada, T., Ogawa, C., Kariyama, K., Morita, M., Nouso, K., et al. (2019). Proposed a simple score for recommendation of scheduled ultrasonography surveillance for hepatocellular carcinoma after Direct Acting Antivirals: Multicenter analysis. *Journal of Gastroenterology and Hepatology, 34*(2), 436–441. https://doi.org/10.1111/jgh.14378.

Hoshida, Y., Villanueva, A., Kobayashi, M., Peix, J., Chiang, D. Y., Camargo, A., et al. (2008). Gene expression in fixed tissues and outcome in hepatocellular carcinoma. *The New England Journal of Medicine, 359*(19), 1995–2004. https://doi.org/10.1056/NEJMoa0804525.

Hoshida, Y., Villanueva, A., Sangiovanni, A., Sole, M., Hur, C., Andersson, K. L., et al. (2013). Prognostic gene expression signature for patients with hepatitis C-related early-stage cirrhosis. *Gastroenterology, 144*(5), 1024–1030. https://doi.org/10.1053/j.gastro.2013.01.021.

Hsu, Y. C., Jun, T., Huang, Y. T., Yeh, M. L., Lee, C. L., Ogawa, S., et al. (2018). Serum M2BPGi level and risk of hepatocellular carcinoma after oral anti-viral therapy in patients with chronic hepatitis B. *Alimentary Pharmacology & Therapeutics, 48*(10), 1128–1137. https://doi.org/10.1111/apt.15006.

Hsu, Y. C., Yip, T. C., Ho, H. J., Wong, V. W., Huang, Y. T., El-Serag, H. B., et al. (2018). Development of a scoring system to predict hepatocellular carcinoma in Asians on anti-virals for chronic hepatitis B. *Journal of Hepatology, 69*(2), 278–285. https://doi.org/10.1016/j.jhep.2018.02.032.

Huang, D. Q., El-Serag, H. B., & Loomba, R. (2021). Global epidemiology of NAFLD-related HCC: Trends, predictions, risk factors and prevention. *Nature Reviews. Gastroenterology & Hepatology, 18*(4), 223–238. https://doi.org/10.1038/s41575-020-00381-6.

Huang, Y. H., Liang, K. H., Chien, R. N., Hu, T. H., Lin, K. H., Hsu, C. W., et al. (2017). A circulating MicroRNA signature capable of assessing the risk of hepatocellular carcinoma in cirrhotic patients. *Scientific Reports, 7*(1), 523. https://doi.org/10.1038/s41598-017-00631-9.

Hughes, D. M., Berhane, S., Emily de Groot, C. A., Toyoda, H., Tada, T., Kumada, T., et al. (2021). Serum levels of α-fetoprotein increased more than 10 years before detection of hepatocellular carcinoma. *Clinical Gastroenterology and Hepatology, 19*(1), 162–170. e164. https://doi.org/10.1016/j.cgh.2020.04.084.

Hung, Y. C., Lin, C. L., Liu, C. J., Hung, H., Lin, S. M., Lee, S. D., et al. (2015). Development of risk scoring system for stratifying population for hepatocellular carcinoma screening. *Hepatology, 61*(6), 1934–1944. https://doi.org/10.1002/hep.27610.

Ioannou, G. N., Green, P., Kerr, K. F., & Berry, K. (2019). Models estimating risk of hepatocellular carcinoma in patients with alcohol or NAFLD-related cirrhosis for risk stratification. *Journal of Hepatology*. https://doi.org/10.1016/j.jhep.2019.05.008.

Islami, F., Miller, K. D., Siegel, R. L., Fedewa, S. A., Ward, E. M., & Jemal, A. (2017). Disparities in liver cancer occurrence in the United States by race/ethnicity and state. *CA: A Cancer Journal for Clinicians, 67*(4), 273–289. https://doi.org/10.3322/caac.21402.

Jee, S. H., Kim, M., Kim, M., Yoo, H. J., Kim, H., Jung, K. J., et al. (2018). Metabolomics profiles of hepatocellular carcinoma in a Korean prospective cohort: The Korean Cancer Prevention Study-II. *Cancer Prevention Research (Philadelphia, Pa.), 11*(5), 303–312. https://doi.org/10.1158/1940-6207.capr-17-0249.

Ji, J., Eggert, T., Budhu, A., Forgues, M., Takai, A., Dang, H., et al. (2015). Hepatic stellate cell and monocyte interaction contributes to poor prognosis in hepatocellular carcinoma. *Hepatology, 62*(2), 481–495. https://doi.org/10.1002/hep.27822.

Jiang, D. K., Sun, J., Cao, G., Liu, Y., Lin, D., Gao, Y. Z., et al. (2013). Genetic variants in STAT4 and HLA-DQ genes confer risk of hepatitis B virus-related hepatocellular carcinoma. *Nature Genetics, 45*(1), 72–75. https://doi.org/10.1038/ng.2483.

Jiang, G., Yu, K., Shao, L., Yu, X., Hu, C., Qian, P., et al. (2015). Association between epidermal growth factor gene +61A/G polymorphism and the risk of hepatocellular carcinoma: A meta-analysis based on 16 studies. *BMC Cancer, 15*, 314. https://doi.org/10.1186/s12885-015-1318-6.

Kim, J. H., Kang, S. H., Lee, M., Youn, G. S., Kim, T. S., Jun, B. G., et al. (2020). Serum myostatin predicts the risk of hepatocellular carcinoma in patients with alcoholic cirrhosis: A multicenter study. *Cancers (Basel), 12*(11). https://doi.org/10.3390/cancers12113347.

Kim, J. H., Kim, Y. D., Lee, M., Jun, B. G., Kim, T. S., Suk, K. T., et al. (2018). Modified PAGE-B score predicts the risk of hepatocellular carcinoma in Asians with chronic hepatitis B on antiviral therapy. *Journal of Hepatology, 69*(5), 1066–1073. https://doi.org/10.1016/j.jhep.2018.07.018.

Kim, H. Y., Lampertico, P., Nam, J. Y., Lee, H. C., Kim, S. U., Sinn, D. H., et al. (2021). An artificial intelligence model to predict hepatocellular carcinoma risk in Korean and Caucasian patients with chronic hepatitis B. *Journal of Hepatology*. https://doi.org/10.1016/j.jhep.2021.09.025.

Kim, N. J., Rozenberg-Ben-Dror, K., Jacob, D. A., Rich, N. E., Singal, A. G., Aby, E. S., et al. (2020). Provider attitudes toward risk-based hepatocellular carcinoma surveillance in patients with cirrhosis in the United States. *Clinical Gastroenterology and Hepatology*. https://doi.org/10.1016/j.cgh.2020.09.015.

Kim, J. H., Sohn, B. H., Lee, H. S., Kim, S. B., Yoo, J. E., Park, Y. Y., et al. (2014). Genomic predictors for recurrence patterns of hepatocellular carcinoma: Model derivation and validation. *PLoS Medicine, 11*(12), e1001770. https://doi.org/10.1371/journal.pmed.1001770.

Kim, H. S., Yu, X., Kramer, J., Thrift, A. P., Richardson, P., Hsu, Y. C., et al. (2021). Comparative performance of risk prediction models for hepatitis B-related hepatocellular carcinoma in the United States. *Journal of Hepatology*. https://doi.org/10.1016/j.jhep.2021.09.009.

King, L. Y., Canasto-Chibuque, C., Johnson, K. B., Yip, S., Chen, X., Kojima, K., et al. (2015). A genomic and clinical prognostic index for hepatitis C-related early-stage cirrhosis that predicts clinical deterioration. *Gut*, *64*(8), 1296–1302. https://doi.org/10.1136/gutjnl-2014-307862.

Kozlitina, J., Smagris, E., Stender, S., Nordestgaard, B. G., Zhou, H. H., Tybjaerg-Hansen, A., et al. (2014). Exome-wide association study identifies a TM6SF2 variant that confers susceptibility to nonalcoholic fatty liver disease. *Nature Genetics*, *46*(4), 352–356. https://doi.org/10.1038/ng.2901.

Kubota, N., Fujiwara, N., & Hoshida, Y. (2020). Clinical and molecular prediction of hepatocellular carcinoma risk. *Journal of Clinical Medicine*, *9*(12). https://doi.org/10.3390/jcm9123843.

Lee, M. H., Lu, S. N., Yuan, Y., Yang, H. I., Jen, C. L., You, S. L., et al. (2014). Development and validation of a clinical scoring system for predicting risk of HCC in asymptomatic individuals seropositive for anti-HCV antibodies. *PLoS One*, *9*(5), e94760. https://doi.org/10.1371/journal.pone.0094760.

Lee, H. W., Park, S. Y., Lee, M., Lee, E. J., Lee, J., Kim, S. U., et al. (2020). An optimized hepatocellular carcinoma prediction model for chronic hepatitis B with well-controlled viremia. *Liver International*, *40*(7), 1736–1743. https://doi.org/10.1111/liv.14451.

Lee, H. W., Yoo, E. J., Kim, B. K., Kim, S. U., Park, J. Y., Kim, D. Y., et al. (2014). Prediction of development of liver-related events by transient elastography in hepatitis B patients with complete virological response on antiviral therapy. *The American Journal of Gastroenterology*, *109*(8), 1241–1249. https://doi.org/10.1038/ajg.2014.157.

Letouze, E., Shinde, J., Renault, V., Couchy, G., Blanc, J. F., Tubacher, E., et al. (2017). Mutational signatures reveal the dynamic interplay of risk factors and cellular processes during liver tumorigenesis. *Nature Communications*, *8*(1), 1315. https://doi.org/10.1038/s41467-017-01358-x.

Liang, K. H., Cheng, M. L., Lo, C. J., Lin, Y. H., Lai, M. W., Lin, W. R., et al. (2020). Plasma phenylalanine and glutamine concentrations correlate with subsequent hepatocellular carcinoma occurrence in liver cirrhosis patients: An exploratory study. *Scientific Reports*, *10*(1), 10926. https://doi.org/10.1038/s41598-020-67971-x.

Liang, K. H., Lai, M. W., Lin, Y. H., Chu, Y. D., Lin, C. L., Lin, W. R., et al. (2021). Plasma interleukin-17 and alpha-fetoprotein combination effectively predicts imminent hepatocellular carcinoma occurrence in liver cirrhotic patients. *BMC Gastroenterology*, *21*(1), 177. https://doi.org/10.1186/s12876-021-01761-1.

Liang, Y., Yang, Z., & Zhong, R. (2012). Primary biliary cirrhosis and cancer risk: A systematic review and meta-analysis. *Hepatology*, *56*(4), 1409–1417. https://doi.org/10.1002/hep.25788.

Liu, J., Ayada, I., Zhang, X., Wang, L., Li, Y., Wen, T., et al. (2021). Estimating global prevalence of metabolic dysfunction-associated fatty liver disease in overweight or obese adults. *Clinical Gastroenterology and Hepatology*. https://doi.org/10.1016/j.cgh.2021.02.030.

Liu, Q., Liu, G., Lin, Z., Lin, Z., Tian, N., Lin, X., et al. (2021). The association of lncRNA SNPs and SNPs-environment interactions based on GWAS with HBV-related HCC risk and progression. *Molecular Genetics & Genomic Medicine*, *9*(2), e1585. https://doi.org/10.1002/mgg3.1585.

Liu, W., Ma, N., Zhao, D., Gao, X., Zhang, X., Yang, L., et al. (2019). Correlation between the DEPDC5 rs1012068 polymorphism and the risk of HBV-related hepatocellular carcinoma. *Clinics and Research in Hepatology and Gastroenterology*, *43*(4), 446–450. https://doi.org/10.1016/j.clinre.2018.12.005.

Liu, J., Tang, W., Budhu, A., Forgues, M., Hernandez, M. O., Candia, J., et al. (2020). A viral exposure signature defines early onset of hepatocellular carcinoma. *Cell, 182*(2), 317–328.e310. https://doi.org/10.1016/j.cell.2020.05.038.

Loomba, R., Friedman, S. L., & Shulman, G. I. (2021). Mechanisms and disease consequences of nonalcoholic fatty liver disease. *Cell, 184*(10), 2537–2564. https://doi.org/10.1016/j.cell.2021.04.015.

Luo, X., Wang, Y., Shen, A., Deng, H., & Ye, M. (2019). Relationship between the rs2596542 polymorphism in the MICA gene promoter and HBV/HCV infection-induced hepatocellular carcinoma: A meta-analysis. *BMC Medical Genetics, 20*(1), 142. https://doi.org/10.1186/s12881-019-0871-2.

Luo, P., Yin, P., Hua, R., Tan, Y., Li, Z., Qiu, G., et al. (2018). A large-scale, multicenter serum metabolite biomarker identification study for the early detection of hepatocellular carcinoma. *Hepatology, 67*(2), 662–675. https://doi.org/10.1002/hep.29561.

Luo, Y. Y., Zhang, H. P., Huang, A. L., & Hu, J. L. (2019). Association between KIF1B rs17401966 genetic polymorphism and hepatocellular carcinoma susceptibility: An updated meta-analysis. *BMC Medical Genetics, 20*(1), 59. https://doi.org/10.1186/s12881-019-0778-y.

Mancebo, A., González-Diéguez, M. L., Cadahía, V., Varela, M., Pérez, R., Navascués, C. A., et al. (2013). Annual incidence of hepatocellular carcinoma among patients with alcoholic cirrhosis and identification of risk groups. *Clinical Gastroenterology and Hepatology, 11*(1), 95–101. https://doi.org/10.1016/j.cgh.2012.09.007.

Mangoud, N. O. M., Ali, S. A., El Kassas, M., & Soror, S. H. (2021). Chitinase 3-like-1, Tolloid-like protein 1, and intergenic gene polymorphisms are predictors for hepatocellular carcinoma development after hepatitis C virus eradication by direct-acting antivirals. *IUBMB Life, 73*(2), 474–482. https://doi.org/10.1002/iub.2444.

Marrero, J. A., Kulik, L. M., Sirlin, C. B., Zhu, A. X., Finn, R. S., Abecassis, M. M., et al. (2018). Diagnosis, staging, and management of hepatocellular carcinoma: 2018 practice guidance by the American Association for the Study of Liver Diseases. *Hepatology, 68*(2), 723–750. https://doi.org/10.1002/hep.29913.

Martinez, M. A., & Franco, S. (2020). Therapy implications of hepatitis C virus genetic diversity. *Viruses, 13*(1). https://doi.org/10.3390/v13010041.

Massarweh, N. N., & El-Serag, H. B. (2017). Epidemiology of hepatocellular carcinoma and intrahepatic cholangiocarcinoma. *Cancer Control, 24*(3). https://doi.org/10.1177/1073274817729245. 1073274817729245.

Matsuura, K., Sawai, H., Ikeo, K., Ogawa, S., Iio, E., Isogawa, M., et al. (2017). Genome-wide association study identifies TLL1 variant associated with development of hepatocellular carcinoma after eradication of hepatitis C virus infection. *Gastroenterology, 152*(6), 1383–1394. https://doi.org/10.1053/j.gastro.2017.01.041.

Mazziotti, G., Sorvillo, F., Morisco, F., Carbone, A., Rotondi, M., Stornaiuolo, G., et al. (2002). Serum insulin-like growth factor I evaluation as a useful tool for predicting the risk of developing hepatocellular carcinoma in patients with hepatitis C virus-related cirrhosis: A prospective study. *Cancer, 95*(12), 2539–2545. https://doi.org/10.1002/cncr.11002.

McGlynn, K. A., Petrick, J. L., & El-Serag, H. B. (2021). Epidemiology of hepatocellular carcinoma. *Hepatology, 73*(Suppl. 1), 4–13. https://doi.org/10.1002/hep.31288.

Meroni, M., Longo, M., Paolini, E., Alisi, A., Miele, L., De Caro, E. R., et al. (2021). The rs599839 A>G variant disentangles cardiovascular risk and hepatocellular carcinoma in NAFLD patients. *Cancers (Basel), 13*(8). https://doi.org/10.3390/cancers13081783.

Miller, K. D., Ortiz, A. P., Pinheiro, P. S., Bandi, P., Minihan, A., Fuchs, H. E., et al. (2021). Cancer statistics for the US Hispanic/Latino population, 2021. *CA: A Cancer Journal for Clinicians*. https://doi.org/10.3322/caac.21695.

Mittal, S., El-Serag, H. B., Sada, Y. H., Kanwal, F., Duan, Z., Temple, S., et al. (2016). Hepatocellular carcinoma in the absence of cirrhosis in United States veterans is associated with nonalcoholic fatty liver disease. *Clinical Gastroenterology and Hepatology, 14*(1), 124–131.e121. https://doi.org/10.1016/j.cgh.2015.07.019.

Moeini, A., Torrecilla, S., Tovar, V., Montironi, C., Andreu-Oller, C., Peix, J., et al. (2019). An immune gene expression signature associated with development of human hepatocellular carcinoma identifies mice that respond to chemopreventive agents. *Gastroenterology, 157*(5), 1383–1397.e1311. https://doi.org/10.1053/j.gastro.2019.07.028.

Morgan, T. R., Mandayam, S., & Jamal, M. M. (2004). Alcohol and hepatocellular carcinoma. *Gastroenterology, 127*(5 Suppl. 1), S87–S96. https://doi.org/10.1053/j.gastro.2004.09.020.

Nakagawa, H., Maeda, S., Yoshida, H., Tateishi, R., Masuzaki, R., Ohki, T., et al. (2009). Serum IL-6 levels and the risk for hepatocarcinogenesis in chronic hepatitis C patients: An analysis based on gender differences. *International Journal of Cancer, 125*(10), 2264–2269. https://doi.org/10.1002/ijc.24720.

Nakagawa, S., Wei, L., Song, W. M., Higashi, T., Ghoshal, S., Kim, R. S., et al. (2016). Molecular liver cancer prevention in cirrhosis by organ transcriptome analysis and lysophosphatidic acid pathway inhibition. *Cancer Cell, 30*(6), 879–890. https://doi.org/10.1016/j.ccell.2016.11.004.

Nault, J. C., Datta, S., Imbeaud, S., Franconi, A., Mallet, M., Couchy, G., et al. (2015). Recurrent AAV2-related insertional mutagenesis in human hepatocellular carcinomas. *Nature Genetics, 47*(10), 1187–1193. https://doi.org/10.1038/ng.3389.

Ono, A., Goossens, N., Finn, R. S., Schmidt, W. N., Thung, S. N., Im, G. Y., et al. (2017). Persisting risk of hepatocellular carcinoma after hepatitis C virus cure monitored by a liver transcriptome signature. *Hepatology, 66*(4), 1344–1346. https://doi.org/10.1002/hep.29203.

Palmer, W. C., & Patel, T. (2012). Are common factors involved in the pathogenesis of primary liver cancers? A meta-analysis of risk factors for intrahepatic cholangiocarcinoma. *Journal of Hepatology, 57*(1), 69–76. https://doi.org/10.1016/j.jhep.2012.02.022.

Papatheodoridis, G., Dalekos, G., Sypsa, V., Yurdaydin, C., Buti, M., Goulis, J., et al. (2016). PAGE-B predicts the risk of developing hepatocellular carcinoma in Caucasians with chronic hepatitis B on 5-year antiviral therapy. *Journal of Hepatology, 64*(4), 800–806. https://doi.org/10.1016/j.jhep.2015.11.035.

Papatheodoridis, G. V., Sypsa, V., Dalekos, G. N., Yurdaydin, C., Van Boemmel, F., Buti, M., et al. (2020). Hepatocellular carcinoma prediction beyond year 5 of oral therapy in a large cohort of Caucasian patients with chronic hepatitis B. *Journal of Hepatology, 72*(6), 1088–1096. https://doi.org/10.1016/j.jhep.2020.01.007.

Poh, Z., Shen, L., Yang, H. I., Seto, W. K., Wong, V. W., Lin, C. Y., et al. (2016). Real-world risk score for hepatocellular carcinoma (RWS-HCC): A clinically practical risk predictor for HCC in chronic hepatitis B. *Gut, 65*(5), 887–888. https://doi.org/10.1136/gutjnl-2015-310818.

Poynard, T., Peta, V., Deckmyn, O., Munteanu, M., Moussalli, J., Ngo, Y., et al. (2019). LCR1 and LCR2, two multi-analyte blood tests to assess liver cancer risk in patients without or with cirrhosis. *Alimentary Pharmacology & Therapeutics, 49*(3), 308–320. https://doi.org/10.1111/apt.15082.

Qin, S., Wang, J., Zhou, C., Xu, Y., Zhang, Y., Wang, X., et al. (2019). The influence of interleukin 28B polymorphisms on the risk of hepatocellular carcinoma among patients with HBV or HCV infection: An updated meta-analysis. *Medicine (Baltimore), 98*(38), e17275. https://doi.org/10.1097/md.0000000000017275.

Qu, C., Wang, Y., Wang, P., Chen, K., Wang, M., Zeng, H., et al. (2019). Detection of early-stage hepatocellular carcinoma in asymptomatic HBsAg-seropositive individuals by liquid biopsy. *Proceedings of the National Academy of Sciences of the United States of America*, 116(13), 6308–6312. https://doi.org/10.1073/pnas.1819799116.

Ren, Z., Li, A., Jiang, J., Zhou, L., Yu, Z., Lu, H., et al. (2019). Gut microbiome analysis as a tool towards targeted non-invasive biomarkers for early hepatocellular carcinoma. *Gut*, 68(6), 1014–1023. https://doi.org/10.1136/gutjnl-2017-315084.

Rich, N. E., Carr, C., Yopp, A. C., Marrero, J. A., & Singal, A. G. (2020). Racial and ethnic disparities in survival among patients with hepatocellular carcinoma in the United States: A systematic review and meta-analysis. *Clinical Gastroenterology and Hepatology*. https://doi.org/10.1016/j.cgh.2020.12.029.

Rich, N. E., Murphy, C. C., Yopp, A. C., Tiro, J., Marrero, J. A., & Singal, A. G. (2020). Sex disparities in presentation and prognosis of 1110 patients with hepatocellular carcinoma. *Alimentary Pharmacology & Therapeutics*, 52(4), 701–709. https://doi.org/10.1111/apt.15917.

Roudot-Thoraval, F. (2021). Epidemiology of hepatitis C virus infection. *Clinics and Research in Hepatology and Gastroenterology*, 45(3), 101596. https://doi.org/10.1016/j.clinre.2020.101596.

Sanyal, A. J., Van Natta, M. L., Clark, J., Neuschwander-Tetri, B. A., Diehl, A., Dasarathy, S., et al. (2021). Prospective study of outcomes in adults with nonalcoholic fatty liver disease. *The New England Journal of Medicine*, 385(17), 1559–1569. https://doi.org/10.1056/NEJMoa2029349.

Schäffer, A. A., Dominguez, D. A., Chapman, L. M., Gertz, E. M., Budhu, A., Forgues, M., et al. (2021). Integration of adeno-associated virus (AAV) into the genomes of most Thai and Mongolian liver cancer patients does not induce oncogenesis. *BMC Genomics*, 22(1), 814. https://doi.org/10.1186/s12864-021-08098-9.

Schwabe, R. F., & Greten, T. F. (2020). Gut microbiome in HCC—Mechanisms, diagnosis and therapy. *Journal of Hepatology*, 72(2), 230–238. https://doi.org/10.1016/j.jhep.2019.08.016.

Sharma, S. A., Kowgier, M., Hansen, B. E., Brouwer, W. P., Maan, R., Wong, D., et al. (2017). Toronto HCC risk index: A validated scoring system to predict 10-year risk of HCC in patients with cirrhosis. *Journal of Hepatology*. https://doi.org/10.1016/j.jhep.2017.07.033.

Shinkai, N., Nojima, M., Iio, E., Matsunami, K., Toyoda, H., Murakami, S., et al. (2018). High levels of serum Mac-2-binding protein glycosylation isomer (M2BPGi) predict the development of hepatocellular carcinoma in hepatitis B patients treated with nucleot(s)ide analogues. *Journal of Gastroenterology*, 53(7), 883–889. https://doi.org/10.1007/s00535-017-1424-0.

Simmons, O. L., Feng, Y., Parikh, N. D., & Singal, A. G. (2019). Primary care provider practice patterns and barriers to hepatocellular carcinoma surveillance. *Clinical Gastroenterology and Hepatology*, 17(4), 766–773. https://doi.org/10.1016/j.cgh.2018.07.029.

Simon, T. G., & Chan, A. T. (2020). Lifestyle and environmental approaches for the primary prevention of hepatocellular carcinoma. *Clinics in Liver Disease*, 24(4), 549–576. https://doi.org/10.1016/j.cld.2020.06.002.

Simon, T. G., Roelstraete, B., Sharma, R., Khalili, H., Hagstrom, H., & Ludvigsson, J. F. (2021). Cancer risk in patients with biopsy-confirmed nonalcoholic fatty liver disease: A population-based cohort study. *Hepatology*, 74(5), 2410–2423. https://doi.org/10.1002/hep.31845.

Singal, A. G., & El-Serag, H. B. (2021). Rational screening approaches for HCC in NAFLD patients. *Journal of Hepatology*. https://doi.org/10.1016/j.jhep.2021.08.028.

Singal, A. G., Manjunath, H., Yopp, A. C., Beg, M. S., Marrero, J. A., Gopal, P., et al. (2014). The effect of PNPLA3 on fibrosis progression and development of hepatocellular carcinoma: A meta-analysis. *The American Journal of Gastroenterology*, *109*(3), 325–334. https://doi.org/10.1038/ajg.2013.476.

Singal, A. G., Mukherjee, A., Elmunzer, B. J., Higgins, P. D., Lok, A. S., Zhu, J., et al. (2013). Machine learning algorithms outperform conventional regression models in predicting development of hepatocellular carcinoma. *The American Journal of Gastroenterology*, *108*(11), 1723–1730. https://doi.org/10.1038/ajg.2013.332.

Singal, A. G., Tiro, J. A., Murphy, C. C., Blackwell, J. M., Kramer, J. R., Khan, A., et al. (2021). Patient-reported barriers are associated with receipt of hepatocellular carcinoma surveillance in a multicenter cohort of patients with cirrhosis. *Clinical Gastroenterology and Hepatology*, *19*(5), 987–995.e981. https://doi.org/10.1016/j.cgh.2020.06.049.

Sinn, D. H., Kang, D., Cho, S. J., Paik, S. W., Guallar, E., Cho, J., et al. (2020). Risk of hepatocellular carcinoma in individuals without traditional risk factors: Development and validation of a novel risk score. *International Journal of Epidemiology*, *49*(5), 1562–1571. https://doi.org/10.1093/ije/dyaa089.

Sinn, D. H., Lee, J. H., Kim, K., Ahn, J. H., Lee, J. H., Kim, J. H., et al. (2017). A novel model for predicting hepatocellular carcinoma development in patients with chronic hepatitis B and Normal alanine aminotransferase levels. *Gut and Liver*, *11*(4), 528–534. https://doi.org/10.5009/gnl16403.

Sohn, W., Cho, J. Y., Kim, J. H., Lee, J. I., Kim, H. J., Woo, M. A., et al. (2017). Risk score model for the development of hepatocellular carcinoma in treatment-naïve patients receiving oral antiviral treatment for chronic hepatitis B. *Clinical and Molecular Hepatology*, *23*(2), 170–178. https://doi.org/10.3350/cmh.2016.0086.

Stepien, M., Keski-Rahkonen, P., Kiss, A., Robinot, N., Duarte-Salles, T., Murphy, N., et al. (2021). Metabolic perturbations prior to hepatocellular carcinoma diagnosis: Findings from a prospective observational cohort study. *International Journal of Cancer*, *148*(3), 609–625. https://doi.org/10.1002/ijc.33236.

Stickel, F., Lutz, P., Buch, S., Nischalke, H. D., Silva, I., Rausch, V., et al. (2020). Genetic variation in HSD17B13 reduces the risk of developing cirrhosis and hepatocellular carcinoma in alcohol misusers. *Hepatology*, *72*(1), 88–102. https://doi.org/10.1002/hep.30996.

Strnad, P., McElvaney, N. G., & Lomas, D. A. (2020). Alpha(1)-antitrypsin deficiency. *The New England Journal of Medicine*, *382*(15), 1443–1455. https://doi.org/10.1056/NEJMra1910234.

Su, F., Weiss, N. S., Beste, L. A., Moon, A. M., Jin, G. Y., Green, P., et al. (2021). Screening is associated with a lower risk of hepatocellular carcinoma-related mortality in patients with chronic hepatitis B. *Journal of Hepatology*, *74*(4), 850–859. https://doi.org/10.1016/j.jhep.2020.11.023.

Suh, B., Park, S., Shin, D. W., Yun, J. M., Yang, H. K., Yu, S. J., et al. (2015). High liver fibrosis index FIB-4 is highly predictive of hepatocellular carcinoma in chronic hepatitis B carriers. *Hepatology*, *61*(4), 1261–1268. https://doi.org/10.1002/hep.27654.

Sung, H., Ferlay, J., Siegel, R. L., Laversanne, M., Soerjomataram, I., Jemal, A., et al. (2021). Global cancer statistics 2020: GLOBOCAN estimates of incidence and mortality worldwide for 36 cancers in 185 countries. *CA: A Cancer Journal for Clinicians*, *71*(3), 209–249. https://doi.org/10.3322/caac.21660.

Tang, S., Zhang, J., Mei, T. T., Guo, H. Q., Wei, X. H., Zhang, W. Y., et al. (2019). Association of TM6SF2 rs58542926 T/C gene polymorphism with hepatocellular carcinoma: A meta-analysis. *BMC Cancer*, *19*(1), 1128. https://doi.org/10.1186/s12885-019-6173-4.

Thilakarathna, W., Rupasinghe, H. P. V., & Ridgway, N. D. (2021). Mechanisms by which probiotic bacteria attenuate the risk of hepatocellular carcinoma. *International Journal of Molecular Sciences, 22*(5). https://doi.org/10.3390/ijms22052606.

Thylur, R. P., Roy, S. K., Shrivastava, A., LaVeist, T. A., Shankar, S., & Srivastava, R. K. (2020). Assessment of risk factors, and racial and ethnic differences in hepatocellular carcinoma. *JGH Open, 4*(3), 351–359. https://doi.org/10.1002/jgh3.12336.

Trépo, E., Caruso, S., Yang, J., Imbeaud, S., Couchy, G., Bayard, Q., et al. (2021). Common genetic variation in alcohol-related hepatocellular carcinoma: A case-control genome-wide association study. *The Lancet Oncology*. https://doi.org/10.1016/s1470-2045(21)00603-3.

Tseng, T. C., Peng, C. Y., Hsu, Y. C., Su, T. H., Wang, C. C., Liu, C. J., et al. (2020). Baseline mac-2 binding protein glycosylation isomer level stratifies risks of hepatocellular carcinoma in chronic hepatitis B patients with oral antiviral therapy. *Liver Cancer, 9*(2), 207–220. https://doi.org/10.1159/000504650.

Verhelst, X., Vanderschaeghe, D., Castéra, L., Raes, T., Geerts, A., Francoz, C., et al. (2017). A glycomics-based test predicts the development of hepatocellular carcinoma in cirrhosis. *Clinical Cancer Research, 23*(11), 2750–2758. https://doi.org/10.1158/1078-0432.ccr-16-1500.

Voulgaris, T., Papatheodoridi, M., Lampertico, P., & Papatheodoridis, G. V. (2020). Clinical utility of hepatocellular carcinoma risk scores in chronic hepatitis B. *Liver International, 40*(3), 484–495. https://doi.org/10.1111/liv.14334.

Wang, C., Hann, H.-W., Ye, Z., Hann, R. S., Wan, S., Ye, X., et al. (2016). Prospective evidence of a circulating microRNA signature as a non-invasive marker of hepatocellular carcinoma in HBV patients. *Oncotarget*. Retrieved from https://www.oncotarget.com/article/9429/.

Wang, B., Rudnick, S., Cengia, B., & Bonkovsky, H. L. (2019). Acute hepatic porphyrias: Review and recent progress. *Hepatology Communications, 3*(2), 193–206. https://doi.org/10.1002/hep4.1297.

Wolf, E., Rich, N. E., Marrero, J. A., Parikh, N. D., & Singal, A. G. (2021). Use of hepatocellular carcinoma surveillance in patients with cirrhosis: A systematic review and meta-analysis. *Hepatology, 73*(2), 713–725. https://doi.org/10.1002/hep.31309.

Wong, V. W., Chan, S. L., Mo, F., Chan, T. C., Loong, H. H., Wong, G. L., et al. (2010). Clinical scoring system to predict hepatocellular carcinoma in chronic hepatitis B carriers. *Journal of Clinical Oncology, 28*(10), 1660–1665. https://doi.org/10.1200/jco.2009.26.2675.

Wong, G. L., Chan, H. L., Wong, C. K., Leung, C., Chan, C. Y., Ho, P. P., et al. (2014). Liver stiffness-based optimization of hepatocellular carcinoma risk score in patients with chronic hepatitis B. *Journal of Hepatology, 60*(2), 339–345. https://doi.org/10.1016/j.jhep.2013.09.029.

Woolen, S. A., Singal, A. G., Davenport, M. S., Troost, J. P., Khalatbari, S., Mittal, S., et al. (2021). Patient preferences for hepatocellular carcinoma surveillance parameters. *Clinical Gastroenterology and Hepatology*. https://doi.org/10.1016/j.cgh.2021.02.024.

World Health Organization. (2021). *Global progress report on HIV, viral hepatitis and sexually transmitted infections, 2021*. Retrieved from http://apps.who.int/iris/bitstream/handle/10665/342813/9789240030992-eng.pdf.

Wu, H. C., Yang, H. I., Wang, Q., Chen, C. J., & Santella, R. M. (2017). Plasma DNA methylation marker and hepatocellular carcinoma risk prediction model for the general population. *Carcinogenesis, 38*(10), 1021–1028. https://doi.org/10.1093/carcin/bgx078.

Xu, J., Zhan, Q., Fan, Y., Lo, E. K. K., Zhang, F., Yu, Y., et al. (2021). Clinical aspects of gut microbiota in hepatocellular carcinoma management. *Pathogens, 10*(7). https://doi.org/10.3390/pathogens10070782.

Yamasaki, K., Tateyama, M., Abiru, S., Komori, A., Nagaoka, S., Saeki, A., et al. (2014). Elevated serum levels of Wisteria floribunda agglutinin-positive human Mac-2 binding protein predict the development of hepatocellular carcinoma in hepatitis C patients. *Hepatology, 60*(5), 1563–1570. https://doi.org/10.1002/hep.27305.

Yamashita, T., Koshikawa, N., Shimakami, T., Terashima, T., Nakagawa, M., Nio, K., et al. (2021). Serum laminin γ2 monomer as a diagnostic and predictive biomarker for hepatocellular carcinoma. *Hepatology, 74*(2), 760–775. https://doi.org/10.1002/hep.31758.

Yang, J. D., Kim, W. R., Coelho, R., Mettler, T. A., Benson, J. T., Sanderson, S. O., et al. (2011). Cirrhosis is present in most patients with hepatitis B and hepatocellular carcinoma. *Clinical Gastroenterology and Hepatology, 9*(1), 64–70. https://doi.org/10.1016/j.cgh.2010.08.019.

Yang, H. I., Sherman, M., Su, J., Chen, P. J., Liaw, Y. F., Iloeje, U. H., et al. (2010). Nomograms for risk of hepatocellular carcinoma in patients with chronic hepatitis B virus infection. *Journal of Clinical Oncology, 28*(14), 2437–2444. https://doi.org/10.1200/jco.2009.27.4456.

Yang, H. I., Yeh, M. L., Wong, G. L., Peng, C. Y., Chen, C. H., Trinh, H. N., et al. (2020). Real-world effectiveness from the Asia Pacific rim liver consortium for HBV risk score for the prediction of hepatocellular carcinoma in chronic hepatitis B patients treated with oral antiviral therapy. *The Journal of Infectious Diseases, 221*(3), 389–399. https://doi.org/10.1093/infdis/jiz477.

Yang, H. I., Yuen, M. F., Chan, H. L., Han, K. H., Chen, P. J., Kim, D. Y., et al. (2011). Risk estimation for hepatocellular carcinoma in chronic hepatitis B (REACH-B): Development and validation of a predictive score. *The Lancet Oncology, 12*(6), 568–574. https://doi.org/10.1016/s1470-2045(11)70077-8.

Yeo, Y. H., & Nguyen, M. H. (2021). Review article: Current gaps and opportunities in HBV prevention, testing and linkage to care in the United States-a call for action. *Alimentary Pharmacology & Therapeutics, 53*(1), 63–78. https://doi.org/10.1111/apt.16125.

Younossi, Z., Tacke, F., Arrese, M., Chander Sharma, B., Mostafa, I., Bugianesi, E., et al. (2019). Global perspectives on nonalcoholic fatty liver disease and nonalcoholic steatohepatitis. *Hepatology, 69*(6), 2672–2682. https://doi.org/10.1002/hep.30251.

Yousafzai, M. T., Bajis, S., Alavi, M., Grebely, J., Dore, G. J., & Hajarizadeh, B. (2021). Global cascade of care for chronic hepatitis C virus infection: A systematic review and meta-analysis. *Journal of Viral Hepatitis, 28*(10), 1340–1354. https://doi.org/10.1111/jvh.13574.

Yu, J. H., Suh, Y. J., Jin, Y. J., Heo, N. Y., Jang, J. W., You, C. R., et al. (2019). Prediction model for hepatocellular carcinoma risk in treatment-naive chronic hepatitis B patients receiving entecavir/tenofovir. *European Journal of Gastroenterology & Hepatology, 31*(7), 865–872. https://doi.org/10.1097/meg.0000000000001357.

Yuan, J. M., Wang, Y., Wang, R., Luu, H. N., Adams-Haduch, J., Koh, W. P., et al. (2021). Serum IL27 in relation to risk of hepatocellular carcinoma in two nested case-control studies. *Cancer Epidemiology, Biomarkers & Prevention, 30*(2), 388–395. https://doi.org/10.1158/1055-9965.epi-20-1081.

Yuen, M. F., Tanaka, Y., Fong, D. Y., Fung, J., Wong, D. K., Yuen, J. C., et al. (2009). Independent risk factors and predictive score for the development of hepatocellular carcinoma in chronic hepatitis B. *Journal of Hepatology, 50*(1), 80–88. https://doi.org/10.1016/j.jhep.2008.07.023.

Zhang, D. Y., Goossens, N., Guo, J., Tsai, M. C., Chou, H. I., Altunkaynak, C., et al. (2016). A hepatic stellate cell gene expression signature associated with outcomes in hepatitis C cirrhosis and hepatocellular carcinoma after curative resection. *Gut, 65*(10), 1754–1764. https://doi.org/10.1136/gutjnl-2015-309655.

Zhang, L., Xu, K., Liu, C., & Chen, J. (2017). Meta-analysis reveals an association between signal transducer and activator of transcription-4 polymorphism and hepatocellular carcinoma risk. *Hepatology Research*, *47*(4), 303–311. https://doi.org/10.1111/hepr.12733.

Zhao, C., & Nguyen, M. H. (2016). Hepatocellular carcinoma screening and surveillance: Practice guidelines and real-life practice. *Journal of Clinical Gastroenterology*, *50*(2), 120–133. https://doi.org/10.1097/mcg.0000000000000446.

Zheng, R., Wang, G., Pang, Z., Ran, N., Gu, Y., Guan, X., et al. (2020). Liver cirrhosis contributes to the disorder of gut microbiota in patients with hepatocellular carcinoma. *Cancer Medicine*, *9*(12), 4232–4250. https://doi.org/10.1002/cam4.3045.

Zhou, A., Tang, L., Zeng, S., Lei, Y., Yang, S., & Tang, B. (2020). Gut microbiota: A new piece in understanding hepatocarcinogenesis. *Cancer Letters*, *474*, 15–22. https://doi.org/10.1016/j.canlet.2020.01.002.

Zhu, M., Lu, T., Jia, Y., Luo, X., Gopal, P., Li, L., et al. (2019). Somatic mutations increase hepatic clonal fitness and regeneration in chronic liver disease. *Cell*, *177*(3), 608–621.e612. https://doi.org/10.1016/j.cell.2019.03.026.

CHAPTER TWO

Inflammatory pathways and cholangiocarcinoma risk mechanisms and prevention

Massimiliano Cadamuro[a] and Mario Strazzabosco[b],*

[a]Department of Molecular Medicine (DMM), University of Padova, Padova, Italy
[b]Liver Center, Department of Internal Medicine, Yale University, New Haven, CT, United States
*Corresponding author: e-mail address: mario.strazzabosco@yale.edu

Contents

1.	Introduction	41
2.	Diseases prodromal to cholangiocarcinoma development	45
	2.1 Primary sclerosing cholangitis	45
	2.2 Congenital hepatic fibrosis/Caroli's disease	47
	2.3 Fluke infestations	50
	2.4 Other risk factors	51
3.	Mechanisms of neoplastic transformation	53
4.	Role of inflammatory cells in modulating CCA malignant features	57
5.	Conclusions	62
	Grant support	62
	Conflict of interest statement	62
	References	62

Abstract

Cholangiocarcinoma (CCA), a neoplasm burdened by a poor prognosis and currently lacking adequate therapeutic treatments, can originate at different levels of the biliary tree, in the intrahepatic, hilar, or extrahepatic area. The main risk factors for the development of CCA are the presence of chronic cholangiopathies of various etiology. To date, the most studied prodromal diseases of CCA are primary sclerosing cholangitis, Caroli's disease and fluke infestations, but other conditions, such as metabolic syndrome, nonalcoholic fatty liver disease and obesity, are emerging as associated with an increased risk of CCA development. In this review, we focused on the analysis of the pro-inflammatory mechanisms that induce the development of CCA and on the role of cells of the immune response in cholangiocarcinogenesis. In very recent times, these cellular mechanisms have been the subject of emerging studies aimed at verifying how the modulation of the inflammatory and immunological responses can have a therapeutic significance and how these can be used as therapeutic targets.

Abbreviations

BilIN	biliary intraepithelial neoplasia
CAFs	cancer-associated fibroblasts
CCA	cholangiocarcinoma
CCL25	Chemokine C-C motif ligand
CD	Caroli's disease
Cdc25A	cell division cycle 25 homolog A
CHF	congenital hepatic fibrosis
COX-2	cyclooxygenase-2
CTGF	connective tissue growth factor
CXCR	C-X-C chemokine receptor
CXCL	C-X-C motif chemokine ligand
dCCA	distal CCA
DC	dendritic cells
DPC4	deleted in pancreatic cancer 4
eCCA	extrahepatic CCA
ECM	extracellular matrix
EGFR	epidermal growth factor receptor
ERK	extracellular-signal-regulated kinase
FPC	fibrocystin
FXR	farnesoid X receptor
GWAS	genome-wide association study
HCC	hepatocellular carcinoma
HDAC	histone deacetylase
HLA	human leukocyte antigen
HR	hazard ratio
iCCA	intrahepatic CCA
iNOS	inducible nitric oxide synthase
iPSC	induced pluripotent stem cell
IFN	interferon
IL	interleukin
IPNB	intraductal papillary neoplasm of the bile duct
MadCAM-1	mucosal vascular address in cell adhesion molecule 1
MAPK	mitogen-activated protein kinases
Mcl1	myeloid cell leukemia 1
MDSC	Myeloid-Derived Suppressor Cells
MMP	metalloproteases
miRNA	micro RNA
mTOR	mammalian target of rapamycin
NAFLD	non-alcoholic fatty liver disease
NASH	non-alcoholic steatohepatitis
NK	Neutral Killer Cells
8-NG	8-nitroguanine
NO	nitric oxide
OR	odds ratio
8-oxoDG	8-oxo-2′-deoxyguanosine
OvESP	*O. viverrini* excretory/secretory products

Ov-GRN-1	*O. viverrini* granulin
Ov-Trx-1	*O. viverrini* thioredoxin
PDGF	platelet-derived growth factors
pCCA	perihilar CCA
PSC	primary sclerosing cholangitis
PCNA	proliferating cell nuclear antigen
PKHD1	polycystic kidney and hepatic disease 1
PI3K	phosphatidylinositol 3-kinases
PKA	protein kinase A
ROS	reactive oxidative species
Smad4	signaling effectors mothers against decapentaplegic protein 4
TLR	Toll-like receptors
TCF/LEF	T cell factor/lymphoid enhancer factor
TNF	tumor necrosis factor
TGF-β	transforming growth factor β
TME	tumor microenvironment
TIL	tumor-infiltrating lymphocytes
TAM	tumor-associated macrophages
TAN	tumor-associated neutrophils
VEGF	vascular endothelial growth factor
VCAM-1	vascular cell adhesion molecule 1
YAP	Yes-associated protein

1. Introduction

Cholangiocarcinoma (CCA) is a highly aggressive neoplasia that can arise from the neoplastic transformation of cholangiocytes in any portion of the biliary system, from the finest intrahepatic ramifications to the main extrahepatic bile ducts. CCA is the most common biliary malignancy and the second most common primary liver cancer after hepatocellular carcinoma (HCC) (Banales et al., 2016, 2020), CCA remains an unmet need as less than 5% of patients survive up to 5 years from the diagnosis, making CCA responsible for 13% of overall cancer-related deaths (Welzel, McGlynn, Hsing, O'Brien, & Pfeiffer, 2006). A number of features contribute to the dismal prognosis of CCA: its insidious presentation due to late onset of the symptoms, lack of reliable clinical, laboratory or instrumental markers, and its early metastatic dissemination. Many patients are diagnosed with metastatic disease at presentation, significantly limiting the possibility of curative treatment.

CCA may arise from any tract of the biliary system and can be anatomically classified as intrahepatic (iCCA), perihilar (pCCA), or distal CCA (dCCA) (Rizvi & Gores, 2013). These site-restricted variants show significant clinical and genomic differences. The pCCA subtype is the most common type, and in the USA, it accounts for approximately 50–60% of all CCAs, followed by dCCA (20–30%) and iCCA (10–20%) (Banales et al., 2020). Globally, CCA has an incidence rate of 0.3–6/100,000 inhabitants per year, with a mortality rate of 1–6/100,000 inhabitants per year (Sarcognato et al., 2021), but its incidence has wide geographical variations, reflecting the heterogeneous distribution of environmental risk factors as well as the genetic predisposition of the different populations (Clements et al., 2020; Banales et al., 2020).

A number of risk factors for CCA have been identified and are listed in Table 1. There are significant difference in terms of carcinogenetic mechanisms according to the type of risk factors involved. CCA often arises in the setting of prolonged biliary inflammation, cholestasis and activation of reparative regenerative hepatic mechanisms, involving cell proliferation (Brindley et al., 2021; Leone, Ali, Weber, Tschaharganeh, & Heikenwalder, 2021). Recent transcriptomic analysis studies distinguished two iCCA profiles, enriched with pro-inflammatory (inflammatory class) and oncogenic pathways (proliferation class), respectively. The "inflammatory" subclass accounts for about 38% of the cases and is characterized by increased expression of immune-related signaling pathways, while the "proliferation" subclass, accounting for about 62% of iCCA cases, is enriched with oncogenic pathways and shows a worse prognosis (Sia et al., 2013). Similarly to iCCA, eCCA could also be differentiated in four subclasses, one of which is characterized by the upregulation of inflammatory mediators (Montal et al., 2020). There are marked differences in the genomic features of CCA, depending on the anatomical location and risk factors. Their discussion is beyond the scope of the current work and the reader is referred to excellent recent reviews (Banales et al., 2020; Cadamuro et al., 2021; Rodrigues et al., 2021; Simbolo et al., 2021).

Here we will focus on the pathways that can be related to the inflammatory changes. The majority of CCA risk factors are associated with chronic inflammatory states and changes in signaling pathways involved in cholangiocyte biology, leading among others to uncontrolled proliferation, evasion of apoptosis, and loss of genome integrity that could eventually carry to cell senescence. Cellular senescence, often the result of chronic inflammation, has been little studied and it is not clear what its significance is in

Table 1 Risk factors for CCA.

Risk factors	Refs.
Primary sclerosing cholangitis	Bjøro, Brandsaeter, Foss, and Schrumpf (2006), Lazaridis and LaRusso (2016), and Aune, Sen, Norat, Riboli, and Folseraas (2021)
Fibrocystic liver diseases: • Congenital hepatic fibrosis • Caroli disease • ARPKD • Choledocal cysts	Khan, Toledano, and Taylor-Robinson (2008), Strazzabosco and Fabris (2012), and Cannito et al. (2018)
Fluke infestation: • Platyhelminthes: *Opistorchis viverrini* and *Opistorchis felineus* • Thrematoda: *Clonorchis sinensis*	Brindley et al. (2021)
Hepatolithiasis	Kim et al. (2015)
Viral chronic hepatitis: • HBV • HCV	Zhou et al. (2012), Matsumoto et al. (2014), Wang, Sheng, Dong, and Qin (2016), and Seo et al. (2020)
Metabolic diseases: • Metabolic syndrome • Obesity • NAFLD/NASH	Welzel et al. (2011), Palmer and Patel (2012), Wongjarupong et al. (2017), and Petrick et al. (2017)
Diabetes mellitus	Tyson and El-Serag (2011) and Clements, Eliahoo, Kim, Taylor-Robinson, and Khan (2020)
Toxins (e.g., alcohol, thorotrast, dioxin)	McGee et al. (2019)
Genetic polymorphisms (e.g., ABCC2, MTHFR, KLRK1)	Tyson and El-Serag (2011) and Clements et al. (2020)

the pathogenesis of CCA even if it is hypothesized that it may be a nodal point in the development of this tumor (Sasaki & Nakanuma, 2016).

A distinguishing feature of CCA is the presence of a rich tumor-reactive stroma that it is believed to dictate the different steps of cholangiocarcinogenesis through the exchange of a complex network of signaling among tumor cells and a number of infiltrating cells, ultimately promoting cell proliferation, survival and genetic and/or epigenetic alterations. In iCCAs and pCCAs, the reactive desmoplastic stroma contains cancer-associated fibroblasts (CAFs) that exhibit an extensive crosstalk with CCA.

Bile acids are retained in cholestasis and may promote cholangiocarcinogenesis through the activation of epidermal growth factor receptor (EGFR), induction of cyclooxygenase-2 (COX-2), myeloid cell leukemia 1 (Mcl-1) and interleukin (IL)-6, and downregulation of farnesoid X receptor (FXR), which favors NF-kB-mediate inflammation (Sirica, 2011; Affo, Yu, & Schwabe, 2017; Cadamuro, Brivio, et al., 2018; Cadamuro, Stecca, et al., 2018). FXR expression was reported to be decreased in human CCA tumors, while the levels of TGR5, another bile acid receptor, were found to be increased in CCA tumors. In turn, IL-6–STAT3 may contribute to mitogenesis by upregulating Mcl-1 or altering *EGFR* promoter methylation (Wehbe, Henson, Meng, Mize-Berge, & Patel, 2006).

Given their relationships with biliary and liver repair, it worth mentioning that developmental pathways involved in biliary development (including Notch, WNT/β-catenin, Hedgehog, Hippo-Yes-associated protein, or YAP, and transforming growth factor β, or TGF-β), play an important role in cholangiocarcinogenesis. The Notch pathway, for example, is known to play a major role in biliary repair, tubulogenesis, fibrosis and maintenance of the stem cell niche and increased Notch activity has been associated with primary liver tumors. Overexpression or aberrant Notch receptor expression has been reported in all types of CCA (Aoki et al., 2016; Wu et al., 2014). Notch signaling also mediates hepatocyte-cholangiocyte transdifferentiation during carcinogenesis in iCCA (Sekiya & Suzuki, 2012; Wang et al., 2018). Furthermore, the WNT–β-catenin signaling pathway is also known to be activated in most CCAs, in part as an effect of the release of Wnt ligands by inflammatory macrophages infiltrating the stroma, but also as a consequence of mutations encoding key components of the canonical WNT–β-catenin signaling pathway (Perugorria et al., 2019). Relationships between β-catenin/YAP/inflammation will be discussed later (see fibropolycystic diseases). YAP is a transcriptional co-activator that could be also modulated by Hippo-independent signals, such changes in extracellular matrix (ECM) composition, stiffness and inflammation (Sugihara, Isomoto, Gores, & Smoot, 2019; Affo et al., 2021; see also Chapter 10). YAP nuclear expression is increased in CCA, and YAP has been shown *in vitro* to be activated in CCA cell lines by IL-6, platelet-derived growth factors (PDGF) and fibroblast growth factor (Rizvi et al., 2016; Smoot et al., 2018; Sugiura et al., 2019).

These pathways are complex and cross talk with each other in diverse ways, depending on specific predisposing conditions and risk factors. To focus more specifically on this concept, we will first describe some known conditions having a strong association with CCA and then discussed the respectively involved inflammatory pathways in more detail.

2. Diseases prodromal to cholangiocarcinoma development

2.1 Primary sclerosing cholangitis

Primary sclerosing cholangitis (PSC) is a rare chronic liver disease in which persistent peribiliary inflammation and progressive fibrosis lead to strictures of the intrahepatic or/and extrahepatic bile ducts. The etiology of PSC is still unknown, but it is well established that inflammatory damage to the biliary epithelium play a causative role (Dyson, Beuers, Jones, Lohse, & Hudson, 2018). The pathogenesis of PSC is that still unidentified, but it is probably the end-results of the interaction between a number of factors acquired and genetic determinants that trigger an inflammatory response in a predisposed individual. Inflammation damages cholangiocytes, which then elicits the onset of persistent inflammatory status sustained by an immune cell infiltrate and fibrotic tissue deposition around the bile ducts, resulting in the typical peribiliary onion-like lesion characteristic of PSC (Lewis, 2017). At the histological level, PSC is further characterized by the presence reactive ductular cholangiocytes, activated fibroblasts, macrophages and neutrophils, and by the deposition of extracellular matrix components, mainly collagen I and fibronectin (Matsumoto et al., 1999). The increasing deposition of fibrotic tissue not only causes strictures in the biliary tree, but also marks disease progression towards biliary cirrhosis and portal hypertension (Dyson et al., 2018).

Evidences of the role of inflammation/immunity in the pathogenesis of PSC is provided by a strong association with the Human Leukocyte Antigen (HLA) complex on chromosome 6, along with the evidence that in 25% of cases PSC is associated with other non-gastrointestinal autoimmune diseases (Karlsen, Folseraas, Thorburn, & Vesterhus, 2017). A genome-wide association study (GWAS) of large cohorts of patients has shown a strong association with HLA (Karlsen et al., 2010; Liu et al., 2013), but more than 20 other genes (e.g., genes of the IL-2 pathway, CARD9, REL, FUT2 and MST1) have been linked with PSC (Liu et al., 2013; Mells, Kaser, & Karlsen, 2013; Melum et al., 2011). However, to date, the combined genetic risk only accounts for less than 10% of the susceptibility to PSC (Karlsen et al., 2017), and the autoimmune nature of the inflammation in PSC remains unproven.

The gut-liver axis is also a factor in the pathogenesis of PSC, not only because of the known association with inflammatory bowel diseases. The innate immune response could be activated by the leakage of pro-inflammatory bacterial products (e.g., lipopolysaccharides) from the intestinal microbiota (Özdirik, Müller, Wree, Tacke, & Sigal, 2021). Furthermore, gut-derived antigens can be presented to the T-cell Receptor (TCR), thus activating T cells and inducing their migration to liver and gut because of the overlapping adhesion molecule profiles of the endothelium in these two organs (e.g., mucosal vascular address in cell adhesion molecule 1 (MadCAM-1), vascular cell adhesion molecule 1 (VCAM-1) and Chemokine C-C motif ligand 25 (CCL25)) (Trivedi & Adams, 2013). The role of Thy17 cells and IL17 is still under investigation, but it could be linked to the translocation of certain bacterial species (Kunzmann et al., 2020).

The composition of gut microbiota in PSC patients is usually altered, with an overall reduction in bacterial diversity and an increase of specific bacterial phyla when compared with a healthy patient. Studies regarding other diseases with altered gut microbiota suggest that the reduced bacterial diversity occurs prior to and independent from clinical disease manifestations, but its role remains unclear (Karlsen, 2016). Recent studies (Bajer et al., 2017; Pontecorvi, Carbone, & Invernizzi, 2016) support the involvement of the gut microbiota in the initiating events of PSC pathogenesis. It is interesting to note that changes in gut microbiota have been linked also to the development of HCC in other liver conditions (Schwabe & Greten, 2020; Zheng, Ran, Zhang, Wang, & Zhou, 2021).

As the disease progresses, chronic injury converges on the common mechanisms of fibrosis, involving hepatic stellate cells, portal myofibroblasts and cholangiocytes accompanied by several other cells involved in immune response (Mederacke et al., 2013). These cells take part in a cross talk that is still largely to be dissected, but cholangiocytes have a key role in driving mesenchymal cells and fibrosis (like in the "onion skin" histological lesions) (Karlsen et al., 2017). As the disease progresses there is an increased risk of developing dominant strictures and of recurrent episodes of bacterial cholangitis. Notably, dominant strictures could be the presentation of CCA (Williamson & Chapman, 2015). Patients with PSC are at increased risk to develop neoplasias, such as colon, and pancreatic cancers, but hepatobiliary malignancies are the most common cause of cancer-related mortality and the most dreadful complication is CCA (Aune et al., 2021; Bjøro et al., 2006; Lazaridis & LaRusso, 2016).

It has been reported that CCA that develops in PSC is driven by *KRAS* and *TP53* mutations in association with *FGFR2* fusion and IDH 1 and 2 mutations, similar to other iCCAs (Banales et al., 2020). Deregulation of micro RNAs (miRNAs) involved in cell proliferation, stemness, and migration, may also play a relevant role; for example, IL-6 reduces the expression of the tumor suppressor NF2, by upregulating miR-let-7a, along with several other miRNAs targeting different transcription factors (Meng et al., 2007).

The persistence of the inflammatory insult in PSC is associated with the secretion of pro-inflammatory mediators, in particular of IL-6, interferon (IFN)γ, and tumor necrosis factor (TNF)α (Pinto, Giordano, Maroni, & Marzioni, 2018), which upregulate the expression in the biliary structures of inducible nitric oxide synthase (iNOS) (Spirlì et al., 2003). This phenomenon leads to the increase in local concentration of nitric oxide (NO), which in turn stimulates infiltrating inflammatory cells to oversecrete various cyto- and chemokines, PDGF, and vascular endothelial growth factor (VEGF) (Cannito et al., 2018; Sirica, 2005). The local accumulation of high concentration of NO_2^- promotes the formation of reactive oxidative species (ROS), such as peroxynitrites ($ONOO^-$) that, being responsible for the peroxidation of lipids, play a major role in the neoplastic transformation, and in nitrosylation and inactivation of several proteins, including enzymes involved in DNA proofreading and repair of single or double-strand breaks in DNA, such as 8-oxo-deoxyguanine DNA glycosylase 1. Furthermore, NO and ROS could generate two mutagenic compounds, 8-oxo-7,8-dihydro-20-deoxyguanosine and 8-nitroguanine (8-NG) (Jaiswal, LaRusso, Shapiro, Billiar, & Gores, 2001). This sequence of events is particularly relevant, as increased iNOS expression and nitrosylation of protein has been also described in bile ducts of patients with early stages of PSC.

2.2 Congenital hepatic fibrosis/Caroli's disease

Congenital hepatic fibrosis (CHF) and Caroli's disease (CD) are two cholangiopathies belonging to the family of the fibropolycystic liver diseases, a group of genetic diseases caused by mutations to the *polycystic kidney and hepatic disease 1* (*PKHD1*) gene, which encodes for the protein fibrocystin

(FPC) (Harris & Torres, 2009; Zerres, Rudnik-Schöneborn, Steinkamm, Becker, & Mücher, 1998). These conditions are all characterized by saccular cyst-like biliary structures in continuity with the biliary tree, accompanied by robust fibro-inflammatory infiltrate preeminently with peribiliary localization, and the development of liver fibrosis (Lasagni, Cadamuro, Morana, Fabris, & Strazzabosco, 2021). Liver disease progresses overtime and 10–40% of the patients presents esophageal varices bleeding or need portosystemic shunting. In adulthood, patients with Caroli's disease suffer from recurrent cholangitis and in 80% of the cases from portal hypertension (PH). No pharmacological therapies are available to relief or block the worsening of this fibroinflammatory liver disease, that could also be causative for CCA in 7–14% of patients with Caroli's disease. (Cannito et al., 2018; Khan et al., 2008; Strazzabosco & Fabris, 2012).

FPC is a transmembrane protein expressed in the primary cilium and in centromeres of epithelial cells of the bile, pancreatic and renal tubules. The role of fibrocystin is not completely known, but appears to be crucial in the differentiation of the ductal epithelium and, if mutated, it induces a maturational arrest of the nascent ductal plate (Veigel et al., 2009). The hepatic phenotype is characterized by a wide spectrum of dysgenetic alterations of the biliary epithelium, including the formation of biliary microhamartomas, which can give rise to segmental dilatations of the bile ducts, both intra and extrahepatic, rather than real cysts. Biliary dysgenesis is associated with an exuberant production of peribiliary fibrosis accompanied by an inflammatory infiltrate, mainly composed by macrophages. This is clinically significant as it is accompanied by the development of PH, and by a non-negligible risk of malignant transformation to CCA (Locatelli et al., 2016; Pech et al., 2016). The mechanisms regulating the development of fibroinflammation in these diseases remain largely unknown.

Biliary cystogenesis in CD/CHF is mediated by altered intracellular signaling, including cAMP and Ca^{2+} (Banales et al., 2009; Rohatgi et al., 2008). The interconnection between these intracellular signals leads to the activation of various pathways involved in the proliferation of biliary structures such as the phosphatidylinositol 3-kinases (PI3K)/AKT/ mammalian target of rapamycin (mTOR) or protein kinase A (PKA)/ RAF/extracellular-signal-regulated kinase (ERK1/2) pathways (Fischer et al., 2009). The activation of these signaling pathways leads to the transcription of the *cell division cycle 25 homolog A (Cdc25A)* gene, which is essential in mediating biliary cystogenesis (Masyuk et al., 2012). It should be noted that Cdc25A is also modulated by epigenetic anomalies; in fact

PCK rats, a model of CD, show the downregulation of miR-15A, whose experimental overexpression induces the repression of Cdc25A and consequently the reduction of the cystic area (Lee et al., 2008). Another epigenetic modulator involved in the pathogenesis of CD/CHF is histone deacetylase (HDAC) 6, an enzyme that plays a key role in cell cycle regulation, and whose overexpression has been demonstrated in cysts in both PCK rats and human patients (Gradilone et al., 2014). The inhibition of its function by ACY-1215, has been shown to induce a significant reduction in the growth of the cysts (Gradilone et al., 2014). The enzymes belonging to the HDAC family are known to modulate the Hippo pathway (Basu, Reyes-Múgica, & Rebbaa, 2013). This is a highly conserved signaling pathway, which orchestrate organ size, cell proliferation, stem cell stability and several other biological functions (Piccolo, Dupont, & Cordenonsi, 2014). Recent studies have shown that biliary cysts of PCK rat and of $Pkhd1^{LSL(-)/LSL(-)}$ mice, both rodent model orthologs to the human disease, display and increased nuclear expression YAP, the main transcription factor of the Hippo pathway, and that this evidence correlate with increased biliary expression of proliferating cell nuclear antigen (PCNA) and of cyclin D1 (Jiang et al., 2017).

One of the most characteristic and concerning traits of CD/CHF is the presence of progressive fibro-inflammation. Recent studies point to the accumulation of β-catenin in the nucleus as a key step in the pathogenesis of CHF/CD. Beta-catenin is a structural protein that is part of the so-called WNT/β-catenin pathway and which is usually retained in the cytoplasm in a non-phosphorylated form linked to the so-called "β-catenin destruction complex." When expressed into the nucleus, β-catenin binds to the T cell factor/lymphoid enhancer factor (TCF/LEF) complex (Nusse & Clevers, 2017) and acts as a transcriptional factor. In FPC-deficient cholangiocytes, β-catenin seems to favor the acquisition of a profibrotic and proinflammatory phenotype and promote the secretion of cytokines such as C-X-C Motif Chemokine Ligand (CXCL) 1, CXCL10, CXCL12. The secretion of these mediators into the pericystic space elicits the recruitment of inflammatory cells into the portal space, and, in particular of M1 and M2 macrophages. Macrophages, by secreting TNFα and TGF-β, in turn, induce the expression of integrin αVβ6, an activator of latent TGF-β, by cholangiocytes, driving fibrosis mediated by myofibroblasts (Locatelli et al., 2016). Furthermore, a recent paper (Tsunoda et al., 2019) demonstrated, using induced pluripotent stem Cell (iPSC) technology, that FPC-deficiency induces the secretion of IL-8 by cholangiocytes, which in turn, through an autocrine loop, stimulates the secretion of connective tissue growth factor (CTGF), a fundamental profibrotic mediator, providing further mechanistic insight into the pathogenesis of this family of diseases. How this intense inflammatory signaling contributes to biliary carcinogenesis is the focus of ongoing investigations.

2.3 Fluke infestations

A major cause of biliary neoplastic transformation is the infestation of the biliary tract with flukes. Liver fluke infestations, in particular of Platyhelminthes, such as *Opistorchis viverrini, Opistorchis felineus*, and of Thrematoda, such as *Clonorchis sinensis*, are endemic in several Asian countries and Eastern Europe (Brindley et al., 2021). *O. viverrini*, is endemic in Thailand, Cambodia and Vietnam and infects from 8 to 10 million people (Sripa, Kaewkes, Intapan, Maleewong, & Brindley, 2010), while *O. felineus* is mainly diffuse in Eastern Europe countries, in particular in Siberia, Ukraine, Belarus, and Kazakhstan (Fedorova et al., 2020). Finally, *C. sinensis*, is diffuse in the rural areas of Korea, Vietnam and China where it is responsible for the infestation of more than 35 million people (Keiser & Utzinger, 2005). The infestation of *O. viverrini* is due to the consumption of undercooked or raw fish, in particular of different species of cyprinids, a fish largely used in all South Asian regions. The first host of *O. viverrini* larvae is represented by freshwater snails, which are eaten by the fish. The metacercariae of the parasites ingested by man through the raw fish proliferate inside the bile ducts of the hosts releasing eggs. Eggs are then dispersed in the water through the stools and develop in the water perpetuating the cycle of infection (Sripa et al., 2010). This infection mechanism is also common to other flukes that infest, in particular, rural populations that do not have healthy drinking water available. Once encysted inside the bile ducts, the parasites develop and induce a strong fibroinflammatory reaction that triggers the development of a chronic cholangiopathy characterized by biliary proliferation, inflammation of the liver, and deposition of peribiliary fibrosis induced by an overt immune response (Qian, Utzinger, Keiser, & Zhou, 2016).

O. viverrini secretes a very large pool of growth factors, cytokine mediators, and other molecules (collectively named *O. viverrini* excretory/secretory products (OvESP)), that promote a chronic cholangiopathy (Young et al., 2010). Among OvESP, *O. viverrini* glutathione S-transferase (OvGST) stimulates cholangiocyte proliferation though AKT and ERK mediated pathways (Daorueang et al., 2012). Furthermore, *O. viverrini* granulin (Ov-GRN-1) and thioredoxin (Ov-Trx-1) can stimulate the proliferation of human cholangiocytes (Smout et al., 2009; Smout, Mulvenna, Jones, & Loukas, 2011; Suttiprapa et al., 2012). The response of biliary tree to the *O. viverrini* infestation is similar to the biliary response to pathogen-associated molecular pattern. In fact, OvESP act by activating the signaling pathway mediated by Toll-like receptors (TLR) 4, which results in nuclearization and consequent MyD88-dependent activation of the transcriptional action of NF-kB (Ninlawan et al., 2010). This signaling cascade leads to the secretion of various mediators, among which the most studied are IL-6 and IL-8, which are involved in stimulating the biliary ductular reaction typical of this liver disease and in the recruitment of neutrophils, respectively (Ninlawan et al., 2010). The TLR4 and TLRT2 pathways are also activated by the infestation of *C. sinensis*. This escalates into the generation of other mediators, including TNFα, IFNγ and several interleukins (1β, 4, 6, 10, and 13), further supporting the secretion of a chronic fibroinflammatory cholangiopathy (Kim et al., 2017; Prueksapanich et al., 2018). During the earliest stages of disease, there is a preponderance of infiltrating M1 macrophages with consequent release of M1-related factors, such as iNOS, TNFα, and CXCL9, while in the later phases there is a shift towards a population of profibrotic and pro-tumorigenic, arginase-1-positive M2 macrophages characterized by hypersecretion of CCL2 (Kim et al., 2017). For these reasons, both *O. viverrini* and *C. sinensis* have been included among the group 1 agents/biological carcinogens by the International Agency for Research on Cancer (IARC) (Brindley et al., 2021).

2.4 Other risk factors

As shown in table 1, several other risk factors related to the development of CCA have been identified. Among them, infection with hepatitis B and C viruses, hepatolithiasis, alcohol consumption, metabolic syndrome, obesity and non-alcoholic fatty liver disease (NAFLD) and non-alcoholic steatohepatitis (NASH). These are all risk factors for chronic liver disease. To date, the mechanisms responsible for the transition from chronic liver damage to tumor formation are unknown (Clements et al., 2020; Tyson & El-Serag, 2011), but in most cases, inflammatory pathways play a central role.

Hepatolithiasis is the presence of intrahepatic biliary stones, and is a well-known risk factor for CCA development in Far Eastern countries (Kim et al., 2015). Notably, hepatolithiasis has been reported to be a prodromal factor for CCA in the 1.6–9.9% of cases (Kim et al., 2015). The presence of stones generates an inflammatory response that induces the dedifferentiation of the biliary epithelium with possible transition to an intraductal papillary neoplasm of the bile duct (IPNB) and biliary intraepithelial neoplasia (BilIN), both of which are considered to be preneoplastic lesions (Aishima, Kubo, Tanaka, & Oda, 2014). Chronic inflammation induces the overexpression of COX-2, which positively modulates prostaglandin E2 that, similar to the increase in NF-kB transcription activity, transforms chronic hepatolithiasis into a chronic proliferating cholangitis (Shoda et al., 2003). The accumulation of gene mutations typically linked to conditions of neoplasia, such as EGFR, p16, deleted in pancreatic cancer 4 (DPC4)/signaling effectors mothers against decapentaplegic protein 4 (Smad4), c-Met, and c-ERBb2 ultimately lead to the development of CCA (Kim et al., 2015).

Hepatotropic virus infection, like HBV and HCV, are also a known cause of liver cancer. While there have been numerous reports on how these viruses induce neoplastic transformation of hepatocytes, favoring the development of HCC, only more recently have HBV and HCV been identified as independent risk factors for the development of iCCA (Matsumoto et al., 2014; Seo et al., 2020; Wang et al., 2016; Zhou et al., 2012). The mechanisms hypothesized for the viral related development of CA are similar to those of the HCC. They rely on the formation of chronic inflammation that induces fibrosis with development of cirrhosis, and subsequently to the accumulation of genetic modifications within cholangiocytes that underlie the sequential appearance of precancerous dysplastic foci, and ultimately to CCA. See at the Introduction concerning the role of Notch in driving the transdifferentiation of tumor cells from hepatocyte-like into cholangiocyte-like in iCCA.

Several studies have shown a correlation between alcohol consumption and smoking with the development of CCA (HR=2.35 and HR=2.15, respectively) (McGee et al., 2019). Furthermore, the association between diabetes, obesity and metabolic syndrome and cholangiocarcinogenesis is becoming clearer (Palmer & Patel, 2012; Vale, Gouveia, Gärtner, and Brindley, 2020). NAFLD is currently the most common liver disease and its incidence and prevalence is expected to reach epidemic proportions with a prevalence of 25% of the general population (Bedogni et al., 2005).

NASH, the inflammatory variant of NAFLD, is a progressive disease in which a sizable proportion of patients develop liver fibrosis, progressing to cirrhosis and primary liver cancer. While most research efforts have been dedicated to study how hepatocellular carcinoma develops from NAFLD/NASH (Michelotti et al., 2021), not much is known about the mechanisms of CCA development in NASH. Recent epidemiological studies have shown that NAFLD is associated with a 3-time increased risk to develop iCCA (Petrick et al., 2017; Wongjarupong et al., 2017). Obesity, a metabolic disorder commonly associated with NAFLD/NASH, is significantly associated with CCA development (OR 1.52, 95 % CI 1.13–1.89) (Li et al., 2014); up to 20% of patients with resectable iCCA having been diagnosed with NASH (Reddy et al., 2013).

3. Mechanisms of neoplastic transformation

Chronic injury induced by inflammation of the biliary epithelia is a premalignant condition leading to CCA. Chronic cholangiopathies share similar carcinogenetic processes. Major mechanisms that support and facilitate the neoplastic transformation in CCA are the proliferation of cholangiocytes and their concomitant acquisition of resistance to apoptosis, together with the accumulation of oncogenic DNA mutations. These mechanisms are closely linked and are self-sustaining. "Reactive" cholangiocytes produce inflammatory mediators that act in an autocrine manner on the cholangiocytes themselves and in a paracrine way on cells of the tumor microenvironment (TME) (Fig. 1) (Banales et al., 2019; Cannito et al., 2018; Fabris, Sato, Alpini, & Strazzabosco, 2021; Sirica et al., 2019).

Cholangiocyte-derived secretion of IL-6 and consequent activation of IL-6/STAT3 signaling stimulates the growth of malignant cells by activating the p44/42 mitogen-activated protein kinases (MAPK) pathways, leading to cell proliferation (Park, Tadlock, Gores, & Patel, 1999), and, at the same time, stimulates the p38 MAPK pathway, which decreases the production of $p21^{WAF1/CIP1}$, a cyclin dependent inhibitor controlling cell cycle (Tadlock & Patel, 2001). Furthermore, the secretion of IL-6 through the stimulation of STAT3 expression and Akt activation induces the overexpression of Mcl-1, a fundamental anti-apoptotic mediator (Isomoto et al., 2005; Kobayashi, Werneburg, Bronk, Kaufmann, & Gores, 2005; Yu et al., 2005). The role of IL-6 in cholangiocarcinogenesis is also related to its ability to induce epigenetic modifications. In fact, IL-6 is able to modulate

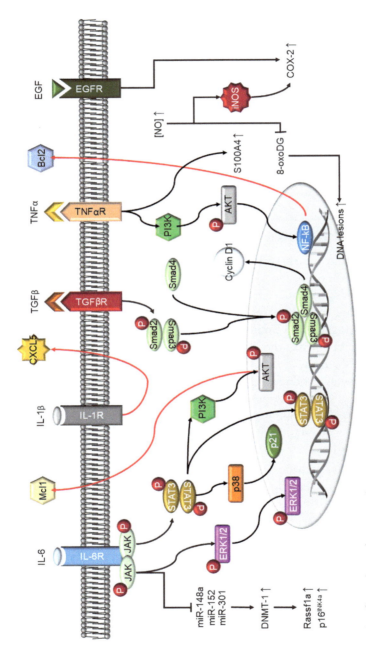

Fig. 1 See figure legend on opposite page.

miRNAs (Braconi, Huang, & Patel, 2010) and to demethylate EGFR promoters, thus favoring their transcription and consequently their proliferative function (Wehbe et al., 2006). IL-6 inhibit the expression of miR-148a, miR-152 and miR-301. DNA methyltransferase-1 (DNMT-1) is a specific target of 148a and miR-152, and due to their absence, increases its activity, thus inducing the methylation and the consequent repression of two tumor suppressors, Rassf1a and p16^{INK4a}, both of which are involved in repressing tumoral transformation (Braconi et al., 2010).

The demethylation of the EGFR promoters hyperactivate the downstream signal pathways of EGFR and induces the expression of COX-2 (Endo, Yoon, Pairojkul, Demetris, & Sirica, 2002; Yoon, Canbay, Werneburg, Lee, & Gores, 2004; Yoon, Gwak, et al., 2004; Yoon, Higuchi, Werneburg, Kaufmann, & Gores, 2002) that promotes cell proliferation and angiogenesis, as well as and inhibits apoptosis (Yoon et al., 2002). In CCA, other mechanisms could modulate COX-2, such as the presence of

Fig. 1 Mediators released by biliary epithelial cells involved in cholangiocancerogenesis. Cholangiocytes respond to signals that mediate responses involved in CCA tumorigenesis, such as proliferative responses, cell survival, chemotactic messages or others involved in the depression of proofreading enzyme activity. Among these effectors, the best known and studied is IL-6 which stimulates biliary proliferation through the ERK1/2 pathway, cellular senescence mediated by p21 overexpression, via p38, and secretion of Mcl1 (red arrow), through the PI3K/Akt pathway. Furthermore, by inhibiting the expression of miR-148a, miR-152, and miR-301, it stimulates, via DNMT1, the expression of the protooncogenes Rassf1a and p16^{INK4a}. IL-1β is involved in the secretion of CXCL5 (red arrow), a potent chemotactic agent for neutrophils. Another mediator of fundamental importance is TGF-β, which, through the nuclearization of the SMAD2/3/4 complex, fuels various responses involved in tumor biology, such as proliferation by stimulating cyclin D1. TNFα is involved in tumorigenesis because it stimulates the secretion of Bcl2 and the expression of S100A4, a protein involved in the metastasis of CCA. Finally, the regulation of COX-2, an enzyme involved in cell proliferation and in protection against apoptosis, is affected by two pathways, on the one hand is stimulated by the hyperactivation of EGFR, on the other hand by the increase of local NO concentration due to the de novo expression of iNOS by extracellular proinflammatory signals. NO can also repress the activity of 8-oxoDG, thus favoring the accumulation of mutations in the DNA. Legend: P, phosphorylation; CCA, cholangiocarcinoma; ERK, extracellular-signal-regulated kinase; IL, interleukin; Mcl1, myeloid cell leukemia 1; PI3K, phosphatidylinositol 3-kinases; miR, microRNA; SMAD, signaling effectors mothers against decapentaplegic protein; TGF-β, transforming growth factor β; TNF, tumor necrosis factor; COX-2, cyclooxygenase-2; EGFR, epidermal growth factor receptor; iNOS, inducible nitric oxide synthase; NO, nitric oxide; 8-oxoDG, 8-Oxo-2'-deoxyguanosine. *Black lines*, intracellular pathways; *red lines*, secretion of soluble proteins.

oxysterols, which can both stabilize COX-2 expression and activate the hedgehog signaling pathway linked to CCA development (El Khatib et al., 2013; Fingas et al., 2011; Nachtergaele et al., 2012; Yoon, Canbay, et al., 2004; Yoon, Gwak, et al., 2004). Another inducer of COX-2 is the excessive local concentration of NO, as a result of the activation of inducible iNOS. As discussed earlier, the excess of NO induces nitrotyrosine stress and inhibits 8-oxo-2'-deoxyguanosine (8-oxoDG) base excision in DNA repair processes, causing the accumulation of oxidative DNA lesions and promoting carcinogenesis (Jaiswal et al., 2001; Jaiswal, LaRusso, Burgart, & Gores, 2000; Spirlì et al., 2003). Fluke infestations exploit similar mechanisms to induce neoplastic transformation of cholangiocytes. The accumulation of metabolites and factors released by the parasites, such as oxysterol or catechol estrogen quinone, stimulate the iNOS-mediated production of reactive oxygen species (ROS) and reactive nitrogen species (RNS). Both compounds not only induce DNA damage, but are also capable of interfering with the activity of DNA proofreading enzymes, such as 8-oxoDG that leads to the accumulation of 8-NG, a product of RNS-induced damage of DNA (Correia da Costa et al., 2014; Ohshima, Sawa, & Akaike, 2006; Vale et al., 2020). Overexpression of iNOS is an indicator of poor prognosis in CCA patients (Suksawat et al., 2017) and its *in vitro* inhibition with N^{ω}-nitro-l-arginine methyl ester hydrochloride leads to a reduced migratory and invasive ability of CCA cell lines (Suksawat et al., 2018).

Another critically important inflammatory mediator released into the TME is IL-1β. This interleukin plays a dual role by stimulating CXCL5 expression on CCA cells (Okabe et al., 2012). The expression of the latter chemokine is in fact able to stimulate the chemotaxis of neutrophils in the tumor area through the PI3K-Akt signaling pathway and to activate cell proliferation via p44/p42 MAPK (Zhou et al., 2014).

TGF-β is a cytokine/growth factor that is also of primary importance in CCA development. This multifunctional cytokine, primarily known as a mediator of organ fibrosis, is heavily involved in anti- and pro-carcinogenetic functions. Paradoxically, in the initial stages of the tumor it acts as a tumor suppressor, whereas in the advanced stages it stimulates various biological functions that support tumor growth and metastasis (Maemura, Natsugoe, & Takao, 2014; Morrison, Parvani, & Schiemann, 2013). In particular, in the most advanced phases of CCA, mutations found in the TFGRβ receptor, upregulate the expression of cyclin D1 via SMAD4 phosphorylation (Zen et al., 2005). A recent work, supporting the importance of TGF-β in cholangiocarcinogenesis has shown that pharmacological inhibition of the TGF-β signaling pathway is able to decrease both the clonogenic potential of tumor cells and the metastasis of CCA (Puthdee et al., 2021).

Finally, another fundamental piece of the puzzle composed by cytokines secreted into the TME is TNFα. TNFα can support the key neoplastic characteristics, not only by stimulating cell survival by activating the Akt/NF-kB/Bcl2 pathway (Luo, Maeda, Hsu, Yagita, & Karin, 2004), but also by increasing metastatization via modulation of S100A4 (Techasen et al., 2014). Moreover, TNFα stimulates CD4 lymphocytes to secrete trophic interleukins such as IL-17 and IL-23 (He et al., 2012; Tang et al., 2013).

4. Role of inflammatory cells in modulating CCA malignant features

As discussed previously, the TME is particularly important for supporting and sustaining the different biological functions of CCA, such as proliferation, metastasis and chemoresistance. The TME is composed of the ECM and of the infiltrating cell types. With regards to the non-cellular component of the TME, desmoplastic tumors, such as CCA, pancreatic adenocarcinoma, and breast cancer are characterized by a marked modification of the normal matrix, which becomes dense and stiff due to the increased deposition of collagen I, III and IV, fibronectin, nidogen, perlecan, and other supporting proteins (Brivio, Cadamuro, Strazzabosco, & Fabris, 2017; Cadamuro, Brivio, et al., 2018; Cadamuro, Stecca, et al., 2018). The increased stiffness provides a mechanical stimulus to the tumor cells, leading to the activation of signaling pathways mediated by mechanoreceptors, such as the Hippo pathway, resulting in a more aggressive tumor phenotype characterized by a marked cell proliferation and increases in chemoresistance and metastasis (Marti et al., 2015; Zanconato, Cordenonsi, & Piccolo, 2019; Zhang et al., 2018).

The TME is infiltrated by a "multiethnic" collection of cells and includes vascular structures (blood and lymphatic), CAFs, neural-derived cells, and cells belonging to innate (i.e., macrophages, granulocytes, and adaptive (T cells) immune responses. In this review, we focus on the cells of the immune response. For the pathogenetic mechanisms mediated in the CCA by other cell types, we refer the reader to other reviews (Banales et al., 2016, 2020; Sirica, Strazzabosco, & Cadamuro, 2021; Vaquero, Aoudjehane, & Fouassier, 2020).

Several recent works of functional genomics confirm the importance of the inflammatory infiltrate. These studies categorized iCCA (Job et al., 2020; Sia et al., 2013; Wang et al., 2022), eCCA (Montal et al., 2020) and mixed HCC-CCA (Nguyen et al., 2021) based on their genomic and transcriptomic fingerprint, identifying initially two classes in iCCA (proliferation and inflammatory) (Sia et al., 2013) and then four subclasses based on the characteristics of TME (immune desert, immunogenic, myeloid and mesenchymal) (Job et al., 2020), four for eCCA (metabolic, proliferation, mesenchymal, and immune) and two classes, further divided into two subclasses, for the mixed HCC-CCA (immune high, IH, and immune low, IL). In iCCA, immunogenic classes are characterized by accumulation of CD4+ CD8+ lymphocytes, CD8RO+ T memory cells, CD20+ B cells, and reduced presence of CD68+ macrophages, as well as express significantly higher levels of IL1β, 6 and 15, C-X-C Chemokine Receptor (CXCR)4, CXCL9 and 13, and IFNγ, with the desert immune class showing the worst prognosis (Job et al., 2020). Similarly to iCCA, in eCCA, the immune class is the one with the better survival outcome, being characterized by an higher representation of CD8+ T cells and CD20+ B cells (Montal et al., 2020). Further confirming the importance of activation of the adaptive response in repressing and containing CCA aggressiveness, IH mixed HCC-CCA, characterized by increased tumoral expression of T lymphocytes and overexpression of chemotactic mediators, such as IL1β, 2, 4, 10, and 33, CCL9, CXCL19, CXCR3, IFNγ, also showed a better survival outcomes as compared to IL neoplasm (Nguyen et al., 2021).

It is interesting to note that these studies collectively demonstrate that CCAs with a high representation of the immune component have a better prognosis. It is worth noting that immune response cells have a dual role in cancer biology. Among cells that tend to contain tumorigenesis are Neutral Killer Cells (NKs), dendritic cells (DCs), and tumor-infiltrating lymphocytes (TILs). Macrophages and neutrophils, due to their plasticity and their ability to change phenotype and biological characteristics, can be either pro- or anti- tumorigenic, while Myeloid-Derived Suppressor Cells (MDSCs) act as tumor-stimulating factors (Fig. 2). For these reasons, a line of research in liver oncology is oriented towards modulating the immune response and the signaling pathways involved in the recruitment of the inflammatory infiltrate. Unfortunately, in CCA these studies are still in their infancy.

As previously mentioned, the cell types infiltrating the TME and capable of containing tumor aggressiveness include NKs, DCs, and TILs. Briefly, NKs are CD3-CD56+ cells capable of eliminating tumor cells via the

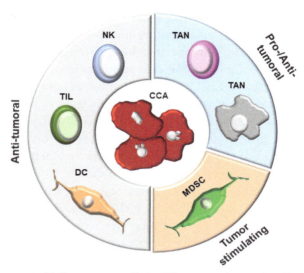

Fig. 2 Involvement of inflammatory cells in CCA progression and malignant behavior. The cells of the immune response are differently involved in tumor progression of the CCA. Dendritic cells (DC), Natural Killer cells (NK) and tumor-associated lymphocytes (TIL), have marked anticancer properties. Conversely, myeloid-derived cells (MDSC), thanks to the ability to inhibit the activation of NK and DC, stimulate tumor malignancy. Finally, cancer-associated macrophages (TAMs) and cancer-associated neutrophils (TANs) can exhibit anti- or pro-tumor behaviors depending on their phenotype.

secretion of lytic enzymes such as perforins, granzyme and proteases, or through an indirect pathway by stimulating the activation on target cells of the proapoptotic pathway mediated by Fas Cell Surface Death Receptor Ligand/TNF-related apoptosis-inducing ligand (FasL/TRAIL) (Banales et al., 2020; Cardoso Alves, Corazza, Micheau, & Krebs, 2021). DCs, on the other hand, acting as professional Antigen Presenting Cells (APCs), are able to activate the TILs in order to stimulate their ability to attack and kill cancer cells (Junking, Grainok, Thepmalee, Wongkham, & Yenchitsomanus, 2017).

Also cells of the adaptive response, i.e. tumor-infiltrating lymphocytes (TILs), either T or B cells, can be endowed with anti-tumor function (Fabris et al., 2019; Kitano et al., 2018). Their interaction within the microenvironment, may oppose tumor growth. On the other hand, MDSCs are myeloid-derived cells with a strong immunosuppressive activity, capable of inhibiting the action of TILs and DCs, thus generating a TME favorable to tumor growth (Bergenfelz & Leandersson, 2020; Job et al., 2020).

As alluded to above, macrophages and neutrophils can act in a dual way. In fact, both can have an inflammatory phenotype (M1 for macrophages and N1 for neutrophils), characterized by anti-tumor properties. Within the TME, however, there are M2 type macrophages or tumor-associated macrophages (TAMs) and N2 neutrophils, or tumor-associated neutrophils (TAN). TAMs are mainly present in the tumor front of the CCA (Raggi et al., 2017; Zhou et al., 2021) and are able to modify the matrix, and stimulate tumor growth thanks to the secretion of different cytokines (such as IL-6, IL-13, IL-34, and TNFα), promote neo-angiogenesis, inhibit the recruitment and activation of T cells, and induce apoptosis of M1-type macrophages (Ge & Ding, 2020). TANs can also secrete a plethora of cyto-chemokines and growth factors, such as VEGF, CXCL1, CXCL2, CXCL6 and CXCL8, having a trophic effect on CCA cells, as well as enzymes capable of modifying the extracellular matrix, such as metalloproteases (MMP) 8 and MMP9 (Fabris et al., 2019; Masucci, Minopoli, & Carriero, 2019; Rimassa, Personeni, Aghemo, & Lleo, 2019).

These data allow a better understanding of the complex and multifaceted biology of the CCA, but also indicate that, at least for some categories of CCA, it is necessary to study more in depth the inflammation that accompanies the tumor and how this can be targeted therapeutically. Currently, the biology and pathogenetic significance of TAMs in CCA represent an important chapter of study because their presence is a hallmark of poorer patient survival outcomes (Sun et al., 2020), probably due to their ability at stimulating cell proliferation, neoangiogenesis, and metastasis through the secretion of various mediators, including VEGF (Zhou, Wang, Lu, et al., 2021). In addition to participating in mechanisms more strictly linked to tumor aggressiveness, VEGF is also able to favor tumor malignancies by stimulating TAM proliferation. TAMs, in fact, have the ability to impede tumor antigen presentation, diminishing the T cell-mediated immune response against the tumor (Rothlin, Ghosh, Zuniga, Oldstone, & Lemke, 2007). Moreover, several studies have shown that the presence and relative quantity of CD4+, CD8+ and CD20+ cells, belonging to the adaptive immune response, are indices of more favorable outcomes in patients with CCA who exhibit lower disease recurrence and an increased overall survival (Goeppert et al., 2013; Loeuillard, Conboy, Gores, & Rizvi, 2019) A clinical trial is actually using Pembrolizumab, a monoclonal antibody anti programmed death (PD)-1, in combination with the anti-VEGF agent Lenvatinib. The rational for this clinical trial is to inhibit the immune suppressive effect of VEGF, thus promoting T cell tumor infiltration and differentiation of CD8+ T cells secreting IFNγ (NCT04550624). A similar approach was used to design another phase Ib/II study, which involves the use of Atezolizumab, a monoclonal antibody inhibiting PD-L1, in combination with Tivozanib, an oral VEGFRs inhibitor to induce activation of CD4+ and CD8+ T cells and DCs (NCT05000294). The stimulation of T cells with antitumor function has also been used in other studies. The phase 1/2, open label, ABILITY study, exploits the use of MDNA11, a modification of IL2 that enhances its action, alone or in combination with checkpoint inhibitor. The use of MDNA11 could therefore lead to an increased activation of naïve CD8+ T-cells and NK cells (NCT05086692). Finally, another phase 2 recruiting study involves the use of Atezolizumab and/or Cobimetinib, a MEK inhibitor, with a monoclonal antibody (CDX-1127) that inhibits CD27, having the ability to activate CD8+ T cells (NCT04941287).

Another cell population potentially important for the containment of CCA is that of DCs, thanks to their known ability to stimulate the activity of T cells (Junking et al., 2017; Panya et al., 2018). An early phase 1 clinical trial involves the infusion of autologous dendritic cells in patients with advanced CCA in combination with Prevnar, a pneumonia vaccine, which aims to further stimulate the immune response (NCT03942328). A second phase 1 study involves the use of CDX-1140, an antibody that targets CD40 alone or in combination with other chemotherapy drugs. CD40 is a key activator of immune response, which is found on dendritic cells, macrophages and B cells, as well as on neoplastic cells (NCT03329950).

NK cells are a fundamental building block for protection from microorganisms and for tumor immune surveillance. Xenograft studies have shown that infusion of NK in athymic mice xenotransplanted with human iCCA cells (HuCCT1) significantly reduces tumor growth in mice (Jung et al., 2018). Given the promising results, two subsequent phase 1/2 studies ("SMT-NK") have been designed involving the infusion of allogenic NK cells in advanced CCA, but the data are not yet available (NCT03358849, NCT03937895). To date, although it has been shown that the accumulation of TAN is an indicator of poorer survival outcomes for both iCCA and eCCA (Kitano et al., 2018; Zhou et al., 2021), and that MDSCs are heavily involved in immune response depression (Fabris et al., 2019; Rimassa et al., 2019), there are currently no human clinical trials that foresee their modulation.

5. Conclusions

Despite a large body of literature that suggest a prominent role of inflammation in CCA associated with certain liver conditions or risk factors, and that the balance of cells of the innate and adaptive immune responses infiltrating the TME are fundamental in modulating the cancer/host balance, little is currently known about the mechanisms that mediate neoplastic transformation in chronic biliary diseases. We have discussed a few examples, such as PSC CD/CHF and fluke infestation, but the association discussed are mostly correlative and relate more to the pathogenesis of the disease than to the specific carcinogenetic mechanism. This is due on one side to the rarity of these conditions and on the other side to the lack of reliable animal and cellular models (Cadamuro, Brivio, et al., 2018; Cadamuro, Stecca, et al., 2018; Mariotti et al., 2019). There is an urgent need to develop animal models of neoplastic transformation more akin to the human conditions to better study how the inflammatory microenvironment and the presence of cells of the immune response interact with the tumor in order to hypothesize new pharmacological approaches. Functional genomic studies will be able to identify druggable inflammatory signature. Analysis of the microenvironment using single cell transcriptomics will be key to understand cell dynamics and crosstalk and their adaptation under treatment. The future certainly looks to be very active for scholars interested in cholangiocarcinoma and inflammation. Meanwhile, efforts should be devoted to advancing the prevention and treatment of known CCA risk factors and predisposing conditions.

Grant support

This project was supported in part by the Yale Liver Center award NIH P30 DK034989 Molecular and Translational core. MS acknowledges the support of grant 5R01DK101528-06.

Conflict of interest statement

M.C. and M.S. have no financial or personal disclosures relevant to the contents of this manuscript.

References

Affo, S., Nair, A., Brundu, F., Ravichandra, A., Bhattacharjee, S., Matsuda, M., et al. (2021). Promotion of cholangiocarcinoma growth by diverse cancer-associated fibroblast subpopulations. *Cancer Cell, 39*, 866–882. e11. https://doi.org/10.1016/j.ccell.2021.03.012.

Affo, S., Yu, L. X., & Schwabe, R. F. (2017). The role of cancer-associated fibroblasts and fibrosis in liver cancer. *Annual Review of Pathology, 12*, 153–186. https://doi.org/10.1146/annurev-pathol-052016-100322.

Aishima, S., Kubo, Y., Tanaka, Y., & Oda, Y. (2014). Histological features of precancerous and early cancerous lesions of biliary tract carcinoma. *Journal of Hepato-Biliary-Pancreatic Sciences, 21*, 448–452. https://doi.org/10.1002/jhbp.71.

Aoki, S., Mizuma, M., Takahashi, Y., Haji, Y., Okada, R., Abe, T., et al. (2016). Aberrant activation of Notch signaling in extrahepatic cholangiocarcinoma: Clinicopathological features and therapeutic potential for cancer stem cell-like properties. *BMC Cancer, 16*, 854. https://doi.org/10.1186/s12885-016-2919-4.

Aune, D., Sen, A., Norat, T., Riboli, E., & Folseraas, T. (2021). Primary sclerosing cholangitis and the risk of cancer, cardiovascular disease, and all-cause mortality: A systematic review and meta-analysis of cohort studies. *Scientific Reports, 11*, 10646. https://doi.org/10.1038/s41598-021-90175-w.

Bajer, L., Kverka, M., Kostovcik, M., Macinga, P., Dvorak, J., Stehlikova, Z., et al. (2017). Distinct gut microbiota profiles in patients with primary sclerosing cholangitis and ulcerative colitis. *World Journal of Gastroenterology, 23*, 4548–4558. https://doi.org/10.3748/wjg.v23.i25.4548.

Banales, J. M., Cardinale, V., Carpino, G., Marzioni, M., Andersen, J. B., Invernizzi, P., et al. (2016). Expert consensus document: Cholangiocarcinoma: Current knowledge and future perspectives consensus statement from the European Network for the Study of Cholangiocarcinoma (ENS-CCA). *Nature Reviews. Gastroenterology & Hepatology, 13*, 261–280. https://doi.org/10.1038/nrgastro.2016.51.

Banales, J. M., Huebert, R. C., Karlsen, T., Strazzabosco, M., LaRusso, N. F., & Gores, G. J. (2019). Cholangiocyte pathobiology. *Nature Reviews. Gastroenterology & Hepatology, 16*, 269–281. https://doi.org/10.1038/s41575-019-0125-y.

Banales, J. M., Marin, J., Lamarca, A., Rodrigues, P. M., Khan, S. A., Roberts, L. R., et al. (2020). Cholangiocarcinoma 2020: The next horizon in mechanisms and management. *Nature Reviews. Gastroenterology & Hepatology, 17*, 557–588. https://doi.org/10.1038/s41575-020-0310-z.

Banales, J. M., Masyuk, T. V., Gradilone, S. A., Masyuk, A. I., Medina, J. F., & LaRusso, N. F. (2009). The cAMP effectors Epac and protein kinase a (PKA) are involved in the hepatic cystogenesis of an animal model of autosomal recessive polycystic kidney disease (ARPKD). *Hepatology, 49*, 160–174. https://doi.org/10.1002/hep.22636.

Basu, D., Reyes-Múgica, M., & Rebbaa, A. (2013). Histone acetylation-mediated regulation of the Hippo pathway. *PLoS One, 8*, e62478. https://doi.org/10.1371/journal.pone.0062478.

Bedogni, G., Miglioli, L., Masutti, F., Tiribelli, C., Marchesini, G., & Bellentani, S. (2005). Prevalence of and risk factors for nonalcoholic fatty liver disease: The Dionysos nutrition and liver study. *Hepatology, 42*, 44–52. https://doi.org/10.1002/hep.20734.

Bergenfelz, C., & Leandersson, K. (2020). The generation and identity of human myeloid-derived suppressor cells. *Frontiers in Oncology, 10*, 109. https://doi.org/10.3389/fonc.2020.00109.

Bjøro, K., Brandsaeter, B., Foss, A., & Schrumpf, E. (2006). Liver transplantation in primary sclerosing cholangitis. *Seminars in Liver Disease, 26*, 69–79. https://doi.org/10.1055/s-2006-933565.

Braconi, C., Huang, N., & Patel, T. (2010). MicroRNA-dependent regulation of DNA methyltransferase-1 and tumor suppressor gene expression by interleukin-6 in human malignant cholangiocytes. *Hepatology, 51*, 881–890. https://doi.org/10.1002/hep.23381.

Brindley, P. J., Bachini, M., Ilyas, S. I., Khan, S. A., Loukas, A., Sirica, A. E., et al. (2021). Cholangiocarcinoma. *Nature Reviews. Disease Primers, 7*, 65. https://doi.org/10.1038/s41572-021-00300-2.

Brivio, S., Cadamuro, M., Strazzabosco, M., & Fabris, L. (2017). Tumor reactive stroma in cholangiocarcinoma: The fuel behind cancer aggressiveness. *World Journal of Hepatology, 9*, 455–468. https://doi.org/10.4254/wjh.v9.i9.455.

Cadamuro, M., Brivio, S., Stecca, T., Kaffe, E., Mariotti, V., Milani, C., et al. (2018). Animal models of cholangiocarcinoma: What they teach us about the human disease. *Clinics and Research in Hepatology and Gastroenterology, 42,* 403–415. https://doi.org/10.1016/j.clinre.2018.04.008.

Cadamuro, M., Lasagni, A., Lamarca, A., Fouassier, L., Guido, M., Sarcognato, S., et al. (2021). Targeted therapies for extrahepatic cholangiocarcinoma: Preclinical and clinical development and prospects for the clinic. *Expert Opinion on Investigational Drugs, 30,* 377–388. https://doi.org/10.1080/13543784.2021.1880564.

Cadamuro, M., Stecca, T., Brivio, S., Mariotti, V., Fiorotto, R., Spirli, C., et al. (2018). The deleterious interplay between tumor epithelia and stroma in cholangiocarcinoma. *Biochimica et Biophysica Acta, Molecular Basis of Disease, 1864,* 1435–1443. https://doi.org/10.1016/j.bbadis.2017.07.028.

Cannito, S., Milani, C., Cappon, A., Parola, M., Strazzabosco, M., & Cadamuro, M. (2018). Fibroinflammatory liver injuries as preneoplastic condition in cholangiopathies. *International Journal of Molecular Sciences, 19,* 3875. https://doi.org/10.3390/ijms19123875.

Cardoso Alves, L., Corazza, N., Micheau, O., & Krebs, P. (2021). The multifaceted role of TRAIL signaling in cancer and immunity. *The FEBS Journal, 288,* 5530–5554. https://doi.org/10.1111/febs.15637.

Clements, O., Eliahoo, J., Kim, J. U., Taylor-Robinson, S. D., & Khan, S. A. (2020). Risk factors for intrahepatic and extrahepatic cholangiocarcinoma: A systematic review and meta-analysis. *Journal of Hepatology, 72,* 95–103. https://doi.org/10.1016/j.jhep.2019.09.007.

Correia da Costa, J. M., Vale, N., Gouveia, M. J., Botelho, M. C., Sripa, B., Santos, L. L., et al. (2014). Schistosome and liver fluke derived catechol-estrogens and helminth associated cancers. *Frontiers in Genetics, 5,* 444. https://doi.org/10.3389/fgene.2014.00444.

Daorueang, D., Thuwajit, P., Roitrakul, S., Laha, T., Kaewkes, S., Endo, Y., et al. (2012). Secreted Opisthorchis viverrini glutathione S-transferase regulates cell proliferation through AKT and ERK pathways in cholangiocarcinoma. *Parasitology International, 61,* 155–161. https://doi.org/10.1016/j.parint.2011.07.011.

Dyson, J. K., Beuers, U., Jones, D., Lohse, A. W., & Hudson, M. (2018). Primary sclerosing cholangitis. *Lancet, 391,* 2547–2559. https://doi.org/10.1016/S0140-6736(18)30300-3.

El Khatib, M., Kalnytska, A., Palagani, V., Kossatz, U., Manns, M. P., Malek, N. P., et al. (2013). Inhibition of hedgehog signaling attenuates carcinogenesis in vitro and increases necrosis of cholangiocellular carcinoma. *Hepatology, 57,* 1035–1045. https://doi.org/10.1002/hep.26147.

Endo, K., Yoon, B. I., Pairojkul, C., Demetris, A. J., & Sirica, A. E. (2002). ERBB-2 overexpression and cyclooxygenase-2 up-regulation in human cholangiocarcinoma and risk conditions. *Hepatology, 36,* 439–450. https://doi.org/10.1053/jhep.2002.34435.

Fabris, L., Perugorria, M. J., Mertens, J., Björkström, N. K., Cramer, T., Lleo, A., et al. (2019). The tumour microenvironment and immune milieu of cholangiocarcinoma. *Liver International, 39,* 63–78. https://doi.org/10.1111/liv.14098.

Fabris, L., Sato, K., Alpini, G., & Strazzabosco, M. (2021). The tumor microenvironment in cholangiocarcinoma progression. *Hepatology, 73,* 75–85. https://doi.org/10.1002/hep.31410.

Fedorova, O. S., Fedotova, M. M., Zvonareva, O. I., Mazeina, S. V., Kovshirina, Y. V., Sokolova, T. S., et al. (2020). Opisthorchis felineus infection, risks, and morbidity in rural Western Siberia, Russian Federation. *PLoS Neglected Tropical Diseases, 14,* e0008421. https://doi.org/10.1371/journal.pntd.0008421.

Fingas, C. D., Bronk, S. F., Werneburg, N. W., Mott, J. L., Guicciardi, M. E., Cazanave, S. C., et al. (2011). Myofibroblast-derived PDGF-BB promotes Hedgehog survival signaling in cholangiocarcinoma cells. *Hepatology, 54,* 2076–2088. https://doi.org/10.1002/hep.24588.

Fischer, D. C., Jacoby, U., Pape, L., Ward, C. J., Kuwertz-Broeking, E., Renken, C., et al. (2009). Activation of the AKT/mTOR pathway in autosomal recessive polycystic kidney disease (ARPKD). *Nephrology, Dialysis, Transplantation, 24*, 1819–1827. https://doi.org/10.1093/ndt/gfn744.

Ge, Z., & Ding, S. (2020). The Crosstalk between tumor-associated macrophages (TAMs) and tumor cells and the corresponding targeted therapy. *Frontiers in Oncology, 10*, 590941. https://doi.org/10.3389/fonc.2020.590941.

Goeppert, B., Frauenschuh, L., Zucknick, M., Stenzinger, A., Andrulis, M., Klauschen, F., et al. (2013). Prognostic impact of tumour-infiltrating immune cells on biliary tract cancer. *British Journal of Cancer, 109*, 2665–2674. https://doi.org/10.1038/bjc.2013.610.

Gradilone, S. A., Habringer, S., Masyuk, T. V., Howard, B. N., Masyuk, A. I., & Larusso, N. F. (2014). HDAC6 is overexpressed in cystic cholangiocytes and its inhibition reduces cystogenesis. *The American Journal of Pathology, 184*, 600–608. https://doi.org/10.1016/j.ajpath.2013.11.027.

Harris, P. C., & Torres, V. E. (2009). Polycystic kidney disease. *Annual Review of Medicine, 60*, 321–337. https://doi.org/10.1146/annurev.med.60.101707.125712.

He, D., Li, H., Yusuf, N., Elmets, C. A., Athar, M., Katiyar, S. K., et al. (2012). IL-17 mediated inflammation promotes tumor growth and progression in the skin. *PLoS One, 7*, e32126. https://doi.org/10.1371/journal.pone.0032126.

Isomoto, H., Kobayashi, S., Werneburg, N. W., Bronk, S. F., Guicciardi, M. E., Frank, D. A., et al. (2005). Interleukin 6 upregulates myeloid cell leukemia-1 expression through a STAT3 pathway in cholangiocarcinoma cells. *Hepatology, 42*, 1329–1338. https://doi.org/10.1002/hep.20966.

Jaiswal, M., LaRusso, N. F., Burgart, L. J., & Gores, G. J. (2000). Inflammatory cytokines induce DNA damage and inhibit DNA repair in cholangiocarcinoma cells by a nitric oxide-dependent mechanism. *Cancer Research, 60*, 184–190.

Jaiswal, M., LaRusso, N. F., Shapiro, R. A., Billiar, T. R., & Gores, G. J. (2001). Nitric oxide-mediated inhibition of DNA repair potentiates oxidative DNA damage in cholangiocytes. *Gastroenterology, 120*, 190–199. https://doi.org/10.1053/gast.2001.20875.

Jiang, L., Sun, L., Edwards, G., Manley, M., Jr., Wallace, D. P., Septer, S., et al. (2017). Increased YAP activation is associated with hepatic cyst epithelial cell proliferation in ARPKD/CHF. *Gene Expression, 17*, 313–326. https://doi.org/10.3727/105221617X15034976037343.

Job, S., Rapoud, D., Dos Santos, A., Gonzalez, P., Desterke, C., Pascal, G., et al. (2020). Identification of four immune subtypes characterized by distinct composition and functions of tumor microenvironment in intrahepatic cholangiocarcinoma. *Hepatology, 72*, 965–981. https://doi.org/10.1002/hep.31092.

Jung, I. H., Kim, D. H., Yoo, D. K., Baek, S. Y., Jeong, S. H., Jung, D. E., et al. (2018). In vivo study of natural killer (NK) cell cytotoxicity against cholangiocarcinoma in a nude mouse model. *In Vivo, 32*, 771–781. https://doi.org/10.21873/invivo.11307.

Junking, M., Grainok, J., Thepmalee, C., Wongkham, S., & Yenchitsomanus, P. T. (2017). Enhanced cytotoxic activity of effector T-cells against cholangiocarcinoma by dendritic cells pulsed with pooled mRNA. *Tumour Biology, 39*. https://doi.org/10.1177/1010428317733367. 1010428317733367.

Karlsen, T. H. (2016). Primary sclerosing cholangitis: 50 years of a gut-liver relationship and still no love? *Gut, 65*, 1579–1581. https://doi.org/10.1136/gutjnl-2016-312137.

Karlsen, T. H., Folseraas, T., Thorburn, D., & Vesterhus, M. (2017). Primary sclerosing cholangitis—A comprehensive review. *Journal of Hepatology, 67*, 1298–1323. https://doi.org/10.1016/j.jhep.2017.07.022.

Karlsen, T. H., Franke, A., Melum, E., Kaser, A., Hov, J. R., Balschun, T., et al. (2010). Genome-wide association analysis in primary sclerosing cholangitis. *Gastroenterology, 138*, 1102–1111. https://doi.org/10.1053/j.gastro.2009.11.046.

Keiser, J., & Utzinger, J. (2005). Emerging foodborne trematodiasis. *Emerging Infectious Diseases, 11*, 1507–1514. https://doi.org/10.3201/eid1110.050614.

Khan, S. A., Toledano, M. B., & Taylor-Robinson, S. D. (2008). Epidemiology, risk factors, and pathogenesis of cholangiocarcinoma. *HPB (Oxford), 10*, 77–82. https://doi.org/10.1080/13651820801992641.

Kim, H. J., Kim, J. S., Joo, M. K., Lee, B. J., Kim, J. H., Yeon, J. E., et al. (2015). Hepatolithiasis and intrahepatic cholangiocarcinoma: A review. *World Journal of Gastroenterology, 21*, 13418–13431. https://doi.org/10.3748/wjg.v21.i48.13418.

Kim, E. M., Kwak, Y. S., Yi, M. H., Kim, J. Y., Sohn, W. M., & Yong, T. S. (2017). Clonorchis sinensis antigens alter hepatic macrophage polarization in vitro and in vivo. *PLoS Neglected Tropical Diseases, 11*, e0005614. https://doi.org/10.1371/journal.pntd.0005614.

Kitano, Y., Okabe, H., Yamashita, Y. I., Nakagawa, S., Saito, Y., Umezaki, N., et al. (2018). Tumour-infiltrating inflammatory and immune cells in patients with extrahepatic cholangiocarcinoma. *British Journal of Cancer, 118*, 171–180. https://doi.org/10.1038/bjc.2017.401.

Kobayashi, S., Werneburg, N. W., Bronk, S. F., Kaufmann, S. H., & Gores, G. J. (2005). Interleukin-6 contributes to Mcl-1 up-regulation and TRAIL resistance via an Akt-signaling pathway in cholangiocarcinoma cells. *Gastroenterology, 128*, 2054–2065. https://doi.org/10.1053/j.gastro.2005.03.010.

Kunzmann, L. K., Schoknecht, T., Poch, T., Henze, L., Stein, S., Kriz, M., et al. (2020). Monocytes as potential mediators of pathogen-induced T-helper 17 differentiation in patients with primary sclerosing cholangitis (PSC). *Hepatology, 72*, 1310–1326. https://doi.org/10.1002/hep.31140.

Lasagni, A., Cadamuro, M., Morana, G., Fabris, L., & Strazzabosco, M. (2021). Fibrocystic liver disease: Novel concepts and translational perspectives. *Translational Gastroenterology and Hepatology, 6*, 26. https://doi.org/10.21037/tgh-2020-04.

Lazaridis, K. N., & LaRusso, N. F. (2016). Primary sclerosing cholangitis. *The New England Journal of Medicine, 375*, 1161–1170. https://doi.org/10.1056/NEJMra1506330.

Lee, S. O., Masyuk, T., Splinter, P., Banales, J. M., Masyuk, A., Stroope, A., et al. (2008). MicroRNA15a modulates expression of the cell-cycle regulator Cdc 25A and affects hepatic cystogenesis in a rat model of polycystic kidney disease. *The Journal of Clinical Investigation, 118*, 3714–3724. https://doi.org/10.1172/JCI34922.

Leone, V., Ali, A., Weber, A., Tschaharganeh, D. F., & Heikenwalder, M. (2021). Liver inflammation and hepatobiliary cancers. *Trends in Cancer, 7*, 606–623. https://doi.org/10.1016/j.trecan.2021.01.012.

Lewis, J. (2017). Pathological patterns of biliary disease. *Clinical Liver Disease, 10*, 107–110. https://doi.org/10.1002/cld.667.

Li, J. S., Han, T. J., Jing, N., Li, L., Zhang, X. H., Ma, F. Z., et al. (2014). Obesity and the risk of cholangiocarcinoma: A meta-analysis. *Tumour Biology, 35*, 6831–6838. https://doi.org/10.1007/s13277-014-1939-4.

Liu, J. Z., Hov, J. R., Folseraas, T., Ellinghaus, E., Rushbrook, S. M., Doncheva, N. T., et al. (2013). Dense genotyping of immune-related disease regions identifies nine new risk loci for primary sclerosing cholangitis. *Nature Genetics, 45*, 670–675. https://doi.org/10.1038/ng.2616.

Locatelli, L., Cadamuro, M., Spirlì, C., Fiorotto, R., Lecchi, S., Morell, C. M., et al. (2016). Macrophage recruitment by fibrocystin-defective biliary epithelial cells promotes portal fibrosis in congenital hepatic fibrosis. *Hepatology, 63*, 965–982. https://doi.org/10.1002/hep.28382.

Loeuillard, E., Conboy, C. B., Gores, G. J., & Rizvi, S. (2019). Immunobiology of cholangiocarcinoma. *JHEP Reports, 1*, 297–311. https://doi.org/10.1016/j.jhepr.2019.06.003.

Luo, J. L., Maeda, S., Hsu, L. C., Yagita, H., & Karin, M. (2004). Inhibition of NF-kappaB in cancer cells converts inflammation- induced tumor growth mediated by TNFalpha to TRAIL-mediated tumor regression. *Cancer Cell*, *6*, 297–305. https://doi.org/10.1016/j.ccr.2004.08.012.

Maemura, K., Natsugoe, S., & Takao, S. (2014). Molecular mechanism of cholangiocarcinoma carcinogenesis. *Journal of Hepato-Biliary-Pancreatic Sciences*, *21*, 754–760. https://doi.org/10.1002/jhbp.126.

Mariotti, V., Cadamuro, M., Spirli, C., Fiorotto, R., Strazzabosco, M., & Fabris, L. (2019). Animal models of cholestasis: An update on inflammatory cholangiopathies. *Biochimica et Biophysica Acta, Molecular Basis of Disease*, *1865*, 954–964. https://doi.org/10.1016/j.bbadis.2018.07.025.

Marti, P., Stein, C., Blumer, T., Abraham, Y., Dill, M. T., Pikiolek, M., et al. (2015). YAP promotes proliferation, chemoresistance, and angiogenesis in human cholangiocarcinoma through TEAD transcription factors. *Hepatology*, *62*, 1497–1510. https://doi.org/10.1002/hep.27992.

Masucci, M. T., Minopoli, M., & Carriero, M. V. (2019). Tumor associated neutrophils. Their role in tumorigenesis, metastasis, prognosis and therapy. *Frontiers in Oncology*, *9*, 1146. https://doi.org/10.3389/fonc.2019.01146.

Masyuk, T. V., Radtke, B. N., Stroope, A. J., Banales, J. M., Masyuk, A. I., Gradilone, S. A., et al. (2012). Inhibition of Cdc25A suppresses hepato-renal cystogenesis in rodent models of polycystic kidney and liver disease. *Gastroenterology*, *142*, 622–633. e4. https://doi.org/10.1053/j.gastro.2011.11.036.

Matsumoto, K., Onoyama, T., Kawata, S., Takeda, Y., Harada, K., Ikebuchi, Y., et al. (2014). Hepatitis B and C virus infection is a risk factor for the development of cholangiocarcinoma. *Internal Medicine*, *53*, 651–654. https://doi.org/10.2169/internalmedicine.53.1410.

Matsumoto, S., Yamamoto, K., Nagano, T., Okamoto, R., Ibuki, N., Tagashira, M., et al. (1999). Immunohistochemical study on phenotypical changes of hepatocytes in liver disease with reference to extracellular matrix composition. *Liver*, *19*, 32–38. https://doi.org/10.1111/j.1478-3231.1999.tb00006.x.

McGee, E. E., Jackson, S. S., Petrick, J. L., Van Dyke, A. L., Adami, H. O., Albanes, D., et al. (2019). Smoking, alcohol, and biliary tract cancer risk: A pooling project of 26 prospective studies. *Journal of the National Cancer Institute*, *111*, 1263–1278. https://doi.org/10.1093/jnci/djz103.

Mederacke, I., Hsu, C. C., Troeger, J. S., Huebener, P., Mu, X., Dapito, D. H., et al. (2013). Fate tracing reveals hepatic stellate cells as dominant contributors to liver fibrosis independent of its aetiology. *Nature Communications*, *4*, 2823. https://doi.org/10.1038/ncomms3823.

Mells, G. F., Kaser, A., & Karlsen, T. H. (2013). Novel insights into autoimmune liver diseases provided by genome-wide association studies. *Journal of Autoimmunity*, *46*, 41–54. https://doi.org/10.1016/j.jaut.2013.07.004.

Melum, E., Franke, A., Schramm, C., Weismüller, T. J., Gotthardt, D. N., Offner, F. A., et al. (2011). Genome-wide association analysis in primary sclerosing cholangitis identifies two non-HLA susceptibility loci. *Nature Genetics*, *43*(1), 17–19. https://doi.org/10.1038/ng.728.

Meng, F., Henson, R., Wehbe-Janek, H., Smith, H., Ueno, Y., & Patel, T. (2007). The microRNA let-7a modulates interleukin-6-dependent STAT-3 survival signaling in malignant human cholangiocytes. *The Journal of Biological Chemistry*, *282*, 8256–8264. https://doi.org/10.1074/jbc.M607712200.

Michelotti, A., de Scordilli, M., Palmero, L., Guardascione, M., Masala, M., Roncato, R., et al. (2021). NAFLD-related hepatocarcinoma: The malignant side of metabolic syndrome. *Cell*, *10*, 2034. https://doi.org/10.3390/cells10082034.

Montal, R., Sia, D., Montironi, C., Leow, W. Q., Esteban-Fabró, R., Pinyol, R., et al. (2020). Molecular classification and therapeutic targets in extrahepatic cholangiocarcinoma. *Journal of Hepatology*, 73, 315–327. https://doi.org/10.1016/j.jhep.2020.03.008.

Morrison, C. D., Parvani, J. G., & Schiemann, W. P. (2013). The relevance of the TGF-β paradox to EMT-MET programs. *Cancer Letters*, 341, 30–40. https://doi.org/10.1016/j.canlet.2013.02.048.

Nachtergaele, S., Mydock, L. K., Krishnan, K., Rammohan, J., Schlesinger, P. H., Covey, D. F., et al. (2012). Oxysterols are allosteric activators of the oncoprotein smoothened. *Nature Chemical Biology*, 8, 211–220. https://doi.org/10.1038/nchembio.765.

Nguyen, C. T., Caruso, S., Maille, P., Beaufrère, A., Augustin, J., Favre, L., et al. (2021). Immune profiling of combined hepatocellular-cholangiocarcinoma reveals distinct subtypes and activation of gene signatures predictive of response to immunotherapy. *Clinical Cancer Research*. https://doi.org/10.1158/1078-0432.CCR-21-1219. doi: 10.1158/1078-0432.CCR-21-1219.

Ninlawan, K., O'Hara, S. P., Splinter, P. L., Yongvanit, P., Kaewkes, S., Surapaitoon, A., et al. (2010). Opisthorchis viverrini excretory/secretory products induce toll-like receptor 4 upregulation and production of interleukin 6 and 8 in cholangiocyte. *Parasitology International*, 59, 616–621. https://doi.org/10.1016/j.parint.2010.09.008.

Nusse, R., & Clevers, H. (2017). Wnt/β-catenin signaling, disease, and emerging therapeutic modalities. *Cell*, 169, 985–999. https://doi.org/10.1016/j.cell.2017.05.016.

Ohshima, H., Sawa, T., & Akaike, T. (2006). 8-nitroguanine, a product of nitrative DNA damage caused by reactive nitrogen species: Formation, occurrence, and implications in inflammation and carcinogenesis. *Antioxidants & Redox Signaling*, 8, 1033–1045. https://doi.org/10.1089/ars.2006.8.1033.

Okabe, H., Beppu, T., Ueda, M., Hayashi, H., Ishiko, T., Masuda, T., et al. (2012). Identification of CXCL5/ENA-78 as a factor involved in the interaction between cholangiocarcinoma cells and cancer-associated fibroblasts. *International Journal of Cancer*, 131, 2234–2241. https://doi.org/10.1002/ijc.27496.

Özdirik, B., Müller, T., Wree, A., Tacke, F., & Sigal, M. (2021). The role of microbiota in primary sclerosing cholangitis and related biliary malignancies. *International Journal of Molecular Sciences*, 22, 6975. https://doi.org/10.3390/ijms22136975.

Palmer, W. C., & Patel, T. (2012). Are common factors involved in the pathogenesis of primary liver cancers? A meta-analysis of risk factors for intrahepatic cholangiocarcinoma. *Journal of Hepatology*, 57, 69–76. https://doi.org/10.1016/j.jhep.2012.02.022.

Panya, A., Thepmalee, C., Sawasdee, N., Sujjitjoon, J., Phanthaphol, N., Junking, M., et al. (2018). Cytotoxic activity of effector T cells against cholangiocarcinoma is enhanced by self-differentiated monocyte-derived dendritic cells. *Cancer Immunology, Immunotherapy*, 67, 1579–1588. https://doi.org/10.1007/s00262-018-2212-2.

Park, J., Tadlock, L., Gores, G. J., & Patel, T. (1999). Inhibition of interleukin 6-mediated mitogen-activated protein kinase activation attenuates growth of a cholangiocarcinoma cell line. *Hepatology*, 30, 1128–1133. https://doi.org/10.1002/hep.510300522.

Pech, L., Favelier, S., Falcoz, M. T., Loffroy, R., Krause, D., & Cercueil, J. P. (2016). Imaging of Von Meyenburg complexes. *Diagnostic and Interventional Imaging*, 97, 401–409. https://doi.org/10.1016/j.diii.2015.05.012.

Perugorria, M. J., Olaizola, P., Labiano, I., Esparza-Baquer, A., Marzioni, M., Marin, J., et al. (2019). Wnt-β-catenin signalling in liver development, health and disease. *Nature Reviews. Gastroenterology & Hepatology*, 16, 121–136. https://doi.org/10.1038/s41575-018-0075-9.

Petrick, J. L., Yang, B., Altekruse, S. F., Van Dyke, A. L., Koshiol, J., Graubard, B. I., et al. (2017). Risk factors for intrahepatic and extrahepatic cholangiocarcinoma in the United States: A population-based study in SEER-Medicare. *PLoS One*, *12*, e0186643. https://doi.org/10.1371/journal.pone.0186643.

Piccolo, S., Dupont, S., & Cordenonsi, M. (2014). The biology of YAP/TAZ: Hippo signaling and beyond. *Physiological Reviews*, *94*, 1287–1312. https://doi.org/10.1152/physrev.00005.2014.

Pinto, C., Giordano, D. M., Maroni, L., & Marzioni, M. (2018). Role of inflammation and proinflammatory cytokines in cholangiocyte pathophysiology. *Biochimica et Biophysica Acta, Molecular Basis of Disease*, *1864*, 1270–1278. https://doi.org/10.1016/j.bbadis.2017.07.024.

Pontecorvi, V., Carbone, M., & Invernizzi, P. (2016). The "gut microbiota" hypothesis in primary sclerosing cholangitis. *Annals of Translational Medicine*, *4*, 512. https://doi.org/10.21037/atm.2016.12.43.

Prueksapanich, P., Piyachaturawat, P., Aumpansub, P., Ridtitid, W., Chaiteerakij, R., & Rerknimitr, R. (2018). Liver fluke-associated biliary tract cancer. *Gut and Liver*, *12*, 236–245. https://doi.org/10.5009/gnl17102.

Puthdee, N., Sriswasdi, S., Pisitkun, T., Ratanasirintrawoot, S., Israsena, N., & Tangkijvanich, P. (2021). The LIN28B/TGF-β/TGFBI feedback loop promotes cell migration and tumour initiation potential in cholangiocarcinoma. *Cancer Gene Therapy*. https://doi.org/10.1038/s41417-021-00387-5. doi:10.1038/s41417-021-00387-5.

Qian, M. B., Utzinger, J., Keiser, J., & Zhou, X. N. (2016). Clonorchiasis. *Lancet*, *387*, 800–810. https://doi.org/10.1016/S0140-6736(15)60313-0.

Raggi, C., Correnti, M., Sica, A., Andersen, J. B., Cardinale, V., Alvaro, D., et al. (2017). Cholangiocarcinoma stem-like subset shapes tumor-initiating niche by educating associated macrophages. *Journal of Hepatology*, *66*, 102–115. https://doi.org/10.1016/j.jhep.2016.08.012.

Reddy, S. K., Hyder, O., Marsh, J. W., Sotiropoulos, G. C., Paul, A., Alexandrescu, S., et al. (2013). Prevalence of nonalcoholic steatohepatitis among patients with resectable intrahepatic cholangiocarcinoma. *Journal of Gastrointestinal Surgery*, *17*, 748–755. https://doi.org/10.1007/s11605-013-2149-x.

Rimassa, L., Personeni, N., Aghemo, A., & Lleo, A. (2019). The immune milieu of cholangiocarcinoma: From molecular pathogenesis to precision medicine. *Journal of Autoimmunity*, *100*, 17–26. https://doi.org/10.1016/j.jaut.2019.03.007.

Rizvi, S., & Gores, G. J. (2013). Pathogenesis, diagnosis, and management of cholangiocarcinoma. *Gastroenterology*, *145*, 1215–1229. https://doi.org/10.1053/j.gastro.2013.10.013.

Rizvi, S., Yamada, D., Hirsova, P., Bronk, S. F., Werneburg, N. W., Krishnan, A., et al. (2016). A Hippo and fibroblast growth factor receptor autocrine pathway in cholangiocarcinoma. *The Journal of Biological Chemistry*, *291*, 8031–8047. https://doi.org/10.1074/jbc.M115.698472.

Rodrigues, P. M., Olaizola, P., Paiva, N. A., Olaizola, I., Agirre-Lizaso, A., Landa, A., et al. (2021). Pathogenesis of cholangiocarcinoma. *Annual Review of Pathology*, *16*, 433–463. https://doi.org/10.1146/annurev-pathol-030220-020455.

Rohatgi, R., Battini, L., Kim, P., Israeli, S., Wilson, P. D., Gusella, G. L., et al. (2008). Mechanoregulation of intracellular Ca2+ in human autosomal recessive polycystic kidney disease cyst-lining renal epithelial cells. *American Journal of Physiology. Renal Physiology*, *294*, F890–F899. https://doi.org/10.1152/ajprenal.00341.2007.

Rothlin, C. V., Ghosh, S., Zuniga, E. I., Oldstone, M. B., & Lemke, G. (2007). TAM receptors are pleiotropic inhibitors of the innate immune response. *Cell, 131*, 1124–1136. https://doi.org/10.1016/j.cell.2007.10.034.

Sarcognato, S., Sacchi, D., Fassan, M., Fabris, L., Cadamuro, M., Zanus, G., et al. (2021). Cholangiocarcinoma. *Pathologica, 113*, 158–169. https://doi.org/10.32074/1591-951X-252.

Sasaki, M., & Nakanuma, Y. (2016). New concept: Cellular senescence in pathophysiology of cholangiocarcinoma. *Expert Review of Gastroenterology & Hepatology, 10*, 625–638. https://doi.org/10.1586/17474124.2016.1133291.

Schwabe, R. F., & Greten, T. F. (2020). Gut microbiome in HCC—Mechanisms, diagnosis and therapy. *Journal of Hepatology, 72*, 230–238. https://doi.org/10.1016/j.jhep.2019.08.016.

Sekiya, S., & Suzuki, A. (2012). Intrahepatic cholangiocarcinoma can arise from Notch-mediated conversion of hepatocytes. *The Journal of Clinical Investigation, 122*, 3914–3918. https://doi.org/10.1172/JCI63065.

Seo, J. W., Kwan, B. S., Cheon, Y. K., Lee, T. Y., Shim, C. S., Kwon, S. Y., et al. (2020). Prognostic impact of hepatitis B or C on intrahepatic cholangiocarcinoma. *The Korean Journal of Internal Medicine, 35*, 566–573. https://doi.org/10.3904/kjim.2018.062.

Shoda, J., Ueda, T., Kawamoto, T., Todoroki, T., Asano, T., Sugimoto, Y., et al. (2003). Prostaglandin E receptors in bile ducts of hepatolithiasis patients and the pathobiological significance for cholangitis. *Clinical Gastroenterology and Hepatology, 1*, 285–296.

Sia, D., Hoshida, Y., Villanueva, A., Roayaie, S., Ferrer, J., Tabak, B., et al. (2013). Integrative molecular analysis of intrahepatic cholangiocarcinoma reveals 2 classes that have different outcomes. *Gastroenterology, 144*, 829–840. https://doi.org/10.1053/j.gastro.2013.01.001.

Simbolo, M., Bersani, S., Vicentini, C., Taormina, S. V., Ciaparrone, C., Bagante, F., et al. (2021). Molecular characterization of extrahepatic cholangiocarcinoma: Perihilar and distal tumors display divergent genomic and transcriptomic profiles. *Expert Opinion on Therapeutic Targets, 25*, 1095–1105. https://doi.org/10.1080/14728222.2021.2013801.

Sirica, A. E. (2005). Cholangiocarcinoma: Molecular targeting strategies for chemoprevention and therapy. *Hepatology, 41*, 5–15. https://doi.org/10.1002/hep.20537.

Sirica, A. E. (2011). The role of cancer-associated myofibroblasts in intrahepatic cholangiocarcinoma. *Nature Reviews. Gastroenterology & Hepatology, 9*, 44–54. https://doi.org/10.1038/nrgastro.2011.222.

Sirica, A. E., Gores, G. J., Groopman, J. D., Selaru, F. M., Strazzabosco, M., Wei Wang, X., et al. (2019). Intrahepatic cholangiocarcinoma: Continuing challenges and translational advances. *Hepatology, 69*, 1803–1815. https://doi.org/10.1002/hep.30289.

Sirica, A. E., Strazzabosco, M., & Cadamuro, M. (2021). Intrahepatic cholangiocarcinoma: Morpho-molecular pathology, tumor reactive microenvironment, and malignant progression. *Advances in Cancer Research, 149*, 321–387. https://doi.org/10.1016/bs.acr.2020.10.005.

Smoot, R. L., Werneburg, N. W., Sugihara, T., Hernandez, M. C., Yang, L., Mehner, C., et al. (2018). Platelet-derived growth factor regulates YAP transcriptional activity via Src family kinase dependent tyrosine phosphorylation. *Journal of Cellular Biochemistry, 119*, 824–836. https://doi.org/10.1002/jcb.26246.

Smout, M. J., Laha, T., Mulvenna, J., Sripa, B., Suttiprapa, S., Jones, A., et al. (2009). A granulin-like growth factor secreted by the carcinogenic liver fluke, Opisthorchis viverrini, promotes proliferation of host cells. *PLoS Pathogens, 5*, e1000611. https://doi.org/10.1371/journal.ppat.1000611.

Smout, M. J., Mulvenna, J. P., Jones, M. K., & Loukas, A. (2011). Expression, refolding and purification of Ov-GRN-1, a granulin-like growth factor from the carcinogenic liver fluke, that causes proliferation of mammalian host cells. *Protein Expression and Purification, 79*, 263–270. https://doi.org/10.1016/j.pep.2011.06.018.

Spirlì, C., Fabris, L., Duner, E., Fiorotto, R., Ballardini, G., Roskams, T., et al. (2003). Cytokine-stimulated nitric oxide production inhibits adenylyl cyclase and cAMP-dependent secretion in cholangiocytes. *Gastroenterology*, *124*, 737–753. https://doi.org/10.1053/gast.2003.50100.

Sripa, B., Kaewkes, S., Intapan, P. M., Maleewong, W., & Brindley, P. J. (2010). Food-borne trematodiases in Southeast Asia epidemiology, pathology, clinical manifestation and control. *Advances in Parasitology*, *72*, 305–350. https://doi.org/10.1016/S0065-308X(10)72011-X.

Strazzabosco, M., & Fabris, L. (2012). Development of the bile ducts: essentials for the clinical hepatologist. *Journal of Hepatology*, *56*, 1159–1170. https://doi.org/10.1016/j.jhep.2011.09.022.

Sugihara, T., Isomoto, H., Gores, G., & Smoot, R. (2019). YAP and the Hippo pathway in cholangiocarcinoma. *Journal of Gastroenterology*, *54*, 485–491. https://doi.org/10.1007/s00535-019-01563-z.

Sugiura, K., Mishima, T., Takano, S., Yoshitomi, H., Furukawa, K., Takayashiki, T., et al. (2019). The expression of Yes-associated protein (YAP) maintains putative cancer stemness and is associated with poor prognosis in intrahepatic cholangiocarcinoma. *The American Journal of Pathology*, *189*, 1863–1877. https://doi.org/10.1016/j.ajpath.2019.05.014.

Suksawat, M., Techasen, A., Namwat, N., Boonsong, T., Titapun, A., Ungarreevittaya, P., et al. (2018). Inhibition of endothelial nitric oxide synthase in cholangiocarcinoma cell lines—A new strategy for therapy. *FEBS Open Bio*, *8*, 513–522. https://doi.org/10.1002/2211-5463.12388.

Suksawat, M., Techasen, A., Namwat, N., Yongvanit, P., Khuntikeo, N., Titapun, A., et al. (2017). Upregulation of endothelial nitric oxide synthase (eNOS) and its upstream regulators in Opisthorchis viverrini associated cholangiocarcinoma and its clinical significance. *Parasitology International*, *66*, 486–493. https://doi.org/10.1016/j.parint.2016.04.008.

Sun, D., Luo, T., Dong, P., Zhang, N., Chen, J., Zhang, S., et al. (2020). CD86+/CD206+ tumor-associated macrophages predict prognosis of patients with intrahepatic cholangiocarcinoma. *PeerJ*, *8*, e 8458. https://doi.org/10.7717/peerj.8458.

Suttiprapa, S., Matchimakul, P., Loukas, A., Laha, T., Wongkham, S., Kaewkes, S., et al. (2012). Molecular expression and enzymatic characterization of thioredoxin from the carcinogenic human liver fluke Opisthorchis viverrini. *Parasitology International*, *61*, 101–106. https://doi.org/10.1016/j.parint.2011.06.018.

Tadlock, L., & Patel, T. (2001). Involvement of p38 mitogen-activated protein kinase signaling in transformed growth of a cholangiocarcinoma cell line. *Hepatology*, *33*, 43–51. https://doi.org/10.1053/jhep.2001.20676.

Tang, Q., Li, J., Zhu, H., Li, P., Zou, Z., & Xiao, Y. (2013). Hmgb1-IL-23-IL-17-IL-6-Stat3 axis promotes tumor growth in murine models of melanoma. *Mediators of Inflammation*, *2013*, 713859. https://doi.org/10.1155/2013/713859.

Techasen, A., Namwat, N., Loilome, W., Duangkumpha, K., Puapairoj, A., Saya, H., et al. (2014). Tumor necrosis factor-α modulates epithelial mesenchymal transition mediators ZEB2 and S100A4 to promote cholangiocarcinoma progression. *Journal of Hepato-Biliary-Pancreatic Sciences*, *21*, 703–711. https://doi.org/10.1002/jhbp.125.

Trivedi, P. J., & Adams, D. H. (2013). Mucosal immunity in liver autoimmunity: A comprehensive review. *Journal of Autoimmunity*, *46*, 97–111. https://doi.org/10.1016/j.jaut.2013.06.013.

Tsunoda, T., Kakinuma, S., Miyoshi, M., Kamiya, A., Kaneko, S., Sato, A., et al. (2019). Loss of fibrocystin promotes interleukin-8-dependent proliferation and CTGF- production of biliary epithelium. *Journal of Hepatology*, *71*, 143–152. https://doi.org/10.1016/j.jhep.2019.02.024.

Tyson, G. L., & El-Serag, H. B. (2011). Risk factors for cholangiocarcinoma. *Hepatology*, *54*, 173–184. https://doi.org/10.1002/hep.24351.

Vale, N., Gouveia, M. J., Gärtner, F., & Brindley, P. J. (2020). Oxysterols of helminth parasites and pathogenesis of foodborne hepatic trematodiasis caused by Opisthorchis and Fasciola species. *Parasitology Research*, *119*, 1443–1453. https://doi.org/10.1007/s00436-020-06640-4.

Vaquero, J., Aoudjehane, L., & Fouassier, L. (2020). Cancer-associated fibroblasts in cholangiocarcinoma. *Current Opinion in Gastroenterology*, *36*, 63–69. https://doi.org/10.1097/MOG.0000000000000609.

Veigel, M. C., Prescott-Focht, J., Rodriguez, M. G., Zinati, R., Shao, L., Moore, C. A., et al. (2009). Fibropolycystic liver disease in children. *Pediatric Radiology*, *39*, 317–421. https://doi.org/10.1007/s00247-008-1070-z.

Wang, J., Dong, M., Xu, Z., Song, X., Zhang, S., Qiao, Y., et al. (2018). Notch2 controls hepatocyte-derived cholangiocarcinoma formation in mice. *Oncogene*, *37*, 3229–3242. https://doi.org/10.1038/s41388-018-0188-1.

Wang, Z., Sheng, Y. Y., Dong, Q. Z., & Qin, L. X. (2016). Hepatitis B virus and hepatitis C virus play different prognostic roles in intrahepatic cholangiocarcinoma: A meta-analysis. *World Journal of Gastroenterology*, *22*, 3038–3051. https://doi.org/10.3748/wjg.v22.i10.3038.

Wang, X. Y., Zhu, W. W., Wang, Z., Huang, J. B., Wang, S. H., Bai, F. M., et al. (2022). Driver mutations of intrahepatic cholangiocarcinoma shape clinically relevant genomic clusters with distinct molecular features and therapeutic vulnerabilities. *Theranostics*, *12*, 260–276. https://doi.org/10.7150/thno.63417.

Wehbe, H., Henson, R., Meng, F., Mize-Berge, J., & Patel, T. (2006). Interleukin-6 contributes to growth in cholangiocarcinoma cells by aberrant promoter methylation and gene expression. *Cancer Research*, *66*, 10517–10524. https://doi.org/10.1158/0008-5472.CAN-06-2130.

Welzel, T. M., Graubard, B. I., Zeuzem, S., El-Serag, H. B., Davila, J. A., & McGlynn, K. A. (2011). Metabolic syndrome increases the risk of primary liver cancer in the United States: A study in the SEER-Medicare database. *Hepatology*, *54*, 463–471. https://doi.org/10.1002/hep.24397.

Welzel, T. M., McGlynn, K. A., Hsing, A. W., O'Brien, T. R., & Pfeiffer, R. M. (2006). Impact of classification of hilar cholangiocarcinomas (Klatskin tumors) on the incidence of intra- and extrahepatic cholangiocarcinoma in the United States. *Journal of the National Cancer Institute*, *98*, 873–875. https://doi.org/10.1093/jnci/djj234.

Williamson, K. D., & Chapman, R. W. (2015). Primary sclerosing cholangitis: A clinical update. *British Medical Bulletin*, *114*, 53–64. https://doi.org/10.1093/bmb/ldv019.

Wongjarupong, N., Assavapongpaiboon, B., Susantitaphong, P., Cheungpasitporn, W., Treeprasertsuk, S., Rerknimitr, R., et al. (2017). Non-alcoholic fatty liver disease as a risk factor for cholangiocarcinoma: A systematic review and meta-analysis. *BMC Gastroenterology*, *17*, 149. https://doi.org/10.1186/s12876-017-0696-4.

Wu, W. R., Zhang, R., Shi, X. D., Zhu, M. S., Xu, L. B., Zeng, H., et al. (2014). Notch1 is overexpressed in human intrahepatic cholangiocarcinoma and is associated with its proliferation, invasiveness and sensitivity to 5-fluorouracil in vitro. *Oncology Reports*, *31*, 2515–2524. https://doi.org/10.3892/or.2014.3123.

Yoon, J. H., Canbay, A. E., Werneburg, N. W., Lee, S. P., & Gores, G. J. (2004). Oxysterols induce cyclooxygenase-2 expression in cholangiocytes: Implications for biliary tract carcinogenesis. *Hepatology*, *39*, 732–738. https://doi.org/10.1002/hep.20125.

Yoon, J. H., Gwak, G. Y., Lee, H. S., Bronk, S. F., Werneburg, N. W., & Gores, G. J. (2004). Enhanced epidermal growth factor receptor activation in human cholangiocarcinoma cells. *Journal of Hepatology*, *41*, 808–814. https://doi.org/10.1016/j.jhep.2004.07.016.

Yoon, J. H., Higuchi, H., Werneburg, N. W., Kaufmann, S. H., & Gores, G. J. (2002). Bile acids induce cyclooxygenase-2 expression via the epidermal growth factor receptor in a human cholangiocarcinoma cell line. *Gastroenterology, 122*, 985–993. https://doi.org/10.1053/gast.2002.32410.

Young, N. D., Campbell, B. E., Hall, R. S., Jex, A. R., Cantacessi, C., Laha, T., et al. (2010). Unlocking the transcriptomes of two carcinogenic parasites, Clonorchis sinensis and Opisthorchis viverrini. *PLoS Neglected Tropical Diseases, 4*, e719. https://doi.org/10.1371/journal.pntd.0000719.

Yu, C., Bruzek, L. M., Meng, X. W., Gores, G. J., Carter, C. A., Kaufmann, S. H., et al. (2005). The role of Mcl-1 downregulation in the proapoptotic activity of the multikinase inhibitor BAY 43-9006. *Oncogene, 24*, 6861–6869. https://doi.org/10.1038/sj.onc.1208841.

Zanconato, F., Cordenonsi, M., & Piccolo, S. (2019). YAP and TAZ: A signalling hub of the tumour microenvironment. *Nature Reviews. Cancer, 19*, 454–464. https://doi.org/10.1038/s41568-019-0168-y.

Zen, Y., Harada, K., Sasaki, M., Chen, T. C., Chen, M. F., Yeh, T. S., et al. (2005). Intrahepatic cholangiocarcinoma escapes from growth inhibitory effect of transforming growth factor-beta 1 by overexpression of cyclin D1. *Laboratory Investigation, 85*, 572–581. https://doi.org/10.1038/labinvest.3700236.

Zerres, K., Rudnik-Schöneborn, S., Steinkamm, C., Becker, J., & Mücher, G. (1998). Autosomal recessive polycystic kidney disease. *Journal of Molecular Medicine, 76*, 303–309. https://doi.org/10.1007/s001090050221.

Zhang, S., Wang, J., Wang, H., Fan, L., Fan, B., Zeng, B., et al. (2018). Hippo cascade controls lineage commitment of liver tumors in mice and humans. *The American Journal of Pathology, 188*, 995–1006. https://doi.org/10.1016/j.ajpath.2017.12.017.

Zheng, Y., Ran, Y., Zhang, H., Wang, B., & Zhou, L. (2021). The microbiome in autoimmune liver diseases: Metagenomic and metabolomic changes. *Frontiers in Physiology, 12*, 715852. https://doi.org/10.3389/fphys.2021.715852.

Zhou, S. L., Dai, Z., Zhou, Z. J., Chen, Q., Wang, Z., Xiao, Y. S., et al. (2014). CXCL5 contributes to tumor metastasis and recurrence of intrahepatic cholangiocarcinoma by recruiting infiltrative intratumoral neutrophils. *Carcinogenesis, 35*, 597–605. https://doi.org/10.1093/carcin/bgt397.

Zhou, M., Wang, C., Lu, S., Xu, Y., Li, Z., Jiang, H., et al. (2021). Tumor-associated macrophages in cholangiocarcinoma: Complex interplay and potential therapeutic target. *eBioMedicine, 67*, 103375. https://doi.org/10.1016/j.ebiom.2021.103375.

Zhou, Z., Wang, P., Sun, R., Li, J., Hu, Z., Xin, H., et al. (2021). Tumor-associated neutrophils and macrophages interaction contributes to intrahepatic cholangiocarcinoma progression by activating STAT3. *Journal for Immunotherapy of Cancer, 9*, e001946. https://doi.org/10.1136/jitc-2020-001946.

Zhou, Y., Zhao, Y., Li, B., Huang, J., Wu, L., Xu, D., et al. (2012). Hepatitis viruses infection and risk of intrahepatic cholangiocarcinoma: Evidence from a meta-analysis. *BMC Cancer, 12*, 289. https://doi.org/10.1186/1471-2407-12-289.

CHAPTER THREE

Causes and functional intricacies of inter- and intratumor heterogeneity of primary liver cancers

Subreen A. Khatib[a,b] and Xin Wei Wang[a,c,]*

[a]Laboratory of Human Carcinogenesis, Center for Cancer Research, National Cancer Institute, Bethesda, MD, United States
[b]Department of Tumor Biology, Lombardi Comprehensive Cancer Center, Georgetown University Medical Center, Washington, DC, United States
[c]Liver Cancer Program, Center for Cancer Research, National Cancer Institute, Bethesda, MD, United States
*Corresponding author: e-mail address: xw3u@nih.gov

Contents

1. Introduction	77
1.1 Liver cancer overview and statistics	77
1.2 Classification of liver cancer types	78
1.3 Etiologies, risk factors, and progression of disease	78
1.4 Therapeutic options and early detection for patients	81
2. Liver cancer heterogeneity	83
2.1 Inter- vs intratumor heterogeneity	83
2.2 Models of tumor heterogeneity	83
2.3 Tools to investigate tumor heterogeneity	84
2.4 Causes and drivers of tumor heterogeneity	86
2.5 Functional implications of tumor heterogeneity	89
2.6 Microenvironmental influences on tumor heterogeneity	90
2.7 Clinical ramifications of tumor heterogeneity	91
3. Spatial architecture and tumor heterogeneity	92
3.1 The need to understand the spatial architecture of tumors	92
3.2 Spatial organization dictates functional heterogeneity	93
3.3 Leveraging spatial techniques to better understand liver cancer heterogeneity	94
3.4 Computational methods for downstream spatial analysis	95
3.5 The interplay between spatial biology and therapeutic development	96
4. Concluding remarks and future perspectives	96
Acknowledgments	97
Author contributions	98
Declaration of interests	98
References	98

Abstract

Tumor heterogeneity is a major feature of primary liver cancers. Defined as the unique genotypic and phenotypic differences of cancer cells within a single tumor (intratumor) or amongst different patients (intertumor), tumor heterogeneity has consistently been linked to worse clinical outcomes in most, if not all, solid tumor types. In particular, liver cancer heterogeneity has been associated with altered immune infiltration, resistance to therapeutics, and worse overall patient survival. Current advancements in single-cell omic technologies have allowed for a deeper understanding and appreciation of the intricate composition and relationships between individual cells within a tumor. These observations have led to the discovery of new cell types in liver cancer, potential new mechanisms of therapy resistance and tumor progression, and new insights into the evolutionary patterns of liver cancer. To better understand the tumor biology of liver cancers and their heterogeneous features, we will begin this chapter on a brief background of liver cancer and then discuss the various etiologies of this disease and how each one can contribute to diverse genomic, transcriptomic, proteomic, and spatial architecture observations. Next, we will go into the specific causes and implications of tumor heterogeneity and end with how understanding the spatial architecture of liver tumors can provide us with new insights and ideas for tumor diversity and therapeutic development.

Abbreviations

ADH	alcohol dehydrogenase
ALDH	aldehyde dehydrogenase
APC	Adenomatous polyposis coli
ATAC	assay for transposase accessible chromatin
CAF	cancer-associated fibroblast
CODEX	CO-Detection by IndEXing
COX-2	cyclooxygenase-2
CSC	cancer stem cell
CTSE	cathepsin
DSP	digital spatial profiler
EMT	epithelial-mesenchymal transition
ECM	extracellular matrix
FDA	Food and Drug Administration
FGFR	fibroblast growth factor receptor
HBV	hepatitis B virus
HCC	hepatocellular carcinoma
HCV	hepatitis C virus
CCA	cholangiocarcinoma
IDH1	isocitrate dehydrogenase 1
IGF-2	insulin-like growth factor 2
iNOS	nitric oxide synthase
KRAS	Kirsten rat sarcoma virus
LAYN	Layilin
MSI	microsatellite instability
NAFLD	nonalcoholic fatty liver disease

NASH	nonalcoholic steatohepatitis
NECTIN2	nectin Cell Adhesion Molecule 2
NF-kB	nuclear factor-kB
OPN	osteopontin
PDGFR	platelet-derived growth factor receptor
PD1	programmed cell death protein 1
PDL1	programmed death ligand 1
RASSF1	RalGDS/AF-6 domain family member 1
ROS	reactive oxygen species
SMAD4	SMAD family member 4
SPP1	secreted phosphoprotein 1
TACE	trans-arterial chemoembolization
TAM	tumor associated macrophages
TIGIT	T cell immunoglobulin and ITIM
TP53	tumor protein p53
TSEI	tumor shannon entropy index
VEGF	vascular endothelial growth factor
VEGFR	vascular endothelial growth factor receptor

1. Introduction
1.1 Liver cancer overview and statistics

Primary liver cancer is one of the few and fastest rising malignancies worldwide, with the highest rates of diagnoses in Southeast Asia. While the overall cancer cases have decreased in U.S. in the past several decades, liver cancer rates are on the rise with an approximate 43,000 new diagnoses every year and an overall 5-year survival rate of 18% (Dasgupta et al., 2020; Siegel, Miller, & Jemal, 2020). This discrepancy may be attributed to most patients being diagnosed at late stages which is caused by a lack of screening tools and diagnostic applications to detect early-stage disease. Moreover, there is no clear molecular driver or mutational profile that distinguishes the onset of primary liver cancer and today's standard of practice does not make biopsies mandatory before the initiation of care. Thus, current tumor samples are often taken post-treatment and/or long after a tumor is formed and possibly spread to other regions of the liver and distant organs, making the model system in which we study liver cancer difficult to decipher major drivers of disease onset and progression. However, with the advancement of many omic technologies and numerous clinical trials underway, it is clear that our understanding of liver cancer is growing and new therapeutic options are on the horizon.

1.2 Classification of liver cancer types

To date, there are two main forms of liver cancer: hepatocellular carcinoma (HCC) and cholangiocarcinoma (CCA). However, multiple studies have shown that within these two forms, there are unique molecular subtypes that can be distinguished by genetic drivers and clinical outcome (Chaisaingmongkol et al., 2017; Shimada et al., 2019). Nevertheless, HCC and CCA are both vastly heterogeneous and fast progressing tumors with 90% of diagnosed cases falling under HCC and the remaining 10% as CCA. Classification of HCC is denoted by malignancy of the hepatocytes while manifestation of CCA arises from the intrahepatic bile duct epithelium. CCA is further divided anatomically by intrahepatic and extrahepatic CCA based on the location of the tumor within the biliary tree (Lazaridis & Gores, 2005). Intrahepatic CCA is present within the hepatic parenchyma and is manifested by mass lesions while extrahepatic CCA is found in large bile ducts and can cause biliary obstruction (Lazaridis & Gores, 2005).

1.3 Etiologies, risk factors, and progression of disease

While the molecular mechanism which causes primary liver cancer is largely unknown, nearly all cases stem from a given risk factor that leads to progression of various steps which include liver injury and inflammation that may result in chronic liver disease and subsequently liver cancer. Currently, there are common risk factors that are attributed to the onset of primary liver cancer. These include cirrhosis, hepatitis B and/or C viral infection (HBV, HCV), alcohol abuse, obesity, nonalcoholic fatty liver disease, ingestion of liver flukes, and exposure to aflatoxin (Fig. 1A). Cirrhosis defined as scarring of the liver results in permanently damaged tissue and impaired liver function. Patients with underlying cirrhosis have a 20- to 30-fold increased risk of developing HCC or CCA and over 40-fold increased mortality rate compared to non-cirrhotic patients (Thiele, Gluud, Fialla, Dahl, & Krag, 2014; Welzel et al., 2007). However, individuals can still develop cancer with a noncirrhotic liver.

Worldwide, the most common risk factor of primary liver cancer includes viral infection of hepatitis B (HBV) and/or C (HBC). HBV accounts for over 50% of HCC cases while HCV accounts for 25% of cases diagnosed each year (Perz, Armstrong, Farrington, Hutin, & Bell, 2006). Across multiple countries, 57% of patients who developed cirrhosis were either infected with HBV or HCV (Perz et al., 2006). HBV infection can lead to a 15- to 20-fold increased risk of developing HCC and a 4- to

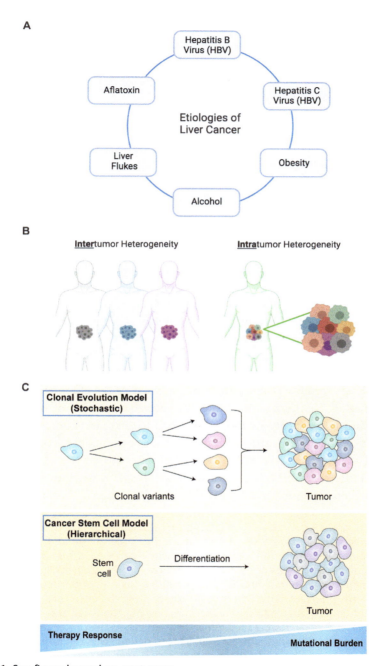

Fig. 1 See figure legend on next page.

6-fold increased risk of developing CCA and this is most common in countries other than the United States. Interestingly, HCV is the leading cause of HCC and CCA in the United States. Those infected with HCV have an estimated 17-fold increased risk of developing HCC or CCA (Donato et al., 2002; Yamamoto et al., 2004).

It is no surprise that substantial alcohol intake is related to the onset of liver cancer as numerous studies have shown this association (Donato et al., 2002; Lieber, 1994; Matsushita & Takaki, 2019; Testino, Leone, & Borro, 2014). Heavy use of alcohol can result in acute/chronic hepatitis or inflammation, fatty liver, and cirrhosis. Upon the intake of alcohol, ethanol is metabolized in the liver and is converted to acetaldehyde by alcohol dehydrogenase (ADH) which is further oxidized to acetate by mitochondrial aldehyde dehydrogenase (ALDH). Research has shown that there exist polymorphisms in the ALDH2 gene that has been associated with HCC development. Acetaldehyde has been shown to be carcinogenic by modulating the DNA repair system to produce certain DNA mutagenic effects. Moreover, the metabolism of ethanol also produces reactive oxygen species (ROS) which can result in increased levels of lipid peroxidation and subsequent mutagenic effects such as p53 mutations (Hu et al., 2002).

Aside from alcohol intake, nonalcoholic fatty liver disease (NAFLD) and nonalcoholic steatohepatitis (NASH), which is the most severe form of NAFLD, are significant indications of liver cancer development. NAFLD is classified as a condition in which an excessive amount of triglycerides accumulate in liver cells leading to steatosis (excess fat). Interestingly, this observation occurs in the absence of excessive alcohol consumption. Current manifestations of NAFLD/NASH have shown to cause liver cell injury, inflammation, and progression towards fibrosis with 10–20% of patients then developing cirrhosis (Huang, El-Serag, & Loomba, 2021).

Fig. 1 Understanding the causes of liver cancer and models of tumor heterogeneity. (A) Etiologies of liver cancer. (B) Tumor heterogeneity is defined as the unique genotypic and phenotypic differences of cancer cells amongst different patients (intertumor) or within a single tumor (intratumor). (C) (Top panel) The clonal evolution model of tumor heterogeneity proposes that a cancer cell originates from one clone and through multiple rounds of replication and division in which new mutations can arise leads to new clonal variants that can form a solid tumor. (Bottom panel) The cancer stem model of tumor heterogeneity proposes that the differences in tumor cells can be linked to their differentiation stage. This model suggests that there is a certain population of cells with indefinite self-renewal capabilities are organized in a hierarchical manner pertinent to the stage of differentiation.

Within the last few decades, prevalence of NAFLD-related HCC has nearly doubled in the United States and is projected to increase 122% by 2030 (Huang et al., 2021). Moreover, in a similar observation of excessive triglycerides, obesity is also another risk factor of liver cancer. Because obesity is associated with the metabolic syndrome, studies have shown these two manifestations as instigators of HCC development.

Furthermore, exposure to aflatoxin, which is produced by the fungi *Aspergillus flavus* and *Aspergillus parasiticus*, in maize and nuts is linked to approximately 5–30% of HCC-related cases worldwide (Liu & Wu, 2010). Studies have shown that aflatoxin gets metabolized in the liver by P450 enzymes which converts the carcinogen into ROS and binds to proteins or DNA, causing aflatoxicosis (acute toxicity) leading to lesions over time and ultimately cancer (Groopman, Kensler, & Wild, 2008). Exposure to aflatoxin and infection of hepatitis B virus can lead to a 30-fold increase risk of liver cancer (Groopman et al., 2008). On the other hand, liver fluke infection is one of the major risk factors for cholangiocarcinoma in Asian countries (Prueksapanich et al., 2018). Liver flukes also known as *Opisthorchis viverrine*, *Clonorchis sinensis*, and *Schistosomiasis japonica* are parasitic trematodes and are often found in the liver of mammals such as humans (Prueksapanich et al., 2018). Various mechanisms have been proposed for the link between liver flukes and cholangiocarcinoma. Physically, these parasites can injure the bile duct epitheliums which can cause bile duct ulcers (Prueksapanich et al., 2018). Mechanistically, liver fluke infection can lead to increased inflammation in the bile duct through upregulation of key mediators involved in the nuclear factor-kB (NF-kB) pathway, nitric oxide synthase (iNOS) induction, and cyclooxygenase (COX-2) stimulation (Prueksapanich et al., 2018). Taken together, the many risk factors that are associated with the onset of liver cancer demonstrates the vast diversity that exists within these tumors and the difficulties of developing targeted therapeutics that can significantly improve patient outcomes.

1.4 Therapeutic options and early detection for patients

Today, there exists few curable treatment options for patients with liver cancer. Selection of the appropriate treatment is often determined by the stage of the tumor, an example of which has been summarized as the Barcelona Clinic Liver Cancer Staging System (Vitale et al., 2011), among many other staging systems. Early-stage tumors typically undergo ablation, resection, or transplantation whereas intermediate stage tumors that cannot be treated with surgery will undergo trans-arterial chemoembolization (TACE). TACE is a procedure that combines embolization (or reduced blood flow to the liver tumor) with direct injection of chemotherapy near the tumor (Raoul et al., 2019). Furthermore, a majority of HCC cases form in cirrhotic livers and are often multifocal, eliminating the option for surgical treatment. Once the tumor reaches to an advanced stage, first-line systemic therapy is given. These options include Sorafenib, which was the first U.S. Food and Drug Administration (FDA)-approved first-line systemic drug for advanced HCC, and Lenvatinib which was developed to combat Sorafenib resistance (Feng, Pan, Kong, & Shu, 2020). They are both oral kinase inhibitors that target the vascular endothelial growth factor receptor (VEGFR)-family and platelet-derived growth factor receptor (PDGFR) and have median overall survival of 13.6 and 12.3 months, respectively (Feng et al., 2020). Various second-line treatments have been approved to treat advanced stage liver cancer which include Regorafenib, Nivolumab, Pembrolizumab, Cabozantinib, and Ramucirumab (Feng et al., 2020). Unfortunately, most cases of liver cancer are diagnosed at advanced or terminal stage which the only option then is supportive care. More specially, intrahepatic CCA, which is often diagnosed at advanced stages where surgical resection and transplantation are not an option, have the choice to undergo TACE or be given the more recently approved FDA drugs called pemigatinib, an fibroblast growth factor receptor (FGFR) inhibitor, and ivosidenib, an isocitrate dehydrogenase 1 (IDH1) inhibitor (Brindley et al., 2021; Fostea, Fontana, Torga, & Arkenau, 2020). These two drugs were identified through molecular profiling of tumor samples that revealed FGFR2 fusions and IDH1 mutations. An additional study identified KRAS, TP53, and SMAD4 mutations in extrahepatic CCA samples (Nakamura et al., 2015). Current approaches are beginning to utilize immunotherapeutic agents to combat targeted therapy resistance. These approaches include the anti-PDL1 antibody, atezolizumab, a PD1 inhibitor, pembrolizumab, and the VEGF-neutralizing antibody, bevacizumab (Brindley et al., 2021; Sangro, Sarobe, Hervás-Stubbs, & Melero, 2021). However, there is a large need to develop technologies and assays that can detect early-stage cancer when the prognosis is favorable. Recently, our lab developed a viral exposure signature that could accurately detect HCC in high-risk patients before a clinical diagnosis using a blood sample (Liu et al., 2020). This signature was much more specific and sensitive than the typical alpha-fetoprotein biomarker that is commonly used to survey liver cancer development. The findings of this study could provide a new screening tool to detect early-stage disease before a primary liver tumor becomes largely heterogeneous and unresponsive to treatment.

2. Liver cancer heterogeneity
2.1 Inter- vs intratumor heterogeneity

Tumor heterogeneity can be classified into two main types, inter- and intratumor heterogeneity. Intertumor heterogeneity is defined as diverse genotypic and phenotypic differences amongst different patients' tumor and intratumor heterogeneity consists of differences found within each tumor (Fig. 1B). With the development of single-cell omic technologies and multi-regional bulk sequencing analysis, extensive heterogeneity has been illustrated in primary liver cancers. Numerous studies have shown that there exist unique cell compositions found within each tumor such as different populations of nonmalignant cells including T cells, macrophages, dendritic cells, natural killer cells, endothelial cells, epithelium cells and various subclonal populations of malignant cells as well as extensive genetic, transcriptomic, and histopathological heterogeneity (Carotenuto et al., 2017; Hsu et al., 1991; Ma et al., 2019, 2021; Tanaka et al., 1993; Xue et al., 2016; Zheng et al., 2017).

2.2 Models of tumor heterogeneity

To better understand how tumor heterogeneity develops, it is important to consider the different models proposed—the clonal evolution model (stochastic) and cancer stem cell model (hierarchical) (Fig. 1C). The clonal evolution model, first defined by Peter Nowell in 1976, proposes that a cancer cell originates from one clone and through multiple rounds of replication and division in which new mutations can arise leads to new clonal variants that can form a solid tumor (Nowell, 1976). Recent studies have illustrated the clonal evolution of liver cancer based on single-cell mutational profiles. For example, Su et al. investigated the association between genetic and phenotypic heterogeneity by dissecting the single-variant clonal structure of five HCC patients (Su et al., 2021). Here, they found a common clonal origin but independent evolutionary patterns amongst different tumor samples and that this unique inter- and intratumor genetic heterogeneity was paralleled with phenotypic heterogeneity based on differential clustering of cells obtained from single-cell RNA sequencing (Su et al., 2021).

The plasticity of cancer cells can also be understood through the cancer stem cell hypothesis, in which differences in tumor cells can be linked to their differentiation stage (Fig. 1C). This model suggests that there is a certain population of cells with indefinite self-renewal capabilities are organized in a hierarchical manner pertinent to the stage of differentiation (Prasetyanti & Medema, 2017). Intra-tumor heterogeneity and cancer stem cells (CSCs) are not mutually exclusive events as research has shown that there exists genetic heterogeneity in populations of cancer stem cells. More specifically, Zheng et al. found heterogeneous expression of CSCs in two HCC cancer cell lines at the single-cell level and that CSC-marker expression associated genes (CD133, CD24, and EpCAM) correlated with HCC prognosis (Zheng et al., 2018). These findings indicate that unique populations of CSCs in HCC may contribute to the biological and transcriptomic heterogeneity of cancer cells within a liver. Furthermore, Ho et al. delineated through single-cell RNAseq of an HCC PDTX (patient-derived tumor xenograft) model rare populations of EpCAM+ cells as contributing to the upregulation of several oncogenes and a unique subclonal population of CD24+/CD44+ cells within the EpCAM+ cells (Ho et al., 2019). Likewise, intratumoral EpCAM+ cancer stem cell heterogeneity within HCC nodules was found to be associated with a higher risk of tumor recurrence nodules (Krause et al., 2020).

2.3 Tools to investigate tumor heterogeneity

With the recent explosion over the past decade in single-cell technologies, such as single-cell RNAseq, single-cell ATACseq (assay for transposase accessible chromatin), and single-cell multiplex immunofluorescence assays (Svensson, Vento-Tormo, & Teichmann, 2018), the field has made great progress in identifying unique cell types, stromal composition, and understanding of the spatial context of liver tumors (Fig. 2). One of the first single-cell RNAseq studies to investigate the cellular landscape of HCC tumors was a study that found distinct subpopulations of T cells, Tregs and exhausted CD8 T cells, to be enriched and that Layilin (LAYN) was linked to the suppressive function of these immune cells in the tumor (Zheng et al., 2017). Additionally, single-cell transcriptomics uncovered stem cell specific subpopulations within HCC patients, in particular CD24+/CD44+ enriched cells within EpCAM+ cells and that the Cathepsin E (CTSE) gene, contributed to this enrichment (Ho et al., 2019). Moreover, Ma et al. discovered tumor cell biodiversity drives microenvironmental reprogramming in hepatocellular and cholangiocarcinoma patients and that high diverse tumors led to worse patient outcomes (Ma et al., 2019). Furthermore, in HBV-associated human hepatocellular carcinoma patients, Ho et al. found that tumor associated macrophages

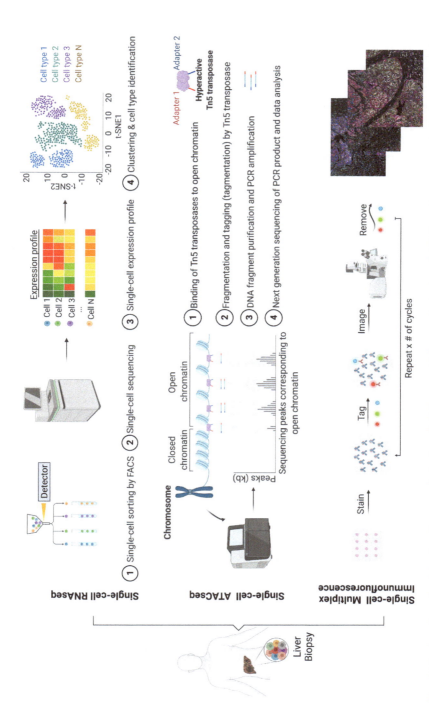

Fig. 2 Single-cell methodologies, (top panel: single-cell RNAseq, middle panel: single-cell ATACseq, bottom panel: single-cell multiplex immunofluorescence) to investigate tumor heterogeneity.

(TAMs) reduce tumor T cell infiltration and that specifically the interaction between T cell Immunoglobulin and ITIM domain-Nectin Cell Adhesion Molecule 2 (TIGIT-NECTIN2) regulates this immunosuppressive environment (Ho et al., 2021). The utility of single-cell RNAseq has allowed for monumental discoveries of unique and previously unknown cellular subtypes that has provided new insights into liver cancer biology and mechanisms of therapeutic resistance.

A relatively new emerging technology, single-cell ATACseq, has garnered interest in the liver cancer field to further investigate the chromatin landscape of primary liver cancers and decipher further mechanisms of tumor heterogeneity. The quantity of studies investigating the chromatin landscape of tumor samples from liver cancer patients is limited due to its recent development and optimization; however, there are studies which have utilized HCC cell lines. For example, Wang et al. conducted single-cell ATACseq in five HCC cell lines to determine whether there were chromatin regions of accessibility that were specifically remodeled towards epithelial-mesenchymal transition (EMT). They found that CDH1 was more open and contained higher gene activity in Huh7 and HepG2 cells (Wang et al., 2020).

Lastly, there has been great progress made in assays and technologies that can investigate single-cell level immunofluorescence detection of numerous genes and/or proteins and their spatial relationships to one another. For example, a few of the available commercial assays include RNAScope®, a multiplex in situ hybridization by Advanced Cell Diagnostics (Wang et al., 2012), CO-Detection by IndEXing (CODEX) of proteins by Akoya Biosciences (Goltsev et al., 2018), GeoMx Digital Spatial Profiler (DSP) by Nanostring to measure mRNA and protein (Merritt et al., 2020), and 10X Genomics Visium of spatial transcriptomics (Ståhl et al., 2016). A further look into these assays was nicely summarized in a recent Nature Methods Review (Lewis et al., 2021) and studies utilizing these technologies to decipher liver tumor heterogeneity is detailed in the next section of this chapter. With all these single-cell technological advancements, understanding the causes and mechanisms of liver cancer heterogeneity can be further achieved.

2.4 Causes and drivers of tumor heterogeneity

Tumor heterogeneity can arise from genomic and non-genomic mechanisms, stem cell influences, and microenvironmental causes that can all lead to functional heterogeneity which is defined as any prosurvival adaptability that cancer cells undergo to maintain their continued replicative and plastic functions (González-Silva, Quevedo, & Varela, 2021; Meacham & Morrison, 2013). Over the past few decades, research has focused on identifying targetable causes of tumor heterogeneity to reduce the burden of cancer progression and metastasis. However, in liver cancer, known drivers are limited which can be contributed to the fact that most liver cancer patients are often diagnosed at late stages where the tumor has progressed, and the diversity of cancer clones is at its maximum. Currently, there is evidence of some drivers of tumor heterogeneity in liver cancer which include mutations, genomic instability, the SPP1 gene, and cell death (Khatib, Pomyen, Dang, & Wang, 2020; Ma et al., 2021; Rao, Asch, & Yamada, 2016; Sung et al., 2019).

2.4.1 Mutational landscape

The mutational burden of tumors has long been associated with tumor heterogeneity. Tumors with higher levels of tumor heterogeneity have been found to have more mutations in tumor suppressor genes and oncogenes compared to tumors with a lower degree of heterogeneity (Sung et al., 2019). The most common mutated genes in liver cancer include TP53, CTNNB1, ARID1A, ARID2, AXIN1, PRS6KA3, VCAM1, CDK14, TERT, MLL4, and CCNE1 (Zhang, 2012). Interestingly, liver cancer has a broad variety of apparent mutations and limited number of encompassing mutations, making identifying therapeutic targets challenging. Liver intratumor heterogeneity can be attested from the diverse mutations found in tumors as well as the distribution of expression. For example, An et al. found that p53 and B-catenin mutated HCC tumors displayed a heterogenous distribution of expression and differentiation within the tumor using histological approaches (An et al., 2001). Another study found intratumoral HCC heterogeneity of mutations within the TP53, CTNNB1, and TERT genes using circulating tumor DNA (Huang et al., 2016).

2.4.2 Genomic instability

A key regulator of initiating tumor cell diversity is genomic instability. Genomic instability is defined as various structural variations in chromosomes that results in increased number of chromosomes, base pair mutations, or microsatellite instability (MSI) (Yao & Dai, 2014). The causes explaining these instabilities have been linked to mutations that cause loss of gene functions that are essential to maintain proper cellular homeostasis and oncogenic induced DNA replication stress. The loss of gene function can lead to a mutator phenotype in which genes that are imperative for DNA repair are mutated and can cause genomic instability, resulting in tumor formation (Sarni & Kerem, 2017). Moreover, the oncogene induced DNA replication model promotes genomic instability through DNA replication stress originating from a high proliferative index. This mechanism puts forth the idea that an oncogene induces instability and tumor development. Thus, the genetically unstable repertoire of tumor cells allows for them to selectively bypass key intracellular signaling such as cell cycle arrest, apoptosis, and immunosurveillance to promote a survival advantage and lead to the emergence of new subclonal populations and potential for increased heterogeneity.

In liver cancer, genomic instability has been found to be caused by mutations, epigenetic dysregulation, or HBV/HCV infection. Studies have shown that viral infections can promote genomic instability by disrupting mitotic regulator proteins, integrating into various sites of the human genome, and inducing chronic inflammation in the liver (Kim et al., 2008; Rao et al., 2016). Further instability found within the chromosome has been linked to aberrant epigenetic changes such as abnormal DNA methylation found within specific genes such as Ras association RalGDS/AF-6 domain family member 1 (RASSF1), Insulin-like growth factor 2 (IGF-2), and Adenomatous polyposis coli (APC) and were associated with poor survival in HCC patients (Villanueva et al., 2015).

2.4.3 SPP1
Recently, our lab uncovered secreted phosphoprotein 1 (SPP1) as a potential modulator of tumor heterogeneity in primary liver cancers. This study utilized pre- and post-treatment immunotherapy samples obtained from HCC and CCA patients (Ma et al., 2021). Single-cell RNAseq analysis revealed a unique repertoire of malignant and nonmalignant cell types with diverse functional subclonal populations. Further analysis revealed SPP1 expression strongly associated with tumor cell evolution and microenvironmental reprogramming. In particular, SPP1 expression was elevated in patients with a greater number of functional clones and in post-treatment samples that contained higher diversity. Interestingly, patients who responded to immunotherapy had lower levels of SPP1. These findings are not surprising as SPP1 which encodes osteopontin (OPN), an extracellular structural protein, and has been found to be associated with the progression of multiple cancer types such as breast, lung, colon, gastric,

ovarian, and liver cancer (Chuang et al., 2012; Fedarko, Jain, Karadag, Van Eman, & Fisher, 2001; Junnila et al., 2010; Liu et al., 2021; Xu et al., 2017; Zeng, Zhou, Wu, & Xiong, 2018). Plausible mechanisms of OPN-induced carcinogenesis include polarization of macrophages to M2 tumor associated macrophages (TAMs) and interaction with vascular endothelial growth factor (VEGF) (Cui et al., 2009; Liu et al., 2021).

2.4.4 Cell death

Another hypothesis our lab has published in reference to one of the causes of tumor heterogeneity is cell death. We demonstrated that an Apoptotic Index that was calculated based on the expression of genes involved in the initiation and execution of apoptosis, a form of programed cell death, was significantly correlated with a Tumor Shannon Entropy Index (TSEI), which is a measurement of tumor diversity, in seven different types of cancer using single-cell RNAseq datasets (Khatib et al., 2020). Specifically, liver cancer illustrated the highest association value and most significant p-value, indicating induced cell death within a liver tumor may further instigate greater levels of intratumor heterogeneity compared to other cancer types. Interestingly, this phenomenon was not observed in liquid cancers, suggesting that a solid tumor ecosystem is needed for cell death to initiate diversity to its nearby surroundings. Moreover, patients that were stratified in high expression of the apoptosis signature displayed worse overall survival compared to low expressing patients. Through these analyses, we proposed the idea that removal of a major clonal population in a tumor through targeted therapy leaves behind a vacant niche that minor and more diverse clones can repopulate and increase the total diversity of the tumor and thus make it much more resistant to therapy (Khatib et al., 2020).

2.5 Functional implications of tumor heterogeneity

To maintain their survival, cancer cells will undergo various methods of adaption to proliferate, evolve, and resist drug intervention in a specified microenvironment linked to a particular etiology as seen in liver cancer. We previously defined this as "functional heterogeneity," in which genomic and nongenomic heterogeneity, stemness heterogeneity, and microenvironmental heterogeneity are all associated with the functional causes of diversity within a liver tumor (Liu, Dang, & Wang, 2018). For example, scRNAseq analysis revealed the functional (increased transcription) and inflammatory role of SLC40A1 and GPNMB in infiltrating TAMs within HCC patients that led to poor prognosis (Zhang et al., 2019). Moreover, Marzioni et al.

investigated the functional heterogeneity of cholangiocytes and found unique morphological differences within small and large intrahepatic ducts, differential secretion in response to hormones, bile acids, and peptides, and a proliferative repertoire in response to injury or toxicity, which together may contribute to the extensive heterogeneity seen in cholangiocarcinoma (Marzioni et al., 2002). Swon et al. found that higher levels of functional genomic complexity, which was defined as the degree of molecular heterogeneity between HCC and CCA tumors, to be associated with TP53 mutations, chromosome instability, and worse overall patient survival (Kwon et al., 2019). Lastly, Payen et al. found diverse subpopulations of cells, specifically GPC3+ and DBH+ hepatic stellate cells, within the extracellular matrix of parenchymal and non-parenchymal liver cells that contribute to its production and organization (Payen et al., 2021). Together, the findings presented above highlight the unique and complex composition of the human liver and how these high levels of heterogeneity can ultimately make primary liver tumors difficult to treat.

2.6 Microenvironmental influences on tumor heterogeneity

The tumor microenvironment plays a substantial role in promoting a heterogeneous population of surrounding stromal cells, macrophages, and epithelial cells as well as remodeling of the extracellular matrix (ECM). Microenvironmental changes can shape the tumor cell phenotype through inducing cellular stress responses and promoting genomic instability. Various mathematical modeling has shown that heterogeneity in the tumor microenvironment may lead to the selection of more aggressive clonal phenotypes (Mumenthaler et al., 2015). Tumor cells and its surrounding environments undergo diverse physiological processes that can lead to heterogeneity such as differences in vasculature and hypoxic regions causing inconsistencies in blood flow and oxygen consumption over time. More specifically, these dynamic processes can vary within the same tumor as certain cells may become more hypoxic than others as well as obtain a greater potential to metastasize through increased vascularization. Hence, it is difficult to account for and measure the magnitude of these variations as cancer cells and the surrounding tissue adapt consistently over time.

In liver cancer, the tumor microenvironment plays a critical role in its survival and progression. Liver cancer stroma consists mainly of three main subclasses of cells: cancer-associated fibroblasts (CAFs), immune and inflammatory cells, and angiogenic cells (Yin et al., 2019). Crosstalk between stromal and tumor cells allows for tumor cell proliferation, migration, invasion and promotes an immunosuppressive microenvironment. We have shown that tumor heterogeneity can reprogram the tumor microenvironment by interacting with VEGF to worsen patient outcomes in both HCC and CCA (Ma et al., 2019). Interestingly, Zhang et al. found that when immune cells in the microenvironment are clustered, three unique HCC subtypes were identified with immunosuppressive, immunodeficient, and immunocompetent characteristics and distinctive chemokine/cytokine compositions and cellular metabolism of tumor cells (Zhang et al., 2019). Furthermore, a high degree of heterogeneity within the immune microenvironment was found to be associated with worse clinical outcome in HCC patients and this degree of immune heterogeneity was correlated with tumor transcriptomic heterogeneity (Nguyen et al., 2021). Specifically, tumors with high levels of immune intratumor heterogeneity were found to be significantly enriched with immunosuppressive and exhausted GB-inactive memory CD4+ T cells, regulatory T cells (Tregs), and Tim-3+ and PD-1+GB- exhausted CD8+ T cells (Nguyen et al., 2021). Lastly, CAFs one of the main subtypes of cells in the microenvironment, contain a heterogeneous mixture of different cell types such as endothelial cells, vascular smooth muscle cells, pericytes, and cancer cells that have undergone EMT, which together can fuel a tumor population with a high degree of heterogeneity and resistance to targeted therapy. In liver cancer, one source of CAFs is hepatic stellate cells, which is one major contributor of causing liver fibrosis and can ultimately lead to a mixture of heterogeneous cell types within the liver (Henderson et al., 2013). Interestingly, a study by Zhang et al. found CD146+ CAFs interacting strongly with malignant cells through the IL-6/IL-6R axis in highly heterogeneous intrahepatic CCA samples which led to advanced disease progression (Zhang et al., 2020).

2.7 Clinical ramifications of tumor heterogeneity

The magnitude of phenotypic heterogeneity can be clinically relevant. Multiple studies have noted that a large amount of intra-tumor heterogeneity is correlated with a worse overall survival in liver cancer patients (Han et al., 2013; Kwon et al., 2019; Ma et al., 2019, 2021; Nguyen et al., 2021). Reasons for these poor prognoses may be due to how clonal diversity drives therapeutic relapse as each tumor cell has a unique signature that allows it to either respond to or resist cancer therapy. This response can also vary over time with treatment as in some cases therapy can eliminate a certain population of cells while others are able to overcome the cellular stresses of cell death and maintain proliferative activity. Tumor heterogeneity can also change during cancer therapy by cancer cells adapting to resistance mechanisms. Determining why these cells become resistant and at what point during treatment they become resistant compared to its neighboring ones has been difficult to assess. Another potential pitfall of cancer therapy is that it introduces greater genomic instability through promoting cell death and yet the cells that survive have evolved to with an increased level of cellular genetic fitness and diversity (Ichim & Tait, 2016). This increased genetic fitness through a selective pressure obtained during therapy gives rise to a multitude of gene expression signatures within individual tumor cells (Venkatesan, Swanton, Taylor, & Costello, 2017). Thus, it is nearly impossible for one standard therapy to successfully treat the diverse population of tumor cells as we must account for the evolutionary mechanisms cancers cells will adapt to maintain their heterogeneous nature.

3. Spatial architecture and tumor heterogeneity

3.1 The need to understand the spatial architecture of tumors

Every day we witness various observations of spatial organization that create positive or effective behaviors in humans, animals, microorganisms, and systems. For example, birds fly in a specific V formation that best suits their needs of travel in an efficient manner that conserves their energy by reducing wind resistance. Moreover, higher order spatial organization is illustrated in every system of the body such as how atriums of the heart are organized to perform the most optimal cardiac output, the gastrointestinal tract is specifically ordered by organ function to digest nutrients, absorb energy, and expel waste, and spinal nerves are coordinately innervated regions of the spine to create movement in our everyday lives. These sophisticated elements of the human body have evolved from the early stages of development into complex and intricately spaced organ systems to maintain human life and survival. What if cancer cells adapted these same principles of higher order spatial organization to orchestrate a mimicry of the human body system to maintain their own life? How can we go about understanding how tumor evolution and its destruction inside a human body is a spatially orchestrated phenomenon?

There is no doubt that spatial organization is a vital component to achieve an effective result or behavior in a variety of diverse experiences.

The same can be said about cancer in which tumor cells will organize themselves in a particular manner to maintain their longevity and survival. Over the past few years, growing interest in dissecting the spatial context of tumors has emerged but deciphering what these new observations means towards our understanding of cancer development, creation of therapeutics, and treating cancer patients remains largely unclear. We have yet to leverage the spatial biology and information garnered from omic technologies into making meaningful conclusions of how the spatial architecture of tumors can influence tumor development, heterogeneity, survival, and patients' response to therapy.

3.2 Spatial organization dictates functional heterogeneity

The functional intricacies of tumor heterogeneity can also be related to the spatial organization of cells within a tumor population. Cancer is a disease in which its growth and survival is perpetrated by the space it is given. Macroscopically, in solid cancers, the tumor resides in an organ which is best fit to maintain its life and microscopically, different cell types are uniquely organized to work together to fuel an environment for carcinogenesis to occur. This is similar to a key biological principle that is evident in our everyday lives such as the concept that structure determines function, meaning the way something is organized and arranged allows for its role or purpose to be fulfilled. These structure-function relationships are a result of natural selection in which the most suitable and advantageous relationships are preserved and passed on to future generations. It may be assumed that because tumors are largely heterogeneous and diverse, their organization is therefore random and chaotic. However, this naïve thinking is nonetheless simple and primitive. Tumors are highly evolved structures that are intricately and selectively designed to maintain their survival and destroy any potential threat towards their existence. Thus, understanding how individual cells (malignant and nonmalignant) within a tumor communicate with one another and their microenvironment in order to survive is imperative to unravel (Fig. 3A). One way to better comprehend these relationships is through the spatial context of all cell types in a tumor. For example, are cancer clones organized in convergent or divergent patterns as illustrated in Fig. 3B? If so, why and how do these relationships influence therapy response? With information like this, we can design better therapeutic options that limits cell-to-cell communication which could lead to the destruction of highly organized tumors.

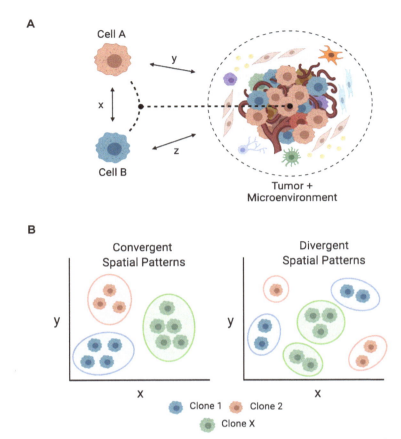

Fig. 3 Understanding the spatial interactions between and organization of cancer clones and may lead to new discoveries of liver tumor biology. (A) The interactions between different cells and their distance to one another and its surrounding tumor microenvironment may fuel tumors to become more heterogeneous and less responsive to treatment. (B) Understanding convergent versus divergent spatial patterns of cancer clones can provide key insights on how tumors organize and respond to targeted therapy.

3.3 Leveraging spatial techniques to better understand liver cancer heterogeneity

Recent studies utilizing spatial omic technologies in liver cancer have begun to dissect the meaning between how tumor cells organize and contribute to treatment failure. In particular, Wu et al. conducted a spatial transcriptomic study of normal to leading-edge to tumor regions of HCC tumors and found that the tumor capsule within the tumor microenvironment contributes to intratumor spatial cluster continuity, transcriptome diversity, and differential immune cell infiltration (Wu et al., 2021). More specifically, the cells found within the tumor capsule, mainly fibroblasts and endothelial cells, act as a barrier in preventing immune cells from infiltrating, which has been observed previously (Rahmanzade, 2020). Furthermore, another study investigated the spatial context of the tumor and immune microenvironment within HCC patients and found three unique histological subtypes that exhibited distinct nuclear features and diverse spatial distribution and relation between tumor cells and infiltrating lymphocytes (Wang et al., 2020). These three histological subtypes were further associated with somatic genomic alterations in the context of aneuploidy, specific molecular pathways such as oxidative phosphorylation and cell cycle progression, and more importantly, independent prognostic differences with subtype two and three having statistically significant worse prognosis compared to subtype one (Wang, Jiang, et al., 2020). Lastly, a HCC proteomic spatial analysis study demonstrated intratumoral proteomic heterogeneity and abundance with differentially expressed proteins (SerpinB3 and SerpinB4) at the center and periphery of the same tumor, further supporting spatially distinct metabolic and functional heterogeneity of liver cancer cells (Buczak et al., 2018). The findings of these studies highlight the importance of incorporating spatial information into deciphering causes of liver carcinogenesis and identify better biomarkers for clinical detection and management of patients' care.

3.4 Computational methods for downstream spatial analysis

Interpreting the functional relevance and clinical implications of the spatial relationships between cells presents a new challenge in our understanding of tumor biology. Currently, publications incorporating the transcriptomic and genomic profiles of cells with its spatial context has been limited in addressing what the functional meaning is behind why malignant and non-malignant cells organize the way they do to fulfill their role in tumor progression and metastasis. These studies have been primarily observational and touch the surface, however future studies can dive deeper into these relationships with the advancement of pathology software and computational modeling. For example, HALO software by Indica Labs has revolutionized pathology analysis at the single-cell level. Their software platform can measure the expression of stained markers, identify unique subclasses of cells, perform spatial analysis, and conduct artificial intelligence and deep learning analysis. Moreover, the CODEX ® Multiplex Analysis Viewer by Akoya Biosciences analyzes cellular segmentation and expression to identify unique cellular neighborhoods by pairwise cell proximity analysis. The data generated from these software and further downstream analysis using computer programming can lead to new insights into understanding the spatial biology of liver tumors.

3.5 The interplay between spatial biology and therapeutic development

Perturbing the life of any unicellular or multicellular organism creates the potential to influence their surrounding environment. That is why it is essential to not only investigate how to destroy cancer cells, but also what happens to its surrounding environment. These events are not mutually exclusive and as evidenced by the complexity of treating cancer, actually work together to maintain the survival and fitness of cancer clones. To do this, it is important to combine single-cell genomics and transcriptomics with the spatial relationships of each individual cell to better understand tumor cell organization. If we can begin to understand the spatial dynamics of a tumor and address key questions on how each cell coordinates with one another and their environment (Fig. 4), then we can therapeutically target the strongest cell-to-cell relationships. With the knowledge we have gathered from the numerous advancements made in single-cell omics studies, it is evident that cancer cells do not work alone but in a coordinated effort with each other and their environment.

4. Concluding remarks and future perspectives

In this chapter, we discussed the latest research in identifying the functional implications of liver cancer heterogeneity with insights into its potential causes, drivers, clinical influence, and spatial architecture. Liver cancer heterogeneity can be classified as one of the greatest challenges of combating tumor development and therapy resistance. With the incidence and mortality rate of liver cancer on the continual rise within the United States and across the world, it is imperative to identify ways to mitigate the heterogeneity of liver tumors and develop therapeutic options that will be more successful in prolonging patients' lives. It is no secret that the extensive heterogeneity present in liver cancer patients makes identifying therapeutic targets challenging. However, with the recent advancement in single-cell

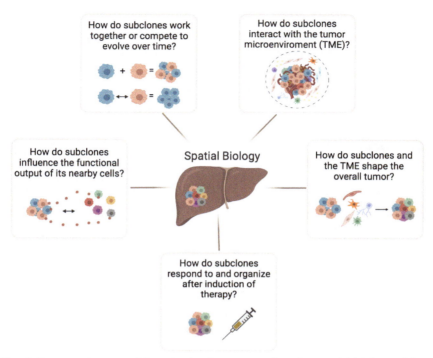

Fig. 4 Key questions to address in the future regarding the spatial biology of liver tumors.

technologies that have led to a deeper understanding of the unique cell types, functions, and spatial organization of liver cancer cells, we are making substantial progress in understanding the etiology, development, and progression of this disease. These efforts have allowed for the discovery of unique subtypes of immune, cancer stem cells, epithelial, progenitor, and hepatic cells and how they contribute to the tumor microenvironment and survival of liver tumor cells. With these findings and new discoveries on the horizon, greater insights in the causes and functional intricacies of liver cancer heterogeneity can be made to help save patient lives.

Acknowledgments

We would like to thank members of Wang Lab/Liver Carcinogenesis Section for thoughtful and critical discussions. This work was supported by grants Z01 BC 010877 and Z01 BC 010876 from the intramural research program of the Center for Cancer Research, National Cancer Institute, Bethesda, MD, USA.

Author contributions

All authors contributed to the writing of the manuscript. S.A.K. generated all figures using Biorender. All authors read, edited, and approved the manuscript.

Declaration of interests

The authors declare no competing interests.

References

An, F.-Q., Matsuda, M., Fujii, H., Tang, R.-F., Amemiya, H., Dai, Y.-M., et al. (2001). Tumor heterogeneity in small hepatocellular carcinoma: Analysis of tumor cell proliferation, expression and mutation of p53 AND β-catenin. *International Journal of Cancer*, *93*(4), 468–474.

Brindley, P. J., Bachini, M., Ilyas, S. I., Khan, S. A., Loukas, A., Sirica, A. E., et al. (2021). Cholangiocarcinoma. *Nature Reviews Disease Primers*, *7*(1), 65.

Buczak, K., Ori, A., Kirkpatrick, J. M., Holzer, K., Dauch, D., Roessler, S., et al. (2018). Spatial tissue proteomics quantifies inter- and intratumor heterogeneity in hepatocellular carcinoma (HCC)*. *Molecular & Cellular Proteomics*, *17*(4), 810–825.

Carotenuto, P., Fassan, M., Pandolfo, R., Lampis, A., Vicentini, C., Cascione, L., et al. (2017). Wnt signalling modulates transcribed-ultraconserved regions in hepatobiliary cancers. *Gut*, *66*(7), 1268–1277.

Chaisaingmongkol, J., Budhu, A., Dang, H., Rabibhadana, S., Pupacdi, B., Kwon, S. M., et al. (2017). Common molecular subtypes among asian hepatocellular carcinoma and cholangiocarcinoma. *Cancer Cell*, *32*(1), 57–70.e53.

Chuang, C. Y., Chang, H., Lin, P., Sun, S. J., Chen, P. H., Lin, Y. Y., et al. (2012). Up-regulation of osteopontin expression by aryl hydrocarbon receptor via both ligand-dependent and ligand-independent pathways in lung cancer. *Gene*, *492*(1), 262–269.

Cui, R., Takahashi, F., Ohashi, R., Yoshioka, M., Gu, T., Tajima, K., et al. (2009). Osteopontin is involved in the formation of malignant pleural effusion in lung cancer. *Lung Cancer*, *63*(3), 368–374.

Dasgupta, P., Henshaw, C., Youlden, D. R., Clark, P. J., Aitken, J. F., & Baade, P. D. (2020). Global trends in incidence rates of primary adult liver cancers: A systematic review and meta-analysis. *Frontiers in Oncology*, *10*, 171.

Donato, F., Tagger, A., Gelatti, U., Parrinello, G., Boffetta, P., Albertini, A., et al. (2002). Alcohol and hepatocellular carcinoma: The effect of lifetime intake and hepatitis virus infections in men and women. *American Journal of Epidemiology*, *155*(4), 323–331.

Fedarko, N. S., Jain, A., Karadag, A., Van Eman, M. R., & Fisher, L. W. (2001). Elevated serum bone sialoprotein and osteopontin in colon, breast, prostate, and lung cancer. *Clinical Cancer Research*, *7*(12), 4060–4066.

Feng, M., Pan, Y., Kong, R., & Shu, S. (2020). Therapy of primary liver cancer. *Innovation (New York, N.Y.)*, *1*(2), 100032.

Fostea, R. M., Fontana, E., Torga, G., & Arkenau, H.-T. (2020). Recent progress in the systemic treatment of advanced/metastatic cholangiocarcinoma. *Cancers*, *12*(9), 2599.

Goltsev, Y., Samusik, N., Kennedy-Darling, J., Bhate, S., Hale, M., Vazquez, G., et al. (2018). Deep profiling of mouse splenic architecture with CODEX multiplexed imaging. *Cell*, *174*(4), 968–981.e915.

González-Silva, L., Quevedo, L., & Varela, I. (2021). Tumor functional heterogeneity unraveled by sc RNA-seq technologies: (Trends in Cancer 6, 13–19, 2020). *Trends Cancer*, *7*(3), 265.

Groopman, J. D., Kensler, T. W., & Wild, C. P. (2008). Protective interventions to prevent aflatoxin-induced carcinogenesis in developing countries. *Annual Review of Public Health*, *29*, 187–203.

Han, D. H., Choi, G. H., Kim, K. S., Choi, J. S., Park, Y. N., Kim, S. U., et al. (2013). Prognostic significance of the worst grade in hepatocellular carcinoma with heterogeneous histologic grades of differentiation. *Journal of Gastroenterology and Hepatology*, *28*(8), 1384–1390.

Henderson, N. C., Arnold, T. D., Katamura, Y., Giacomini, M. M., Rodriguez, J. D., McCarty, J. H., et al. (2013). Targeting of αv integrin identifies a core molecular pathway that regulates fibrosis in several organs. *Nature Medicine*, *19*(12), 1617–1624.

Ho, D. W.-H., Tsui, Y.-M., Chan, L.-K., Sze, K. M.-F., Zhang, X., Cheu, J. W.-S., et al. (2021). Single-cell RNA sequencing shows the immunosuppressive landscape and tumor heterogeneity of HBV-associated hepatocellular carcinoma. *Nature Communications*, *12*(1), 3684.

Ho, D. W.-H., Tsui, Y.-M., Sze, K. M.-F., Chan, L.-K., Cheung, T.-T., Lee, E., et al. (2019). Single-cell transcriptomics reveals the landscape of intra-tumoral heterogeneity and stemness-related subpopulations in liver cancer. *Cancer Letters*, *459*, 176–185.

Hsu, H. C., Chiou, T. J., Chen, J. Y., Lee, C. S., Lee, P. H., & Peng, S. Y. (1991). Clonality and clonal evolution of hepatocellular carcinoma with multiple nodules. *Hepatology*, *13*(5), 923–928.

Hu, W., Feng, Z., Eveleigh, J., Iyer, G., Pan, J., Amin, S., et al. (2002). The major lipid peroxidation product, trans-4-hydroxy-2-nonenal, preferentially forms DNA adducts at codon 249 of human p53 gene, a unique mutational hotspot in hepatocellular carcinoma. *Carcinogenesis*, *23*(11), 1781–1789.

Huang, D. Q., El-Serag, H. B., & Loomba, R. (2021). Global epidemiology of NAFLD-related HCC: Trends, predictions, risk factors and prevention. *Nature Reviews. Gastroenterology & Hepatology*, *18*(4), 223–238.

Huang, A., Zhang, X., Zhou, S.-L., Cao, Y., Huang, X.-W., Fan, J., et al. (2016). Detecting circulating tumor DNA in hepatocellular carcinoma patients using droplet digital PCR is feasible and reflects intratumoral heterogeneity. *Journal of Cancer*, *7*(13), 1907–1914.

Ichim, G., & Tait, S. W. (2016). A fate worse than death: Apoptosis as an oncogenic process. *Nature Reviews. Cancer*, *16*(8), 539–548.

Junnila, S., Kokkola, A., Mizuguchi, T., Hirata, K., Karjalainen-Lindsberg, M.-L., Puolakkainen, P., et al. (2010). Gene expression analysis identifies over-expression of CXCL1, SPARC, SPP1, and SULF1 in gastric cancer. *Genes, Chromosomes and Cancer*, *49*(1), 28–39.

Khatib, S., Pomyen, Y., Dang, H., & Wang, X. W. (2020). Understanding the cause and consequence of tumor heterogeneity. *Trends in Cancer*, *6*(4), 267–271.

Kim, S., Park, S. Y., Yong, H., Famulski, J. K., Chae, S., Lee, J. H., et al. (2008). HBV X protein targets hBubR1, which induces dysregulation of the mitotic checkpoint. *Oncogene*, *27*(24), 3457–3464.

Krause, J., von Felden, J., Casar, C., Fründt, T. W., Galaski, J., Schmidt, C., et al. (2020). Hepatocellular carcinoma: Intratumoral EpCAM-positive cancer stem cell heterogeneity identifies high-risk tumor subtype. *BMC Cancer*, *20*(1), 1130.

Kwon, S. M., Budhu, A., Woo, H. G., Chaisaingmongkol, J., Dang, H., Forgues, M., et al. (2019). Functional genomic complexity defines intratumor heterogeneity and tumor aggressiveness in liver cancer. *Scientific Reports*, *9*(1), 16930.

Lazaridis, K. N., & Gores, G. J. (2005). Cholangiocarcinoma. *Gastroenterology*, *128*(6), 1655–1667.

Lewis, S. M., Asselin-Labat, M.-L., Nguyen, Q., Berthelet, J., Tan, X., Wimmer, V. C., et al. (2021). Spatial omics and multiplexed imaging to explore cancer biology. *Nature Methods*, *18*(9), 997–1012.

Lieber, C. S. (1994). Alcohol and the liver: 1994 update. *Gastroenterology*, *106*(4), 1085–1105.

Liu, J., Dang, H., & Wang, X. W. (2018). The significance of intertumor and intratumor heterogeneity in liver cancer. *Experimental & Molecular Medicine*, *50*(1), e416.

Liu, J., Tang, W., Budhu, A., Forgues, M., Hernandez, M. O., Candia, J., et al. (2020). A viral exposure signature defines early onset of hepatocellular carcinoma. *Cell*, *182*(2), 317–328.e310.

Liu, Y., & Wu, F. (2010). Global burden of aflatoxin-induced hepatocellular carcinoma: A risk assessment. *Environmental Health Perspectives*, *118*(6), 818–824.

Liu, L., Zhang, R., Deng, J., Dai, X., Zhu, X., Fu, Q., et al. (2021). Construction of TME and identification of crosstalk between malignant cells and macrophages by SPP1 in hepatocellular carcinoma. *Cancer Immunology, Immunotherapy*, *71*(1), 121–136.

Ma, L., Hernandez, M. O., Zhao, Y., Mehta, M., Tran, B., Kelly, M., et al. (2019). Tumor cell biodiversity drives microenvironmental reprogramming in liver cancer. *Cancer Cell*, *36*(4), 418–430.e416.

Ma, L., Wang, L., Khatib, S. A., Chang, C.-W., Heinrich, S., Dominguez, D. A., et al. (2021). Single-cell atlas of tumor cell evolution in response to therapy in hepatocellular carcinoma and intrahepatic cholangiocarcinoma. *Journal of Hepatology*, *75*(6), 1397–1408.

Marzioni, M., Glaser, S. S., Francis, H., Phinizy, J. L., LeSage, G., & Alpini, G. (2002). Functional heterogeneity of cholangiocytes. *Seminars in Liver Disease*, *22*(03), 227–240.

Matsushita, H., & Takaki, A. (2019). Alcohol and hepatocellular carcinoma. *BMJ Open Gastroenterology*, *6*(1), e000260.

Meacham, C. E., & Morrison, S. J. (2013). Tumour heterogeneity and cancer cell plasticity. *Nature*, *501*(7467), 328–337.

Merritt, C. R., Ong, G. T., Church, S. E., Barker, K., Danaher, P., Geiss, G., et al. (2020). Multiplex digital spatial profiling of proteins and RNA in fixed tissue. *Nature Biotechnology*, *38*(5), 586–599.

Mumenthaler, S. M., Foo, J., Choi, N. C., Heise, N., Leder, K., Agus, D. B., et al. (2015). The impact of microenvironmental heterogeneity on the evolution of drug resistance in cancer cells. *Cancer Informatics*, *14*(Suppl. 4), 19–31.

Nakamura, H., Arai, Y., Totoki, Y., Shirota, T., Elzawahry, A., Kato, M., et al. (2015). Genomic spectra of biliary tract cancer. *Nature Genetics*, *47*(9), 1003–1010.

Nguyen, P. H. D., Ma, S., Phua, C. Z. J., Kaya, N. A., Lai, H. L. H., Lim, C. J., et al. (2021). Intratumoural immune heterogeneity as a hallmark of tumour evolution and progression in hepatocellular carcinoma. *Nature Communications*, *12*(1), 227.

Nowell, P. C. (1976). The clonal evolution of tumor cell populations. *Science*, *194*(4260), 23–28.

Payen, V. L., Lavergne, A., Alevra Sarika, N., Colonval, M., Karim, L., Deckers, M., et al. (2021). Single-cell RNA sequencing of human liver reveals hepatic stellate cell heterogeneity. *JHEP Reports*, *3*(3), 100278.

Perz, J. F., Armstrong, G. L., Farrington, L. A., Hutin, Y. J. F., & Bell, B. P. (2006). The contributions of hepatitis B virus and hepatitis C virus infections to cirrhosis and primary liver cancer worldwide. *Journal of Hepatology*, *45*(4), 529–538.

Prasetyanti, P. R., & Medema, J. P. (2017). Intra-tumor heterogeneity from a cancer stem cell perspective. *Molecular Cancer*, *16*(1), 41.

Prueksapanich, P., Piyachaturawat, P., Aumpansub, P., Ridtitid, W., Chaiteerakij, R., & Rerknimitr, R. (2018). Liver fluke-associated biliary tract cancer. *Gut and Liver*, *12*(3), 236–245.

Rahmanzade, R. (2020). Redefinition of tumor capsule: Rho-dependent clustering of cancer-associated fibroblasts in favor of tensional homeostasis. *Medical Hypotheses*, *135*, 109425.

Rao, C. V., Asch, A. S., & Yamada, H. Y. (2016). Frequently mutated genes/pathways and genomic instability as prevention targets in liver cancer. *Carcinogenesis*, *38*(1), 2–11.

Raoul, J. L., Forner, A., Bolondi, L., Cheung, T. T., Kloeckner, R., & de Baere, T. (2019). Updated use of TACE for hepatocellular carcinoma treatment: How and when to use it based on clinical evidence. *Cancer Treat Reviews*, *72*, 28–36.

Sangro, B., Sarobe, P., Hervás-Stubbs, S., & Melero, I. (2021). Advances in immunotherapy for hepatocellular carcinoma. *Nature Reviews. Gastroenterology & Hepatology*, *18*(8), 525–543.

Sarni, D., & Kerem, B. (2017). Oncogene-induced replication stress drives genome instability and tumorigenesis. *International Journal of Molecular Sciences*, *18*(7), 1339.

Shimada, S., Mogushi, K., Akiyama, Y., Furuyama, T., Watanabe, S., Ogura, T., et al. (2019). Comprehensive molecular and immunological characterization of hepatocellular carcinoma. *EBioMedicine*, *40*, 457–470.

Siegel, R. L., Miller, K. D., & Jemal, A. (2020). Cancer statistics, 2020. *CA Cancer Journal of Clinicians*, *70*(1), 7–30.

Ståhl, P. L., Salmén, F., Vickovic, S., Lundmark, A., Navarro, J. F., Magnusson, J., et al. (2016). Visualization and analysis of gene expression in tissue sections by spatial transcriptomics. *Science*, *353*(6294), 78–82.

Su, X., Zhao, L., Shi, Y., Zhang, R., Long, Q., Bai, S., et al. (2021). Clonal evolution in liver cancer at single-cell and single-variant resolution. *Journal of Hematology & Oncology*, *14*(1), 22.

Sung, J.-Y., Shin, H.-T., Sohn, K.-A., Shin, S.-Y., Park, W.-Y., & Joung, J.-G. (2019). Assessment of intratumoral heterogeneity with mutations and gene expression profiles. *PLoS One*, *14*(7), e0219682.

Svensson, V., Vento-Tormo, R., & Teichmann, S. A. (2018). Exponential scaling of single-cell RNA-seq in the past decade. *Nature Protocols*, *13*(4), 599–604.

Tanaka, S., Toh, Y., Adachi, E., Matsumata, T., Mori, R., & Sugimachi, K. (1993). Tumor progression in hepatocellular carcinoma may be mediated by p53 mutation. *Cancer Research*, *53*(12), 2884–2887.

Testino, G., Leone, S., & Borro, P. (2014). Alcohol and hepatocellular carcinoma: A review and a point of view. *World Journal of Gastroenterology*, *20*(43), 15943–15954.

Thiele, M., Gluud, L. L., Fialla, A. D., Dahl, E. K., & Krag, A. (2014). Large variations in risk of hepatocellular carcinoma and mortality in treatment naïve hepatitis B patients: Systematic review with meta-analyses. *PLoS One*, *9*(9), e107177.

Venkatesan, S., Swanton, C., Taylor, B. S., & Costello, J. F. (2017). Treatment-induced mutagenesis and selective pressures sculpt cancer evolution. *Cold Spring Harbor Perspectives in Medicine*, *7*(8), a026617.

Villanueva, A., Portela, A., Sayols, S., Battiston, C., Hoshida, Y., Méndez-González, J., et al. (2015). DNA methylation-based prognosis and epidrivers in hepatocellular carcinoma. *Hepatology*, *61*(6), 1945–1956.

Vitale, A., Morales, R. R., Zanus, G., Farinati, F., Burra, P., Angeli, P., et al. (2011). Barcelona clinic liver cancer staging and transplant survival benefit for patients with hepatocellular carcinoma: A multicentre, cohort study. *Lancet Oncology*, *12*(7), 654–662.

Wang, F., Flanagan, J., Su, N., Wang, L. C., Bui, S., Nielson, A., et al. (2012). RNAscope: A novel in situ RNA analysis platform for formalin-fixed, paraffin-embedded tissues. *Journal of Molecular Diagnostics*, *14*(1), 22–29.

Wang, H., Jiang, Y., Li, B., Cui, Y., Li, D., & Li, R. (2020). Single-cell spatial analysis of tumor and immune microenvironment on whole-slide image reveals hepatocellular carcinoma subtypes. *Cancers (Basel)*, *12*(12).

Wang, S., Xie, J., Zou, X., Pan, T., Zhuang, Z., Wang, Z., et al. (2020). Single-cell Multi-omics reveal heterogeneity and metastasis potential in different liver cancer cell lines. *bioRxiv*. 2020.2011.2003.367532.

Welzel, T. M., Graubard, B. I., El-Serag, H. B., Shaib, Y. H., Hsing, A. W., Davila, J. A., et al. (2007). Risk factors for intrahepatic and extrahepatic cholangiocarcinoma in the

United States: A population-based case-control study. *Clinical Gastroenterology and Hepatology: The Official Clinical Practice Journal of the American Gastroenterological Association*, 5(10), 1221–1228.

Wu, R., Guo, W., Qiu, X., Wang, S., Sui, C., Lian, Q., et al. (2021). Comprehensive analysis of spatial architecture in primary liver cancer. *bioRxiv*. 2021.2005.2024.445446.

Xu, C., Sun, L., Jiang, C., Zhou, H., Gu, L., Liu, Y., et al. (2017). SPP1, analyzed by bioinformatics methods, promotes the metastasis in colorectal cancer by activating EMT pathway. *Biomedicine & Pharmacotherapy*, 91, 1167–1177.

Xue, R., Li, R., Guo, H., Guo, L., Su, Z., Ni, X., et al. (2016). Variable intra-tumor genomic heterogeneity of multiple lesions in patients with hepatocellular carcinoma. *Gastroenterology*, 150(4), 998–1008.

Yamamoto, S., Kubo, S., Hai, S., Uenishi, T., Yamamoto, T., Shuto, T., et al. (2004). Hepatitis C virus infection as a likely etiology of intrahepatic cholangiocarcinoma. *Cancer Science*, 95(7), 592–595.

Yao, Y., & Dai, W. (2014). Genomic instability and cancer. *Journal of Carcinogenesis & Mutagenesis*, 5, 1000165.

Yin, Z., Dong, C., Jiang, K., Xu, Z., Li, R., Guo, K., et al. (2019). Heterogeneity of cancer-associated fibroblasts and roles in the progression, prognosis, and therapy of hepatocellular carcinoma. *Journal of Hematology & Oncology*, 12(1), 101.

Zeng, B., Zhou, M., Wu, H., & Xiong, Z. (2018). SPP1 promotes ovarian cancer progression via Integrin β1/FAK/AKT signaling pathway. *OncoTargets and Therapy*, 11, 1333–1343.

Zhang, Z. (2012). Genomic landscape of liver cancer. *Nature Genetics*, 44(10), 1075–1077.

Zhang, Q., He, Y., Luo, N., Patel, S. J., Han, Y., Gao, R., et al. (2019). Landscape and dynamics of single immune cells in hepatocellular carcinoma. *Cell*, 179(4), 829–845.e820.

Zhang, Q., Lou, Y., Yang, J., Wang, J., Feng, J., Zhao, Y., et al. (2019). Integrated multiomic analysis reveals comprehensive tumour heterogeneity and novel immunophenotypic classification in hepatocellular carcinomas. *Gut*, 68(11), 2019–2031.

Zhang, M., Yang, H., Wan, L., Wang, Z., Wang, H., Ge, C., et al. (2020). Single-cell transcriptomic architecture and intercellular crosstalk of human intrahepatic cholangiocarcinoma. *Journal of Hepatology*, 73(5), 1118–1130.

Zheng, H., Pomyen, Y., Hernandez, M. O., Li, C., Livak, F., Tang, W., et al. (2018). Single-cell analysis reveals cancer stem cell heterogeneity in hepatocellular carcinoma. *Hepatology*, 68(1), 127–140.

Zheng, C., Zheng, L., Yoo, J.-K., Guo, H., Zhang, Y., Guo, X., et al. (2017). Landscape of infiltrating T cells in liver cancer revealed by single-cell sequencing. *Cell*, 169(7), 1342–1356.e1316.

CHAPTER FOUR

Implications of genetic heterogeneity in hepatocellular cancer

Akanksha Suresh and Renumathy Dhanasekaran*

Division of Gastroenterology and Hepatology, Stanford University, Stanford, CA, United States
*Corresponding author: e-mail address: dhanaser@stanford.edu

Contents

1. Introduction — 104
2. Approaches to studying heterogeneity in HCC — 106
3. Molecular heterogeneity in HCC — 110
   - 3.1 Tumor heterogeneity — 110
   - 3.2 Cancer cell heterogeneity — 112
   - 3.3 Immune cell heterogeneity — 114
   - 3.4 Stromal heterogeneity — 116
4. Clinical implications of genetic heterogeneity in HCC — 118
   - 4.1 Challenges in stratification of HCC — 118
   - 4.2 Challenges in developing biomarkers for HCC — 120
5. Therapeutic implications of genetic heterogeneity in HCC — 122
   - 5.1 Personalized medicine — 122
   - 5.2 Novel targets for therapies and combination therapies — 123
6. Current perspectives and future directions — 125

Acknowledgements — 127
Grant support — 127
Conflict of interest statement — 128
References — 128

Abstract

Hepatocellular carcinoma (HCC) exhibits a remarkable degree of heterogeneity, not only at an inter-patient level but also between and within tumors in the same patient. The advent of next-generation sequencing (NGS)-based technologies has allowed the creation of high-resolution atlases of HCC. This review outlines recent findings from genomic, epigenomic, transcriptomic, and proteomic sequencing that have yielded valuable insights into the spatial and temporal heterogeneity of HCC. The high heterogeneity of HCC has both clinical and therapeutic implications. The challenges in prospectively validating molecular classifications for HCC either for prognostication or for prediction of therapeutic response are partly due to the immense heterogeneity in HCC. Moreover, the heterogeneity of HCC tumors combined with the lack of commonly mutated, druggable targets severely limits treatment options for HCC. Recently, immune checkpoint inhibitors and combination therapies have shown promise for advanced HCC, while T cell therapies and vaccines are currently being investigated. Yet, immunotherapies show benefit only in a limited subset of patients, making it imperative to decipher tumor heterogeneity in HCC in order to enable optimal patient selection. This review summarizes the cutting-edge research on heterogeneity in HCC and explores the implications of heterogeneity on stratifying patients and developing biomarkers and therapies for HCC.

Abbreviations

CAF	cancer-associated fibroblast
CTC	circulating tumor cell
ctDNA	circulating tumor DNA
EMT	epithelial-mesenchymal transformation
HCC	hepatocellular carcinoma
ICI	immune checkpoint inhibitor
NGS	next-generation sequencing
PDX	patient-derived xenograft
SNV	single-nucleotide variation
TGFβ	transforming growth factor-β
TKI	tyrosine kinase inhibitor
TLS	tertiary lymphoid structure

1. Introduction

Liver cancer causes over 800,000 deaths annually and is the third leading cause of cancer-related mortality in the world (Llovet et al., 2021; Sung et al., 2021). Furthermore, the incidence of liver cancer is on the rise in the Americas and Europe (Petrick et al., 2020). In the US, for instance, HCC incidence rose by 4% from 1978 to 2012 (Petrick et al., 2020). Hepatocellular carcinoma (HCC), the most common form of liver cancer, represents an average of 75–85% of all liver cancer cases (Thun, Linet, Cerhan, Haiman, & Schottenfeld, 2017). Patients with HCC can have variable clinical outcomes depending on the stage of tumor diagnosis and severity of the underlying liver disease. In general, patients with advanced HCC have a dismal prognosis with 5-year survival ranging from 15% to 20% (Siegel, Miller, & Jemal, 2020).

A key factor that complicates the diagnosis and treatment of HCC is tumor heterogeneity. Relative to other tumors, HCC exhibits a remarkable degree of heterogeneity (Khatib, Pomyen, Dang, & Wang, 2020). This heterogeneity makes HCC extremely resistant to most conventional chemotherapy. In 2007, the tyrosine kinase inhibitor, sorafenib, was approved for use in advanced HCC. Today, more treatment options exist for patients with advanced HCC in addition to sorafenib including regorafenib, lenvatinib, cabozantinib, and ramucirumab. However, all these options still provide limited clinical benefits. In 2020, the combination of atezolizumab (an antibody against PDL-1) and bevacizumab (an antibody against VEGF) was approved for patients with unresectable HCC (Finn, Qin, et al., 2020). Yet, this combination immunotherapy beat the median progression-free survival of sorafenib by only 2.5 months (Finn, Qin, et al., 2020; Sukowati, El-Khobar, & Tiribelli, 2021). Thus, there is a vast unmet need to develop novel therapies for HCC. We believe that a deeper understanding of tumor heterogeneity is necessary to develop successful therapeutic solutions and improve clinical outcomes for patients with HCC.

The complexity of HCC arises from several sources including diverse etiologies, varying geographical trends, and also molecular heterogeneity. Several different etiological risk factors are strongly associated with HCC development like chronic hepatitis B, chronic hepatitis C, alcohol-related or nonalcoholic fatty liver disease, aflatoxin exposure, hemochromatosis, and other causes of cirrhosis (Ghouri, Mian, & Rowe, 2017; Fujiwara, Friedman, Goossens, & Hoshida, 2018). The specific etiologies of HCC also drive the geographical trends in HCC incidence and outcomes in different parts of the world. For instance, HCC incidence is high in East and Southeast Asia due to the high prevalence of the hepatitis B virus (HBV) (Jiang, Han, Zhao, & Zhang, 2021; Yang et al., 2019). Even with falling HBV rates accompanied by a decline in HCC in these regions, liver cancer incidence remains very high (Dasgupta et al., 2020). On the other hand, metabolic risk factors for HCC like obesity and diabetes rates are increasing in many parts of the world including the US. In 2016, an alarming 1.9 billion adults worldwide were classified as overweight or obese, and this is a major contributor to the rising incidence of liver cancer (Sun & Karin, 2012; World Health Organization, 2018). Thus, disparate etiological and geographical factors contribute to the heterogeneity seen in HCC incidence.

Molecular heterogeneity is a critical area of research in HCC and will be the focus of the rest of this chapter. HCC tumor progression frequently occurs in the context of inflammation and fibrosis in a cirrhotic liver

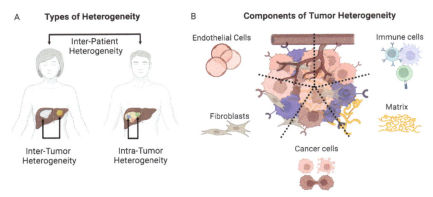

Fig. 1 Types and components of tumor heterogeneity in hepatocellular carcinoma (HCC). (A) The different types of heterogeneity in HCC are shown. Inter-patient—molecular differences between tumors of different patients, inter-tumor—variations between different tumors in the same patient, and intra-tumor heterogeneity—variations across different regions of an individual tumor in the same patient. (B) The different components of heterogeneity in HCC are shown. Cancer-cell intrinsic variations, differences in tumor immune cell infiltrates, and changes in endothelial cells, fibroblasts, and the matrix.

(Llovet et al., 2021; Villanueva, 2019). This environment is permissive to synchronous and metachronous tumor-initiating events and acts as a catalyst for molecular heterogeneity (Dhanasekaran, Bandoh, & Roberts, 2016). Molecular heterogeneity occurs on different levels. 1. on the interpatient level, there are molecular variations between tumors in different patients, and 2. on the intrapatient level, there are differences between tumors in the same patient. Moreover, intratumorally, there are variations in different regions within one tumor in a given patient. Further, a tumor is not just a mass of genetically mutated cancer cells; it is infiltrated by a variety of immune cells, structurally supported by the extracellular matrix, and nourished with oxygen and nutrients by the vasculature. These components, therefore, are also important considerations in the discussion of HCC molecular heterogeneity (Fig. 1).

2. Approaches to studying heterogeneity in HCC

Prior to the advancement of next-generation sequencing technologies, experimental schemes were largely hypothesis-driven and mostly only allowed investigators to confirm or refute a hypothesis. However, this approach limited our exploration of tumor heterogeneity because it necessitated the targets of interest to be known *a priori*. The advent of next-generation sequencing-based technologies for multi-omic and single-cell analysis has greatly accelerated a discovery-based understanding of tumor heterogeneity. Genome, epigenome, transcriptome, and proteome sequencing of serial and multi-region samples has allowed us to create high-resolution atlases of tumors and yielded several valuable insights into the spatial and temporal heterogeneity of HCC.

An important approach to studying heterogeneity in HCC is single-cell sequencing. In this approach, bulk tumors are dissociated into single cells, and DNA or RNA is extracted and sequenced from each cell. Additionally, the study of single-cell epigenetics is made possible by technologies like DNA methylome sequencing and transposase-accessible chromatin with sequencing (ATAC-seq). Today, the profiling of tens of thousands of cells is both technically feasible and relatively cost-effective. Thus, there is an explosion of research performing single-cell analysis in different cancers. Single-cell technology is powerful because of the resolution it offers. Single-cell RNA sequencing, for example, allows the investigation of the transcriptome of individual cancer, immune and stromal cells, and the tracking of subclones and their evolution. Several of the studies discussed in Section 3 leveraged single-cell sequencing for these purposes.

A drawback of single-cell sequencing is that it cannot capture the spatial distribution of different cell types. Given that paracrine signaling operates over short distances of less than 200 μm, knowing regional distributions of sequenced cells is critical in disentangling crosstalk between cells (Longo, Guo, Ji, & Khavari, 2021). Spatial transcriptomics offers a way to gain transcriptional information while maintaining tissue architecture. Two major techniques are used for spatial transcriptomics: high-plex RNA imaging and spatial barcoding. While high-plex RNA imaging has single-cell resolution and a greater depth of sequencing, spatial barcoding has a lower depth per transcript. On the other hand, spatial barcoding can be used on a larger field of view and has a greater transcriptome coverage compared to high-plex RNA imaging (Longo et al., 2021). The study by Brady et al exemplifies the strength of spatial transcriptomics in deciphering tumor heterogeneity. The researchers sampled different primary and metastatic tumor lesions in patients with prostate cancer. They found significant intrapatient heterogeneity when comparing tumor lesions within each patient, including in the gene expression of the therapeutically relevant, *NCAM1* for which antibodies and CAR-T cells exist. This showed that metastatic lesions in prostate cancer are not monoclonal and have distinct mutation profiles. Within a given tumor lesion, too, spatially distinct regions did not always have similar gene expression profiles. One metastasis, for instance, had two separate phenotypes within it that would have been misattributed by a bulk sequencing approach (Brady et al., 2021).

Liquid biopsies are a growing area of interest in the study of tumor heterogeneity. The term 'liquid biopsy' refers to analyzing tumors through circulating tumor cells (CTCs), which are cells that break off from tumors and enter circulation, or circulating tumor DNA (ctDNA), which is a type of fragmented DNA in blood circulation that is derived from tumors. Liquid biopsies offer many benefits over traditional needle biopsies which have long been used to detect genomic alterations in HCC. Needle biopsies are an invasive approach, requiring sampling from the tumor, while liquid biopsies are comparatively noninvasive and only require a blood draw. The lower risk profile of liquid biopsies means it is feasible to repeat them over time to track temporal changes and use them in surveillance for recurrence. Further, because needle biopsies sample from one tumor lesion at a singular point in time, they may fail to reflect tumor burden and fully capture genomic changes across the tumor (Swanton, 2012; Zhang et al., 2021). Liquid biopsies, on the other hand, measure CTCs or ctDNA that enter the bloodstream from all major tumor regions. Thus, liquid biopsies capture genetic changes from a majority of the tumor rather than from a specific region. Cai et al investigated the utility of ctDNA as a biomarker in HCC and found the ctDNA adequately reflected tumor subclonal mutations, thus capturing tumor heterogeneity (Cai et al., 2017). The researchers studied mutations in 574 cancer genes from tissue and plasma samples of three patients. Overall, over 200 subclonal mutations were discovered in the three patients. Intrapatient heterogeneity was also found: only 50–80% of subclonal mutations were overlapping between the primary tumor and portal vein tumor thrombus in each patient. Notwithstanding, the researchers reported that over 98% of detected subclonal mutations were also detected in ctDNA, indicating the ability of ctDNA to capture tumor heterogeneity (Cai et al., 2017).

Sun et al. investigated the spatial and temporal features of CTCs in the blood (Sun et al., 2018). The study used blood drawn from the peripheral vein, peripheral artery, hepatic veins, infrahepatic inferior vena cava, and portal vein of patients prior to tumor resection and during follow-up. A majority of CTCs at sites close to the tumor were epithelial while a majority of CTCs in systemic blood circulation had activated epithelial-mesenchymal transition (EMT). The total number of CTCs in the hepatic vein blood correlated with EMT activation in the primary tumor. Follow-up analysis showed that an increased number of CTCs in the peripheral vein or peripheral artery indicated an increased risk of intrahepatic recurrence (Sun et al., 2018). Overall, this study demonstrated the importance of the location of liquid biopsy and showcased the potential of CTCs in guiding HCC clinical progress. However, despite the promise of CTC, the lack of specific biomarkers to detect CTCs remains a challenge in HCC. Many studies have used the epithelial cell adhesion molecule (EpCAM) as a marker, however, this alone may not be ideal to detect CTCs that undergo EMT and lose their epithelial phenotype (Maravelia et al., 2021; Plaks, Koopman, & Werb, 2013). Combinations of biomarkers like EpCAM with hepatocyte-specific asialoglycoprotein receptor (ASGPR) and Twist1 have also been explored in other studies (Li et al., 2014; Sun et al., 2018). Another obstacle to the utility of CTCs is that CTCs levels can be particularly low in patients with small or early tumors, making this technique difficult to use in detection or early-disease monitoring (Chen, Zhong, Tan, Wang, & Feng, 2020; Mocan et al., 2020).

Generation of patient-derived primary cell lines, organoids, and xenografts from resected or biopsied HCC tumors offers another way to study HCC tumor heterogeneity. Gao et al performed multi-region sampling on 55 regions from 10 patients and maintained the isolated cells in culture. Whole-exome sequencing, copy-number analysis, and high-throughput screening were performed. The researchers discovered a high degree of intratumoral heterogeneity, with around 40% of mutations being heterogeneous. They also discovered that cells from some subregions had genetic aberrations in *FGF19, DDR2, PDGFRA,* and *TOP1,* and showed sensitivity to the application of a therapeutic treatment. However, subregions from the same tumor without these mutations did not show sensitivity. This experiment shows an in vitro approach to study HCC tumor heterogeneity and drug responses (Gao et al., 2017).

Organoids are models where cancer cells are cultured as three-dimensional structures (Tuveson & Clevers, 2019). Zhao et al generated seven hepatobiliary tumor organoids and studied them using single-cell RNA sequencing (Zhao et al., 2021). The researchers noted that there was inter-organoid transcriptional and therapeutic heterogeneity. One of the organoids, HC272, for example, showed enrichment of HIF-1, MAPK, and PI3K-Akt signaling pathways compared to the other organoids. Consequently, this organoid also displayed resistance to the eleven tyrosine kinase inhibitors (TKIs) tested in the study, while other organoids showed sensitivity to one or more TKIs (Zhao et al., 2021).

Patient-derived xenografts have also been investigated to identify biomarkers, study tumor progression, and conduct personalized drug screens. Blumer et al validated that PDXs generated from biopsies of Grade III or IV HCC preserved tumor markers, gene signatures, and copy-number alterations from the original tumor over 6 generations of retransplantation (Blumer et al., 2019). Hu et al. studied the effects of three drugs in sixteen PDX models. They discovered inter-individual heterogeneity as well as intra-individual differences in sensitivity to the agents tested. One of their findings was that impaired sorafenib response was correlated with low MAP3K1 expression in patients. The investigators further studied this finding in their PDX model and cell lines and found that MAP3K1 overexpression resulted in a more robust response to sorafenib treatment while MAP3K1 inhibition increased cell proliferation during sorafenib treatment (Hu et al., 2020). Taken together, these results indicate the ability of organoids and PDXs to recapitulate tumor biology and study drug responses (see also chapter "Molecular therapeutic targets for cholangiocarcinoma: Present challenges and future possibilities" by Anderson et al.) (Fig. 2).

3. Molecular heterogeneity in HCC

Tumors are composed of more than just clusters of cancer cells: they are complex quasi-organs that are infiltrated by immune cells, structurally supported by the extracellular matrix, and supplied with oxygen and nutrients from the vasculature. All of these components contribute to molecular heterogeneity in HCC tumors. Here, we discuss three contributors to molecular heterogeneity: cancer cells, immune cells, and stromal cells.

3.1 Tumor heterogeneity

Next-generation sequencing (NGS) has revolutionized our understanding of cancer genetics. DNA sequencing of bulk HCC tumors has revealed that they have a median of 50–70 protein-altering mutations and an average of 2–6 driver mutations across all tumor stages (Schulze et al., 2015; The Cancer Genome Atlas Research Network, 2017; Nault et al., 2020; Ahn et al., 2014). The most prevalent driver gene mutations are in the *TERT* promoter, *TP53*, *CTNNB1*, *AXIN1*, *ARID1A*, and *ARID2* which cause changes in the activation of several pathways including telomere maintenance, P53 cell regulation, Wnt/β-catenin, Akt/mTOR, MAP kinase, and oxidative stress (Ahn et al., 2014; The Cancer Genome Atlas Research Network, 2017; Fujimoto et al., 2012; Guichard et al., 2012; Nault et al., 2013; Schulze et al., 2015; Totoki et al., 2014). Some driver gene mutations tend to co-occur, suggesting

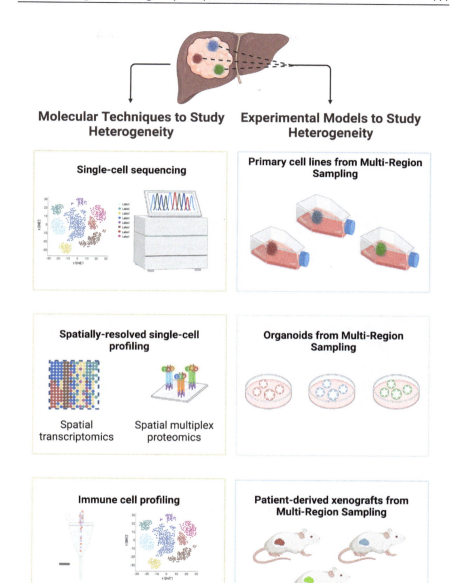

Fig. 2 Molecular techniques (left) and experimental models (right) to study tumor heterogeneity in hepatocellular carcinoma (HCC) are shown. On the left, extracts from the tumor or liquid biopsy can be used to perform single-cell sequencing, spatial transcriptomics and multiplex proteomics, and flow and mass cytometry. On the right, multi-region samples from tumors can be used to establish primary cell lines, organoids, and patient-derived xenografts in immunocompromised mice.

a synergistic effect in tumorigenesis and tumor progression such as *CTNNB1* and *TERT* promoter mutations with *ARID2* and *NFE2L2* mutations while others like *AXIN1* and *TP53* mutations are mutually exclusive suggesting redundancy or disadvantageous effects when co-occurring (Schulze et al., 2015). These findings highlight the variety of mutations and mutation combinations that are present in different tumors that contribute to inter-patient heterogeneity in HCC.

On an intrapatient level too, there can be incredible heterogeneity between tumors in the same patient (Xue et al., 2016). Exome and whole-genome sequencing of 43 lesions from 10 patients with HCC revealed that, in general, intrahepatic metastases and tumor thrombi tended to have different mutations and copy number variations compared to the primary lesion. However, the percentage of shared mutations between lesions in a given patient varied widely from 8% to 97%. This broad range of intrapatient heterogeneity suggests that a single lesion may not always sufficiently capture the genomic features of HCC in every patient (Xue et al., 2016).

In about half of patients with recurrent HCC, the primary and recurrent tumors have dissimilar genetic features, according to a study by Ding et al. (2019). By sequencing 824 HCC-relevant genes in multi-region samples from 41 matched primary and recurrent tumor pairs, the investigators found that recurrent HCC fell into two broad categories: multicentric and progressive. Multicentric HCC comprised 48% of recurrent tumors and had an independent lineage from the primary tumor. Progressive HCC, on the other hand, comprised the remaining 52% of recurrent tumors, shared several mutations with the primary tumor, and were derived from the same clonal lineage. The recurrence time between these two categories also appeared distinct—though not statistically significant—with the multicentric type generally recurring after 2 years and progressive type recurring earlier. Thus, the similarity between the primary and recurrent HCC tumors is another contributor to interpatient and intrapatient heterogeneity (Ding et al., 2019).

3.2 Cancer cell heterogeneity

Cancer cells are the primary contributor to tumor heterogeneity because of their high degree of genome instability. Several studies have explored cancer cell heterogeneity present in HCC tumors using single-cell sequencing and/or multi-region sampling.

Guo et al used single-cell DNA sequencing to study copy number alterations (CNAs), which occur due to genome instability and contribute to cancer cell heterogeneity (Guo et al., 2022). The percentage of amplifications and deletions in the genome of cells from a given patient varied widely with standard deviations between 1% and 30%. There were also differences between patients with the average percentage of genomic alterations ranging from 15% to 60% and the percentage of non-diploid cells ranging from 34% to 94%. These findings indicated a variable degree of intrapatient and interpatient heterogeneity in cancer cell genomes (Guo et al., 2022).

Single-cell sequencing studies using multi-region sampling have further revealed the extraordinary diversity of cancer cells present in HCC tumors. Ling et al performed whole-genome and whole-exome sequencing on 286 regions from a 35 mm, histopathological grade III HCC tumor (Ling et al., 2015). Their results revealed a high degree of genetic diversity in the tumor with over 100 million estimated coding region mutations, including 6 driver gene mutations and 209 single-nucleotide variations (SNVs). A portion of the cancer cells with no driver gene mutations was discovered that defined a whopping 20 distinct cell clones in the whole-exome sequenced regions. Thus, the study reported extreme heterogeneity in the HCC tumor, much higher than would be predicted by Darwinian evolution (Ling et al., 2015).

Zhai et al also used single-cell sequencing of multi-regional samples to investigate the evolution and spatial organization of intratumoral heterogeneity in HCC (Zhai et al., 2017). A series of sections were sampled along the diameter of nine early HCC tumors and either whole-genome or whole-exome sequenced. The authors found that observed variability within each tumor increased substantially with an increasing number of sampled regions, suggesting that single samples of HCC tumors likely fail to fully capture tumor heterogeneity. Phylogenetic trees for each tumor were constructed using variations in somatic mutations across sections and showed that sections from one end of the tumor on the central axis were consistently more similar than those from the other end. This led to the suggestion that HCC ancestral clones may be found in the tumor center, with new clades arising as the tumor expands outwards (Zhai et al., 2017).

Wu et al. performed high-resolution spatial transcriptomics to study 3 distinct regions–nontumor, leading-edge, and tumor–in seven HCC patients (Wu et al., 2021). One aspect of the study investigated cancer stem cell (CSC) niches and found that the distribution of CSCs varies from the leading-edge to the tumor and constituted diverse cells and activated pathways. The PROM1+ and CD47+ CSC niche, for example, was found to increase in abundance from the leading-edge to the tumor and the portal vein thrombus and was implicated in immunosuppression and vascular metastasis. The preservation of spatial information in the study allowed the investigation of CSC niche composition and distribution within tumors (Wu et al., 2021). Overall, the aforementioned studies reveal a rich diversity of cancer cells in HCC tumors both on the interpatient and intrapatient levels.

3.3 Immune cell heterogeneity

HCC tumors vary widely in their immune subpopulation frequency and distribution, the activation and functional states of immune cells, and the overall degree of immune infiltration. With the recent approval of combination immunotherapies for HCC, atezolizumab/bevacizumab and nivolumab/ipilimumab, there is a renewed interest in immune cells in the tumor microenvironment.

Zhang et al. analyzed immune heterogeneity in HCC using mass cytometry (Zhang, Lou, et al., 2019). They identified 40 distinct clusters of immune cells within the tumor microenvironment and noted differences between immune environments in their eight patients and between lesions within the same tumor in each patient. Certain immune cell populations were positively correlated such as PD-L1+ macrophages with PD-1+ CD4+ T cells, and dendritic cells with CD11b+ CD4+ T cells. The immune environments were clustered into three categories with distinct cell frequencies, gene expression of immune suppressive genes and metabolomes. Their study of the immune landscape of tumors revealed notable interpatient and intrapatient immune heterogeneity (Zhang, Lou, et al., 2019).

The immune environment is dynamically regulated through the stages of tumor progression. Song et al. used single-cell RNA sequencing of immune cells in hepatitis B/C-related HCC tumors and found 29 unique immune cell subsets of myeloid, lymphoid, and NK cells that shifted between functional states and interacted with other subsets (Song et al., 2020). Effector CD8+ T cells showcased distinct transcriptomes and cytotoxicity that varied from early to advanced HCC. A subset of M2 macrophages with high expression of CCL18 and CREM was elevated in patients with advanced HCC (Song et al., 2020). The findings of this paper provide insight into the changes in the HCC immune environment with tumor progression.

Several studies have studied specific immune cell populations and their functional states in HCC tumors. Losic et al. investigated T cell clonality in samples from spatially different regions of HCC tumors (Losic et al.,

2020). Using a combination of DNA, RNA, and T cell receptor (TCR) sequencing, they found significant differences in the number of unique T cell expansions, and density of immune cells in different regions of tumors. They also found that some regions within a given tumor had tertiary lymphoid structures while other regions did not, demonstrating another instance of intratumoral immune heterogeneity. They determined that tumor neoantigens primarily recruit tumor-infiltrating lymphocytes (B and T cells). Immune checkpoint genes such as CTLA4 and CD274 also tended to be upregulated in tumor cells in regions with more tumor-infiltrating lymphocytes, suggesting regional differences in immunosuppressive effect on T cell activation (Losic et al., 2020).

T cell heterogeneity was further explored by Hung et al who investigated the impact of tumor methionine metabolism on T cell exhaustion in HCC. They used ATAC-seq to uncover the chromatin accessibility of T cells during activation and treatment with two products of methionine metabolism. In their results, they show that the products of tumor methionine metabolism induce significant changes in chromatin in T cells that inhibit effector T cell function and potentially contribute to exhaustion (Hung et al., 2021).

Immune cell distribution and properties were explored by Zhang et al. by RNA sequencing of immune cells at five regions: tumor, adjacent liver, hepatic lymph node, blood, and ascites (Zhang, He, et al., 2019). The study found a population of LAMP3+ dendritic cells had the potential to migrate from the tumor to the lymph node and expressed immune-modulating ligands that could regulate various lymphocytes. Macrophages with distinct transcriptional states were discovered, and the expression of *SLC40A1* and *GPNMB* in tumor-associated macrophages (TAMs) was found to play an inflammatory role in the tumor. The origins of myeloid and lymphoid cells in the ascites were traced to the tumor and blood, respectively. These findings reveal the diversity of immune cell functions in the regions in and around the tumor (Zhang, He, et al., 2019).

The heterogeneity of the HCC immune microenvironment has implications on patient prognosis. Kurebayashi et al. studied the immune microenvironment in 919 regions from 158 HCC tumors and classified them into three subtypes: immune-high, immune-mid and immune-low. The immune-high subtype was characterized by increased B cell and T cell infiltration. Importantly, they found that high-grade tumors of the immune high subtype had a significantly better prognosis (Kurebayashi et al., 2018). In a similar vein, Calderaro et al. investigated intratumoral tertiary lymphoid structures (TLSs) in HCC. TLSs are sites for immune response initiation and maintenance present in nonlymphoid tissue. The presence of TLSs within HCC tumors was found to be associated with a decreased risk of early HCC recurrence in two cohorts consisting of 273 and 225 tumors respectively (Calderaro et al., 2019). The above studies showcase the incredible heterogeneity in immune cell frequency, distribution, and activation in HCC.

3.4 Stromal heterogeneity

Stromal cells, including cancer-associated fibroblasts (CAFs) and endothelial cells, make up important components of the tumor, though their contribution to HCC tumor heterogeneity remains an active area of investigation.

CAFs in the HCC microenvironment display heterogeneity in tissue distribution, cellular origin, and function. CAFs are often present in the HCC fibrous septum, fibrous capsule, and hepatic blood sinusoids (Yin et al., 2019). CAFs originate from a variety of cells including hepatic stellate cells, mesenchymal stem cells, and cancer cells. Secretory products from cancer cells can cause these various cell types to take on a CAF role. A growing body of evidence suggests that CAFs play a pro-tumor role in HCC. Liu et al showed that CAFs secrete chemokines including CCL2, CCL5, CCL7, and CXCL16 which promote cancer cell migration, invasion, and epithelial-mesenchymal transformation (EMT) in vitro, and promote HCC metastasis in vivo through the activation of the hedgehog (Hh) and transforming growth factor-β (TGFβ) pathways (Liu et al., 2016). Some CAFs induce the chemotaxis of immune cells such as neutrophils, dendritic cells, and monocytes to the tumor and promote immunosuppressive phenotypes through IL6-STAT3 signaling (Cheng et al., 2018). Thus, CAFs display heterogeneity in the tumor through their presence in a variety of tumor regions, having multiple cellular origins, and activating different pro-tumor and immunomodulatory signaling pathways.

The role of fibroblasts in HCC tumorigenesis from non-alcoholic steatohepatitis (NASH)-related fibrosis has been explored (Asakawa et al., 2019). RNA sequencing was used to study the gene expression profiles of fibroblasts in a NASH mouse model of fibrosis. Compared to a chemical-induced fibrosis model, fibroblasts in the NASH model showed a robust upregulation in cancer-related pathways including fibroblast growth factor 9 (FGF9) expression. Follow-up experiments validated the role of FGF9 in promoting tumorigenesis to drive HCC in the NASH fibrotic liver

(Asakawa et al., 2019). This study suggests that there may be heterogeneity in the transcriptional state and roles of fibroblasts in HCC of different etiologies, such as in NASH-driven HCC.

Overall, however, the heterogeneity of CAF phenotypes and functions in HCC remains to be thoroughly investigated. Studies from other cancers indicate that the functional states and spatial distribution of CAFs are important contributors to tumor heterogeneity (see also chapter "Cancer-associated fibroblasts in intrahepatic cholangiocarcinoma progression and therapeutic resistance" by Affò et al.). Sebastian et al explored the molecular heterogeneity of CAFs in triple-negative breast cancer in a mouse model using single-cell RNA sequencing and found six distinct subpopulations such as (i) myofibroblastic CAFs enriched in contractile proteins, (ii) inflammatory CAFs with increased expression of inflammatory cytokines, and (iii) CAFs expressing major histocompatibility complex (MHC) class II, which are generally found on antigen-presenting cells (Sebastian et al., 2020). The spatial distribution of CAFs in pancreatic cancer was explored by a study that found that the myofibroblast CAF population is frequently in regions close to the pancreatic tumor, while inflammatory CAFs are found in regions distant from the tumor, though TGF-β or JAK/STAT signaling can facilitate their interconversion (Biffi et al., 2019; Öhlund et al., 2017). The differing transcriptional states and spatial distribution of different CAF subclasses suggest that they may have differing contributions to tumorigenesis and tumor progression. Comparing HCC CAF phenotypes and distribution in different patients and between lesions in a given patient will further elucidate their contribution to interpatient and intrapatient heterogeneity.

Endothelial cells (ECs) which typically line blood vessels are another source of heterogeneity in HCC. Sharma et al performed single-cell RNA sequencing on HCC tumors and discovered eleven unique clusters of ECs including a cluster of CD9+ ECs found primarily in the adjacent liver, a cluster of phospholipid phosphatase 3-positive (PLPP3+) ECs present in the peripheral tumor, and a cluster of plasmalemma vesicle-associated protein-positive (PLVAP+) ECs present in the core tumor (Sharma et al., 2020). This study sheds light on the rich diversity of spatially distributed endothelial cells present within HCC tumors. The roles of these EC subpopulations in HCC pathophysiology remain to be explored. However, some EC subpopulations are likely pro-tumorigenic: ECs are known to engage in crosstalk with cancer cells that contribute to epithelial to mesenchymal transition, cancer stem cell phenotype, and metastasis (Alsina-Sanchis, Mülfarth, & Fischer, 2021; Maishi et al., 2016; McCoy et al., 2019).

The abundance of tumor vasculature, which is composed of ECs, shows interpatient differences and impacts HCC prognosis. Several studies have reported the biomarker potential of vessels that encapsulated-tumor clusters (VETCs)–detected through immunostaining for vascular ECs (Fang et al., 2015, 2019; Renne et al., 2020). Renne et al report the presence of VETCs in only a subset of patients, about 19%, where their presence was associated with several factors including tumor sizes greater than 5 cm, a poorly differentiated phenotype, microvascular invasion, and early recurrence (Renne et al., 2020). Overall, this study exemplifies the interpatient heterogeneity in HCC vasculature and the role of vascular endothelial cells in HCC pathophysiology and prognostication.

4. Clinical implications of genetic heterogeneity in HCC
4.1 Challenges in stratification of HCC

The understanding and classification of tumors into subtypes based on molecular or mutational signatures has improved patient outcomes for several cancers. The advances in single-cell sequencing technologies carry the promise of deepening our understanding of the characteristics of, and interactions between, cancer, immune, and stromal cells. NGS technologies offer high-plex information at a significantly higher sensitivity than traditional methods like immunohistochemistry, and this could, in turn, facilitate the identification of molecular signatures for distinct classes of HCC.

Several studies have used transcriptome analysis to categorize HCC tumors into subtypes based on hierarchical clustering or correlations with clinical variables (Boyault et al., 2007; Chiang et al., 2008; Hoshida et al., 2009, 2010; Lachenmayer et al., 2012). The analysis of multiple, genome-wide studies reveals that HCC tumors can be categorized into two major classes: proliferative and non-proliferative (Boyault et al., 2007; The Cancer Genome Atlas Research Network, 2017; Chiang et al., 2008; Hoshida et al., 2009; Lee et al., 2004). The proliferative class consists of tumors of a more aggressive and poorly-differentiated phenotype. These tumors showcase high chromosomal instability, global DNA hypomethylation, TP53 inactivation, and *CCND1* and *FGF19* amplification. Pro-survival pathways are frequently activated including cell cycle, MET, RAS-MAPK, and mTOR (Boyault et al., 2007; Calvisi et al., 2007; Chiang et al., 2008; Hoshida et al., 2010; Lee et al., 2004). The proliferative class can be further subdivided into the Wnt-TGFβ subclass, characterized by an exhausted immune response, and the progenitor subclass, consisting of hepatic progenitor marker overexpression and ERK hyperphosphorylation (Boyault et al., 2007; Calderaro et al., 2017; Hoshida et al., 2009; Sia et al., 2017).

The non-proliferative class consists of tumors with Wnt/β-catenin pathway activation, higher genome stability, and a less aggressive, more differentiated phenotype (The Cancer Genome Atlas Research Network, 2017; Hoshida et al., 2009, 2010; Lee et al., 2004). However, even within tumors of this class, there is a significant degree of heterogeneity and tumors can be subdivided into those with *CTNNB1* mutations and "G4" subclasses. The *CTNNB1* subclass show activation of the Wnt/β-catenin pathway, *TERT* promoter mutations, and *CDKN2A* and *CDH1* promoter hypermethylation (The Cancer Genome Atlas Research Network, 2017; Chiang et al., 2008; Xu et al., 2018). Tumors of this subclass tend to be "cold", with reduced T cell infiltration and an increased presence of regulatory T cells (Calderaro et al., 2017; Sia et al., 2017). The "G4" subclass shows activation of the IL6/JAK-STAT pathway and tends to be "hot", with an active immune response (Boyault et al., 2007; Calderaro et al., 2017; Chiang et al., 2008; Sia et al., 2017). Tumors of the non-proliferative class have alternatively been grouped depending on their metabolic functions into periportal and perivenous subclasses. The periportal subclass expresses the *HNF4A* gene signature and is associated with the least likelihood of early recurrence and the highest survival rate (Chan, Tsui, Ho, & Ng, 2021; Désert et al., 2017).

Another approach has been to use cell deconvolution methods on transcriptomic data from HCC tumors. This approach has allowed the identification of the proportion of different immune cells within HCC and yielded two immune classes of HCC, immune-high and immune-low (Sia et al., 2017). About 25% of HCC tumors fall in the immune-high class and show macrophage, T cell, cytotoxic cell, and tertiary lymphoid structure enrichment and have elevated PD1 signaling. The immune-high subclass can be further divided into those that show an active or exhausted T cell response, with the exhausted response subclass showing TGFβ1-mediated gene regulation and immunosuppression.

HCC is yet to have a molecular classification system to prospectively stratify patients and guide clinical care. A central challenge to validating and implementing a classification system is that the current guidelines for HCC diagnosis and treatment do not require tissue biopsies. This limits the systematic collection of tissue samples and the correlation of molecular features with clinical variables.

4.2 Challenges in developing biomarkers for HCC

Biomarkers can serve as useful tools to guide clinical care for patients with HCC. One area of interest has been the discovery of biomarkers that predict response to therapies, particularly to tyrosine kinase inhibitors which constitute a major portion of the available therapies for patients with advanced HCC. Sorafenib is a multikinase inhibitor that is used as a first-line treatment in patients with advanced HCC. Fang et al. report that primary tumors with CD34+ vessels that encapsulate tumor clusters (VETC) pattern show a better response to sorafenib treatment upon HCC recurrence than tumors without VETC pattern (Fang et al., 2019). Other factors including *FGF3/FGF4* amplification, *VEGFA* amplification, VEGFR overexpression, and Mapk14-Atf2 signaling elevation have also been associated with better sorafenib response (Arao et al., 2013; Horwitz et al., 2014; Peng et al., 2014; Rudalska et al., 2014). Harding et al performed NGS on 81 tumors from patients with advanced HCC who received sorafenib treatment. They found that the activation of the P13K-mTOR pathway in patients receiving sorafenib treatment was associated with poorer outcomes including shorter median progression-free survival and shorter median overall survival (Harding et al., 2019). Similarly, another study found that a 9-marker myeloid signature predicted sorafenib efficacy in recurrent HCC (Wu et al., 2020). Despite these findings, there is no validated biomarker that is currently used in patient care to predict sorafenib response. Similarly, a retrospective analysis revealed that 5 proteins and 9 miRNAs present in plasma could predict response to regorafenib, a tyrosine kinase inhibitor that is used as second-line treatment (Teufel et al., 2019). However, this finding also needs validation in a prospective cohort. Phase III of the REACH-2 study validated the first predictive biomarker for an HCC drug that is usable in clinical practice. It was found that high serum alpha-fetoprotein is associated with sensitivity to ramucirumab (Zhu, Kang, et al., 2019).

Immune checkpoint inhibitors have shown major benefits in the treatment of several cancers. Given that HCC arises in the context of inflammation and immunosuppression, immunotherapies are an exciting therapeutic prospect. Despite these facts, a sizable portion of HCC patients does not respond to ICIs. For example, in a phase II trial for nivolumab, a PD-1 inhibitor, the objective response rate was only 20%. Further, tumoral PD-L1 expression did not correlate with the effectiveness of PD-1 inhibition (El-Khoueiry et al., 2017). This suggests a need to identify biomarkers that can predict benefit from PD-1 inhibition, or other ICIs. One study found that a higher proportion of CD38+ cells in the tumor correlated with better response to PD-1 inhibition, while another study showed that the upregulation of indoleamine 2,3-dioxygenase (IDO) was associated with poorer response to PD-1 inhibition (Brown et al., 2018; Garnelo et al., 2017). It is also possible that the immune classification of HCC, discussed in Section 4.1, could predict response to PD-1 therapy. The "G4" subclass, which has high immune infiltration and enriched PD1 signaling, is likely to be more responsive to ICI, especially PD-1 inhibition. However, the association between this HCC subclass and ICI response is yet to be confirmed. On the other hand, the *CTNNB1* subclass, which shows activation of the Wnt/β-catenin pathway and has reduced immune infiltration, is likely to be resistant to ICI. Indeed, a study found that in patients receiving PD-1 inhibition, Wnt/β-catenin pathway mutations were associated with poorer outcomes including shorter median progression-free survival and shorter median overall survival (Harding et al., 2019). For the combination therapy of PD-1/PD-L1 inhibition, many studies across different cancers have validated the biomarker potential of tumor mutational burden, a measure of the total number of somatic mutations per megabase. These studies point to the association between a high tumor mutational burden and greater benefit from PD-1/PD-L1 inhibition (Zhu, Zhang, et al., 2019; Forschner et al., 2019; Lee, Samstein, Valero, Chan, & Morris, 2020; Chen et al., 2019). Several factors including genetic heterogeneity, activated pathways, tumor microenvironment, systemic immunity status and metastases can impact the efficacy of immunotherapy (Siu et al., 2018; Sukowati et al., 2021). However, there is still much progress to be made in identifying predictive biomarkers for response to immune checkpoint inhibitor therapies for patients with HCC.

Tumor biopsy remains the most reliable source for identifying biomarkers. However, given the difficulty in obtaining single and longitudinal tissue biopsies as in HCC, a liquid biopsy is a lucrative, non-invasive intervention that allows temporal and some degree of spatial monitoring of tumor heterogeneity. A study by Kim et al. points to the ability of ctDNA to reflect tumor heterogeneity and predict prognosis in HCC. The researchers first investigated common SNVs in ctDNA using a digital droplet PCR panel of 2924 SNVs in 69 genes. There was considerable inter-patient heterogeneity in detected SNVs but the researchers determined that the top four most frequent SNVs in ctDNA from HCC patients were *MLH1, STK11, PTEN*, and *CTNNB1*. The biomarker potential of these four candidates was then validated in a separate cohort of 62 patients. The researchers found that the presence of *MLH1* SNV and elevated ctDNA levels together predicted poor overall survival (Kim et al., 2020). Liquid biopsies hold incredible potential to aid the management and treatment of many cancers. However, this method has a long way to go before it can reach the bedside, including the need to improve sensitivity and specificity (Rebouissou & Nault, 2020).

5. Therapeutic implications of genetic heterogeneity in HCC

Tumor heterogeneity has important implications for predicting response to therapy. In general, tumors with a high degree of heterogeneity do not respond well to therapies because the selection pressure can lead to the expansion of resistant sub-clones or the emergence of new drug-resistant clones. HCC tumors are highly heterogeneous, as described in detail in Section 3. The study of the genomic landscape of HCC has revealed that HCC tumors often present with loss-of-function mutations in tumor-suppressor genes like *P53*, *AXIN1*, *ARID1A*, and *TSC1/2* (The Cancer Genome Atlas Research Network, 2017; Guichard et al., 2012; Ho et al., 2017). While gain-of-function mutations result in activated pathways and present clear potential targets for therapy, loss-of-function mutations are more difficult to target therapeutically. Though a subset of patients has aberrant activation of pathways from *TERT* promoter and *CTNNB1* mutations, these targets are largely considered undruggable (Chan et al., 2021). Thus, the high heterogeneity of HCC tumors combined with the lack of commonly mutated, druggable targets severely limits treatment options in HCC. For patients with early or intermediate stage HCC, local radio- or chemo-embolization, surgical resection, and liver transplantation are curative treatment options. However, for patients with advanced HCC that is inoperable, first-line therapy includes protein kinase inhibitors like sorafenib and lenvatinib which have limited survival benefits, high toxicity, and a high likelihood of resistance (Kudo et al., 2018; Llovet et al., 2008). There is a need to not only identify more therapeutic targets but also to select appropriate therapies for each patient's specific tumor biology.

5.1 Personalized medicine

Affected pathways and druggable targets for individual patients have the potential to be discovered using a personalized medicine approach where the genomic and transcriptomic alterations from biopsy- or liquid biopsy-obtained cancer cells are investigated. Such findings can be used to select appropriate anti-tumor therapies. While individual patients may have druggable mutations, the use of bulk sequencing or single location biopsy approaches may make it difficult to disentangle whether the mutation is shared by cancer cells in different tumor regions. Thus, it is likely that truncal or driver mutations need to be targeted to account for otherwise heterogeneous mutations between subclones (Lohr et al., 2014; McGranahan et al., 2015).

While a personalized medicine approach could help select therapies for individual patients, several approaches have been suggested for treatment regimens to address tumor heterogeneity. McQuerry et al suggest that a strategy could be to sequentially cycle different treatments for tumors consisting of heterogeneous subclones (McQuerry, Chang, Bowtell, Cohen, & Bild, 2017). In patients with multiple tumor lesions, intrapatient intertumor heterogeneity can present as a mixed response to therapy with some tumors shrinking and others growing (McQuerry et al., 2017). In such cases, there may be a benefit in obtaining multiple and/or serial biopsies from a patient in order to select appropriate treatments. However, HCC, relative to other cancers, is unique in that there is currently no clinical indication to perform a tumor biopsy to confirm an HCC diagnosis (Dhanasekaran, 2021). This somewhat hinders the ability to obtain tissue samples to study tumor progression and evolution in the context of treatment response. An alternate approach could be to use liquid biopsy instead, though the low concentration of ctDNA and CTCs and the high sensitivity required for their detection may be limiting factors (Maravelia et al., 2021; Mocan et al., 2020).

5.2 Novel targets for therapies and combination therapies

The liver is a highly vascularized organ that is exposed to many antigens due to its physiological functions (Thomson & Knolle, 2010). This feature requires the liver to be immunologically tolerant, a characteristic that may impede anti-tumor immunity. In HCC following viral hepatitis infection, there is immune exhaustion as seen by an exhausted T cell phenotype that may further impede an anti-tumor immune response (Harding, El Dika, & Abou-Alfa, 2016). This unique immune environment makes immunotherapies, and specifically immune checkpoint inhibitors (ICIs) an exciting prospect for the treatment of HCC. Nivolumab and pembrolizumab are PD-1 receptor inhibitors that were both shown to have benefits for HCC patients as second-line treatment following sorafenib failure or toxicity (El-Khoueiry et al., 2017; Finn, Ryoo, et al., 2020; Zhu et al.,

2018). However, as with most ICIs, nivolumab and pembrolizumab only show benefits for a limited subset of patients. These therapies had a 15-20% rate of objective remissions with only a 1-5% rate of complete response (Sangro et al., 2017; Zhu et al., 2018). Further, a study found that tumoral PD-L1 expression did not correlate with the effectiveness of PD-1 inhibition by nivolumab (El-Khoueiry et al., 2017). Thus, there is a pressing need to develop molecular classifications so that patients can be matched to therapies that they are more likely to benefit from.

Surgical resection, radio- or chemo-embolization, and liver transplantation are curative treatment options that remain the standard of care for patients with early or intermediate-stage HCC. An ongoing pilot study is investigating the use of perioperative immunotherapy as part of curative treatment for patients with resectable HCC (Kaseb et al., 2019). The clinical outcomes of perioperative immunotherapy remain to be seen, however, mass cytometry analysis of pre- and post-treatment lesions reveals an increase in two subsets of CD8 + T cells. One subset showed robust anti-tumor immunity with a twofold increase in CD45RO and a threefold increase in granzyme B which are markers of T cell activation. The other subset possessed immune-modulatory and suppressive phenotypes with the elevation of Foxp3 and PD-L1 expression (Kaseb et al., 2019). The immune profiling of patients in this study could prove to be a powerful tool for stratifying patient responses and aid prospective patient selection for perioperative immunotherapy in the future.

Single-agent therapies have not been able to bypass the median overall survival of less than one year (Abou-Alfa & Venook, 2013). As a result, combination therapies have been evaluated in different combinations. High intra-tumoral heterogeneity in patients with HCC may also be an indication for combination therapy over single-agent therapies (Barcena-Varela & Lujambio, 2021). Some of the currently investigated combination therapies include VEGF inhibitors with ICIs, TKIs with ICIs, and ICIs with other ICIs. The rationale for the combination of VEGF inhibitors with ICIs is as follows. It has been shown that anti-VEGF therapies inhibit angiogenesis, inducing hypoxia in the tumor, which in turn upregulates the immune checkpoint protein PD-L1 (Kimura et al., 2018). Further, VEGF inhibition may improve tumor-specific T cell activity (Noman et al., 2014). The combination of VEGF inhibitors and PD-1/PD-L1 inhibitors has already been shown to be beneficial in lung and genitourinary cancers, and likewise as first-line treatment in HCC as well through the combination of atezolizumab, a PD-L1 inhibitor, and bevacizumab, an anti-VEGF agent (Finn, Qin, et al., 2020;

Reck et al., 2019; Rini et al., 2019). Further, combinations of immunotherapies such as anti-PD1/PD-L1 and CTLA4 are also being shown to be clinically beneficial in HCC (Sangro, Sarobe, Hervás-Stubbs, & Melero, 2021).

Chimeric antigen receptor (CAR) engineered T cell therapy may be an upcoming treatment option for HCC. Preclinical studies have shown the adoptive glypican-3 (GPC3) CAR T therapy in patient-derived xenografts (PDX) models slowed down tumor growth in tumors with high expression of GPC3 (Jiang et al., 2016). Other targets such as alpha-fetoprotein (AFP), human TERT, and melanoma antigen gene (MAGE3) are also being investigated. Early phase clinical trials for patients with HCC are currently identifying different CAR targets (NCT02905188, NCT03884751, NCT03980288, NCT03993743). Other immune-based therapies such as adoptive T cell transfer, vaccination and virotherapy, and the combination of locoregional therapies with ICIs are also being studied (Sangro et al., 2021).

While many treatment options and combinations emerge, there is still a need to fully understand how to sequence treatments. Further, many treatments only show responses in a subset of patients, likely due to inter-patient and intra-tumor heterogeneity. Thus, it is important to develop biomarkers to predict response to the array of treatments available, so that the most appropriate therapy can be selected. Achieving this necessitates the molecular profiling of patients in clinical trials so that clinical endpoints can be determined for distinct subgroups and reanalyzed in the future as more evidence emerges from basic and clinical research studies (Fig. 3).

6. Current perspectives and future directions

HCC exhibits a remarkable degree of heterogeneity not just in etiology and geographic prevalence, but also in the molecular signatures within and between tumors. The advent of next-generation sequencing-based technologies for multi-omic and single-cell analysis has greatly accelerated the understanding of tumor heterogeneity. Genome, epigenome, transcriptome, and proteome sequencing of serial and multi-region samples has allowed us to create high-resolution atlases of tumors and yielded several valuable insights into the spatial and temporal heterogeneity of HCC. However, an obstacle in the implementation of NGS sequencing for HCC tumors is that there is currently no clinical indication to biopsy an HCC tumor for diagnosis or treatment. This limits the systematic access to HCC tumor samples, and researchers who wish to study tumors must

Translational Implications of Heterogeneity in HCC

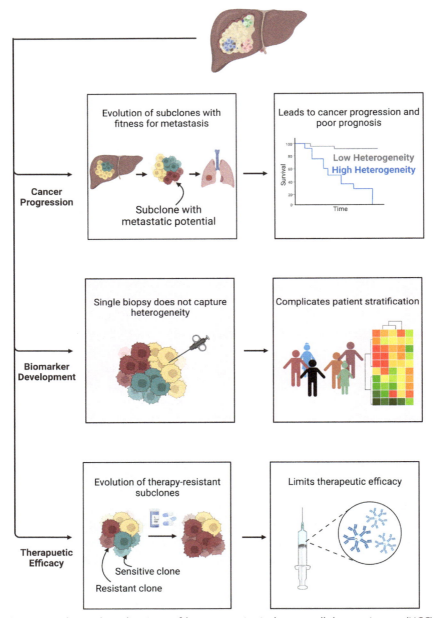

Fig. 3 Translational implications of heterogeneity in hepatocellular carcinoma (HCC). HCC heterogeneity promotes cancer progression—multiple subclones are present and the evolution of subclones with metastatic potential can lead to metastasis. Consequently, patients with high heterogeneity have a greater degree of cancer progression and poorer prognoses. Heterogeneity impedes biomarker development—a single tissue biopsy often doesn't capture tumor heterogeneity, and this limited insight into the tumor complicates patient stratification and biomarker development. Heterogeneity limits therapeutic efficacy—therapies can apply selective pressure on the tumor, leading to the evolution of resistant subclones.

continue collecting tumor samples through approved research protocols with patient consent. Liquid biopsies are a developing area of interest and are a non-invasive approach to studying and monitoring tumor cells in circulation. However, issues related to sensitivity, specificity, and validity in larger cohorts need to be addressed before liquid biopsy can be adopted in clinical practice.

NGS sequencing has paved the way for the development of several classification systems for HCC. Tumors have been classified based on driver mutations, pathway activation, and immune infiltration. However, a classification system is yet to be approved for use in clinical care, whether to predict prognosis or decide a treatment course. Additional validation of classification systems is awaited for these tools to be translated.

Expanding therapeutic options for advanced HCC and developing guidelines to select appropriate therapies for each patient's tumor biology remains an active area of research. Tumor heterogeneity is postulated to be a major reason for the failure of multiple therapies in HCC clinical trials. While there is no direct evidence to support this claim in HCC, experimental data have linked heterogeneity to drug resistance in multiple cancers (Dagogo-Jack & Shaw, 2018; Turner & Reis-Filho, 2012). Thus, there is a pressing need to explicitly study and account for tumor heterogeneity both in research and in clinical trials. For instance, research protocols that biopsy tumors before and after treatment could help identify tumor evolution and therapy resistance. Additionally, stratifying patients in clinical trials by genetic signatures or other biomarkers could help identify therapies that may provide benefit to patients with specific tumor biology, even if the average benefit across all patients is minimal.

Thus, single cell sequencing and spatial biology techniques have allowed the creation of high-resolution tumor landscapes that are ushering in a new era of personalized medicine. We believe that deciphering the molecular heterogeneity of HCC will ultimately lead to the discovery of novel biomarkers and therapeutic strategies, and thus improve the outcomes of patients with HCC.

Acknowledgements
All figures were created with BioRender.com.

Grant support
None.

Conflict of interest statement

A.S. and R.D. have no financial or personal disclosures relevant to the contents of this manuscript.

References

Abou-Alfa, G. K., & Venook, A. P. (2013). The antiangiogenic ceiling in hepatocellular carcinoma: does it exist and has it been reached? *The Lancet Oncology*, *14*(7), e283–e288.

Ahn, S.-M., Jang, S. J., Shim, J. H., Kim, D., Hong, S.-M., Sung, C. O., et al. (2014). Genomic portrait of resectable hepatocellular carcinomas: implications of RB1 and FGF19 aberrations for patient stratification. *Hepatology*, *60*(6), 1972–1982.

Alsina-Sanchis, E., Mülfarth, R., & Fischer, A. (2021). Control of Tumor Progression by Angiocrine Factors. *Cancers*, *13*(11).

Arao, T., Ueshima, K., Matsumoto, K., Nagai, T., Kimura, H., Hagiwara, S., et al. (2013). FGF3/FGF4 amplification and multiple lung metastases in responders to sorafenib in hepatocellular carcinoma. *Hepatology*, *57*(4), 1407–1415.

Asakawa, M., Itoh, M., Suganami, T., Sakai, T., Kanai, S., Shirakawa, I., et al. (2019). Upregulation of cancer-associated gene expression in activated fibroblasts in a mouse model of non-alcoholic steatohepatitis. *Scientific Reports*, *9*(1), 19601.

Barcena-Varela, M., & Lujambio, A. (2021). The Endless Sources of Hepatocellular Carcinoma Heterogeneity. *Cancers*, *13*(11).

Biffi, G., Oni, T. E., Spielman, B., Hao, Y., Elyada, E., Park, Y., et al. (2019). IL1-Induced JAK/STAT Signaling Is Antagonized by TGFβ to Shape CAF Heterogeneity in Pancreatic Ductal Adenocarcinoma. *Cancer Discovery*, *9*(2), 282–301.

Blumer, T., Fofana, I., Matter, M. S., Wang, X., Montazeri, H., Calabrese, D., et al. (2019). Hepatocellular Carcinoma Xenografts Established From Needle Biopsies Preserve the Characteristics of the Originating Tumors. *Hepatology Communications*, *3*(7), 971–986.

Boyault, S., Rickman, D. S., de Reyniès, A., Balabaud, C., Rebouissou, S., Jeannot, E., et al. (2007). Transcriptome classification of HCC is related to gene alterations and to new therapeutic targets. *Hepatology*, *45*(1), 42–52.

Brady, L., Kriner, M., Coleman, I., Morrissey, C., Roudier, M., True, L. D., et al. (2021). Inter- and intra-tumor heterogeneity of metastatic prostate cancer determined by digital spatial gene expression profiling. *Nature Communications*, *12*(1), 1426.

Brown, Z. J., Yu, S. J., Heinrich, B., Ma, C., Fu, Q., Sandhu, M., et al. (2018). Indoleamine 2,3-dioxygenase provides adaptive resistance to immune checkpoint inhibitors in hepatocellular carcinoma. *Cancer Immunology, Immunotherapy: CII*, *67*(8), 1305–1315.

Cai, Z.-X., Chen, G., Zeng, Y.-Y., Dong, X.-Q., Lin, M.-J., Huang, X.-H., et al. (2017). Circulating tumor DNA profiling reveals clonal evolution and real-time disease progression in advanced hepatocellular carcinoma. *International Journal of Cancer. Journal International Du Cancer*, *141*(5), 977–985.

Calderaro, J., Couchy, G., Imbeaud, S., Amaddeo, G., Letouzé, E., Blanc, J.-F., et al. (2017). Histological subtypes of hepatocellular carcinoma are related to gene mutations and molecular tumour classification. *Journal of Hepatology*, *67*(4), 727–738.

Calderaro, J., Petitprez, F., Becht, E., Laurent, A., Hirsch, T. Z., Rousseau, B., et al. (2019). Intra-tumoral tertiary lymphoid structures are associated with a low risk of early recurrence of hepatocellular carcinoma. *Journal of Hepatology*, *70*(1), 58–65.

Calvisi, D. F., Ladu, S., Gorden, A., Farina, M., Lee, J.-S., Conner, E. A., et al. (2007). Mechanistic and prognostic significance of aberrant methylation in the molecular pathogenesis of human hepatocellular carcinoma. *The Journal of Clinical Investigation*, *117*(9), 2713–2722.

Chan, L.-K., Tsui, Y.-M., Ho, D. W.-H., & Ng, I. O.-L. (2021). Cellular heterogeneity and plasticity in liver cancer. *Seminars in Cancer Biology*.

Chen, H., Chong, W., Teng, C., Yao, Y., Wang, X., & Li, X. (2019). The immune response-related mutational signatures and driver genes in non-small-cell lung cancer. *Cancer Science, 110*(8), 2348–2356.

Chen, F., Zhong, Z., Tan, H.-Y., Wang, N., & Feng, Y. (2020). The significance of circulating tumor cells in patients with hepatocellular carcinoma: Real-time monitoring and moving targets for cancer therapy. *Cancers, 12*(7).

Cheng, Y., Li, H., Deng, Y., Tai, Y., Zeng, K., Zhang, Y., et al. (2018). Cancer-associated fibroblasts induce PDL1+ neutrophils through the IL6-STAT3 pathway that foster immune suppression in hepatocellular carcinoma. *Cell Death & Disease, 9*(4), 422.

Chiang, D. Y., Villanueva, A., Hoshida, Y., Peix, J., Newell, P., Minguez, B., et al. (2008). Focal gains of VEGFA and molecular classification of hepatocellular carcinoma. *Cancer Research, 68*(16), 6779–6788.

Dagogo-Jack, I., & Shaw, A. T. (2018). Tumour heterogeneity and resistance to cancer therapies. *Nature Reviews. Clinical Oncology, 15*(2), 81–94.

Dasgupta, P., Henshaw, C., Youlden, D., Clark, P., Aitken, J., & Baade, P. (2020). Global trends in incidence rates of primary adult liver cancers: A systematic review and meta-analysis. *Frontiers in Oncology, 10*.

Désert, R., Rohart, F., Canal, F., Sicard, M., Desille, M., Renaud, S., et al. (2017). Human hepatocellular carcinomas with a periportal phenotype have the lowest potential for early recurrence after curative resection. *Hepatology, 66*(5), 1502–1518.

Dhanasekaran, R. (2021). Deciphering tumor heterogeneity in hepatocellular carcinoma (HCC)-multi-omic and singulomic approaches. *Seminars in Liver Disease, 41*(1), 9–18.

Dhanasekaran, R., Bandoh, S., & Roberts, L. R. (2016). Molecular pathogenesis of hepatocellular carcinoma and impact of therapeutic advances. *F1000Research, 5*.

Ding, X., He, M., Chan, A. W. H., Song, Q. X., Sze, S. C., Chen, H., et al. (2019). Genomic and epigenomic features of primary and recurrent hepatocellular carcinomas. *Gastroenterology, 157*(6). 1630–1645.e6.

El-Khoueiry, A. B., Sangro, B., Yau, T., Crocenzi, T. S., Kudo, M., Hsu, C., et al. (2017). Nivolumab in patients with advanced hepatocellular carcinoma (CheckMate 040): an open-label, non-comparative, phase 1/2 dose escalation and expansion trial. *The Lancet, 389*(10088), 2492–2502.

Fang, J.-H., Xu, L., Shang, L.-R., Pan, C.-Z., Ding, J., Tang, Y.-Q., et al. (2019). Vessels that encapsulate tumor clusters (VETC) pattern is a predictor of sorafenib benefit in patients with hepatocellular carcinoma. *Hepatology, 70*(3), 824–839.

Fang, J.-H., Zhou, H.-C., Zhang, C., Shang, L.-R., Zhang, L., Xu, J., et al. (2015). A novel vascular pattern promotes metastasis of hepatocellular carcinoma in an epithelial-mesenchymal transition-independent manner. *Hepatology, 62*(2), 452–465.

Finn, R. S., Qin, S., Ikeda, M., Galle, P. R., Ducreux, M., Kim, T.-Y., et al. (2020). Atezolizumab plus Bevacizumab in Unresectable Hepatocellular Carcinoma. *The New England Journal of Medicine, 382*(20), 1894–1905.

Finn, R. S., Ryoo, B.-Y., Merle, P., Kudo, M., Bouattour, M., Lim, H. Y., et al. (2020). Pembrolizumab as second-line therapy in patients with advanced hepatocellular carcinoma in KEYNOTE-240: A randomized, double-blind, phase III trial. *Journal of Clinical Oncology: Official Journal of the American Society of Clinical Oncology, 38*(3), 193–202.

Forschner, A., Battke, F., Hadaschik, D., Schulze, M., Weißgraeber, S., Han, C.-T., et al. (2019). Tumor mutation burden and circulating tumor DNA in combined CTLA-4 and PD-1 antibody therapy in metastatic melanoma - results of a prospective biomarker study. *Journal for Immunotherapy of Cancer, 7*(1), 180.

Fujimoto, A., Totoki, Y., Abe, T., Boroevich, K. A., Hosoda, F., Nguyen, H. H., et al. (2012). Whole-genome sequencing of liver cancers identifies etiological influences on mutation patterns and recurrent mutations in chromatin regulators. *Nature Genetics*, *44*(7), 760–764.

Fujiwara, N., Friedman, S., Goossens, N., & Hoshida, Y. (2018). Risk factors and prevention of hepatocellular carcinoma in the era of precision medicine. *Journal of Hepatology*, *68*(3), 526–549.

Gao, Q., Wang, Z.-C., Duan, M., Lin, Y.-H., Zhou, X.-Y., Worthley, D. L., et al. (2017). Cell culture system for analysis of genetic heterogeneity within hepatocellular carcinomas and response to pharmacologic agents. *Gastroenterology*, *152*(1). 232–242.e4.

Garnelo, M., Tan, A., Her, Z., Yeong, J., Lim, C. J., Chen, J., et al. (2017). Interaction between tumour-infiltrating B cells and T cells controls the progression of hepatocellular carcinoma. *Gut*, *66*(2), 342–351.

Ghouri, Y. A., Mian, I., & Rowe, J. H. (2017). Review of hepatocellular carcinoma: Epidemiology, etiology, and carcinogenesis. *Journal of Carcinogenesis*, *16*(1), 1.

Guichard, C., Amaddeo, G., Imbeaud, S., Ladeiro, Y., Pelletier, L., Maad, I. B., et al. (2012). Integrated analysis of somatic mutations and focal copy-number changes identifies key genes and pathways in hepatocellular carcinoma. *Nature Genetics*, *44*(6), 694–698.

Guo, L., Yi, X., Chen, L., Zhang, T., Guo, H., Chen, Z., et al. (2022). Single-cell DNA sequencing reveals punctuated and gradual clonal evolution in hepatocellular carcinoma. *Gastroenterology*, *162*(1), 238–252.

Harding, J. J., El Dika, I., & Abou-Alfa, G. K. (2016). Immunotherapy in hepatocellular carcinoma: Primed to make a difference? *Cancer*, *122*(3), 367–377.

Harding, J. J., Nandakumar, S., Armenia, J., Khalil, D. N., Albano, M., Ly, M., et al. (2019). Prospective genotyping of hepatocellular carcinoma: Clinical implications of next-generation sequencing for matching patients to targeted and immune therapies. *Clinical Cancer Research: An Official Journal of the American Association for Cancer Research*, *25*(7), 2116–2126.

Ho, D. W. H., Chan, L. K., Chiu, Y. T., Xu, I. M. J., Poon, R. T. P., Cheung, T. T., et al. (2017). TSC1/2 mutations define a molecular subset of HCC with aggressive behaviour and treatment implication. *Gut*, *66*(8), 1496–1506.

Horwitz, E., Stein, I., Andreozzi, M., Nemeth, J., Shoham, A., Pappo, O., et al. (2014). Human and mouse VEGFA-amplified hepatocellular carcinomas are highly sensitive to sorafenib treatment. *Cancer Discovery*, *4*(6), 730–743.

Hoshida, Y., Nijman, S. M. B., Kobayashi, M., Chan, J. A., Brunet, J.-P., Chiang, D. Y., et al. (2009). Integrative transcriptome analysis reveals common molecular subclasses of human hepatocellular carcinoma. *Cancer Research*, *69*(18), 7385–7392.

Hoshida, Y., Toffanin, S., Lachenmayer, A., Villanueva, A., Minguez, B., & Llovet, J. M. (2010). Molecular classification and novel targets in hepatocellular carcinoma: recent advancements. *Seminars in Liver Disease*, *30*(1), 35–51.

Hu, B., Li, H., Guo, W., Sun, Y.-F., Zhang, X., Tang, W.-G., et al. (2020). Establishment of a hepatocellular carcinoma patient-derived xenograft platform and its application in biomarker identification. *International Journal of Cancer. Journal International Du Cancer*, *146*(6), 1606–1617.

Hung, M. H., Lee, J. S., Ma, C., Diggs, L. P., Heinrich, S., Chang, C. W., et al. (2021). Tumor methionine metabolism drives T-cell exhaustion in hepatocellular carcinoma. *Nature Communications*, *12*(1), 1455.

Jiang, Y., Han, Q., Zhao, H., & Zhang, J. (2021). The mechanisms of HBV-induced hepatocellular carcinoma. *Journal of Hepatocellular Carcinoma*, *8*, 435–450.

Jiang, Z., Jiang, X., Chen, S., Lai, Y., Wei, X., Li, B., et al. (2016). Anti-GPC3-CAR T cells suppress the growth of tumor cells in patient-derived xenografts of hepatocellular carcinoma. *Frontiers in Immunology*, *7*, 690.

Kaseb, A. O., Vence, L., Blando, J., Yadav, S. S., Ikoma, N., Pestana, R. C., et al. (2019). Immunologic correlates of pathologic complete response to preoperative immunotherapy in hepatocellular carcinoma. *Cancer Immunology Research*, 7(9), 1390–1395.

Khatib, S., Pomyen, Y., Dang, H., & Wang, X. W. (2020). Understanding the cause and consequence of tumor heterogeneity. *Trends in Cancer Research*, 6(4), 267–271.

Kim, S. S., Eun, J. W., Choi, J.-H., Woo, H. G., Cho, H. J., Ahn, H. R., et al. (2020). MLH1 single-nucleotide variant in circulating tumor DNA predicts overall survival of patients with hepatocellular carcinoma. *Scientific Reports*, 10(1), 17862.

Kimura, T., Kato, Y., Ozawa, Y., Kodama, K., Ito, J., Ichikawa, K., et al. (2018). Immunomodulatory activity of lenvatinib contributes to antitumor activity in the Hepa1-6 hepatocellular carcinoma model. *Cancer Science*, 109(12), 3993–4002.

Kudo, M., Finn, R. S., Qin, S., Han, K.-H., Ikeda, K., Piscaglia, F., et al. (2018). Lenvatinib versus sorafenib in first-line treatment of patients with unresectable hepatocellular carcinoma: A randomised phase 3 non-inferiority trial. *The Lancet*, 391(10126), 1163–1173.

Kurebayashi, Y., Ojima, H., Tsujikawa, H., Kubota, N., Maehara, J., Abe, Y., et al. (2018). Landscape of immune microenvironment in hepatocellular carcinoma and its additional impact on histological and molecular classification. *Hepatology*, 68(3), 1025–1041.

Lachenmayer, A., Alsinet, C., Savic, R., Cabellos, L., Toffanin, S., Hoshida, Y., et al. (2012). Wnt-pathway activation in two molecular classes of hepatocellular carcinoma and experimental modulation by sorafenib. *Clinical Cancer Research: An Official Journal of the American Association for Cancer Research*, 18(18), 4997–5007.

Lee, J.-S., Chu, I.-S., Heo, J., Calvisi, D. F., Sun, Z., Roskams, T., et al. (2004). Classification and prediction of survival in hepatocellular carcinoma by gene expression profiling. *Hepatology*, 40(3), 667–676.

Lee, M., Samstein, R. M., Valero, C., Chan, T. A., & Morris, L. G. T. (2020). Tumor mutational burden as a predictive biomarker for checkpoint inhibitor immunotherapy. *Human Vaccines & Immunotherapeutics*, 16(1), 112–115.

Li, J., Chen, L., Zhang, X., Zhang, Y., Liu, H., Sun, B., et al. (2014). Detection of circulating tumor cells in hepatocellular carcinoma using antibodies against asialoglycoprotein receptor, carbamoyl phosphate synthetase 1 and pan-cytokeratin. *PLoS One*, 9(4), e96185.

Ling, S., Hu, Z., Yang, Z., Yang, F., Li, Y., Lin, P., et al. (2015). Extremely high genetic diversity in a single tumor points to prevalence of non-Darwinian cell evolution. *Proceedings of the National Academy of Sciences of the United States of America*, 112(47), E6496–E6505.

Liu, J., Chen, S., Wang, W., Ning, B.-F., Chen, F., Shen, W., et al. (2016). Cancer-associated fibroblasts promote hepatocellular carcinoma metastasis through chemokine-activated hedgehog and TGF-β pathways. *Cancer Letters*, 379(1), 49–59.

Llovet, J. M., Kelley, R. K., Villanueva, A., Singal, A. G., Pikarsky, E., Roayaie, S., et al. (2021). Hepatocellular carcinoma. *Nature Reviews. Disease Primers*, 7(1), 6.

Llovet, J. M., Ricci, S., Mazzaferro, V., Hilgard, P., Gane, E., Blanc, J.-F., et al. (2008). Sorafenib in advanced hepatocellular carcinoma. *The New England Journal of Medicine*, 359(4), 378–390.

Lohr, J. G., Stojanov, P., Carter, S. L., Cruz-Gordillo, P., Lawrence, M. S., Auclair, D., et al. (2014). Widespread genetic heterogeneity in multiple myeloma: implications for targeted therapy. *Cancer Cell*, 25(1), 91–101.

Longo, S. K., Guo, M. G., Ji, A. L., & Khavari, P. A. (2021). Integrating single-cell and spatial transcriptomics to elucidate intercellular tissue dynamics. *Nature Reviews. Genetics*, 22(10), 627–644.

Losic, B., Craig, A. J., Villacorta-Martin, C., Martins-Filho, S. N., Akers, N., Chen, X., et al. (2020). Intratumoral heterogeneity and clonal evolution in liver cancer. *Nature Communications*, *11*(1), 291.

Maishi, N., Ohba, Y., Akiyama, K., Ohga, N., Hamada, J.-I., Nagao-Kitamoto, H., et al. (2016). Tumour endothelial cells in high metastatic tumours promote metastasis via epigenetic dysregulation of biglycan. *Scientific Reports*, *6*, 28039.

Maravelia, P., Silva, D. N., Rovesti, G., Chrobok, M., Stål, P., Lu, Y.-C., et al. (2021). Liquid biopsy in hepatocellular carcinoma: Opportunities and challenges for immunotherapy. *Cancers*, *13*(17).

McCoy, M. G., Nyanyo, D., Hung, C. K., Goerger, J. P., R Zipfel, W., Williams, R. M., et al. (2019). Endothelial cells promote 3D invasion of GBM by IL-8-dependent induction of cancer stem cell properties. *Scientific Reports*, *9*(1), 9069.

McGranahan, N., Favero, F., de Bruin, E. C., Birkbak, N. J., Szallasi, Z., & Swanton, C. (2015). Clonal status of actionable driver events and the timing of mutational processes in cancer evolution. *Science Translational Medicine*, *7*(283), 283ra54.

McQuerry, J. A., Chang, J. T., Bowtell, D. D. L., Cohen, A., & Bild, A. H. (2017). Mechanisms and clinical implications of tumor heterogeneity and convergence on recurrent phenotypes. *Journal of Molecular Medicine*, *95*(11), 1167–1178.

Mocan, T., Simão, A. L., Castro, R. E., Rodrigues, C. M. P., Słomka, A., Wang, B., et al. (2020). Liquid biopsies in hepatocellular carcinoma: Are we winning? *Journal of Clinical Medicine Research*, *9*(5).

Nault, J. C., Mallet, M., Pilati, C., Calderaro, J., Bioulac-Sage, P., Laurent, C., et al. (2013). High frequency of telomerase reverse-transcriptase promoter somatic mutations in hepatocellular carcinoma and preneoplastic lesions. *Nature Communications*, *4*, 2218.

Nault, J.-C., Martin, Y., Caruso, S., Hirsch, T. Z., Bayard, Q., Calderaro, J., et al. (2020). Clinical Impact of Genomic Diversity From Early to Advanced Hepatocellular Carcinoma. *Hepatology*, *71*(1), 164–182.

Noman, M. Z., Desantis, G., Janji, B., Hasmim, M., Karray, S., Dessen, P., et al. (2014). PD-L1 is a novel direct target of HIF-1α, and its blockade under hypoxia enhanced MDSC-mediated T cell activation. *The Journal of Experimental Medicine*, *211*(5), 781–790.

Öhlund, D., Handly-Santana, A., Biffi, G., Elyada, E., Almeida, A. S., Ponz-Sarvise, M., et al. (2017). Distinct populations of inflammatory fibroblasts and myofibroblasts in pancreatic cancer. *The Journal of Experimental Medicine*, *214*(3), 579–596.

Peng, S., Wang, Y., Peng, H., Chen, D., Shen, S., Peng, B., et al. (2014). Autocrine vascular endothelial growth factor signaling promotes cell proliferation and modulates sorafenib treatment efficacy in hepatocellular carcinoma. *Hepatology*, *60*(4), 1264–1277.

Petrick, J. L., Florio, A. A., Znaor, A., Ruggieri, D., Laversanne, M., Alvarez, C. S., et al. (2020). International trends in hepatocellular carcinoma incidence, 1978–2012. *International Journal of Cancer. Journal International Du Cancer*, *147*(2), 317–330.

Plaks, V., Koopman, C. D., & Werb, Z. (2013). Cancer. Circulating tumor cells. *Science*, *341*(6151), 1186–1188.

Rebouissou, S., & Nault, J.-C. (2020). Advances in molecular classification and precision oncology in hepatocellular carcinoma. *Journal of Hepatology*, *72*(2), 215–229.

Reck, M., Mok, T. S. K., Nishio, M., Jotte, R. M., Cappuzzo, F., Orlandi, F., et al. (2019). Atezolizumab plus bevacizumab and chemotherapy in non-small-cell lung cancer (IMpower150): key subgroup analyses of patients with EGFR mutations or baseline liver metastases in a randomised, open-label phase 3 trial. *The Lancet. Respiratory Medicine*, *7*(5), 387–401.

Renne, S. L., Woo, H. Y., Allegra, S., Rudini, N., Yano, H., Donadon, M., et al. (2020). Vessels encapsulating tumor clusters (VETC) is a powerful predictor of aggressive hepatocellular carcinoma. *Hepatology*, *71*(1), 183–195.

Rini, B. I., Powles, T., Atkins, M. B., Escudier, B., McDermott, D. F., Suarez, C., et al. (2019). Atezolizumab plus bevacizumab versus sunitinib in patients with previously untreated metastatic renal cell carcinoma (IMmotion151): a multicentre, open-label, phase 3, randomised controlled trial. *The Lancet*, *393*(10189), 2404–2415.

Rudalska, R., Dauch, D., Longerich, T., McJunkin, K., Wuestefeld, T., Kang, T.-W., et al. (2014). In vivo RNAi screening identifies a mechanism of sorafenib resistance in liver cancer. *Nature Medicine*, *20*(10), 1138–1146.

Sangro, B., Melero, I., Yau, T., Hsu, C., Kudo, M., Kim, T.-Y., et al. (2017). *Nivolumab in sorafenib-naive and-experienced patients with advanced hepatocellular carcinoma (HCC): Survival, hepatic safety, and biomarker assessments in CheckMate 040*.

Sangro, B., Sarobe, P., Hervás-Stubbs, S., & Melero, I. (2021). Advances in immunotherapy for hepatocellular carcinoma. *Nature Reviews. Gastroenterology & Hepatology*, *18*(8), 525–543.

Schulze, K., Imbeaud, S., Letouzé, E., Alexandrov, L. B., Calderaro, J., Rebouissou, S., et al. (2015). Exome sequencing of hepatocellular carcinomas identifies new mutational signatures and potential therapeutic targets. *Nature Genetics*, *47*(5), 505–511.

Sebastian, A., Hum, N. R., Martin, K. A., Gilmore, S. F., Peran, I., Byers, S. W., et al. (2020). Single-cell transcriptomic analysis of tumor-derived fibroblasts and normal tissue-resident fibroblasts reveals fibroblast heterogeneity in breast cancer. *Cancers*, *12*(5).

Sharma, A., Seow, J. J. W., Dutertre, C.-A., Pai, R., Blériot, C., Mishra, A., et al. (2020). Onco-fetal reprogramming of endothelial cells drives immunosuppressive macrophages in hepatocellular carcinoma. *Cell*, *183*(2). 377–394.e21.

Sia, D., Jiao, Y., Martinez-Quetglas, I., Kuchuk, O., Villacorta-Martin, C., Castro de Moura, M., et al. (2017). Identification of an immune-specific class of hepatocellular carcinoma, based on molecular features. *Gastroenterology*, *153*(3), 812–826.

Siegel, R. L., Miller, K. D., & Jemal, A. (2020). Cancer statistics, 2020. *CA: a Cancer Journal for Clinicians*, *70*(1), 7–30.

Siu, E. H.-L., Chan, A. W.-H., Chong, C. C.-N., Chan, S. L., Lo, K.-W., & Cheung, S. T. (2018). Treatment of advanced hepatocellular carcinoma: immunotherapy from checkpoint blockade to potential of cellular treatment. *Translational Gastroenterology and Hepatology*, *3*, 89.

Song, G., Shi, Y., Zhang, M., Goswami, S., Afridi, S., Meng, L., et al. (2020). Global immune characterization of HBV/HCV-related hepatocellular carcinoma identifies macrophage and T-cell subsets associated with disease progression. *Cell Discovery*, *6*(1), 90.

Sukowati, C. H. C., El-Khobar, K. E., & Tiribelli, C. (2021). Immunotherapy against programmed death-1/programmed death ligand 1 in hepatocellular carcinoma: Importance of molecular variations, cellular heterogeneity, and cancer stem cells. *World Journal of Stem Cells*, *13*(7), 795–824.

Sun, Y.-F., Guo, W., Xu, Y., Shi, Y.-H., Gong, Z.-J., Ji, Y., et al. (2018). Circulating tumor cells from different vascular sites exhibit spatial heterogeneity in epithelial and mesenchymal composition and distinct clinical significance in hepatocellular carcinoma. *Clinical Cancer Research: An Official Journal of the American Association for Cancer Research*, *24*(3), 547–559.

Sun, B., & Karin, M. (2012). Obesity, inflammation, and liver cancer. *Journal of Hepatology*, *56*(3), 704–713.

Sung, H., Ferlay, J., Siegel, R. L., Laversanne, M., Soerjomataram, I., Jemal, A., et al. (2021). Global Cancer Statistics 2020: GLOBOCAN estimates of incidence and mortality worldwide for 36 cancers in 185 countries. *CA: a Cancer Journal for Clinicians*, *71*(3), 209–249.

Swanton, C. (2012). Intratumor heterogeneity: Evolution through space and time. *Cancer Research*, *72*(19), 4875–4882.

Teufel, M., Seidel, H., Köchert, K., Meinhardt, G., Finn, R. S., Llovet, J. M., et al. (2019). Biomarkers associated with response to regorafenib in patients with hepatocellular carcinoma. *Gastroenterology*, *156*(6), 1731–1741.

The Cancer Genome Atlas Research Network. (2017). Comprehensive and Integrative Genomic Characterization of Hepatocellular Carcinoma. *Cell*, *169*(7), 1327–1341.

Thomson, A. W., & Knolle, P. A. (2010). Antigen-presenting cell function in the tolerogenic liver environment. *Nature Reviews. Immunology*, *10*(11), 753–766.

Thun, M., Linet, M. S., Cerhan, J. R., Haiman, C. A., & Schottenfeld, D. (2017). *Cancer Epidemiology and Prevention*. Oxford University Press.

Totoki, Y., Tatsuno, K., Covington, K. R., Ueda, H., Creighton, C. J., Kato, M., et al. (2014). Trans-ancestry mutational landscape of hepatocellular carcinoma genomes. *Nature Genetics*, *46*(12), 1267–1273.

Turner, N. C., & Reis-Filho, J. S. (2012). Genetic heterogeneity and cancer drug resistance. *The Lancet Oncology*, *13*(4), e178–e185.

Tuveson, D., & Clevers, H. (2019). Cancer modeling meets human organoid technology. *Science*, *364*(6444), 952–955.

Villanueva, A. (2019). Hepatocellular Carcinoma. *The New England Journal of Medicine*, *380*, 1450–1462.

World Health Organization. (2018). *Noncommunicable diseases country profiles 2018*. https://apps.who.int/iris/bitstream/handle/10665/274512/9789241514620-eng.pdf.

Wu, R., Guo, W., Qiu, X., Wang, S., Sui, C., Lian, Q., et al. (2021). Comprehensive analysis of spatial architecture in primary liver cancer. *Science Advances*, *7*(51). eabg3750.

Wu, C., Lin, J., Weng, Y., Zeng, D.-N., Xu, J., Luo, S., et al. (2020). Myeloid signature reveals immune contexture and predicts the prognosis of hepatocellular carcinoma. *The Journal of Clinical Investigation*, *130*(9), 4679–4693.

Xu, X., Tao, Y., Shan, L., Chen, R., Jiang, H., Qian, Z., et al. (2018). The role of MicroRNAs in hepatocellular carcinoma. *Journal of Cancer*, *9*(19), 3557–3569.

Xue, R., Li, R., Guo, H., Guo, L., Su, Z., Ni, X., et al. (2016). Variable intra-tumor genomic heterogeneity of multiple lesions in patients with hepatocellular carcinoma. *Gastroenterology*, *150*(4), 998–1008.

Yang, J. D., Hainaut, P., Gores, G. J., Amadou, A., Plymoth, A., & Roberts, L. R. (2019). A global view of hepatocellular carcinoma: Trends, risk, prevention and management. *Nature Reviews Gastroenterology & Hepatology*, *16*, 589–604.

Yin, Z., Dong, C., Jiang, K., Xu, Z., Li, R., Guo, K., et al. (2019). Heterogeneity of cancer-associated fibroblasts and roles in the progression, prognosis, and therapy of hepatocellular carcinoma. *Journal of Hematology & Oncology*, *12*(1), 101.

Zhai, W., Lim, T. K.-H., Zhang, T., Phang, S.-T., Tiang, Z., Guan, P., et al. (2017). The spatial organization of intra-tumour heterogeneity and evolutionary trajectories of metastases in hepatocellular carcinoma. *Nature Communications*, *8*, 4565.

Zhang, Q., He, Y., Luo, N., Patel, S. J., Han, Y., Goa, R., et al. (2019). Landscape and dynamics of single immune cells in hepatocellular carcinoma. *Cell*, *179*(4), 829–845.

Zhang, Q., Lou, Y., Yang, J., Wang, J., Feng, J., Zhao, Y., et al. (2019). Integrated multiomic analysis reveals comprehensive tumour heterogeneity and novel immunophenotypic classification in hepatocellular carcinomas. *Gut*, *68*(11), 2019–2031.

Zhang, Y., Liu, Z., Ji, K., Li, X., Wang, C., Ren, Z., et al. (2021). Clinical application value of circulating cell-free DNA in hepatocellular carcinoma. *Frontiers in Molecular Biosciences*, *8*, 736330.

Zhao, Y., Li, Z.-X., Zhu, Y.-J., Fu, J., Zhao, X.-F., Zhang, Y.-N., et al. (2021). Single-cell transcriptome analysis uncovers intratumoral heterogeneity and underlying mechanisms for drug resistance in hepatobiliary tumor organoids. *Advancement of Science*, *8*(11), e2003897.

Zhu, A. X., Finn, R. S., Edeline, J., Cattan, S., Ogasawara, S., Palmer, D., et al. (2018). Pembrolizumab in patients with advanced hepatocellular carcinoma previously treated with sorafenib (KEYNOTE-224): a non-randomised, open-label phase 2 trial. *The Lancet Oncology, 19*(7), 940–952.

Zhu, A. X., Kang, Y.-K., Yen, C.-J., Finn, R. S., Galle, P. R., Llovet, J. M., et al. (2019). Ramucirumab after sorafenib in patients with advanced hepatocellular carcinoma and increased α-fetoprotein concentrations (REACH-2): a randomised, double-blind, placebo-controlled, phase 3 trial. *The Lancet Oncology, 20*(2), 282–296.

Zhu, J., Zhang, T., Li, J., Lin, J., Liang, W., Huang, W., et al. (2019). Association between tumor mutation burden (TMB) and outcomes of cancer patients treated with PD-1/PD-L1 inhibitions: A meta-analysis. *Frontiers in Pharmacology, 10*, 673.

CHAPTER FIVE

Understanding the genetic basis for cholangiocarcinoma

Mikayla A. Schmidt* and Lewis R. Roberts
Division of Gastroenterology and Hepatology, Mayo Clinic College of Medicine and Science, Rochester, MN, United States
*Corresponding author: e-mail address: schmidt.mikayla2@mayo.edu

Contents

1. Introduction	138
2. Germline mutations in risk factors associated with cholangiocarcinoma	141
2.1 Primary sclerosing cholangitis	142
2.2 Obesity	143
2.3 Diabetes	144
2.4 Gallstones	145
2.5 NAFLD	145
3. Overview of germline mutations and variants associated with cholangiocarcinoma	146
4. Key candidate gene studies of CCA risk variants	147
5. Pathogenic germline alteration (PGA) sequencing studies	154
6. Conclusion	160
Conflict of interest statement	160
Grant support	160
References	160

Abstract

Cholangiocarcinoma is associated with several different risk factors, many of which have known genetic associations. Advances in our understanding of the human genome have translated to the development of gene specific and whole genome assays for identifying gene variants and other alterations associated with cancer development. An improved understanding of the inherited genetic variants associated with risk of cholangiocarcinoma has the potential to improve our understanding of the basic biology of cholangiocarcinoma, enhance the performance of risk stratification models for identifying individuals at highest risk for cholangiocarcinoma, and identifying genetic variants associated with predisposition to cholangiocarcinoma in families with multiple affected individuals. It is increasingly recognized that major cancer-causing mutations or other gene alterations associated with familial risk of multiple cancers can also occur as germline events in individuals with apparently sporadically occurring cancer. In this chapter we review the major risk factors for cholangiocarcinoma as well as known gene variants associated with these risk factors, gene variants that have been associated with cholangiocarcinoma as the result of interrogation of candidate genes known to be associated with putative cancer causing pathways in cholangiocarcinoma, as well as the prevalence of major cancer causing genetic aberrations shown to be inherited in the germline of patients with sporadically developed cholangiocarcinoma. There has not yet been any large-scale genome wide association study of cholangiocarcinoma, and the results from such a study are eagerly anticipated.

Abbreviations

α1AT	alpha-1-antitypsin
APC	adenomatous polyposis coli
ATM	ataxia-telangiectasia kinase
BER	base excision repair pathway
BAP1	BRCA1-associated protein 1
cMOAT	canalicular multispecific organic anion transporter 1
CCA	cholangiocarcinoma
dCCA	distal cholangiocarcinoma
FH	fumarate hydratase
GWAS	genome-wide association study
HBV	hepatitis B infection
HCV	hepatitis C
iCCA	intrahepatic cholangiocarcinoma
IL-6	interleukin 6
MICA	major histocompatibility complex class I chain-related molecule A
MIFT	microphthalmia inducing transcription factor
MRP2	multidrug resistance-associated protein 2
MST1	macrophage stimulation 1
MUTYH	mutY DNA glycosylase
NAFLD	non-alcoholic fatty liver disease
NASH	non-alcoholic steatohepatitis
NAT2	N-Acetyltransferase 2
OGG1	8-oxoguanine glycosylase 1
pCCA	perihilar/hilar cholangiocarcinoma
PALB2	partner and localizer of BRCA2
PGAs	pathogenic germline alterations
PPARG	peroxisome proliferator-activated receptor gamma
PSC	primary sclerosing cholangitis
PTGS2	prostaglandin-endoperoxide synthase 2

1. Introduction

Cholangiocarcinoma (CCA) is a rare malignancy occurring in the epithelial cells of the biliary tree. It is generally diagnosed at a late stage, and consequently has a poor prognosis. For people diagnosed with CCA, the median survival is currently 24 months and only 10% survive 5 years post diagnosis (Rizvi & Gores, 2013). CCA is the second most common primary malignancy of the liver following hepatocellular carcinoma and accounts for 10–25% of primary hepatic malignancies (Gatto et al., 2010) and approximately 3% of all gastrointestinal cancers (Shaib & El-Serag, 2004). CCA is classified anatomically into three sub-groups, designated as intrahepatic cholangiocarcinoma (iCCA), perihilar/hilar cholangiocarcinoma (pCCA), and distal cholangiocarcinoma (dCCA). Perihilar/hilar CCA and distal CCA are sometimes grouped together as extrahepatic (eCCA). It is typically reported that 60% of cholangiocarcinoma cases are perihilar, 30% are distal and 10% are intrahepatic (Khan & Dageforde, 2019) however, recent population-based studies suggest that the breakdown is closer to 40% iCCA, 40% pCCA, and 20% dCCA (Yang et al., 2012). CCA is 1.5 times more likely to affect men than women and the onset is uncommon before age 40, with CCA being most commonly diagnosed in the seventh decade. The incidence of CCA is twice as high in Asians as in whites or blacks. Due to differences in diagnostic coding and approaches to classification of CCA, global epidemiological data is not completely comparable or reliable. Global geographic differences in CCA rates are apparent, with higher incidences of CCA in Eastern parts of the world and less frequent in Western countries (Khan, Tavolari, & Brandi, 2019). The overall age-standardized incidence rate in the United States is reported as 1.4 persons per 100,000, but some smaller, more detailed studies suggest the rates may be higher (Yang et al., 2012). The countries with the lowest incidences on CCA are Spain (0.5/100,000), Switzerland (0.5/100,000), Canada (0.4/100,000), Australia (0.4/100,000), New Zealand (0.34/100,000), Puerto Rico (0.4/100,000), Costa Rica (0.3/100,000) and Israel (0.3/100,000). Northeast Thailand has the highest incidence of CCA, with 85 cases per 100,000, while the incidence is much lower in North and Central Thailand at 14.5 cases per 100,000. South Thailand has even lower rates at 5.7 cases per 100,000. Variations of incidence within subregions of countries also occur, likely a result of varying environmental risk factors (Khan, Toledano, & Taylor-Robinson, 2008). A number of genetic studies have been performed, associating some genetic variants with the pathogenesis of CCA.

Known risk factors account for less than 50% of CCA cases in Western countries, with most being of unknown etiology. Risk factors for CCA also vary geographically. Identified risk factors include liver fluke infestation, chronic viral hepatitis, primary sclerosing cholangitis, obesity (Li et al., 2014), diabetes (type I and type II), [6], gallstones, and non-alcoholic fatty liver disease (Wongjarupong et al., 2017). These risk factors are highlighted in Fig. 1.

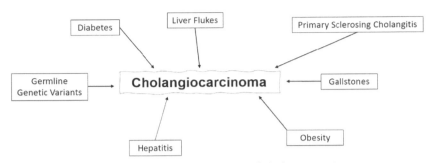

Fig. 1 Risk factors associated with pathogenesis of cholangiocarcinoma.

The liver flukes *Opisthorchis viverrini*, Clonorchis sinenis and, rarely, Schistosoma japonica, are risk factors for CCA. Opisthorchis is commonly found in Thailand, Laos, Vietnam, and Cambodia, while Chlonorchis if common in rural areas of Korea and China (Braconi & Patel, 2010). It is predicted that the colonization of these parasites in the biliary tree causes chronic inflammation (Gatto et al., 2010). In a case-control study in the Khon Kaen province of North Thailand, of 76 people diagnosed with CCA, 38% were found to have *O. viverrini* eggs in their stool. Additionally, 53 of 108 participants with CCA reported eating nitrite-containing foods sometimes (24%) or often (25%), suggesting a correlation (Poomphakwaen et al., 2009). Many toxins have been predicted to be associated with CCA development, including the radiological contrast agent, Thorotrast, exposure to which has been shown to increase CCA risk by 3- fold and the chlorinated organic solvents 1,2 dichloropropane and dichloromethane, which are used in offset printing (Lipshutz, Brennan, & Warren, 2002).

Chronic viral hepatitis is also associated with risk of CCA development and is a potential reason for geographical variation. The core protein of hepatitis C virus (HCV) has the ability to alter proliferation of cells and apoptosis in cholangiocarcinoma (Braconi & Patel, 2010). Patients in the United States with HCV-induced hepatitis have a four-fold higher risk of developing CCA when compared to the general population. In addition, a population-based study of United State veterans found the HCV-positive cohort to have a two-fold increased risk of iCCA. Another recognized risk factor is hepatitis B infection (HBV). The association among CCA and HBV is stronger in Asian countries (OR: 6.0) than Western countries (OR: 4.0) (Zhou et al., 2012). In a case-controlled study conducted in Korea, 13.5% of 622 CCA patients tested positive for HBV (Lee et al., 2008).

2. Germline mutations in risk factors associated with cholangiocarcinoma

Evidence suggests that malignancies in the bile ducts are a result of complex interactions between environmental factors as previously described and inherited genetic variation. Significant risk factors and some of their associated genetic variants are highlighted in Table 1.

Table 1 Genetic variants associated with risk factors for cholangiocarcinoma.

Risk factor	Gene	SNP	References
PSC[a]	HLA-B	rs3099844	Mosaad (2015)
	MST1	rs3197999	Melum et al. (2011)
	BCL2L11	rs6720394	
	FOXP1	rs80060485	
	CCDC88B	rs663743	
	CLEC16A	rs725613	
	UBASH3A	rs1893592	
Obesity	ADCY3	rs6545814	Loos and Yeo (2022)
	AGRP		
	BDNF	rs925946	
	KSR2	rs56214831	
	LEP	rs10487505	
	LEPR	rs11208659	
	MC4R	rs17782313	
	MRAP2		
	NTRK2	rs10868215	
	PCSK1	rs6235	
	PHIP		
	POMC	rs6545975	
	SH2B1	rs7498665	
	SIM1	rs6907240	

Continued

Table 1 Genetic variants associated with risk factors for cholangiocarcinoma.—cont'd

Risk factor	Gene	SNP	References
Diabetes	PTPN22		Ilonen, Lempainen, and Veijola (2019)
	PPARG	rs1801282	Willer et al. (2007)
	KCNJ11	rs5219	
Gallstones	ABCG8	rs4953023	Shulenin et al. (2001)
	ABCG5	rs4299376	
		rs6544718	
		rs6720173	
NAFLD[a]	GCKR	27,594,741	Beer et al. (2009)
	PNPLA3	rs738409	Dai, Liu, Li, Zhou, and He (2019)

[a]PSC, primary sclerosing cholangitis; NAFLD, non-alcoholic fatty liver disease.

2.1 Primary sclerosing cholangitis

Patients with primary sclerosing cholangitis (PSC), a chronic inflammatory disorder of the intrahepatic and extrahepatic bile ducts characterized by bile duct fibrosis and strictures, have a 9% 10-year risk of developing cholangiocarcinoma (Rebholz, Krawczyk, & Lammert, 2018). In a multicenter study of five European countries with median follow-up of 18 years, 50% of the CCA cases were diagnosed within 1 year of PSC diagnosis, suggesting that the development of CCA might have triggered the diagnosis of PSC (Gatto et al., 2010). Approximately 30 germline variants have been linked to PSC through genome-wide association studies (GWAS) (Karlsen et al., 2010).

An initial discovery GWAS of 285 Norwegian PSC patients, followed by a replication phase of 766 PSC patients revealed a strong association of variants at chromosome 6p21, specifically rs3099844, located near the HLA-B chromosome region (OR: 4.8, CI: 3.5–6.4, $p = 2.6 \times 10^{-26}$). HLA-B is a part of the HLA gene family. HLA genes are important in the induction and regulation of immune responses and are commonly associated with inflammatory and autoimmune disorders (Mosaad, 2015).

Other gene variations associated with PSC development are in MST1, BCL2L11 FOXP1, CCDC88B, CLEC16A and UBASH3A (Melum et al., 2011). The SNP rs3197999 of MST1 on chromosome 2 was associated with PSC development (OR: 1.39, 95%, $p = 1.1 \times 10^{-16}$) and rs6720394 on

BCL2L11 of chromosome 3 (OR: 1.29, 95%, $p = 4.1 \times 10^{-8}$) in both a discovery and replication phase GWAS MST1 encodes for macrophage-stimulating protein, a hepatocyte growth-factor like protein that plays a central role in innate immunity (Mosaad, 2015). BCL2L11 encodes Bcl-2-like protein 11 (BIM). This protein is pro-apoptotic and regulates normal cell homeostasis through cell death and survival (Melum et al., 2011).

rs80060485, another PSC associated SNP (OR: 1.44 95%, CI: 1.32–1.58, $P = 2.62 \times 10^{-15}$) identified in a GWAS of 2871 PSC patients, is on chromosome 3, located within the coding region of the FOXP1 gene (Ji et al., 2017). FOXP1 belongs to a subclass of the forkhead box (FOX) transcription group. FOXP2 can form heterodimers resulting in transcriptional repression. It is most known to be associated with development of the heart, lung, immune system, and spinal motor neurons. FOXP2 is also thought to act as a tumor suppressor and is mutated in several malignancies (Bacon & Rappold, 2012). This GWAS also identified mutations in the Coiled Coil Domain Containing Protein 88B (CCDC88B) gene at SNP rs663743 (OR: 1.20, 95%, CI: 1.14–1.26, $P = 2.24 \times 10^{-13}$) (Ji et al., 2017). Studies in mice have found CCDC88B to be important in dendritic cell-dependent inflammatory and immune reactions (Olivier et al., 2020). The third statistically significantly mutated SNP identified in this study of PSC patients is rs725613 (OR: 1.14–1.26, 95%, $P = 9.52 \times 10^{-13}$) located in the candidate gene C-type lectin domain family 16 (CLEC16A). The function of this gene is incompletely understood, but it is highly expressed in dendritic cells, natural killer cells, and B lymphocytes and is also associated with an increased risk of multiple sclerosis and type I diabetes (Booth et al., 2009). Polymorphisms at SNP rs1893592 (OR: 1.22, 95%, CI: 1.15–1.29, $P = 2.19 \times 10^{-12}$) which is part of the encoding region for Ubiquitin-associated and SH3 domain-containing protein A (UBASH3A) are also associated with PSC [20]. Mutations in this protein have been associated with several autoimmune disorders, suggesting a role in autoimmunity (Ge, Paisie, Chen, & Concannon, 2019).

2.2 Obesity

The results of two meta-analyses showed an association between obesity and CCA. One meta-analysis showed a more significant relationship between intrahepatic CCA and obesity (OR: 1.56, CI: 95%). More than 1000 loci significant in obesity traits have been identified (Rebholz et al., 2018). A candidate gene study initially identified polymorphisms in LEP and

PCKS1. Following this, more than 60 GWAS have been performed on obese populations. Numerous genome wide linkage studies and GWAS studies have been performed, mainly in Europeans (Şahın et al., 2013), and the significant gene variants identified are summarized in Table 1.

A major pathway associated with obesity is the leptin-melanocortin pathway. Leptin, a hormone involved in regulation of body weight, is transcribed from the LEP gene. Nine of the genes highlighted in Table 1 (AGRP, LEPR, MCR4R, MRAP2, PCKS1, PHIP, POMC and SH21B and SIMI1) are involved in the melanocortin system, which consists of hormonal and neuropeptidergic pathways. Two important functions of these pathways are energy homeostasis and adrenal development (Yeo et al., 2021). Leptin decreases with food deprivation and increases when food is consumed. When non-mutated, leptin circulates within adipose tissue at an appropriate level, promoting long term energy storage (Loos & Yeo, 2022). Variants in MC4R are among the most important gene polymorphisms leading to early onset obesity. Along with hunger, MC4R plays a role in regulating food preference. Those with risk variants of this gene are more likely to crave food with higher fat and sucrose content, further increasing their risk of obesity (van der Klaauw et al., 2015).

2.3 Diabetes

A meta-analysis of five cohort studies found an association between type I diabetes (RR: 1.97, CI: 95%, OR: 1.57–2.46), and intrahepatic and extrahepatic cholangiocarcinoma (RR: 1.6, CI: 95%, OR: 1.29–2.05) (Jing et al., 2012). Despite associations, it remains unclear if diabetes is a direct risk factor of CCA or a secondary association with non-alcoholic fatty liver disease (NAFLD) and obesity. Diabetes may also increase the risk of biliary stones and eCCA (Khan et al., 2019).

There are over 65 genetic variants associated with the development of type I and type II diabetes that have been replicated. In type 1 diabetes human leukocyte antigen (HLA) variations account for 40–50% of genetic risk. The HLA complex coded genes are clustered on chromosome 6, which is the most important region in the human genome in relation to infection, autoimmunity, and adaptive and innate immunity (Noble & Valdes, 2011).

Besides variants in the HLA region, there is also a significant associated allele discrepancy in type I diabetes in a SNP on PTPN22 allele 1858T (OR: 1.83, CI: 95%, $P \leq 0.001$), which was identified through a candidate gene study in a North American population (Ilonen et al., 2019). PTPN22 encodes a lymphocyte protein tyrosine phosphatase which is a strong suppressor of T-cell activation; the SNP causes hyper-activity of T cells. Thus, the genetic mutation in PTPN22 can cause a stronger than needed immune response against autoantigens after immune stress such as a viral infection (Bottini et al., 2004), resulting in reduced self-antigen tolerance and the destruction of pancreatic beta cells (Sharp, Abdulrahim, Naser, & Naser, 2015).

A replication study of previously reported type II diabetes associated SNPs in Finland confirmed significantly associated SNP variants in the peroxisome proliferator-activated receptor gamma gene (PPARG), rs1801282 (OR: 1.30, CI: 95%, $P = 0.0019$, and KCNJ11, rs5219 (OR: 1.22, CI: 95%, $P = 0.0019$) (Willer et al., 2007). PPARG transcribes a transcription factor involved accumulation of triglycerides, glucose-induced insulin secretion and adipocyte differentiation (Rao, Keating, Chen, & Parkington, 2012). The KCNJ11 gene provides encodes produce subunits of the ATP-sensitive potassium (K-ATP) channel. These channels are found embedded in the membranes of insulin secreting pancreatic beta cells. Further, the K-ATP channels close and open according to the glucose level in the bloodstream allowing for the release of insulin out of beta cells. This process helps to control blood sugar; therefore, it is impactful in diabetes (Rao et al., 2012).

2.4 Gallstones

The presence of gallstones is associated with CCA. Gallstone formation in the gallbladder or biliary tract is a result of abnormal regulation of cholesterol solubilization or levels of bilirubin (Lee, Keane, & Pereira, 2015). A genetic polymorphism of the ABCG8 gene, a hepatobiliary cholesterol transporter, has been associated with the development of gallstones and replicated in numerous studies. This transporter alters bile composition and secretion. A meta-analysis of 3216 individuals associated SNPs rs6544718 ($p = 2.1 \times 10^{-14}$) and rs6720173 ($p = 3.8 \times 10^{-12}$) in the ABCG5 gene with the development of gallstones. This gene is a part of the ATP-binding cassette (ABC) transporters which are responsible for transporting various molecules across intra and extra cellular membranes. The protein transcribed by ABCG5 limits intestinal absorption and promotes biliary secretion of sterols. Risk variants can cause a buildup of sterol in the liver (Shulenin et al., 2001).

2.5 NAFLD

NAFLD, characterized by accumulation of fat in the liver that is unrelated to alcohol consumption, is now the most common liver disease in the world. NAFLD is frequently found in individuals with diabetes or obesity due to its pathogenic relationship to insulin resistance, which promotes triglyceride accumulation in the liver. NAFLD commonly induces inflammation, designated non-alcoholic steatohepatitis (NASH), which can result in cirrhosis of the liver (Smith & Adams, 2011). In analysis of 2092 intrahepatic CCA and 2981 extrahepatic cases, NAFLD was associated with a 3-fold increased risk of CCA (extrahepatic—OR: 2.93, CI: 2.42–3.55; intrahepatic—OR: 3.52, CI: 2.87–4.32) (Petrick et al., 2017).

Underlying genetic predispositions have been found to be significantly associated with NAFLD development. 21 studies including 14,266 subjects with NAFLD found a significant association of the phospholipase domain containing 3 (PNPLA3) gene SNP rs738409. Expression of PNPLA3 is influenced by multiple factors such as obesity, insulin, diet, and glucose levels. GG and GC genotypes, compared to CC wild type alleles, were most common in NAFLD patients (Dai et al., 2019).

A GWAS of 592 individuals diagnosed with NAFLD and 1405 healthy controls determined statistical significance at SNP rs780094 (OR: 1.45, CI: 95%, $p = 2.59 \times 10^{-9}$) (Speliotes et al., 2011) in the glucokinase regulatory protein producing gene (GCKR). This gene functions to control triglyceride and fasting plasma glucose levels in healthy individuals. Thus, mutations in this gene can cause dysregulation of glucose levels, leading to diabetes and promoting NAFLD (Beer et al., 2009).

3. Overview of germline mutations and variants associated with cholangiocarcinoma

Though the pathogenesis of CCA is poorly understood, several genes have been associated with risk of CCA development via candidate gene studies and more recently in targeted sequencing studies of pathogenic germline alterations (PGAs). A candidate gene study is a gene association study in which a genetic variant is selected for testing based on biological knowledge of the disease. Using cases and controls, individual or up to tens or perhaps a few hundred variants are genotyped, typically using a polymerase chain reaction-based test. The required sample sizes in candidate gene studies are often relatively low, allowing more easy determination of associations; however, they have the disadvantage that they depend on a priori knowledge of the mechanistic biology association with risk of the disease under study.

A variation of candidate gene studies are studies of the frequency of known, highly penetrant cancer-causing gene mutations in cohorts of patients with apparently sporadic cancers. These genes, such as ATM,

BRCA1, and TP53, have strong associations with cancer development and were previously presumed to be primarily associated with cancer development in patients with a family history of particular cancers. Pathogenic germline alterations (PGAs) targeted sequencing is another technique used to identify gene variants using a germline variant analysis. This analysis uses an array of genes that are known to increase the risk of numerous cancers in individuals and compares the variant frequency to individuals without cancer. Multiple recent studies have shown that a substantial minority, typically 5–15% of patients with apparently sporadic cancers bear a major cancer-causing gene in their germline (Kwon & Goate, 2000).

With significant advances in sequencing and genotyping technology, genome-wide association studies are gaining popularity in the quest to identify previously unknown genetic risk variants associated with disease. Unlike a candidate gene study, a GWAS can simultaneously test the associations of variants in hundreds of thousands of gene variants located within or adjacent to thousands of genes or within intergenic regions using a small amount of isolated DNA with no assumptions about the underlying genes of interest. Microarray technology is utilized which assays 500,000–4 million SNPs simultaneously on a chip. The arrayed SNPs are selected from sequences with high genetic variability within the genome. Due to the phenomenon of linkage disequilibrium, there can be a high correlation between nearby variants such that the alleles at neighboring polymorphic variants not included as a SNP in an array can be imputed.

GWAS are typically designed as large case-controlled studies and if adequately sized, they have the ability to scan the entire genome for variants that are associated with increased risk of cancer. To account for the multiple comparisons that are performed, typically examining a much larger number of genomic locations than the number of samples in the study, variants associated with a disease in GWAS have to reach a very stringent threshold of statistical significance to be considered real, typically $p < 5 \times 10^{-8}$ (Attia, 2021). Presently, only candidate gene studies and targeted PGA sequencing studies have been performed on CCA. No large scale GWAS of CCA has been published to date.

4. Key candidate gene studies of CCA risk variants

This section will provide an overview of the germline genetic polymorphisms identified as potentially or definitively associated with CCA thus far. Key candidate gene studies are highlighted in Table 2.

Table 2 Gene variants associated with cholangiocarcinoma through genome wide association and candidate gene studies.

Gene	Variant	Number of cases (%)	Number of controls (%)	Crude OR	Adjusted OR	P-value	Reference	Ethnic group/country
OGG1	Ser326Cys Ser/Ser	19 (12.7) iCCA	37 (24.7)	1.0			Ding et al. (2015)	China
	Ser/Cys	71 (47.3)	73 (48.6)	1.9 (1.0–3.6)		0.06		
	Cys/Cys	60 (40.0)	40 (26.7)	2.9 (1.5–5.8)		0.03		
	Ser allele	109 (36.3)	147 (49.0)	1.0				
	Cys allele	191 (63.7)	153 (51.0)	1.7 (1.2–2.3)		0.002		
IL-6R	rs2228145 (also called rs8192284) Ala358Asp A48892C AA	60 (73) O-viverrini related CCA	37 (46)	1.0			Prayong et al. (2014)	Thailand
	AC	17 (24)	37 (46)	0.3 (0.1–0.6)		0.0003		
	CC	2 (3)	6 (8)	0.2 (0.2–1.25)		0.04		
	A allele	137 (87)	111 (69)	1.0				
	C allele	21 (13)	49 (31)	0.4 (0.2–0.6)		0.0002		
α1AT	rs28929474 C allele	349 (96)	688 (98)	1.0			Mihalache et al. (2011)	Caucasian
	T allele	15 (4)	12 (2)	2.5 (1.1–5.3)		0.02		

Gene	Variant	Genotype			OR (95% CI)	p-value	Reference	Population
MST1	rs3197999	GG			1.0		Krawczyk et al. (2013)	Caucasian
		AA			2.1 (1.1–3.8)	0.02		
ABCC2	c.3972 T allele		73 (39)	60 (26)	1.83 (1.09–3.08)	0.026	Hoblinger, Grunhage, Sauerbruch, and Lammert (2009)	Caucasian
MYH	rs3219476	TT	25 (42.4)	26 (26.0)	1.0		Ding et al. (2015)	China
		TG	20 (33.9)	58 (58)	0.359 (0.17–0.758)	0.006		
		GG	14 (23.7)	16 (16.0)	0.91 (0.369–2.246)	0.838		
		TG+GG	34 (57.6)	74 (74)	0.478 (0.241–0.946)	0.033		
	rs3219472	GG	19 (32.2)	47 (47)	1.0			China
		GA	19 (32.2)	47 (47)	0.664 (0.326–1.351)	0.258		
		AA	12 (20.3)	7 (7)	2.816 (0.992–7.999)	0.047		
		GA+AA	31 (52.5)	54 (54)	0.943 (0.495–1.797)	0.859		
NKG2D	rs11053781	GA	49			0.008	Melum et al. (2011)	Scandinavian PSC-CCA
NAT2	*6B		6 (1)	22 (5)	0.28 (0.12–0.69)	0.008	Prawan, Kukongviriyapan, Tassaneeyakul, Pairojkul, and Bhudhisawasdi (2005)	Thailand
	*7A		9 (2)	28 (6)	0.33 (0.16–0.70)	0.004		
	*13		8 (2)	24 (5)	0.35 (0.16–0.77)	0.008		

A candidate gene, case-controlled study was performed in a Chinese population genotyping 150 cases of CCA and 150 controls, focused on the 8-oxoguanine glycosylase 1 (OGG1) gene, located at chromosome 3p25. It is known that cells are constantly adapting to changing intracellular conditions driven by environmental factors. These changes stress the cell and can result in DNA damage. DNA repair processes are required to return the cell to its original functional state. OGG1 plays an important role in these repair processes, specifically encoding a protein that initiates the base excision repair pathway (BER). Alterations of this gene have been shown to increase the risk of several cancers, particularly in BRCA1 and BRCA2 mutation carriers (Xiangmin Ding et al., 2015). Individuals with the GG (Cys/Cys) genotype at Ser326Cys (rsl052133) at chromosome 3p25 had a higher risk of intrahepatic CCA (OR: 2.9; 95% CI = 1.1–22.4) compared to those with the wild-type CC genotype (Ding et al., 2015). An additional study in China including 150 iCCA patients and 150 controls observed polymorphisms in OGG1 at codon 326, specifically the replacement of serine with cysteine. They observed that Ser/Cys was associated with a 1.894-fold increase in iCCA (OR: 1.894, 95% CI = 0.994–3.597) and the Cys/Cys genotype was associated with a 2.942-fold increased risk of iCCA (OR: 2.924, 95% CI 1.475–5.780) (Ding et al., 2015).

The MUTHY gene produces the enzyme MYH glycosylase, another key protein in BER. MYH is involved in the repair of DNA by correcting errors that occur during DNA replication in preparation for cell division. For instance, adenine normally pairs with thymine (A:T) and guanine pairs with cytosine (G:C). During normal cellular activities, guanine sometimes becomes altered by reactive oxygen and nitrogen species to 8-oxoguanine, which can pair with adenine instead of cytosine, leading to a G:C to T:A transversion after replication. MYH glycosylase fixes this error by BER, so mutations do not accumulate in the DNA and lead to tumor formation (Dallosso et al., 2008). Using a candidate gene approach, two SNPS in MYH were analyzed in a case-control study of 10 CCA cases and 100 controls in China. Risk of CCA was associated with rs3219472 in genotype G/G by 2.816-fold compared with individuals carrying the MYH A/A genotype (OR: 2.816, 95% CI = 0.992–7.999, $P = 0.047$). The other SNP, rs3219476, was found to be statistically significant in the genotype variants T/G (OR: 0.359, 95%, CI: 0.17–0.758, $P = 0.006$) compared to the wild-type TT (Ding et al., 2015).

Susceptibility to opisthorchiasis-linked CCA, due to a type of liver fluke, was studied using a candidate gene approach in 79 CCA patients and

80 healthy controls in Thailand. Irritation and inflammation induced by infestation of the bile ducts by the fluke causes enhanced expression of some cytokines, specifically interleukin-6 (IL-6), a glycoprotein with pro and anti- inflammatory properties which produces an antibacterial acute phase response in the liver. Typically, IL-6 is found at low concentrations in human blood but the levels rise dramatically during inflammation. Enhancement of IL-6 receptor (IL-6R) expression on CD4 + T cells occurs upon infection with chronic hepatitis B, a risk factor for CCA (Wolf, Rose-John, & Garbers, 2014). Binding of IL-6 to membrane bound IL-6R, causes induction of intracellular signaling pathways. IL-6R is only expressed on hepatocytes and certain subpopulations of leukocytes. Examination of SNP rs2228145 in exon 9 of the IL-6R gene, showed that carriers of the A allele had a significantly increased risk of CCA development. In contrast, allele C carriers had a significant reduction in CCA risk (OR: 0.27, 95%, CI: 0.13–0.56, $P=0.0001$) (Prayong et al., 2014).

Alpha-1-antitypsin (α1AT) deficiency is a hereditary genetic disorder in which the α1AT protein is not released from the liver at normal rates. As a result, it accumulates in the liver, aggregating in toxic, insoluble, intracellular inclusions in the endoplasmic reticulum of hepatocytes, leading to inflammation, liver fibrosis and cirrhosis. This deficiency is the most common genetic associated cause of liver disease in children. It is also a frequently overlooked cause of liver disease in adults. α1AT is a protein with many functions such as maintenance of protease-antiprotease homeostasis, allowing for enough neutrophils to be present for host defense while also providing antiprotease function to prevent host tissue injury (Mitchell & Khan, 2017). The most severe and common genetic polymorphism is the Z variant, in which 2 alleles are mutated. This variant results in protein misfolding and an accumulation of Z-α1AT polymers in the endoplasmic reticulum of hepatocytes leading to cell death (Bouchecareilh, 2014). Hepatocellular carcinoma has been associated with α1AT. In a candidate gene study to assess effect of the Z variant on risk of CCA in whites from Germany and Romania, 182 CCA cases and 350 controls were enrolled. The SNP rs28929474 Z variant allele carriers were more common in CCA patients, specifically eCCA, compared to the controls (OR: 2.46, 95% CIL 1.14–5.32, $P=0.018$) (Mihalache et al., 2011).

The Macrophage Stimulation 1 (MST1) or is a hepatocyte growth factor (HGF)-like gene encodes a glycoprotein designated MSP that activates murine resident peritoneal macrophages. MSP is mainly synthesized hepatocytes and circulates as a plasma protein. The MSP pathway is associated with pathophysiological conditions such as inflammation and cancer. MSP has both stimulatory and inhibitory effects on macrophages. It increases macrophage spreading, phagocytosis and cytokine production, but also inhibits lipopolysaccharide-induced production of mediators of inflammation (Kawaguchi, Orikawa, Baba, Fukushima, & Kataoka, 2009). Krawczyk et al., used a European cohort of 233 CCA patients and 355 healthy controls to investigate the polymorphisms of the SNP, rs3197999, in the MST1 gene which was previously associated with PSC. A significant association was observed between the homozygous variant carriers, AA, and CCA, specifically extrahepatic CCA (OR: 2.04, 95%, CI: 1.09–3.84, $P=0.022$) (Krawczyk et al., 2013).

Cancer susceptibility has been associated with SNP variations in the natural killer cell (NK) receptor, NKG2D. NKG2D is encoded by KLRK1 (killer cell lectin like receptor K1) gene which is in the NK-gene complex (NKC). NKG2D is activated by numerous ligands which serve as indicators for cellular stress. It functions in both the innate and adaptive immunity and determines the activation of cell-mediated cytotoxicity in natural killer cells. NK cells constitute a key part of the innate immune system and plays a vital role in antitumor and antiviral mechanisms. A key ligand for NKG2D is major histocompatibility complex class I chain-related molecule A (MICA). The 5.1 allele of MICA has been associated with the risk of developing PSC, a known risk factor for CCA. In a sequencing study of seven SNPs in the NKG2D gene in 365 PSC patients, 49 of whom had PSC-CCA, and 368 healthy controls, variants in rs2617167 and rs11053781 were associated with an increase in CCA risk in patients with PSC (OR: 2.3, 95% CI: 1.5–3.7, $P=0.002$; OR: 2.1, 95% CI: 1.3–3.3, $P=0.01$, respectively) (Melum et al., 2011).

HOTAIR is an oncogenic long non-coding RNA transcribed from the antisense strand of the HOXC gene group (Tang & Hann, 2018) which regulates genes at transcriptional, epigenetic and posttranscriptional levels. Overexpression has been shown to reduce the effectiveness of some tumor suppressor genes (Mozdarani, Ezzatizadeh, & Rahbar Parvaneh, 2020). A candidate gene study in a Greek population of 122 CCA patients and 165 healthy controls investigated the significance of three Hox transcript antisense intergenics (HOTAIR) SNP polymorphisms on CCA growth. Two of the three SNPS, rs920778 and rs7958904, did not show a statistically significant association, however, the SNP rs4759314 showed significant results suggesting an association with development of CCA in genotypes AG (OR: 3.13, CI: 95%, $P=0.0004$) and GG (OR: 12.31, CI: 95%, $P=0.005$). Thus, the G allele of this SNP may be associated with a higher risk of CCA (Lampropoulou et al., 2021).

Prostaglandin-endoperoxide synthase 2 (PTGS2) also known as cyclooxygenase 2 (COX2) is encoded by the PTGS2 gene. In humans it is one of two cyclooxygenases. It is involved in the conversion of arachidonic acid to prostaglandin H2, an important precursor of prostacyclin, which is expressed in inflammation, as well as lipoxins and epi-lipoxins, which are potent anti-inflammatory agents. PTGS2 (COX-2) is overexpressed in many cancers. Individuals carrying the C allele (either TC or CC genotype) of the PTGS2 gene SNP rs5275 had a 1.8-fold risk (OR: 1.2–2.7, CI: 95%, $P=0.003$) of bile duct cancer (defined as malignancies of the gallbladder, extrahepatic bile ducts and ampulla of Vater), compared to healthy controls (Sakoda et al., 2005).

Xenobiotic metabolism is the breakdown of xenobiotics that enter the human body through exposure to drugs, environmental chemicals, and carcinogens. The N-Acetyltransferase 2 (NAT2) gene encodes key enzyme involved in xenobiotic metabolism. NAT2 also plays a crucial role against oxidative stress-mediated reactive oxygen species protection. The frequency of gene variants of the NAT2 differs among racial groups (Srivastava, Aggarwal, & Singh, 2020). In a case-controlled, candidate gene study of 213 CCA patients in Thailand, there was a significantly decreased risk of CCA associated with mutant alleles in NAT2, *6B (OR: 0.28, CI: 95%, $P=0.004$) *7A (OR: 0.33, CI: 95%, $P=0.003$), and NAT2*13 (OR: 0.35, CI: 95%, $P=0.008$) (Prawan et al., 2005).

The multidrug resistance-associated protein 2 (MRP2), also called canalicular multispecific organic anion transporter 1 (cMOAT) or ATP-binding cassette sub-family C member 2, encoded by the ABCC2 gene, is an ATP-binding transporter that is expressed in hepatocytes and cholangiocytes (Jedlitschky, Hoffmann, & Kroemer, 2006). ABCC2 functions to transport and remove toxins such as drugs, toxicants, heavy metal anions, and conjugates of lipophilic substances with glutathione, glucuronide and sulfate (Yin & Zhang, 2011). A candidate-gene study of 60 CCA patients and 73 healthy Caucasian controls was performed to evaluate the association of alleles in the c.3972C>T variant with risk of CCA. The T allele was significantly correlated with an increased risk of CCA (OR: 1.95, CI: 95%, $P=0.026$) (Hoblinger et al., 2009).

5. Pathogenic germline alteration (PGA) sequencing studies

Most recently, germline mutations in cholangiocarcinoma were found utilizing a PGA sequencing approach in two studies outlined in Table 3. The first study utilized a panel of 88 known genes associated with cancer development. 104 patients with cholangiocarcinoma were included. 12 germline alterations were found in intrahepatic CCA individuals and five PGAs in extrahepatic CCA individuals. Six mutations were found to have high penetrance in the following genes: APC, ATM, BAP1, BRCA1, BRCA2, PABL2, PMS2 (Maynard et al., 2020). Another PGA targeted sequencing study in Japan used an array containing 146 cancer-predisposing genes. Fifteen of the 146 (10%) Japanese patients were found to have germline variants in one of the targeted cancer predisposing genes (Wardell et al., 2018). These variants are highlighted in Table 3 and will be further reviewed.

Adenomatous polyposis coli (APC) is a negative WNT signaling pathway regulator. WNT pathway activation is initiated by Wnt proteins that pass signals into a cell through cell surface receptors (Komiya & Habas, 2008). APC has many roles in humans. It acts as a tumor-suppressor by preventing cell proliferation or in a non-regulated way via transcriptional activation. It also mediates cellular apoptosis, stabilizes microtubules to assist in cell-to-cell adhesion, regulated cell migration, and promotes cell differentiation (Hankey, Frankel, & Groden, 2018). APC helps to ensure that the number of chromosomes in a cell is correct after cell division (Akiyama, 1996). It is an important component of the cytoplasmic protein complex that destroys beta-catenin and prevents its overaccumulation in the cell, which promotes carcinogenesis. Mutations in APC are strongly associated with risk of gastro-intestinal malignancies (Minde, Anvarian, Rüdiger, & Maurice, 2011).

The ataxia-telangiectasia kinase (ATM) protein is located primarily in the nucleus of cells, where it regulates the rate of cell growth and division. ATM-Chk2 and ATR-Chk1 pathways are activated by DNA double-strand breaks (DSBs) and single-stranded DNA respectively. When an ATM protein identifies a broken DNA strand, it initiates repair by inducing enzymes that aid in the repair, allowing maintenance of key cellular functions.

The BRCA1-associated protein 1 (BAP1) functions as a tumor suppressor, (Louie & Kurzrock, 2020) deubiquitinating enzyme, regulating cell

Table 3 Gene variants in cholangiocarcinoma patient germlines using selected cancer-predisposing genes to determine range and frequency of pathogenic germline alterations (PGAs).

Gene	Nucleotide alteration	Penetrance	Classification of CCA	Ethnic group	Reference
APC	c.3920T>A	Low	Intrahepatic	Ashkenazi Jewish	Maynard et al. (2020)
			Extrahepatic	Ashkenazi Jewish	
	c.3920T>A	Low	Intrahepatic	Ashkenazi Jewish	Wardell et al. (2018)
	c.3920T>A	Low	Extrahepatic	Ashkenazi Jewish	
ATM	c.3669_3670insTAG	Moderate	Intrahepatic	White	Maynard et al. (2020)
	c.7878_7882delTTATA		Distal	Japanese	
	c.3669_3670insTAG	Moderate	Intrahepatic	White	
BAP1	c.784-1G>A	High	Intrahepatic	African American	
BRCA1	c.68_69delAG	High	Intrahepatic	Ashkenazi Jewish	
	c.2411_2412del	High	Extrahepatic	Asian	
	c.3593_3594insATAG		Intrahepatic	Japanese	
	c.47T>A		Intrahepatic	Japanese	
	c.68_69delAG	High	Intrahepatic	Ashkenazi Jewish	Wardell et al. (2018)
	c.2411_2412del	High	Extrahepatic	Asian	

Continued

Table 3 Gene variants in cholangiocarcinoma patient germlines using selected cancer-predisposing genes to determine range and frequency of pathogenic germline alterations (PGAs).—cont'd

Gene	Nucleotide alteration	Penetrance	Classification of CCA	Ethnic group	Reference
BRCA2	c.4131_4132insTGAGGA	High	Intrahepatic	Asian	Maynard et al. (2020)
	c.7558C>t		Perhilar	Japanese	
	c.5574_5577delAATT		Distal	Japanese	
	c.8303T>C		Intrahepatic	Japanese	
	c.9090insA		Intrahepatic	Japanese	
	c.68_69delAG	High	Intrahepatic	Ashkenazi Jewish	Wardell et al. (2018)
	c.3215T>G	High	Intrahepatic	Asian	
FH	c.1431_1433dupAAA	Recessive	Intrahepatic	White	Maynard et al. (2020)
			Intrahepatic	White	Wardell et al. (2018)
			Extrahepatic	White	Maynard et al. (2020)
	c.1431_1433dupAAA		Extrahepatic	White	Wardell et al. (2018)
MITF	c.952G>A	Moderate	Intrahepatic	White	Maynard et al. (2020)
		Moderate	Intrahepatic	White	Wardell et al. (2018)

Gene	Variant	Level	Location	Ethnicity	Reference
MUTYH	c.1187G>A	Low	Intrahepatic	White	Maynard et al. (2020)
	c.1437_1439delGGA	Low	Intrahepatic	White	
	c.312C>A	Low	Extrahepatic	White	
	c.1187G>A	Low	Intrahepatic	White	Wardell et al. (2018)
	c.1437_1439delGGA	Low	Intrahepatic	White	
	c.312C>A	Low	Extrahepatic	White	
PALB2	c.3549C>G	High	Intrahepatic	White	Maynard et al. (2020)
	c.3549C>G	High	Intrahepatic	White	Wardell et al. (2018)
PMS2	c.943C>T	High	Intrahepatic	White	
	c.943C>T	High	Intrahepatic	White	Wardell et al. (2018)
RAD51D	c.298C>T		Distal	Japanese	Maynard et al. (2020)
	c.964−2 A>T		Distal	Japanese	
MLH1	c.430C>T		Distal	Japanese	
MSH2	c.1076+1G>A		Intrahepatic	Japanese	
POLD1	c.1775+1T>G		Distal	Japanese	
POLE	c.779delG		Perhilar	Japanese	

cycle control, gene transcription, DNA damage repair, cell differentiation, apoptosis, and cell metabolism. Gene variants of BAP1 are associated with many cancers (Liu & Lu, 2020).

The protein encoded by the BRCA2 gene also plays a key role in tumor suppression. It aids in homologous repair of double strand DNA breaks. Consequently, mutations in BRCA2 can lead to instability of chromosomes as the defect leads to lack of recognition and repair of double-strand DNA breaks through the DSB repair pathway (Yang et al., 2002).

A specific mutation in Fumarate Hydratase (FH), c.1431_1433dupAAA, was identified in two cases of intrahepatic CCA and two cases of extrahepatic CCA in PCG studies. The FH protein encodes fumarase, which converts fumarate to malate in the Krebs cycle. Mutations in the FH gene result in metabolic dysregulation, modulating the cell's ability to efficiently utilize oxygen.

A protein encoding gene, Microphthalmia Inducing Transcription Factor (MIFT) nucleotide variant, c.952G>A (p.Glu318Lys), has been identified in the germlines of two patients with intrahepatic CCA (Maynard et al., 2020; Wardell et al., 2018). The protein product is a master regulator of extracellular signals, including those triggered by alpha-MSH and c-kit ligand. Interestingly, MITF immunoreactivity has been observed in GLI-1 positive cholangiocarcinomas (Nooron, Ohba, Takeda, Shibahara, & Chiabchalard, 2017). This variant of the gene decreases the SUMOylation of MITF, resulting in altered transcription of target genes and increased colony forming potential. It is associated with an increased risk of melanoma and renal cell carcinoma (Kawakami & Fisher, 2017).

MUTYH (mutY DNA glycosylase) is found in the nucleus and mitochondria where is transcribes the enzyme MYH glycosylase. This enzyme is involved in the repair of DNA by correcting errors during DNA replication. It specifically serves as a base excision repair mechanism when oxidated DNA damage occurs and guanine incorrectly pairs with adenine instead of cytosine (Church, Heald, Burke, & Kalady, 2012).

Partner and localizer of BRCA2 (PALB2), encoded by the PALB2 gene, is a genome maintenance protein that interacts with BRCA1. PALB2 acts in conjunction with BRCA2, serving as a bridge for molecules to connect with the BRCA complex, which as previously mentioned is vital in homologous repair of DNA double stranded breaks as part of the DSM repair pathway. PALB2 specifically. Due to this role, PALB2 acts a tumor suppressor, participating in regulation of a cell's genome (Wu et al., 2020). Mutations in this protein has been strongly associated with various cancers such as breast, pancreatic and ovarian (Wu et al., 2020).

PMS1 homolog 2 (PMS2), located on chromosome 7, encodes a key protein in the system that corrects DNA mismatches such as insertions and deletions of alleles that can occur during homologous recombination and DNA replication. The protein works with the MLH1 gene to form heterodimers that remove the mismatched DNA. Variations in PSM2 have been strongly associated with Lynch syndrome due to DNA recombination errors which result in abnormal cell division and accumulation of these defective cells (Kasela, Nyström, & Kansikas, 2019).

An additional resource in homologous recombination and DNA repair is the RAD51D protein coding gene. It is commonly included in PGA sequencing studies as mutations have been strongly associated with carcinogenesis (Pittman, Weinberg, & Schimenti, 1998). Large intestine, colon, rectum and ovarian cancers have been strongly associated with dysfunctions of this gene.

SNP differences in mutL homolog1 (MLH1) located on chromosome 3 result in DNA mismatching as its key role involves the DNA mismatch repair system, further tumor suppression (Barnetson et al., 2008). MLH1 works with the previously mentioned mismatch repair gene, PMS2, to form a two-protein complex, dimer. When MLH1 is not expresses appropriately, it does not interact with PMS2 resulting in the degradation of PMS2. In opposition, MLH1 can persist without the presence of PMS2. In addition to DNA mismatch repair functionality, MLH1 localizes sites of crossing over in meiotic chromosomes (Truninger et al., 2005).

MSH2 is another one of the seven key proteins in the cellular DNA mismatch repair system. In this process MSH2 forms two heterodimers that bind to identified DNA mismatches and initiate the repair system. It is specifically known to be involved in transcription-coupled repair (Mellon, Rajpal, Koi, Boland, & Champe, 1996), base excision repair and homologous recombination (Pitsikas, Lee, & Rainbow, 2007).

Polymerase delta 1 (POLD1) encodes a large catalytic subunit of the DNA polymerase delta complex (Prindle & Loeb, 2012). This complex plays a major role in excision repair following UV irradiation. The POLE gene, DNA polymerase epsilon, is required for synthesis of the leading DNA strands at the replication fork, where it binds near the replication origins and moves along with the replication fork. Similar to other pathogenic germline alterations in cancers, these genes play crucial roles in DNA replication and DNA repair (Ogi et al., 2010).

6. Conclusion

In summary, genetic studies using both multigene cancer panels and candidate gene studies, have increased our knowledge of gene variants in the risk factors associated with CCA and CCA pathogenesis. After common variants and genes of which they are located on are identified, their physiological role in carcinogenesis can be further analyzed and characterized, leading to a better understanding of the process of biliary carcinogenesis. Genetic testing may lead to better stratification of CCA risk and may help determine patients who require the most aggressive surveillance. As noted in Table 2, the cohorts of genetic variant studies in CCA have small sample sizes. The small cohorts of genetic variant studies produce limited information and therefore, larger cohort studies are needed. Further detection of the important risk variants will greatly benefit the identification of risk for CCA, molecular pathogenesis and treatment options. There is currently a large multinational, multicenter effort underway to further identify and characterize genetic variants associated with risk of CCA through genome wide genotyping and sequencing studies supported collaboratively by the International Cholangiocarcinoma Research Network (ICRN).

Conflict of interest statement

M.A.S. has no financial or personal disclosures relevant to the contents of this manuscript. L.R.R. reports grants from NIH (CA186566 and CA210964), Mayo Clinic, The Cholangiocarcinoma Foundation, Bayer, Boston Scientific, Exact Sciences, Fujifilm Medical Sciences, Gilead Sciences, Glycotest Inc., PSC Partners Seeking a Cure, Redhill Biopharma, and TARGET PharmaSolutions during the conduct of the study; other support from MedEd Design LLC, Pontifax, Global Life Science Consulting, The Lynx Group, AstraZeneca, Bayer, Eisai, Exact Sciences, and GRAIL Inc.

Grant support

None.

References

Akiyama, T. (1996). The APC gene. *Nihon Rinsho, 54*(4), 955–959.

Maynard, H., Stadler, Z. K., Berger, M. F., Solit, D. B., Ly, M., Lowery, M. A., et al. (2020). Germline alterations in patients with biliary tract cancers: A spectrum of significant and previously underappreciated findings. *Cancer, 126*(9), 1995–2002. https://doi.org/10.1002/cncr.32740.

Attia, J. (2021). *Genetic association and GWAS studies: Principles and applications.* Retrieved from https://www.uptodate.com/contents/genetic-association-and-gwas-studies-principles-and-applications.

Bacon, C., & Rappold, G. A. (2012). The distinct and overlapping phenotypic spectra of FOXP1 and FOXP2 in cognitive disorders. *Human Genetics, 131*(11), 1687–1698. https://doi.org/10.1007/s00439-012-1193-z.

Barnetson, R. A., Cartwright, N., van Vliet, A., Haq, N., Drew, K., Farrington, S., et al. (2008). Classification of ambiguous mutations in DNA mismatch repair genes identified in a population-based study of colorectal cancer. *Human Mutation, 29*(3), 367–374. https://doi.org/10.1002/humu.20635.

Beer, N. L., Tribble, N. D., McCulloch, L. J., Roos, C., Johnson, P. R. V., Orho-Melander, M., et al. (2009). The P446L variant in GCKR associated with fasting plasma glucose and triglyceride levels exerts its effect through increased glucokinase activity in liver. *Human Molecular Genetics, 18*(21), 4081–4088. https://doi.org/10.1093/hmg/ddp357.

Booth, D. R., Heard, R. N., Stewart, G. J., Goris, A., Dobosi, R., Dubois, B., et al. (2009). The expanding genetic overlap between multiple sclerosis and type I diabetes. *Genes and Immunity, 10*(1), 11–14. https://doi.org/10.1038/gene.2008.83.

Bottini, N., Musumeci, L., Alonso, A., Rahmouni, S., Nika, K., Rostamkhani, M., et al. (2004). A functional variant of lymphoid tyrosine phosphatase is associated with type I diabetes. *Nature Genetics, 36*(4), 337–338. https://doi.org/10.1038/ng1323.

Bouchecareilh, M. (2014). Le déficit en alpha-1-antitrypsine. *Medical Science (Paris), 30*(10), 889–895. https://doi.org/10.1051/medsci/20143010016.

Braconi, C., & Patel, T. (2010). Cholangiocarcinoma: New insights into disease pathogenesis and biology. *Infectious Disease Clinics of North America, 24*(4), 871–vii. https://doi.org/10.1016/j.idc.2010.07.006.

Church, J., Heald, B., Burke, C., & Kalady, M. (2012). Understanding MYH-associated neoplasia. *Diseases of the Colon & Rectum, 55*(3). Retrieved from https://journals.lww.com/dcrjournal/Fulltext/2012/03000/Understanding_MYH_Associated_Neoplasia.19.aspx.

Dai, G., Liu, P., Li, X., Zhou, X., & He, S. (2019). Association between PNPLA3 rs738409 polymorphism and nonalcoholic fatty liver disease (NAFLD) susceptibility and severity: A meta-analysis. *Medicine (Baltimore), 98*(7), e14324. https://doi.org/10.1097/md.0000000000014324.

Dallosso, A. R., Dolwani, S., Jones, N., Jones, S., Colley, J., Maynard, J., et al. (2008). Inherited predisposition to colorectal adenomas caused by multiple rare alleles of MUTYH but not OGG1, NUDT1, NTH1 or NEIL 1, 2 or 3. *Gut, 57*(9), 1252–1255. https://doi.org/10.1136/gut.2007.145748.

Ding, X., Wang, K., Wu, Z., Yao, A., Li, J., Jiao, C., et al. (2015). The Ser326Cys polymorphism of hOGG1 is associated with intrahepatic cholangiocarcinoma susceptibility in a Chinese population. *International Journal of Clinical and Experimental Medicine, 8*(9), 16294–16300. Retrieved from https://pubmed.ncbi.nlm.nih.gov/26629147. https://www.ncbi.nlm.nih.gov/pmc/articles/PMC4659035/.

Gatto, M., Bragazzi, M. C., Semeraro, R., Napoli, C., Gentile, R., Torrice, A., et al. (2010). Cholangiocarcinoma: Update and future perspectives. *Digestive and Liver Disease, 42*(4), 253–260. https://doi.org/10.1016/j.dld.2009.12.008.

Ge, Y., Paisie, T. K., Chen, S., & Concannon, P. (2019). UBASH3A regulates the synthesis and dynamics of TCR–CD3 complexes. *The Journal of Immunology, 203*(11), 2827–2836. https://doi.org/10.4049/jimmunol.1801338.

Hankey, W., Frankel, W. L., & Groden, J. (2018). Functions of the APC tumor suppressor protein dependent and independent of canonical WNT signaling: Implications for therapeutic targeting. *Cancer Metastasis Reviews, 37*(1), 159–172. https://doi.org/10.1007/s10555-017-9725-6.

Hoblinger, A., Grunhage, F., Sauerbruch, T., & Lammert, F. (2009). Association of the c.3972C>T variant of the multidrug resistance-associated protein 2 gene (MRP2/ABCC2) with susceptibility to bile duct cancer. *Digestion, 80*(1), 36–39. https://doi.org/10.1159/000212990.

Ilonen, J., Lempainen, J., & Veijola, R. (2019). The heterogeneous pathogenesis of type 1 diabetes mellitus. *Nature Reviews Endocrinology, 15*(11), 635–650. https://doi.org/10.1038/s41574-019-0254-y.
Jedlitschky, G., Hoffmann, U., & Kroemer, H. K. (2006). Structure and function of the MRP2 (ABCC2) protein and its role in drug disposition. *Expert Opinion on Drug Metabolism & Toxicology, 2*(3), 351–366. https://doi.org/10.1517/17425255.2.3.351.
Ji, S.-G., Juran, B. D., Mucha, S., Folseraas, T., Jostins, L., Melum, E., et al. (2017). Genome-wide association study of primary sclerosing cholangitis identifies new risk loci and quantifies the genetic relationship with inflammatory bowel disease. *Nature Genetics, 49*(2), 269–273. https://doi.org/10.1038/ng.3745.
Jing, W., Jin, G., Zhou, X., Zhou, Y., Zhang, Y., Shao, C., et al. (2012). Diabetes mellitus and increased risk of cholangiocarcinoma: A meta-analysis. *European Journal of Cancer Prevention, 21*(1), 24–31. https://doi.org/10.1097/CEJ.0b013e3283481d89.
Karlsen, T. H., Franke, A., Melum, E., Kaser, A., Hov, J. R., Balschun, T., et al. (2010). Genome-wide association analysis in primary sclerosing cholangitis. *Gastroenterology, 138*(3), 1102–1111. https://doi.org/10.1053/j.gastro.2009.11.046.
Kasela, M., Nyström, M., & Kansikas, M. (2019). PMS2 expression decrease causes severe problems in mismatch repair. *Human Mutation, 40*(7), 904–907. https://doi.org/10.1002/humu.23756.
Kawaguchi, M., Orikawa, H., Baba, T., Fukushima, T., & Kataoka, H. (2009). Hepatocyte growth factor activator is a serum activator of single-chain precursor macrophage-stimulating protein. *The FEBS Journal, 276*(13), 3481–3490. https://doi.org/10.1111/j.1742-4658.2009.07070.x.
Kawakami, A., & Fisher, D. E. (2017). The master role of microphthalmia-associated transcription factor in melanocyte and melanoma biology. *Laboratory Investigation, 97*(6), 649–656. https://doi.org/10.1038/labinvest.2017.9.
Khan, A. S., & Dageforde, L. A. (2019). Cholangiocarcinoma. *Surgical Clinics of North America, 99*(2), 315–335. https://doi.org/10.1016/j.suc.2018.12.004.
Khan, S. A., Tavolari, S., & Brandi, G. (2019). Cholangiocarcinoma: Epidemiology and risk factors. *Liver International, 39*(S1), 19–31. https://doi.org/10.1111/liv.14095.
Khan, S. A., Toledano, M. B., & Taylor-Robinson, S. D. (2008). Epidemiology, risk factors, and pathogenesis of cholangiocarcinoma. *HPB: The Official Journal of the International Hepato Pancreato Biliary Association, 10*(2), 77–82. https://doi.org/10.1080/13651820801992641.
Komiya, Y., & Habas, R. (2008). Wnt signal transduction pathways. *Organogenesis, 4*(2), 68–75. https://doi.org/10.4161/org.4.2.5851.
Krawczyk, M., Höblinger, A., Mihalache, F., Grünhage, F., Acalovschi, M., Lammert, F., et al. (2013). Macrophage stimulating protein variation enhances the risk of sporadic extrahepatic cholangiocarcinoma. *Digestive and Liver Disease, 45*(7), 612–615. https://doi.org/10.1016/j.dld.2012.12.017.
Kwon, J. M., & Goate, A. M. (2000). The candidate gene approach. *Alcohol Research & Health, 24*(3), 164–168.
Lampropoulou, D. I., Laschos, K., Aravantinos, G., Georgiou, K., Papiris, K., Theodoropoulos, G., et al. (2021). Association between homeobox protein transcript antisense intergenic ribonucleic acid genetic polymorphisms and cholangiocarcinoma. *World Journal of Clinical Cases, 9*(8), 1785–1792. https://doi.org/10.12998/wjcc.v9.i8.1785.
Lee, J. Y., Keane, M. G., & Pereira, S. (2015). Diagnosis and treatment of gallstone disease. *Practitioner, 259*(1783), 15–19 (12).
Lee, T. Y., Lee, S. S., Jung, S. W., Jeon, S. H., Yun, S. C., Oh, H. C., et al. (2008). Hepatitis B virus infection and intrahepatic cholangiocarcinoma in Korea: A case-control study. *American Journal of Gastroenterology, 103*(7), 1716–1720. https://doi.org/10.1111/j.1572-0241.2008.01796.x.

Li, J.-S., Han, T.-J., Jing, N., Li, L., Zhang, X.-H., Ma, F.-Z., et al. (2014). Obesity and the risk of cholangiocarcinoma: A meta-analysis. *Tumor Biology*, *35*(7), 6831–6838. https://doi.org/10.1007/s13277-014-1939-4.

Lipshutz, G. S., Brennan, T. V., & Warren, R. S. (2002). Thorotrast-induced liver neoplasia: A collective review. *Journal of the American College of Surgeons*, *195*(5), 713–718. https://doi.org/10.1016/S1072-7515(02)01287-5.

Liu, Y., & Lu, L.-Y. (2020). BRCA1 and homologous recombination: Implications from mouse embryonic development. *Cell & Bioscience*, *10*(1), 49. https://doi.org/10.1186/s13578-020-00412-4.

Loos, R. J. F., & Yeo, G. S. H. (2022). The genetics of obesity: From discovery to biology. *Nature Reviews Genetics*, *23*(2), 120–133. https://doi.org/10.1038/s41576-021-00414-z.

Louie, B. H., & Kurzrock, R. (2020). BAP1: Not just a BRCA1-associated protein. *Cancer Treatment Reviews*, *90*, 102091. https://doi.org/10.1016/j.ctrv.2020.102091.

Mellon, I., Rajpal, D. K., Koi, M., Boland, C. R., & Champe, G. N. (1996). Transcription-coupled repair deficiency and mutations in human mismatch repair genes. *Science*, *272*(5261), 557–560. https://doi.org/10.1126/science.272.5261.557.

Melum, E., Franke, A., Schramm, C., Weismüller, T. J., Gotthardt, D. N., Offner, F. A., et al. (2011). Genome-wide association analysis in primary sclerosing cholangitis identifies two non-HLA susceptibility loci. *Nature Genetics*, *43*(1), 17–19. https://doi.org/10.1038/ng.728.

Mihalache, F., Höblinger, A., Grünhage, F., Krawczyk, M., Gärtner, B. C., Acalovschi, M., et al. (2011). Heterozygosity for the alpha1-antitrypsin Z allele may confer genetic risk of cholangiocarcinoma. *Alimentary Pharmacology & Therapeutics*, *33*(3), 389–394. https://doi.org/10.1111/j.1365-2036.2010.04534.x.

Minde, D. P., Anvarian, Z., Rüdiger, S. G., & Maurice, M. M. (2011). Messing up disorder: How do missense mutations in the tumor suppressor protein APC lead to cancer? *Molecular Cancer*, *10*, 101. https://doi.org/10.1186/1476-4598-10-101.

Mitchell, E. L., & Khan, Z. (2017). Liver disease in alpha-1 antitrypsin deficiency: Current approaches and future directions. *Current Pathobiology Reports*, *5*(3), 243–252. https://doi.org/10.1007/s40139-017-0147-5.

Mosaad, Y. M. (2015). Clinical role of human leukocyte antigen in health and disease. *Scandinavian Journal of Immunology*, *82*(4), 283–306. https://doi.org/10.1111/sji.12329.

Mozdarani, H., Ezzatizadeh, V., & Rahbar Parvaneh, R. (2020). The emerging role of the long non-coding RNA HOTAIR in breast cancer development and treatment. *Journal of Translational Medicine*, *18*(1), 152. https://doi.org/10.1186/s12967-020-02320-0.

Noble, J. A., & Valdes, A. M. (2011). Genetics of the HLA region in the prediction of type 1 diabetes. *Current Diabetes Reports*, *11*(6), 533–542. https://doi.org/10.1007/s11892-011-0223-x.

Nooron, N., Ohba, K., Takeda, K., Shibahara, S., & Chiabchalard, A. (2017). Dysregulated expression of MITF in subsets of hepatocellular carcinoma and cholangiocarcinoma. *Tohoku Journal of Experimental Medicine*, *242*(4), 291–302. https://doi.org/10.1620/tjem.242.291.

Ogi, T., Limsirichaikul, S., Overmeer, R. M., Volker, M., Takenaka, K., Cloney, R., et al. (2010). Three DNA polymerases, recruited by different mechanisms, carry out NER repair synthesis in human cells. *Molecular Cell*, *37*(5), 714–727. https://doi.org/10.1016/j.molcel.2010.02.009.

Olivier, J.-F., Fodil, N., Al Habyan, S., Gopal, A., Artusa, P., Mandl, J. N., et al. (2020). CCDC88B is required for mobility and inflammatory functions of dendritic cells. *Journal of Leukocyte Biology*, *108*(6), 1787–1802. https://doi.org/10.1002/JLB.3A0420-386R.

Petrick, J. L., Yang, B., Altekruse, S. F., Van Dyke, A. L., Koshiol, J., Graubard, B. I., et al. (2017). Risk factors for intrahepatic and extrahepatic cholangiocarcinoma in the United States: A population-based study in SEER-Medicare. *PLoS One*, *12*(10), e0186643. https://doi.org/10.1371/journal.pone.0186643.

Pitsikas, P., Lee, D., & Rainbow, A. J. (2007). Reduced host cell reactivation of oxidative DNA damage in human cells deficient in the mismatch repair gene hMSH2. *Mutagenesis*, *22*(3), 235–243. https://doi.org/10.1093/mutage/gem008.

Pittman, D. L., Weinberg, L. R., & Schimenti, J. C. (1998). Identification, characterization, and genetic mapping of Rad51d, a new mouse and human RAD51/RecA-related gene. *Genomics*, *49*(1), 103–111. https://doi.org/10.1006/geno.1998.5226.

Poomphakwaen, K., Promthet, S., Kamsa-Ard, S., Vatanasapt, P., Chaveepojnkamjorn, W., Klaewkla, J., et al. (2009). Risk factors for cholangiocarcinoma in Khon Kaen, Thailand: A nested case-control study. *Asian Pacific Journal of Cancer Prevention*, *10*(2), 251–258.

Prawan, A., Kukongviriyapan, V., Tassaneeyakul, W., Pairojkul, C., & Bhudhisawasdi, V. (2005). Association between genetic polymorphisms of CYP1A2, arylamine N-acetyltransferase 1 and 2 and susceptibility to cholangiocarcinoma. *European Journal of Cancer Prevention*, *14*(3). Retrieved from https://journals.lww.com/eurjcancerprev/Fulltext/2005/06000/Association_between_genetic_polymorphisms_of.8.aspx.

Prayong, P., Mairiang, E., Pairojkul, C., Chamgramol, Y., Mairiang, P., Bhudisawasdi, V., et al. (2014). An interleukin-6 receptor polymorphism is associated with opisthorchiasis-linked cholangiocarcinoma risk in Thailand. *Asian Pacific Journal of Cancer Prevention*, *15*(13), 5443–5447. https://doi.org/10.7314/apjcp.2014.15.13.5443.

Prindle, M. J., & Loeb, L. A. (2012). DNA polymerase delta in DNA replication and genome maintenance. *Environmental and Molecular Mutagenesis*, *53*(9), 666–682. https://doi.org/10.1002/em.21745.

Rao, J. R., Keating, D. J., Chen, C., & Parkington, H. C. (2012). Adiponectin increases insulin content and cell proliferation in MIN6 cells via PPARγ-dependent and PPARγ-independent mechanisms. *Diabetes, Obesity and Metabolism*, *14*(11), 983–989. https://doi.org/10.1111/j.1463-1326.2012.01626.x.

Rebholz, C., Krawczyk, M., & Lammert, F. (2018). Genetics of gallstone disease. *European Journal of Clinical Investigation*, *48*(7), e12935. https://doi.org/10.1111/eci.12935.

Rizvi, S., & Gores, G. J. (2013). Pathogenesis, diagnosis, and management of cholangiocarcinoma. *Gastroenterology*, *145*(6), 1215–1229. https://doi.org/10.1053/j.gastro.2013.10.013.

Şahin, S., Rüstemoğlu, A., Tekcan, A., Taşliyurt, T., Güven, H., & Yığıt, S. (2013). Investigation of associations between obesity and LEP G2548A and LEPR 668A/G polymorphisms in a Turkish population. *Disease Markers*, *35*(6), 673–677. https://doi.org/10.1155/2013/216279.

Sakoda, L. C., Gao, Y.-T., Chen, B. E., Chen, J., Rosenberg, P. S., Rashid, A., et al. (2005). Prostaglandin-endoperoxide synthase 2 (PTGS2) gene polymorphisms and risk of biliary tract cancer and gallstones: A population-based study in Shanghai, China. *Carcinogenesis*, *27*(6), 1251–1256. https://doi.org/10.1093/carcin/bgi314.

Shaib, Y., & El-Serag, H. B. (2004). The epidemiology of cholangiocarcinoma. *Seminars in Liver Disease*, *24*(2), 115–125. https://doi.org/10.1055/s-2004-828889.

Sharp, R. C., Abdulrahim, M., Naser, E. S., & Naser, S. A. (2015). Genetic variations of PTPN2 and PTPN22: Role in the pathogenesis of type 1 diabetes and Crohn's disease. *Frontiers in Cellular and Infection Microbiology*, *5*, 95. https://doi.org/10.3389/fcimb.2015.00095.

Shulenin, S., Schriml, L. M., Remaley, A. T., Fojo, S., Brewer, B., Allikmets, R., et al. (2001). An ATP-binding cassette gene (ABCG5) from the ABCG (white) gene subfamily maps to human chromosome 2p21 in the region of the Sitosterolemia locus. *Cytogenetics and Cell Genetics*, *92*(3–4), 204–208. https://doi.org/10.1159/000056903.

Smith, B. W., & Adams, L. A. (2011). Non-alcoholic fatty liver disease. *Critical Reviews in Clinical Laboratory Sciences*, *48*(3), 97–113. https://doi.org/10.3109/10408363.2011.596521.

Speliotes, E. K., Yerges-Armstrong, L. M., Wu, J., Hernaez, R., Kim, L. J., Palmer, C. D., et al. (2011). Genome-wide association analysis identifies variants associated with non-alcoholic fatty liver disease that have distinct effects on metabolic traits. *PLoS Genetics*, *7*(3), e1001324. https://doi.org/10.1371/journal.pgen.1001324.

Srivastava, D. S. L., Aggarwal, K., & Singh, G. (2020). Is NAT2 gene polymorphism associated with vitiligo? *Indian Journal of Dermatology, 65*(3), 173–177. https://doi.org/10.4103/ijd.IJD_388_18.

Tang, Q., & Hann, S. S. (2018). HOTAIR: An oncogenic long non-coding RNA in human Cancer. *Cellular Physiology and Biochemistry, 47*(3), 893–913. https://doi.org/10.1159/000490131.

Truninger, K., Menigatti, M., Luz, J., Russell, A., Haider, R., Gebbers, J.-O., et al. (2005). Immunohistochemical analysis reveals high frequency of PMS2 defects in colorectal Cancer. *Gastroenterology, 128*(5), 1160–1171. https://doi.org/10.1053/j.gastro.2005.01.056.

van der Klaauw, A., Keogh, J., Henning, E., Stephenson, C., Trowse, V. M., Fletcher, P., et al. (2015). Role of melanocortin signalling in the preference for dietary macronutrients in human beings. *Lancet, 385*(Suppl 1), S12. https://doi.org/10.1016/s0140-6736(15)60327-0.

Wardell, C. P., Fujita, M., Yamada, T., Simbolo, M., Fassan, M., Karlic, R., et al. (2018). Genomic characterization of biliary tract cancers identifies driver genes and predisposing mutations. *Journal of Hepatology, 68*(5), 959–969. https://doi.org/10.1016/j.jhep.2018.01.009.

Willer, C. J., Bonnycastle, L. L., Conneely, K. N., Duren, W. L., Jackson, A. U., Scott, L. J., et al. (2007). Screening of 134 single nucleotide polymorphisms (SNPs) previously associated with type 2 diabetes replicates association with 12 SNPs in nine genes. *Diabetes, 56*(1), 256–264. https://doi.org/10.2337/db06-0461.

Wolf, J., Rose-John, S., & Garbers, C. (2014). Interleukin-6 and its receptors: A highly regulated and dynamic system. *Cytokine, 70*(1), 11–20. https://doi.org/10.1016/j.cyto.2014.05.024.

Wongjarupong, N., Assavapongpaiboon, B., Susantitaphong, P., Cheungpasitporn, W., Treeprasertsuk, S., Rerknimitr, R., et al. (2017). Non-alcoholic fatty liver disease as a risk factor for cholangiocarcinoma: A systematic review and meta-analysis. *BMC Gastroenterology, 17*(1), 149. https://doi.org/10.1186/s12876-017-0696-4.

Wu, S., Zhou, J., Zhang, K., Chen, H., Luo, M., Lu, Y., et al. (2020). Molecular mechanisms of PALB2 function and its role in breast Cancer management. *Frontiers in Oncology, 10*. https://doi.org/10.3389/fonc.2020.00301.

Yang, H., Jeffrey, P. D., Miller, J., Kinnucan, E., Sun, Y., Thoma, N. H., et al. (2002). BRCA2 function in DNA binding and recombination from a BRCA2-DSS1-ssDNA structure. *Science, 297*(5588), 1837–1848. https://doi.org/10.1126/science.297.5588.1837.

Yang, J. D., Kim, B., Sanderson, S. O., Sauver, J. S., Yawn, B. P., Larson, J. J., et al. (2012). Biliary tract cancers in Olmsted County, Minnesota, 1976-2008. *American Journal of Gastroenterology, 107*(8), 1256–1262. https://doi.org/10.1038/ajg.2012.173.

Yeo, G. S. H., Chao, D. H. M., Siegert, A. M., Koerperich, Z. M., Ericson, M. D., Simonds, S. E., et al. (2021). The melanocortin pathway and energy homeostasis: From discovery to obesity therapy. *Molecular Metabolism, 48*, 101206. https://doi.org/10.1016/j.molmet.2021.101206.

Yin, J., & Zhang, J. (2011). Multidrug resistance-associated protein 1 (MRP1/ABCC1) polymorphism: from discovery to clinical application. *Zhong Nan Da Xue Xue Bao. Yi Xue Ban, 36*(10), 927–938. https://doi.org/10.3969/j.issn.1672-7347.2011.10.002.

Zhou, Y., Zhao, Y., Li, B., Huang, J., Wu, L., Xu, D., et al. (2012). Hepatitis viruses infection and risk of intrahepatic cholangiocarcinoma: Evidence from a meta-analysis. *BMC Cancer, 12*, 289. https://doi.org/10.1186/1471-2407-12-289.

CHAPTER SIX

Novel insights into molecular and immune subtypes of biliary tract cancers

Emily R Bramel and Daniela Sia*

Division of Liver Diseases, Liver Cancer Program, Department of Medicine, Tisch Cancer Institute, Icahn School of Medicine at Mount Sinai, New York, NY, United States
*Corresponding author: e-mail address: daniela.sia@mssm.edu

Contents

1. Introduction	170
2. The mutational landscape of biliary tract cancers	173
2.1 Genetic subgroups of cholangiocarcinoma	173
2.2 New findings for GBC	177
3. Multi-omics classifications of intrahepatic cholangiocarcinoma	179
3.1 Integrative molecular classifications	179
3.2 Immune microenvironment-based subtypes	181
4. Molecular classifications of other biliary tract cancers	185
4.1 Molecular classification of extrahepatic CCA	185
4.2 Molecular and immune subtypes of gallbladder cancer	187
5. Therapeutic implications	189
6. Conclusions and future perspectives	192
Acknowledgments	193
Grant support	193
Conflict of interest statement	193
References	193

Abstract

Biliary tract cancers (BTCs), which include cholangiocarcinoma (CCA) and gallbladder cancer (GBC), are heterogenous malignancies characterized by distinct molecular features often associated with specific clinical traits and/or outcomes. Such complex molecular heterogeneity, both within each BTC subtype and between distinct subtypes, poses a great challenge to personalized medicine. Recent technological advances have allowed the integration of multiple -*omics* derived from large cohorts of patients with distinct solid cancers to ultimately design stratification algorithms for prognostic prediction or more efficient treatment allocation. In this regard, although BTCs lag behind other tumors when it comes to our understanding of their molecular complexity, over the past decade, tremendous efforts have been made to generate supervised or unsupervised molecular classifications. As a result, CCAs and GBCs can be assigned to distinct molecular and/or prognostic classes. Notably, the discovery of biologically relevant subgroups of tumors harboring frequent targetable alterations (i.e., mutations in *IDH1*, FGFR2 fusion proteins) holds important therapeutic implications for BTCs, particularly iCCA. Furthermore, the recent application of single cell-based technologies or more conservative (and less precise) "virtual microdissection" algorithms to isolate signals derived from the immune and stromal cells has identified the first microenvironment-based classes. In this chapter, we will review the molecular and immune classes of BTCs, with a particular focus on their clinical implications.

Abbreviations

2-HG	2-hydroxyglutarate
AKT	RAC-alpha serine/threonine-protein kinase
ALK	anaplastic lymphoma kinase
ARID1A	AT-rich interaction domain 1A
ATM	ataxia telangiectasia mutated
ATR	ataxia telangiectasia and Rad3-related protein
BAP1	BRCA1-associated protein-1
BLM	BLM RecQ like helicase
BRAC1/2	breast cancer gene
BRAF	B-Raf proto-oncogene
BRD9	bromodomain-containing protein 9
BTC	biliary tract cancer
BTLA	B- and T-lymphocyte attenuator
CAF	cancer-associated fibroblasts
CCA	cholangiocarcinoma
CD133	CD133 antigen or prominin
CD3	cluster of differentiation 3
CD4	cluster of differentiation 4
CD8	cluster of differentiation 8
CDK	cyclin-dependent kinase
CDK12	cyclin-dependent kinase 12
CDKNA/B	cyclin-dependent kinase inhibitor 2A/B
CNA	copy number alterations
CTLA4	cytotoxic T-lymphocyte-associated protein 4
CTNNB1	catenin beta 1
dCCA	distal CCA
DDR	DNA damage repair
DDR	DNA damage response
eCCA	extrahepatic CCA
EGF	epidermal growth factor
EGFR	Epidermal Growth Factor Receptor
ELF3	ETS-related transcription factor
EMT	epithelial mesenchymal transition
ERBB2	Erb-B2 Receptor Tyrosine Kinase 2
EZH2	Enhancer of zeste homolog 2
FANCA	FA Complementation Group A
FANCD2	FA Complementation Group D2
FDA	food and drug administration

FGF2	fibroblast growth factor 2
FGFR2	fibroblast growth factor receptor 2
FOLFOX	5-fluorouracil, leucovorin and oxaliplatin
GBC	gallbladder cancer
HAS2	Hyaluronan synthase 2
HBsAg	hepatitis B surface antigen
HBV	hepatitis B virus
HCC	hepatocellular carcinoma
HCV	hepatitis C virus
HER2	human epidermal growth factor receptor 2
HGF	Hepatocyte Growth Factor
HNF4A	hepatocyte nuclear factor-4 alpha
iCCA	intrahepatic CCA
ICGC	International Cancer Genome Consortium
ICI	immune checkpoint inhibitor
ICOS	Inducible T Cell Costimulator
IDH1	Isocitrate dehydrogenase 1
IL-17	interleukin 17
IL6	interleukin 6
KI67	Marker Of Proliferation Ki-67
KMT2C	Lysine Methyltransferase 2C
KRAS	Kirsten rat sarcoma virus
LAG3	Lymphocyte-activation gene 3
MAPK	Mitogen activated protein kinases
MDK	midkine
MDSC	myeloid derived suppressor cells
MET	mesenchymal–epithelial transition
MHC I/II	major histocompatibility complex I/II
MLH1	MutL Homolog 1
MLL3	mixed-lineage leukemia protein 3
MMR	mismatch repair
MSH2	MutS homolog 2
MSH6	MutS homolog 6
MTOR	mechanistic target of rapamycin kinase
MYC	master regulator of cell cycle entry and proliferative metabolism
NCAM	neural cell adhesion molecule
NCT	national clinical trial
NFKB	nuclear factor kappa light chain enhancer of activated B cells
NGS	next-generation sequencing
NOTCH1	Notch homolog 1
NRAS	neuroblastoma RAS viral oncogene homolog
NTRK1	neurotrophic receptor tyrosine kinase 1
OCT4	octamer-binding transcription factor 4
ONCOKB	MSK's precision oncology knowledge base
PALB2	partner and localizer of BRCA2
PARP	poly-ADP ribose polymerase
PBRM1	polybromo 1
pCCA	perihilar CCA

PD1	programmed cell death protein 1
PDGFR	platelet-derived growth factor receptors
PD-L2	PDCD1LG2 programmed cell death 1 ligand 2
PIK3CA	Phosphatidylinositol-4,5-Bisphosphate 3-Kinase Catalytic Subunit Alpha
POLD1	gene polymerase delta 1
POLE	DNA polymerase epsilon
PRKCA	protein kinase C alpha
PRKCB	protein kinase C beta
PRKDC	protein kinase, DNA-activated, catalytic subunit
PSC	primary sclerosis cholangitis
RABGAP1L	RAB GTPase activating protein 1 like
RAD50	RAD50 double-strand break repair protein
RAS	rat sarcoma virus
RB1	RB transcriptional corepressor 1
ROS1	ROS proto-oncogene 1
SMAD4	SMAD family member 4
SNP	single nucleotide polymorphism
SOS1	Son of Sevenless Homolog 1
SRC	SRC Proto-Oncogene
STAT3	Signal transducer and activator of transcription 3
T cells	thymus cells
TCGA	the cancer genome Atlas
TES	targeted-exome sequencing
TGFBR2	transforming growth factor, beta receptor II
TGFβ	transforming growth factor beta
TME	tumor microenvironment
TP53	tumor protein 53
TPPP	tubulin polymerization-promoting protein
TRK	tropomyosin receptor kinase
VEGF	vascular endothelial growth factor
VEGFR	vascular endothelial growth factor receptor
WES	whole-exome sequencing
WNT	wingless/integrated

1. Introduction

The term biliary tract cancers (BTCs) refers to a spectrum of uncommon and highly aggressive epithelial malignancies with limited treatment options (Valle, Lamarca, Goyal, Barriuso, & Zhu, 2017). BTCs can arise anywhere along the biliary tree (cholangiocarcinomas, CCAs), or from the cells constituting the gallbladder (gallbladder cancer, GBC). According to their anatomical site of origin, CCAs can be divided into intrahepatic (iCCA) and extrahepatic (eCCA), with the latter being further subdivided into perihilar and distal CCA (Banales et al., 2020). iCCA arises from the small ducts within the liver, and accounts for the second most common liver cancer, after hepatocellular carcinoma (HCC). Perihilar CCA (pCCA) develops between the second order of the bile ducts and the cystic duct, while distal CCA (dCCA) arises anywhere in the common bile duct between the cystic duct and the ampulla of Vater (Banales et al., 2020). In addition to their anatomical origin, risk factors and outcome differ between BTC subtypes; worse overall survival rates have been recently reported for patients with GBC compared to those with other BTCs (McNamara et al., 2020). Furthermore, while most cases of CCA occur sporadically (Bridgewater et al., 2014; Valle et al., 2016), risk factors for GBC primarily revolve around the presence of cholecystolithiasis which can be found in up to 90% of all patients (Valle et al., 2016). For both iCCA and eCCA, well established risk factors have been identified in a fraction of cases and include primary sclerosing cholangitis (PSC), biliary cysts, and fluke infections, with the latter representing the dominant risk factor in Southeast Asia (Rizvi & Gores, 2013). Risk factors associated with HCC (i.e., chronic HBV and HCV infections, cirrhosis, etc.) have also been associated with an increased risk for iCCA, but not eCCA (Tyson & El-Serag, 2011; Valle et al., 2016). The diagnosis of BTCs is challenging with clinical presentation varying considerably based on the location of the lesion (Benavides et al., 2015; Valle et al., 2017). Due to its asymptomatic onset, two thirds of cases are diagnosed when the tumor has already metastasized and systemic therapy with gemcitabine and cisplatin represents the first-line option, with a survival of ~9–12 months (Valle et al., 2010). In patients whose disease has progressed on gemcitabine and cisplatin, the ABC-06 study has recently demonstrated benefit of folinic acid, 5-fluorouracil, and oxaliplatin (FOLFOX) in the second line setting (Lamarca et al., 2021). Based on the results of this study, FOLFOX has become the standard second-line therapy for advanced CCA. However, no systemic therapy is curative and outcomes for CCA and GBC remain dismal. Ultimately, novel therapies are required, particularly at advanced stages.

Despite significant differences in prognosis, genetic landscape, and underlying tumor biology, the distinct BTC subtypes are frequently grouped together in clinical trials to achieve adequate sample size. This extensive inter-patient heterogeneity has long been recognized as a major contributor to the failure of targeted therapies in unselected patient populations, ultimately suggesting that molecular profiling of the tumor may be critical to enable new tailored strategies.

Overall, BTCs are rare (~3% of all gastrointestinal cancers), although steadily increasing incidence rates have been reported over the past 40 years (Banales et al., 2016; Bray et al., 2018). The limited availability of fresh human tissue samples has long hampered translational efforts in this disease. The first molecular subtyping endeavors were published in the early 2010s and mostly based on transcriptomic profiling of iCCA (Andersen et al., 2012; Oishi et al., 2012; Sia et al., 2013). More recently, there has been renewed focus in all BTCs, mostly due to increasing incidence and mortality rates. This, combined with the decrease of the once prohibitive costs of next-generation sequencing (NGS), has made it possible to generate genomic profiles at a significantly reduced cost in different types of human material. As a result, multi-institutional collaborations and international consortia including TCGA and ICGC have assembled collections of BTCs and provided comprehensive genomic analyses. In particular, the application of NGS has led to the discovery of a subset of CCAs harboring FGFR2 fusions or *IDH1* mutations, ultimately paving the way for precision medicine in this malignancy with selective FGFR or IDH inhibitors (Abou-Alfa, Macarulla, et al., 2020; Abou-Alfa, Sahai, et al., 2020; Javle et al., 2021). At the same time, the first integrative multi-omics classifications have been proposed. Although these classifications have not been implemented yet in clinical practice, they have greatly contributed to elucidating the molecular basis of this complex disease; however, a lot more remains to be done. As immune checkpoint inhibitors (ICIs) become frontline treatment options in several cancers (Ribas & Wolchok, 2018), multi omics-based subtyping efforts have been complicated by the emerging need of integrating elements of the immune microenvironment to develop robust algorithms able to predict response and resistance to immunotherapy (Liu & Mardis, 2017). This has opened new avenues to immunogenomic analyses of large cohorts using algorithms to virtually isolate immune signals or more sophisticated single cell-based technologies. Considering the poor responses observed in unselected BTC patients treated with the monoclonal antibody directed anti-PD1, pembrolizumab, (Piha-Paul et al., 2020), these new endeavors are becoming critical.

In this chapter, we provide a thorough review of the genomic and immunogenomic classifications of CCA and other BTCs, with particular emphasis on their current and future clinical implications.

2. The mutational landscape of biliary tract cancers

Over the past decade, the application of NGS has significantly impacted the management of BTC, particularly iCCA, where clinically-relevant subgroups have being identified based on the recurrence of specific structural aberrations (i.e., mutations, chromosomal rearrangements, etc.) with important differences according to the anatomical site, underlying risk factors and country of origin (Chaisaingmongkol et al., 2017; Chan-On et al., 2013; Farshidfar et al., 2017; Goeppert et al., 2019; Jiao et al., 2013; Jusakul et al., 2017; Wardell et al., 2018). In the following paragraphs, we will describe the mutational profile of the distinct BTC subtypes with a focus on those subgroups associated with prognosis and better response to targeted therapies (Fig. 1).

2.1 Genetic subgroups of cholangiocarcinoma

The first takeaway from whole-exome sequencing (WES) and targeted-exome sequencing (TES) applied to large cohorts of CCAs is that a

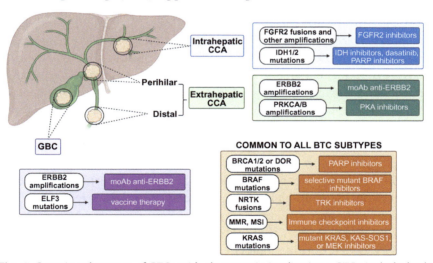

Fig. 1 Genetic subgroups of BTCs with therapeutic implications. BTCs include both CCAs and GBCs. CCAs include intrahepatic and extrahepatic CCA. Extrahepatic tumors can be further sub-classified into perihilar and distal tumors. In the figure, the distinct anatomic location of each BTC subtype is indicated along with targetable molecular alterations. Both emerging and established therapeutic options are indicated.

significant fraction of tumors harbor potentially targetable genetic alterations. A comprehensive WES study conducted on 239 BTCs identified potentially druggable targets in around 40% of CCA patients; the most recurrent occurred in *IDH1/2, FGFR1–3, ALK, EGFR, PIK3CA, BRAF* and *BRCA1/2* (Nakamura et al., 2015). Similarly, the genomic analysis of 412 BTCs identified 32 frequently mutated genes with around 58% of patients harboring one or more actionable alterations (Wardell et al., 2018). In another analysis including only CCAs, of whom 158 (81%) were iCCAs and 37 (19%) were eCCAs), 47% of patients were found to harbor genetic aberrations with potential therapeutic implications, including *IDH1* mutations (22%) and *FGFR2* rearrangements (9%), as well as additional low frequency actionable mutations (*PIK3CA*, 3%; *NRAS*, 2%, *ERBB2*, 1%) (Lowery et al., 2018), confirming the importance of molecular profiling for this malignancy. Despite some variability in frequencies, all these studies confirmed *IDH* mutations and *FGFR2* rearrangements as the most recurrent targetable alterations in CCA. In terms of clinical and biological relevance, these alterations have been more frequently detected in iCCA while they are rarely found in eCCA.

IDH1–2 mutations occur in around 10–25% of CCAs. The hotspots typically mutated in *IDH1* include R132C, R132G, or R132L, while *IDH2* mutations typically include R172W, R172M or R172S. Mechanistically, mutated *IDH* proteins convert alpha ketoglutarate to 2-hydroxyglutarate (2HG) that ultimately suppresses HNF4A, leading to attenuated hepatocyte differentiation, expansion of liver progenitor cells, and development of CCA in presence of additional drivers (i.e., *KRAS* mutations) (Saha et al., 2014). The analysis of 52 iCCAs elegantly illustrated how *IDH* mutant tumors have distinct genomic features and cluster separately from other iCCA tumors in terms of DNA methylation and genomic copy number alterations (CNAs) (Goeppert, Toth, et al., 2019). In particular, the *IDH* mutant group showed a highly disrupted genome, characterized by frequent deletions of chromosome arms 3p and 6q, and hypermethylation. In line with these results, TGCA analysis described *IDH* mutant tumors as characterized by low chromatin modifier gene expression, *ARID1A* hypermethylation, and increased mitochondrial DNA copy number (Farshidfar et al., 2017). Finally, genomic alterations of a large cohort of 496 iCCA patients using three classifier genes (*IDH*, *KRAS*, *TP53*) identified unique mutational signatures, co-mutation profiles and enriched pathways for each subgroup (Nepal et al., 2018). In this study, the *IDH* group displayed the greatest number of differentially methylated regions and the lowest number of somatic CNAs. In terms of signaling pathways, the *IDH* group was enriched in metabolic pathways, which is consistent with the known involvement of *IDH1* in metabolic processes, and better prognosis. The *KRAS* group was highly enriched in immune-related pathways, whereas MAPK, WNT and p53 signaling were enriched in the *TP53* group. By performing a high-throughput screening using a drug library of 525 late-stage trial and/or Food and Drug Administration (FDA)–approved compounds, the authors identified differential sensitivity to specific compounds suggesting that such molecular stratification could be applied to predict therapeutics sensitivity in patients (Nepal et al., 2018).

FGFR2 gene rearrangements account for approximately 15% of all CCA cases (Lowery et al., 2018), although discrepancies have been detected across cohorts (2–45%) (Arai et al., 2014; Borad et al., 2014; Lowery et al., 2018; Nakamura et al., 2015; Ross et al., 2014; Sia et al., 2015; Wu et al., 2013). To date, more than 60 fusion partners have been identified for *FGFR2* rearrangements (Silverman et al., 2021). Regardless of gene identity, the fusion partners allow FGFR2 to oligomerize without the presence of FGF ligands, leading to enhanced cell proliferation and invasion (Wu et al., 2013). Genomic characterization of 138 CCA patients with *FGFR2 rearrangements* enrolled in the FIGHT-202 clinical trial testing the FGFR inhibitor, pemigatinib, identified a total of 140 *FGFR2* fusions with 2 patients showing 2 *FGFR2* rearrangements each (Silverman et al., 2021). Confirming previous results, patients were more likely to show a unique rearrangement partner, whereas only 16% of patients shared a rearrangement partner. *FGFR2* fusions most frequently occurred intra-chromosomally on chromosome 10 (53%) and the most frequent was the *FGFR2–BICC1* (Wu et al., 2013), accounting for 28% of all *FGFR2* rearrangements. *TP53* and *KRAS* alterations were observed less frequently in *FGFR2*-rearranged patients, whereas 63% of *FGFR2*-rearranged patients had co-alterations in a well-known tumor-suppressor gene, including *BAP1, CDKN2A/B, PBRM1,* and *ARID1A* (Silverman et al., 2021). Presence of *FGFR2* fusions and *IDH* mutations has recently been reported more frequently in small duct iCCAs while large duct iCCAs present a genetic profile closer to eCCA (Akita et al., 2019; Graham et al., 2014). Interestingly, *FGFR2* fusions have also been found less frequently in fluke-associated CCAs and are mutually exclusive with mutations in *KRAS, BRAF,* and *FGFR* (Kongpetch et al., 2020).

Besides *FGFR2* fusions, additional genetic differences have been observed between fluke-associated CCA and fluke-negative tumors

(Chan-On et al., 2013; Ong et al., 2012). For example, the comparative analysis of the mutational profiles of 108 liver-fluke infected cases with 101 non-liver fluke-related tumors identified a higher mutation burden and more frequent *TP53* and *SMAD4* mutations in fluke-infected CCAs, whereas *IDH1/2* and *BAP1* mutations were more frequent in fluke-negative tumors (Chan-On et al., 2013). A recent integrative analysis reinforced the genetic impact of distinct risk factors with the identification of fluke-positive clusters showing elevated mutation rates, more frequent *ERBB2* amplifications, *TP53* mutations, and hypermethylation of CpG islands (Jusakul et al., 2017). In line with previous studies, fluke-negative clusters showed high numbers of CNAs, mutations in *IDH1/2* and *BAP, FGFR2* fusions, and hypermethylation of CpG shores (Jusakul et al., 2017). Finally, the comprehensive molecular profiling of 164 Chinese and 283 US patients with iCCA identified important differences between Asian and Western populations that could be attributed to variations in the underlying liver disease (Cao et al., 2020). In this study, Chinese patients had significantly higher frequency of *TP53* as well as *KMT2C, BRCA1/2, DDR, TGFBR2,* and *RB1,* whereas mutations in *IDH1/2, BAP1,* and *CDKN2A/B* were more dominant in the Western cohort (Cao et al., 2020). Consistent with a different tumor profile between Eastern and Western CCA patients, Chaisaingmongkol et al. clustered Asian iCCAs into four subtypes that shared features with HCC but less so with Caucasian patients (Chaisaingmongkol et al., 2017). In Western countries, PSC is one of the most well known risk factors for CCA. A recent study demonstrated that, regardless of the anatomical site of origin, PSC-related BTCs show a molecular profile similar to eCCA with high frequency of *TP53* (36%), *KRAS* (28%), *CDKN2A* (15%), and *SMAD4* (11%), as well as potentially druggable mutations (e.g., *HER2/ERBB2*) (Goeppert et al., 2020).

KRAS and *TP53* mutations have been described in both iCCA and eCCA, although recent studies have reported higher prevalence of *KRAS* (38%) and *TP53* (35%) mutations in eCCA (Montal et al., 2020; Wardell et al., 2018). Both genes play a key role in driving CCA, and their prognostic value has been extensively explored over the years (Sia, Villanueva, Friedman, & Llovet, 2017). Recently, a large study of 412 BTCs confirmed that both mutations negatively affect patient prognosis (Wardell et al., 2018). These results have been further confirmed in a large cohort of 412 iCCAs where *TP53, KRAS,* and *CDKN2A* alterations were identified as independent prognostic factors when controlling for clinical and pathologic variables, disease stage, and treatment (Boerner et al., 2021). Interestingly, *TP53*-mutated CCAs are more likely to be HBsAg-seropositive, whereas *KRAS* mutations have been described nearly exclusively in HBsAg-seronegative CCA patients (Zou et al., 2014). Other genetic alterations showing different distribution according to the anatomical location occur in *BAP1* and *ARID1A*, which are found more frequently in iCCA, whereas novel fusions in *PRKACA* and *PRKABC*, mutations in *ARID1B*, and *ERBB* pathway alterations are mostly found in eCCA cases (Lowery et al., 2018; Nakamura et al., 2015) with the latter representing the most relevant for therapeutic intervention. In this regard, the genomic characterization of 1863 CCA cases (1615 iCCAs and 248 ECCAs) through NGS revealed *ERBB* alterations in 4.2% of iCCAs and 10% of eCCAs with the majority harboring a point mutation in *ERBB2* (Jacobi et al., 2021).

Other clinically relevant alterations, although relatively rare, include *NTRK1* fusions which have been described in 0.67% of patients with BTC (Boileve et al., 2021), with the fusion *RABGAP1L–NTRK1* being identified in iCCA (Ross et al., 2014). TRK inhibitors represent a therapeutic opportunity for these patients as well as those patients with *ROS1* and *ALK* fusions (Ross et al., 2014). Finally, frequency of mismatch repair deficiency (MMR) and/or microsatellite instability (MSI) is of great clinical relevance in BTC patients since it has been associated with high response rates to immunotherapy across multiple tumor types (Le et al., 2015, 2017). Unfortunately, MMR is a rare event in CCA with an overall incidence of 3% in iCCA (Bonneville et al., 2017; Isa et al., 2002; Momoi et al., 2001; Weinberg et al., 2019; Yoshida et al., 2000), and 2% in eCCA (Goeppert et al., 2019; Kim et al., 2003; Weinberg et al., 2019).

2.2 New findings for GBC

Although molecular profiling efforts for GBCs trail behind those for CCA, in the past few years an increasing number of studies have emerged, with the two largest cohorts published thus far including up to 157 (Li et al., 2019) and 224 (Nepal et al., 2021) GBCs samples, respectively. Tumor profiles in these studies are from patients in Korea, India and Chile, countries where GBC incidence is the highest (Rawla, Sunkara, Thandra, & Barsouk, 2019). Prior to these large cohort sequencing efforts, the mutational landscape of GBC had been analyzed only in a few samples (Jiao et al., 2013; Li et al., 2014; Nakamura et al., 2015). Among these small studies, WES of 9 Caucasian GBCs followed by TES in 8 additional cases identified

TP53 mutations (63%) as the most recurrent, followed by mutations in several chromatin remodeling genes (*PBRM1*, 25%; *BAP1*, 13% and *ARID1A*, 13%) (Jiao et al., 2013). Two studies of 57 (Li et al., 2014) and 79 (Nakamura et al., 2015) Asian GBCs, using a mix of WES and TES, confirmed the driving role of *TP53*, which was found mutated in 43–47% of patients, and identified ERBB signaling, including *EGFR*, *ERBB2*, *ERBB3*, *ERBB4* and their downstream genes, as the most extensively mutated pathway, affecting overall 37% of the GBC samples. In a recent follow-up study of 157 GBCs, the authors identified *ERBB2* and *ERBB3* mutations at a frequency of 7–8% and confirmed that patients harboring these mutations exhibited poorer prognosis (Li et al., 2019). The most recent mutational analysis was conducted in 190 GBC samples and, unlike previous studies, only reported 5.4% mutations in *ERBB2*, although it identified aberrant methylation of both *ERBB2* and *ERBB3* suggesting another layer of regulation for these drivers. Pandey et al. described mutations in the *CTNNB1* gene in 12% of GBCs (Pandey et al., 2020) but other studies failed to confirm such results, while *KRAS* mutations have been overall identified in most studies (4–8%). Other genes typically mutated in CCA (i.e., *FGFR2* fusions, *IDH* mutations) were not found altered in GBC. A recurrent fusion protein occurring between the cytoskeletal gene, tubulin polymerization-promoting protein (*TPPP*), and the chromatin remodeler, bromodomain-containing protein 9 (*BRD9*) has been recently reported in 15% of GBCs although its functional consequences or therapeutic implications are still unknown (Nepal et al., 2021).

The ampulla of Vater represents an anatomical structure characterized by the crossroad of three distinct epithelia: biliary, intestinal, and ductal pancreatic. Adenocarcinomas arising in the ampulla of Vater, albeit not strictly grouped with BTCs, can exhibit biliary, intestinal, pancreatic, or mixed features, and represent an opportunity to understand the biology and cell of origin of all periampullary malignancies. Interestingly, genomic profiling of 172 ampullary carcinomas identified a mutational landscape highly resembling that of GBCs with frequent mutations in *TP53* (48%), *ERBB3* (11%), *ERBB2* (12%), and *ELF3* (12%), (Yachida et al., 2016); the latter gene has been described as mutated in 3–21% of GBCs with the majority being frame-shift, stop gained and essential splice-site mutations (73%). *ELF3* mutations were found more frequently in Korean (31%; 28/91) and Chilean GBC patients (22%; 2/9) compared to patients from India (7%) and frequently co-occurred with *TP53* mutations (Pandey et al., 2020). Frame-shift mutations in *ELF3* resulted in the highest number of predicted neoantigens which could represent potential candidates for vaccine therapy (Pandey et al., 2020).

3. Multi-omics classifications of intrahepatic cholangicoarcinoma

In recent years, molecular classifications of many cancer types have informed therapy choices, provided insight into actionable targets, and improved patient outcomes. Considering the genetic and clinical differences observed between BTC tumors, increasing efforts have tried to generate molecular classifications specific for each subtype, although initial studies analyzed CCA samples regardless of their anatomic site of origin. In all these efforts, both dCCA and pCCA were significantly underrepresented compared to iCCA. In the following section, we will review these studies focusing particularly on iCCA.

3.1 Integrative molecular classifications

Evidence of the distinct molecular profile between iCCA and eCCA had already transpired since the first proposed molecular classification (Andersen et al., 2012). The study analyzed the transcriptomic profile of 104 surgically resected CCA samples collected from patients in Australia, Europe, and the United States, including 36 eCCAs and 68 iCCAs. Transcriptome-wide survival association analysis identified 2 main prognostic classes with most of eCCA tumors clustering together in the poor prognosis subgroup (75% of tumors) (Andersen et al., 2012). In 2015, Nakamura et al. molecularly applied WES and RNA-seq to 239 BTCs, including 137 iCCAs, 74 eCCAs and 28 GBCs. Unsupervised clustering of gene expression levels identified four molecular subgroups associated with specific driver genes and distinct prognosis. Interestingly, cluster 1 consisted only of eCCA samples whereas cluster three included only iCCAs with enrichment of mutations in *BAP1*, *IDH1* and *NRAS* as well as *FGFR2* fusions (Nakamura et al., 2015). Similarly, an integrative analysis of 489 CCAs (133 Fluke-Pos and 356 Fluke-Neg cases) was conducted using different genomic platforms including WGS ($n=71$ cases), WES sequencing ($n=200$), high-depth TES ($n=188$), SNP array copy-number profiling ($n=175$), array-based DNA methylation profiling ($n=138$), and array-based expression profiling ($n=118$). Integrative clustering defined four

CCA clusters: clusters 1 and 2 (fluke-positive CCAs) were enriched in eCCA samples with *ERBB2* amplifications and *TP53* mutations, whereas clusters 3 and 4 (fluke-negative CCAs) were mostly iCCA with high copy-number alterations and *PD-1/PD-L2* expression, or epigenetic mutations (*IDH1/2, BAP1*) and *FGFR/PRKA*-related gene rearrangements (Jusakul et al., 2017). This evidence, albeit circumstantial, suggest that CCAs arising from the distinct anatomical sites represent different molecular entities with distinct genetic, molecular and clinical characteristics.

In an attempt to generate a molecular classification specific for iCCA, our group conducted an integrative genomic analysis of 149 iCCAs and identified two main biological classes, termed the Inflammation and Proliferation classes (Sia et al., 2013). The Inflammation class (~38% of iCCA) was associated with good survival and low recurrence rates compared to the Proliferation class (~62% of the cohort). The Inflammation class was enriched in inflammatory features such as chemokine signaling, overexpression of cytokines of the Th2 subtype (*IL-4, IL-10*), enrichment of the IL-17 family, and STAT3 activation (Sia et al., 2013). On the other hand, the Proliferation class was enriched in activation of multiple receptor tyrosine kinase (RTK) pathways such as EGF, RAS, AKT, PDGFR, VEGFR, and MAPK (Sia et al., 2013). There was also found to be frequent *BRAF/RAS, KRAS*, and *EGFR* mutations, NOTCH1 pathway activation, and deletions of the Hippo pathway locus (Sia et al., 2013). Similar to Sia et al.'s classification, the study conducted by Andersen et al. clustered CCAs in 2 groups defined by "good" and "poor" prognosis although as mentioned above, this study included both iCCAs and eCCAs. Interestingly, the poor prognosis cluster had an enrichment of *KRAS/BRAF* mutations, a deregulation of *ERK1/2*, and activation of multiple RTKs, corroborating findings that the subset of CCA with the poorest survival are enriched in RTK signaling (Andersen et al., 2012; Sia et al., 2013). In line with the Proliferation and Inflammation classification, a subset of iCCA with stem cell-like features was encompassed in the Proliferation class (Oishi et al., 2012). These tumors were also associated with poor survival and had an upregulation of TGFβ and NFKB signaling pathways as well as stem cell related genes (i.e., *Oct4, CD133, NANOG, NCAM*), (Oishi et al., 2012). These tumors also had an epithelial-mesenchymal transition (EMT) phenotype potentiated by miR-200c. Several other studies have confirmed the existence of subsets of iCCA with activation of inflammatory pathways or RTKs enrichment. For example, among the four clusters identified by Jusakul et al., two were predominantly iCCA, with Cluster 3 showing an enrichment of immune checkpoint molecules PD-1, PD-L1, and BTLA, and an antigen cross-presentation signature. On the other hand, Cluster 4 was enriched in RTK signaling, exhibited mutations in *BAP1* and *IDH1/2*, *FGFR* alterations, and increased *PI3K* signaling (Jusakul et al., 2017). Similarly, another study found a subclass of iCCA enriched for inflammatory pathways and mutations in *TP53* and *KRAS*. This subclass was associated with poor survival, though the study is somewhat limited by a small cohort ($n = 30$) (Ahn et al., 2019). Taken together, these data demonstrate across multiple cohorts and stratification methods that distinct subsets of iCCA with a more inflammatory versus proliferative phenotype can present unique molecular aberrations and patient outcomes (Fig. 2, as also extensively reviewed in (Sia et al., 2017)).

3.2 Immune microenvironment-based subtypes

Though many studies have aimed to characterize genomic and molecular aberrations of iCCA, few studies have investigated components of the tumor microenvironment (TME) and their prognostic value (Fig. 3). In one of the first studies to classify iCCA based on elements of the TME, tumors were stratified into four different subtypes based on immune infiltration and patient outcome: Immune desert (I1), Immunogenic (I2), Myeloid (I3), and Mesenchymal (I4), (Job et al., 2020). The immune desert cluster, accounting for 48% of iCCA, was defined by no immune activation or presence of MHC class I/II, and an upregulation of metabolism, cell growth, and both FGFR and WNT signaling (Job et al., 2020). The Immunogenic cluster (9%), presenting the best outcome, had high recruitment of innate and adaptive immune cells including CD4+ and CD8+, and activation of immune checkpoint pathways (Job et al., 2020). The myeloid cluster (13%) was defined by strong expression of monocyte-derived signatures, myeloid cell types, high expression of CD69, and low expression of lymphoid signatures. Finally, the cluster associated with the worst survival, Mesenchymal (28%), was enriched for mesenchymal features and strong expression of activated fibroblasts (Job et al., 2020). Similarly, Carapeto et al. described four immune-based clusters of iCCA with ranging levels of immune infiltration based on immune checkpoint-directed immunohistochemistry of 96 iCCA samples. Group 2, described as immune "hot," had increased chemokine signaling, expression of surface markers CD3, CD4, and CD8a, and immune checkpoint molecules PD-1, ICOS, and LAG3 (Carapeto et al., 2021). Groups 3 and 4 consisted of immune cold tumors

iCCA Molecular Classes

Characteristics	Proliferation Class	Inflammation Class
Subtype	Progenitor-like iCCA	
Signaling Pathways	KRAS/RAF/ERK, IGF1R, EGFR, NOTCH, HER2, MET, PI3K/AKT/mTOR	Inflammatory pathways (Interleukins/chemokines), STAT3 activation
Prognostic Signatures	Poor prognostic signatures (i.e. G3, S1-2, Cluster A, stroma, iCCA-like HCC); Stem cell like iCCA	
DNA Methylation	Hyper-methylation	
Oncogenic Mutations	KRAS, BRAF, EGFR; IDH1/2; Chromatin remodeling genes (ARID1A, BAP1, PBMR1), TP53	
Structural Aberrations	Chr 11q13 amplif. (CCND1, FGF19); Chr14q22.1 (SAV1) Focal deletion; FGFR2 translocations	
Clinical Features	Intra-neural invasion; Moderate/poorly differentiated; Poor prognosis (survival/recurrence)	Well differentiated; Good prognosis

Fig. 2 Tumor-based classification of iCCA. According to the gene expression profile of the tumor, 2 main classes of iCCA have been identified, named Proliferation and Inflammation. The Proliferation class is characterized by chromosome instability, activation of oncogenic pathways, mutations in driver genes, and poor prognosis. The Proliferation class encompasses a subset of tumors with progenitor cell-like features, mutations in IDH1/2, and hypermethylation. Tumors of the Inflammation subclass show overexpression of cytokines and less aggressive clinical behavior. *Reprint of Fig. 5 from Sia, D., Villanueva, A., Friedman, S. L., and Llovet, J. M. (2017). Liver cancer cell of origin, molecular class, and effects on patient prognosis. Gastroenterology 152, 745–761.*

with low overall survival. Interestingly, both T-cell and immune checkpoint markers were found enriched at the tumor margins compared to the tumor center (Carapeto et al., 2021).

Though these studies provided insight into immune cell populations and which types of iCCAs are most likely to respond to ICIs, they do not provide integrative information detailing the genomic and transcriptomic landscape of these tumors to inform choices of potential combination strategies

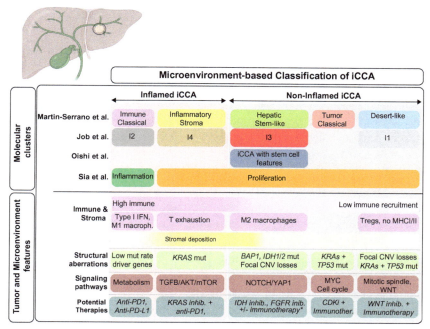

Fig. 3 TME-based molecular classification of iCCA. Main characteristics and overlap of the recently proposed microenvironment-based classes of iCCA. Each class has unique tumor-, stroma- and immune microenvironment-related features. At the bottom of the figure, candidate combination strategies with therapies aimed at targeting both the tumor and the surrounding microenvironment are indicated. These therapies are recommended based on the molecular characterization of each class and need to be thoroughly validated in prospective clinical trials. The * refers to immunotherapy strategies targeting the depletion of MDSC or TAMs.

with other targeted therapies. In this regard, our group recently provided a tumor microenvironment-based classification of iCCA, also called Stroma Tumor Immune Microenvironment (STIM classification) (Martin-Serrano et al., 2021). This classification defined five clusters of iCCA, encompassing both inflamed (35%) and non-inflamed (65%) phenotypes (Fig. 3). The inflamed profiles encompasses both the Immune Classical (10%) and the Inflammatory stroma (25%) classes; both clusters are enriched in gamma delta T cells, CD8 + T cells, and immune checkpoint molecules but differ in terms of stromal infiltration, ECM stiffness and cancer-associated fibroblasts (CAF) subtypes (Martin-Serrano et al., 2021). Of the non-inflamed profiles, the Desert-like (20%) showed the highest tumor purity and lowest immune and stromal infiltration, with the presence of immunosuppressive FOXP3+ T regulatory cells. Interestingly, a non-inflamed profile, the Hepatic stem-like class, highly resembled the previously described iCCA subtype with stem cell features (Oishi et al., 2012). While the Hepatic stem-like class showed low immune and stromal infiltration, its immune landscape was characterized by a high presence of macrophages type 2. The last of the non-inflamed tumors, the Tumor Classical class (12%), was defined by high expression of cholangiocyte markers and cell cycle related pathways (Martin-Serrano et al., 2021). Each class also had distinct mutational aberrations, with the Inflammatory Stroma having frequent *KRAS* mutations, and the Tumor Classical and Desert-like classes showing a co-occurrence of *TP53* and *KRAS* mutations. Consistent with previous studies characterizing the subtype of iCCA with stem cell features, the Hepatic Stem-like was enriched in *IDH1/2* and *BAP1* mutations. Each STIM cluster had differences in survival, with the Hepatic Stem-like having the best probability of survival and Tumor Classical the worst. Transcriptomic overlap with our previously reported molecular classes (Sia et al., 2013) identified transcriptomic similarity between the Inflammation class and the Immune classical, whereas all other classes were mostly enriched in the Proliferation class (Martin-Serrano et al., 2021). Similarities could also be noted between the STIM classification and the immune subtypes proposed by Job et al. (Fig. 3).

In addition to tumor cells, the iCCA TME has many distinct cell types, such as immune cells, extracellular matrix proteins, and fibroblasts. Cancer-associated fibroblasts (CAFs) have been widely studied, and are generally associated with poor survival in iCCA (Chuaysri et al., 2009). CAFs promote iCCA tumor growth through induction of cytokines (IL-6, TGFβ) and growth factors (VEGF, FGF2) (Tamma et al., 2019). Given that inflamed iCCA tumors have an enrichment of fibroblasts and a malignant-ECM transcriptional program that is dependent on TGFβ signaling in CAFs, stratifying the types of CAFs present in the iCCA TME has some prognostic value and could better inform treatment options (Job et al., 2020; Martin-Serrano et al., 2021). Recent single cell RNA-sequencing based studies of iCCA have identified multiple subtypes of CAFs (Affo et al., 2021; Tamma et al., 2019). Distinct hepatic stellate cell-derived populations of CAFs include myofibroblastic and inflammatory CAFs, which make up much of the CAF population and promote iCCA progression though HAS2 and HGF-MET, respectively (Affo et al., 2021). Another large population of CAFs, vascular CAFs have shown high microvascular signatures and levels of IL-6, which enhance malignancy and promote tumor progression of iCCA by IL-6-induced epigenetic alteration of EZH2 (Zhang et al., 2020). A smaller population, antigen presenting CAFs, were implicated in response to IFNγ in addition to antigen presentation to CD4+ T cells via MHC-II (Zhang et al., 2020).

Considering the complexity of the iCCA TME, classifications that have stratified immune cell and stromal components have given new insight into therapeutic strategies for iCCA based on level of immune and stromal infiltration and the molecular pathways implicated. Initial trials with the anti-PD1 therapies have been discouraging (Piha-Paul et al., 2020). However, identifying patients with immune-hot tumors, or combining immunotherapies with targeted therapies based on patients' mutational status may prove to be more effective and sensitize tumors to ICI. Preclinical data from Martin-Serrano et al. suggest that the treatment of KRAS-mutant mice, which recapitulate the Inflammatory Stroma STIM class, with a KRAS-SOS1 inhibitor sensitized mice to anti-PD1 therapy (Martin-Serrano et al., 2021).

4. Molecular classifications of other biliary tract cancers

Compared to iCCA, only a handful of molecular classifications have been proposed for eCCA and GBC, although the number is expected to rise in the near future. In the following paragraphs, we will review the most recent efforts for each BTC subtype.

4.1 Molecular classification of extrahepatic CCA

Due to the low number of eCCA samples analyzed thus far in large international initiatives, only one molecular classification has been proposed for eCCA (Montal et al., 2020). In this study, the unsupervised transcriptomic-based analysis of 189 eCCAs (76% pCCA and 24% dCCA) from Western countries identified four distinct molecular classes, named based on biological features of tumoral cells and their TME (Fig. 4). The Metabolic class (19%) showed a hepatocyte-like phenotype with activation of the transcription factor HNF4A, enrichment in gene signatures related to bile acid metabolism, and high infiltration of gamma delta T cells. Interestingly, this subgroup showed some overlap with the CTNNB1 (Chiang et al., 2008) and S3 classes of HCC (Hoshida et al., 2009), known to have a hepatocyte-like phenotype. The Proliferation class (23%) was enriched in patients with dCCA and was characterized by

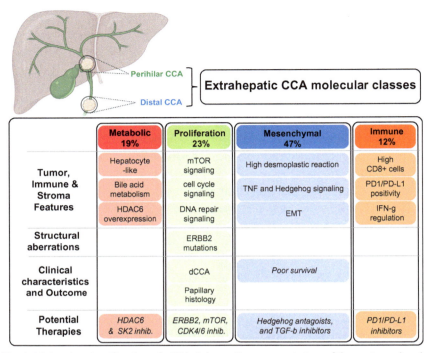

Fig. 4 Molecular classification of eCCA. Schematic representation of the main molecular and clinical characteristics of the recently described four classes of eCCA (Montal et al., 2020). Thus far, this represents the only classification specific for eCCA tumors, including both perihilar CCA (pCCA) and distal CCA (dCCA). At the bottom of the figure, candidate targeted therapies with a potential benefit in each molecular class are indicated. Recommendation of treatment is based on molecular data at transcriptomic and protein level. Phase 2/3 clinical trials to investigate these findings are needed to confirm the efficacy of the proposed targeted therapies.

enrichment of MYC targets, high Ki67$^+$, *ERBB2* mutations/amplifications, and activation of mTOR signaling. Additionally, CD4 + T cells, specifically T helper 17 cells, were the predominant immune cell populations infiltrating the TME of this class. Cell cycle, cyclin-dependent kinase (CDK), and mTOR inhibitors such as everolimus were identified as potential drugs able to reverse this phenotype in vitro. The Mesenchymal class was the most prevalent (47%) and was defined by signatures of EMT, TGFβ signaling, and poor overall survival; these results were consistent with the intense fibrotic reaction detected in these tumors upon pathological analysis and in silico gene expression deconvolution. Of relevance, the extracellular matrix protein periostin, produced by activated CAFs, was the top overexpressed gene in the Mesenchymal class; its overexpression was independent from the degree of CAF infiltration, implying that CAF activation, moreso than abundance of CAFs, dictates the mesenchymal phenotype of this subtype. Finally, the Immune class (11%) showed higher lymphocyte infiltration, overexpression of PD1/PD-L1, and molecular features associated with a better response to ICIs. Direct comparison of eCCA with our previously described iCCA molecular classes (Sia et al., 2013) identified significant similarities only in the proliferation classes, further highlighting the importance of anatomical location in the molecular subtyping of CCA. Indeed, no similarities were detected between the Immune class of eCCA and the "Inflammation class" of iCCA (Sia et al., 2013) nor between the Immune class of eCCA and the more recently reported inflamed classes of iCCA, Immune classical and Inflammatory Stroma (Martin-Serrano et al., 2021). Surprisingly, the only molecular similarity was identified between the Metabolic class of eCCA and the STIM Immune classical of iCCA, with both molecular classes being defined by enrichment of metabolic pathways (Martin-Serrano et al., 2021). The *IDH* mutant-enriched subtype of iCCA identified by TCGA (Farshidfar et al., 2017), displaying high mitochondrial and low chromatin modifier gene expression, was not detected in eCCA tumors. These observations add further evidence supporting iCCA and eCCA as two different molecular entities for clinical decision-making purposes.

4.2 Molecular and immune subtypes of gallbladder cancer

To date, only one integrative study has been conducted in GBC with the goal of identifying genomic subgroups. In this study, matched transcriptomes, DNA methylomes, and somatic CNAs derived from 45 patients were analyzed to explore molecular subtypes (Nepal et al., 2021). Transcriptome-wide survival association analysis identified a 95-gene signature that stratified patients into three subtypes with distinct histopathological features and oncogenic profiles (Fig. 5). Subtypes 1 and 3 included tumors with poor prognoses and advanced stage, whereas subtype 2 included tumors with longer survival, a gastric mucous-producing histomorphology, smoking history, and enrichment for the *TPPP-BRD9* gene rearrangement. No other clinical associations were observed. Similarly, mutations in chromatin remodelers, *TP53* mutations, or other structural aberrations were equally distributed across the three subtypes. The poor survival subtypes were enriched in MAPK and TGFβ signaling,

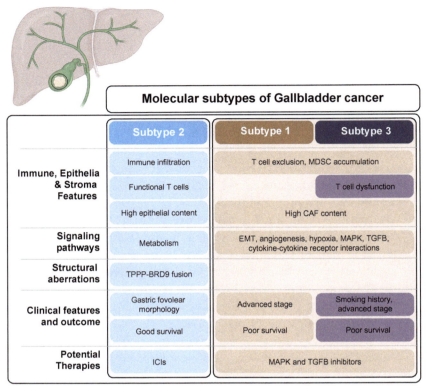

Fig. 5 Molecular subgroups of GBC. Overview of the 3 subtypes of GBCs based on the recent integrative analysis of 190 samples (Nepal et al., 2021). Transcriptome-wide survival association analysis identified a 95-gene signature that stratified all patients into 3 subtypes. The 2 poor-survival subtypes, subtypes 1 and 3, were associated with adverse clinicopathologic features, an immunosuppressive microenvironment, and T cell dysfunction, whereas the good-survival subtype showed the opposite features. The newly identified fusion gene TPPP-BRD9 was enriched in the subtype 2, although no statistical significance was achieved.

EMT, hypoxia, and angiogenesis. Immune deconvolution using xCELL did not identify meaningful differences in terms of immune composition across the distinct subtypes, although endothelial cells and MDSCs were elevated in the poor prognosis subtype 3 along with the gene expression of immune checkpoint inhibitors (i.e., *PD1*, *PD-L1* and *CTLA4*) and T cell dysfunction (Nepal et al., 2021). Another study attempted to identify immune-based classes of GBCs by applying the deconvolution tool xCell to bulk RNA-sequencing data of 115 GBCs from patients in Korea, India and Chile (Pandey et al., 2020). The analysis identified five distinct clusters with cluster 1 showing significantly higher levels of $CD8^+$ T cells and expression of *LAG-3* and *PD-L1*. Unlike the other classification, these immune-based clusters did not show differences in terms of prognosis. Similarly, mutations did not differ across the subtypes. Of interest, cluster 4 showed a higher level of endothelial cell signature and patients in the highest endothelial cell quartile showed a significant reduction in survival, potentially resembling patients of the subtype 3 described in Nepal et al. Interestingly, single cell RNA-seq of 13 GBCs recently linked GBCs with ERBB pathway mutations to reduced anti-cancer immunity and worse survival (Zhang et al., 2021). In this study, the authors observed that ERBB pathway mutated tumors harbored larger populations of specific epithelial cell subtype and expressed higher levels of secreted midkine (MDK), which promoted immunosuppressive macrophage differentiation (Zhang et al., 2021).

5. Therapeutic implications

Multi-omics, molecular, and immune classifications of CCA or GBC have largely contributed to elucidating the molecular basis of these complex malignancies. Unfortunately, due to the lack of validation in prospective clinical trials and the discouraging inter- and intra-tumoral heterogeneity, no classifications have been implemented in clinical practice. While their clinical utility remains unclear, the discovery of frequent targetable alterations has successfully paved the way to precision therapeutic interventions for BTCs. As an example, molecular profiling of 195 CCA patients identified actionable genetic aberrations classified as level 3 using the OncoKB database (Chakravarty et al., 2017) in 39% of patients, and led to biomarker-directed therapy or clinical trial enrollment in 16% of them. Among these patients, 13 received the IDH1 inhibitor ivosidenib, 6 received FGFR inhibitors, and 2 received HER2-directed therapy (Lowery et al., 2018). Of relevance, 64% of the patients being treated with targeted therapy had evidence of response or clinical benefit to treatment, highlighting the potential of molecular testing and enrichment-based trial design to improve the outcome of these patients (Lowery et al., 2018).

Undoubtedly, *IDH1* mutations and *FGFR2* rearrangements represent the most relevant and frequent aberrations identified thus far in CCA. Their groundbreaking discovery has recently led to the FDA approval of ivosidenib for CCAs harboring *IDH1* mutations and the accelerated approval of the first FGFR inhibitors, pemigatinib and infigratinib, for patients with *FGFR2* aberrations. The approval of ivosedinib was granted based on the results of the phase 3 trial ClarIDHy, reporting a median progression free survival of 2.7 months in the ivosidenib arm vs 1.4 months in the placebo group (Abou-Alfa, Macarulla, et al., 2020). A new phase 1 study testing ivosidenib in combination with chemotherapy (NCT04088188), and a phase 2 trial testing ivosidenib with nivolumab (NCT04056910) are currently ongoing. Concurrently, other selective IDH1 inhibitors, including IDH305 (NCT02381886), HMPL-306 (NCT04762602) and FT-2102 (NCT03684811), are being tested. Another agent currently being investigated in a phase 2 study in *IDH*-mutated CCA patients is the multi-tyrosine kinase inhibitor dasatinib (NCT02428855). This study finds its rationale in a recent report showing susceptibility of an *IDH*-mutated orthotopic CCA model to the dasatinib-mediated inhibition of the *SRC* gene (Saha et al., 2016). Finally, it has been suggested that the accumulation of 2-HG produced by mutant *IDH* could sensitize tumors to PARP inhibitors such as olaparib and niraparib (Sulkowski et al., 2017). Based on these premises, 3 phase 2 studies testing olaparib, alone or in combination with immunotherapy, are underway in CCA patients with *IDH* mutations (NCT03212274, NCT03878095, NCT03991832).

The use of PARP inhibitors has been recently FDA-approved for certain solid tumors harboring mutations in *BRCA1/2* or homologous recombination repair (HRR) deficiency due to aberrations in other components of the DNA damage repair (DDR) machinery (Brown, O'Carrigan, Jackson, & Yap, 2017; Konecny & Kristeleit, 2016; Lee, Ledermann, & Kohn, 2014). While *BRAC1/2* mutations are overall rare in BTCs (1–7%) (Nakamura et al., 2015; Spizzo et al., 2020; Terashima et al., 2019), germline or somatic mutations in the DDR pathway have been described in up to 64% of BTC patients (Chae et al., 2019; Ross et al., 2015). Similarly, the analysis of DDR mutations in 760 GBCs identified 14% of mutations in direct DDR genes (*ATM, ATR, BRCA1, BRCA2, FANCA, FANCD2, MLH1, MSH2, MSH6, PALB2, POLD1, POLE, PRKDC, and RAD50*) and up to 63% of mutations in caretaker DDR genes (*BAP1, CDK12, MLL3, TP53,* and *BLM*). Nonetheless, it is important to consider that little consensus currently exists on methods for testing deficiency of the DDR machinery and which mutations are considered actual DDR alterations. As a result, extreme variation in frequency of alterations in DDR genes has been reported thus far (Chae et al., 2019). Notably, a phase 1 study is ongoing with niraparib combined with the anti-angiogenic anlotinib in HRR gene-mutated advanced solid tumors, including CCA (NCT04764084) (Abdel-Wahab et al., 2020).

FGFR2 rearrangements represent the second most relevant alterations reported in CCA. As mentioned above, 2 FGFR inhibitors Abou-Alfa, Sahai, et al., 2020; Javle et al., 2021) have received FDA accelerated approval pending the results of the ongoing phase 3 studies (NCT03773302, NCT03656536) testing these inhibitors in first line CCAs with *FGFR2* fusions. Other FGFR inhibitors, such as ARQ087 (NCT01752920), RLY-4008 (NCT04526106) and Debio1347 (NCT01948297) are currently being tested in clinical trials. Unfortunately, a recent analysis described mechanisms of acquired resistance to FGFR inhibition, including the occurrence of secondary mutations in the kinase domain of *FGFR2* (Goyal et al., 2017, 2019). In this study, the irreversible FGFR inhibitor TAS-120 showed efficacy in 4 patients with advanced FGFR2 fusion–positive iCCA previously treated with BGJ398 and Debio1347, suggesting a greater sustained activity of TAS-120 against multiple *FGFR2* mutations conferring acquired resistance (Goyal et al., 2019). TAS120 is currently being tested versus gemcitabine-cisplatin chemotherapy as first-line treatment in patients with advanced CCA harboring FGFR2 rearrangements (FOENIX-CCA3, NCT04093362).

Despite their driving role and high mutation frequency in all BTC subtypes, both *KRAS* and *TP53* have been traditionally considered undruggable. Recently, the development of novel, potent selective *KRAS* inhibitors has led to the FDA approval of sotorasib, the first $KRAS^{G12C}$ inhibitor, for *KRAS* mutant lung cancer (Skoulidis et al., 2021), while other selective KRAS inhibitors are being investigated in clinical trials (Moore, Rosenberg, McCormick, & Malek, 2020). These advances bring new hope to the significant fraction of BTC patients harboring such alterations, as also supported by our recent preclinical study showing benefit in *KRAS*-mutant iCCA models receiving the KRAS-SOS1 inhibitor BI3406 (Hofmann et al., 2021) in combination with anti-PD1 therapy (Martin-Serrano et al., 2021). Consistently, previous preclinical studies in other cancers have also suggested that KRAS inhibition with sotorasib in immunocompetent murine models results in a pro-inflammatory TME and produces durable responses in combination with PD1 blockade (Canon et al., 2019). Future clinical trials testing selective *KRAS* inhibitors in combination with ICIs in those BTCs carrying mutations in this gene are anxiously awaited.

Among the alterations identified more frequently in eCCA and GBC, those affecting the ERBB signaling pathway are the most relevant from the therapeutic perspective despite occurring in less than 10% of patients

(Fig. 1). In a recent study, the identification of *ERBB/EGFR* aberrations in 3 eCCA patients led to personalized therapy with 2 patients receiving neratinib and 1 patient treated with a chemotherapy-trastuzumab combination; interestingly, clinical benefit was reported in all patients (Jacobi et al., 2021). Based on these premises, zanidatamab, a bispecific antibody that can simultaneously bind two non-overlapping epitopes of HER2, is being tested in a phase 2 study in patients with advanced or metastatic HER2-amplified BTCs (HERIZON-BTC-01, NCT04466891).

Other clinically relevant alterations have been described at low frequency and include fusion genes involving *NTRK* (4%), *ROS1* (8–9%), *ALK* (3%), and the serine/threonine kinases *PRKACA* and *PRKACB* (1% of eCCA). Due to the low incidence of these alterations, BTC patients are more likely to be enrolled in basket trials (NCT02568267) and may benefit from TRK inhibitors such entrectinib (Rolfo et al., 2015) and larotrectinib (Drilon et al., 2018). In this regard, in August 2019, the FDA announced the approval of entrectinib for the treatment of adult and adolescent patients with solid tumors harboring NTRK gene fusions (Marcus et al., 2021). The approval was based on the results from the integrated analysis of the Phase 2 STARTRK-2, Phase 1 STARTRK-1 and Phase 1 ALKA-372-001 trials, and data from the Phase 1/2 STARTRK-NG study testing entrectinib efficacy in several solid tumor types, including CCA (Doebele et al., 2020; Drilon et al., 2020). Overall, a durable overall response rate of 57%, including a complete response rate of 7%, was reported among 54 entrectinib-treated patients enrolled in one of 3 single-arm clinical trials (Marcus et al., 2021).

Finally, given the poor results observed in BTCs treated with ICIs as monotherapy, there has been a growing interest in identifying predictors of response and resistance. In this regard, in a phase 2 trial investigating the efficacy of pembrolizumab in MMR deficient solid tumors, 4 CCA patients were enrolled, with 1 complete response and 3 stable diseases (Le et al., 2017). Unfortunately, MMR deficiency occurs in <5% of all BTCs, highlighting the critical need of integrative immunogenomics analyses for better patient stratification. Although the first studies have emerged, further investigation is needed for such classifications to be implemented in clinical practice.

6. Conclusions and future perspectives

Overall, while the generation of integrative molecular classifications and personalized approaches to BTCs continue to evolve, the importance of molecular profiling remains undeniable. In just a few years, the discovery of subgroups of patients with *FGFR2* fusions or *IDH* mutations has led to the FDA approval of the first targeted therapies for this malignancy, with many more selective inhibitors currently being tested in molecular-based stratified approaches. Therefore, this suggests that the implementation of rigorous molecular testing and enrichment-based trial designs hold promise to improve the outcomes of these patients. In this regard, several studies have demonstrated that the molecular landscape of solid tumors, including BTCs, could be easily accessed via circulating tumor DNA (Deveson et al., 2021; Nakamura et al., 2020), an approach that, if consistently applied in clinical practice, could overcome the need of tissue biopsies. Finally, emerging evidence suggests that, despite extensive genomic, transcriptomic, and epigenomic heterogeneity, the immune profile among multiple tumors within a patient is relatively less heterogeneous (Chen et al., 2021); hence, immune-based stratification of patients could facilitate prediction to immunotherapy. While further analysis is needed to corroborate such preliminary findings, studies combining bulk sequencing technologies and single cell sequencing of tumors derived from patients enrolled in immunotherapy-based clinical trials are desperately needed. These data would not only provide insights into the clinical utility of current immunogenomics classifications, but also provide additional biomarkers for further patient stratification.

Acknowledgments
Figures were created with BioRender.com

Grant support
E.B. and D. S. are supported by the Tisch Cancer Institute, Icahn School of Medicine at Mount Sinai.

Conflict of interest statement
E. B. and D. S. have no financial or personal disclosures relevant to the contents of this manuscript.

References
Abdel-Wahab, R., Yap, T. A., Madison, R., Pant, S., Cooke, M., Wang, K., et al. (2020). Genomic profiling reveals high frequency of DNA repair genetic aberrations in gallbladder cancer. *Scientific Reports, 10*, 22087.
Abou-Alfa, G. K., Macarulla, T., Javle, M. M., Kelley, R. K., Lubner, S. J., Adeva, J., et al. (2020). Ivosidenib in IDH1-mutant, chemotherapy-refractory cholangiocarcinoma (ClarIDHy): A multicentre, randomised, double-blind, placebo-controlled, phase 3 study. *The Lancet Oncology, 21*, 796–807.

Abou-Alfa, G. K., Sahai, V., Hollebecque, A., Vaccaro, G., Melisi, D., Al-Rajabi, R., et al. (2020). Pemigatinib for previously treated, locally advanced or metastatic cholangiocarcinoma: A multicentre, open-label, phase 2 study. *The Lancet Oncology*, *21*, 671–684.

Affo, S., Nair, A., Brundu, F., Ravichandra, A., Bhattacharjee, S., Matsuda, M., et al. (2021). Promotion of cholangiocarcinoma growth by diverse cancer-associated fibroblast subpopulations. *Cancer Cell*, *39*, 866–882, (e811).

Ahn, K. S., O'Brien, D., Kang, Y. N., Mounajjed, T., Kim, Y. H., Kim, T. S., et al. (2019). Prognostic subclass of intrahepatic cholangiocarcinoma by integrative molecular-clinical analysis and potential targeted approach. *Hepatology International*, *13*, 490–500.

Akita, M., Sofue, K., Fujikura, K., Otani, K., Itoh, T., Ajiki, T., et al. (2019). Histological and molecular characterization of intrahepatic bile duct cancers suggests an expanded definition of perihilar cholangiocarcinoma. *HPB: The Official Journal of the International Hepato Pancreato Biliary Association*, *21*, 226–234.

Andersen, J. B., Spee, B., Blechacz, B. R., Avital, I., Komuta, M., Barbour, A., et al. (2012). Genomic and genetic characterization of cholangiocarcinoma identifies therapeutic targets for tyrosine kinase inhibitors. *Gastroenterology*, *142*(4).

Arai, Y., Totoki, Y., Hosoda, F., Shirota, T., Hama, N., Nakamura, H., et al. (2014). Fibroblast growth factor receptor 2 tyrosine kinase fusions define a unique molecular subtype of cholangiocarcinoma. *Hepatology*, *59*, 1427–1434.

Banales, J. M., Cardinale, V., Carpino, G., Marzioni, M., Andersen, J. B., Invernizzi, P., et al. (2016). Expert consensus document: Cholangiocarcinoma: Current knowledge and future perspectives consensus statement from the European network for the study of cholangiocarcinoma (ENS-CCA). *Nature Reviews. Gastroenterology & Hepatology*, *13*, 261–280.

Banales, J. M., Marin, J. J. G., Lamarca, A., Rodrigues, P. M., Khan, S. A., Roberts, L. R., et al. (2020). Cholangiocarcinoma 2020: The next horizon in mechanisms and management. *Nature Reviews. Gastroenterology & Hepatology*, *17*, 557–588.

Benavides, M., Anton, A., Gallego, J., Gomez, M. A., Jimenez-Gordo, A., La Casta, A., et al. (2015). Biliary tract cancers: SEOM clinical guidelines. *Clinical & Translational Oncology*, *17*, 982–987.

Boerner, T., Drill, E., Pak, L. M., Nguyen, B., Sigel, C. S., Doussot, A., et al. (2021). Genetic determinants of outcome in intrahepatic cholangiocarcinoma. *Hepatology*, *74*, 1429–1444.

Boileve, A., Verlingue, L., Hollebecque, A., Boige, V., Ducreux, M., & Malka, D. (2021). Rare cancer, rare alteration: The case of NTRK fusions in biliary tract cancers. *Expert Opinion on Investigational Drugs*, *30*, 401–409.

Bonneville, R., Krook, M. A., Kautto, E. A., Miya, J., Wing, M. R., Chen, H. Z., et al. (2017). Landscape of microsatellite instability across 39 cancer types. *JCO Precision Oncology*, *2017*.

Borad, M. J., Champion, M. D., Egan, J. B., Liang, W. S., Fonseca, R., Bryce, A. H., et al. (2014). Integrated genomic characterization reveals novel, therapeutically relevant drug targets in FGFR and EGFR pathways in sporadic intrahepatic cholangiocarcinoma. *PLoS Genetics*, *10*, e1004135.

Bray, F., Ferlay, J., Soerjomataram, I., Siegel, R. L., Torre, L. A., & Jemal, A. (2018). Global cancer statistics 2018: GLOBOCAN estimates of incidence and mortality worldwide for 36 cancers in 185 countries. *CA: A Cancer Journal for Clinicians*, *68*, 394–424.

Bridgewater, J., Galle, P. R., Khan, S. A., Llovet, J. M., Park, J. W., Patel, T., et al. (2014). Guidelines for the diagnosis and management of intrahepatic cholangiocarcinoma. *Journal of Hepatology*, *60*, 1268–1289.

Brown, J. S., O'Carrigan, B., Jackson, S. P., & Yap, T. A. (2017). Targeting DNA repair in cancer: Beyond PARP inhibitors. *Cancer Discovery*, *7*, 20–37.

Canon, J., Rex, K., Saiki, A. Y., Mohr, C., Cooke, K., Bagal, D., et al. (2019). The clinical KRAS(G12C) inhibitor AMG 510 drives anti-tumour immunity. *Nature*, *575*, 217–223.

Cao, J., Hu, J., Liu, S., Meric-Bernstam, F., Abdel-Wahab, R., Xu, J., et al. (2020). Intrahepatic cholangiocarcinoma: Genomic heterogeneity between eastern and Western patients. JCO precis. *Oncologia*, *4*.

Carapeto, F., Bozorgui, B., Shroff, R. T., Chagani, S., Solis Soto, L., Foo, W. C., et al. (2021). The immunogenomic landscape of resected intrahepatic cholangiocarcinoma. *Hepatology*.

Chae, H., Kim, D., Yoo, C., Kim, K. P., Jeong, J. H., Chang, H. M., et al. (2019). Therapeutic relevance of targeted sequencing in management of patients with advanced biliary tract cancer: DNA damage repair gene mutations as a predictive biomarker. *European Journal of Cancer*, *120*, 31–39.

Chaisaingmongkol, J., Budhu, A., Dang, H., Rabibhadana, S., Pupacdi, B., Kwon, S. M., et al. (2017). Common molecular subtypes among Asian hepatocellular carcinoma and cholangiocarcinoma. *Cancer Cell*, *32*(57–70), e53.

Chakravarty, D., Gao, J., Phillips, S. M., Kundra, R., Zhang, H., Wang, J., et al. (2017). OncoKB: A precision oncology Knowledge Base. *JCO Precision Oncology*, *2017*.

Chan-On, W., Nairismagi, M. L., Ong, C. K., Lim, W. K., Dima, S., Pairojkul, C., et al. (2013). Exome sequencing identifies distinct mutational patterns in liver fluke-related and non-infection-related bile duct cancers. *Nature Genetics*, *45*, 1474–1478.

Chen, S., Xie, Y., Cai, Y., Hu, H., He, M., Liu, L., et al. (2021). Multiomic analysis reveals comprehensive tumor heterogeneity and distinct immune subtypes in multifocal intrahepatic cholangiocarcinoma. *Clinical Cancer Research*.

Chiang, D. Y., Villanueva, A., Hoshida, Y., Peix, J., Newell, P., Minguez, B., et al. (2008). Focal gains of VEGFA and molecular classification of hepatocellular carcinoma. *Cancer Research*, *68*, 6779–6788.

Chuaysri, C., Thuwajit, P., Paupairoj, A., Chau-In, S., Suthiphongchai, T., & Thuwajit, C. (2009). Alpha-smooth muscle actin-positive fibroblasts promote biliary cell proliferation and correlate with poor survival in cholangiocarcinoma. *Oncology Reports*, *21*, 957–969.

Deveson, I. W., Gong, B., Lai, K., LoCoco, J. S., Richmond, T. A., Schageman, J., et al. (2021). Evaluating the analytical validity of circulating tumor DNA sequencing assays for precision oncology. *Nature Biotechnology*, *39*, 1115–1128.

Doebele, R. C., Drilon, A., Paz-Ares, L., Siena, S., Shaw, A. T., Farago, A. F., et al. (2020). Entrectinib in patients with advanced or metastatic NTRK fusion-positive solid tumours: Integrated analysis of three phase 1-2 trials. *The Lancet Oncology*, *21*, 271–282.

Drilon, A., Laetsch, T. W., Kummar, S., DuBois, S. G., Lassen, U. N., Demetri, G. D., et al. (2018). Efficacy of Larotrectinib in TRK fusion-positive cancers in adults and children. *The New England Journal of Medicine*, *378*, 731–739.

Drilon, A., Siena, S., Dziadziuszko, R., Barlesi, F., Krebs, M. G., Shaw, A. T., et al. (2020). Entrectinib in ROS1 fusion-positive non-small-cell lung cancer: Integrated analysis of three phase 1-2 trials. *The Lancet Oncology*, *21*, 261–270.

Farshidfar, F., Zheng, S., Gingras, M. C., Newton, Y., Shih, J., Robertson, A. G., et al. (2017). Integrative genomic analysis of cholangiocarcinoma identifies distinct IDH-mutant molecular profiles. *Cell Reports*, *18*, 2780–2794.

Goeppert, B., Folseraas, T., Roessler, S., Kloor, M., Volckmar, A. L., Endris, V., et al. (2020). Genomic characterization of cholangiocarcinoma in primary sclerosing cholangitis reveals therapeutic opportunities. *Hepatology*, *72*, 1253–1266.

Goeppert, B., Roessler, S., Renner, M., Singer, S., Mehrabi, A., Vogel, M. N., et al. (2019). Mismatch repair deficiency is a rare but putative therapeutically relevant finding in non-liver fluke associated cholangiocarcinoma. *British Journal of Cancer*, *120*, 109–114.

Goeppert, B., Toth, R., Singer, S., Albrecht, T., Lipka, D. B., Lutsik, P., et al. (2019). Integrative analysis defines distinct prognostic subgroups of intrahepatic cholangiocarcinoma. *Hepatology, 69*, 2091–2106.

Goyal, L., Saha, S. K., Liu, L. Y., Siravegna, G., Leshchiner, I., Ahronian, L. G., et al. (2017). Polyclonal secondary FGFR2 mutations drive acquired resistance to FGFR inhibition in patients with FGFR2 fusion–positive cholangiocarcinoma. *Cancer Discovery, 7*, 252–263.

Goyal, L., Shi, L., Liu, L. Y., Fece De La Cruz, F., Lennerz, J. K., Raghavan, S., et al. (2019). TAS-120 overcomes resistance to ATP-competitive FGFR inhibitors in patients with FGFR2 fusion–positive intrahepatic cholangiocarcinoma. *Cancer Discovery, 9*, 1064–1079.

Graham, R. P., Barr Fritcher, E. G., Pestova, E., Schulz, J., Sitailo, L. A., Vasmatzis, G., et al. (2014). Fibroblast growth factor receptor 2 translocations in intrahepatic cholangiocarcinoma. *Human Pathology, 45*, 1630–1638.

Hofmann, M. H., Gmachl, M., Ramharter, J., Savarese, F., Gerlach, D., Marszalek, J. R., et al. (2021). BI-3406, a potent and selective SOS1-KRAS interaction inhibitor, is effective in KRAS-driven cancers through combined MEK inhibition. *Cancer Discovery, 11*, 142–157.

Hoshida, Y., Nijman, S. M., Kobayashi, M., Chan, J. A., Brunet, J. P., Chiang, D. Y., et al. (2009). Integrative transcriptome analysis reveals common molecular subclasses of human hepatocellular carcinoma. *Cancer Research, 69*, 7385–7392.

Isa, T., Tomita, S., Nakachi, A., Miyazato, H., Shimoji, H., Kusano, T., et al. (2002). Analysis of microsatellite instability, K-ras gene mutation and p53 protein overexpression in intrahepatic cholangiocarcinoma. *Hepato-Gastroenterology, 49*, 604–608.

Jacobi, O., Ross, J. S., Goshen-Lago, T., Haddad, R., Moore, A., Sulkes, A., et al. (2021). ERBB2 pathway in biliary tract carcinoma: Clinical implications of a targetable pathway. *Oncology Research and Treatment, 44*, 20–27.

Javle, M., Roychowdhury, S., Kelley, R. K., Sadeghi, S., Macarulla, T., Weiss, K. H., et al. (2021). Infigratinib (BGJ398) in previously treated patients with advanced or metastatic cholangiocarcinoma with FGFR2 fusions or rearrangements: Mature results from a multicentre, open-label, single-arm, phase 2 study. Lancet. *Gastroenterología y Hepatología*.

Jiao, Y., Pawlik, T. M., Anders, R. A., Selaru, F. M., Streppel, M. M., Lucas, D. J., et al. (2013). Exome sequencing identifies frequent inactivating mutations in BAP1, ARID1A and PBRM1 in intrahepatic cholangiocarcinomas. *Nature Genetics, 45*, 1470–1473.

Job, S., Rapoud, D., Dos Santos, A., Gonzalez, P., Desterke, C., Pascal, G., et al. (2020). Identification of four immune subtypes characterized by distinct composition and functions of tumor microenvironment in intrahepatic cholangiocarcinoma. *Hepatology, 72*, 965–981.

Jusakul, A., Cutcutache, I., Yong, C. H., Lim, J. Q., Huang, M. N., Padmanabhan, N., et al. (2017). Whole-genome and epigenomic landscapes of etiologically distinct subtypes of cholangiocarcinoma. *Cancer Discovery, 7*, 1116–1135.

Kim, S. G., Chan, A. O., Wu, T. T., Issa, J. P., Hamilton, S. R., & Rashid, A. (2003). Epigenetic and genetic alterations in duodenal carcinomas are distinct from biliary and ampullary carcinomas. *Gastroenterology, 124*, 1300–1310.

Konecny, G. E., & Kristeleit, R. S. (2016). PARP inhibitors for BRCA1/2-mutated and sporadic ovarian cancer: Current practice and future directions. *British Journal of Cancer, 115*, 1157–1173.

Kongpetch, S., Jusakul, A., Lim, J. Q., Ng, C. C. Y., Chan, J. Y., Rajasegaran, V., et al. (2020). Lack of targetable FGFR2 fusions in endemic fluke-associated cholangiocarcinoma. *JCO Global Oncology, 6*, 628–638.

Lamarca, A., Palmer, D. H., Wasan, H. S., Ross, P. J., Ma, Y. T., Arora, A., et al. (2021). Second-line FOLFOX chemotherapy versus active symptom control for advanced biliary tract cancer (ABC-06): A phase 3, open-label, randomised, controlled trial. *The Lancet Oncology, 22*, 690–701.

Le, D. T., Durham, J. N., Smith, K. N., Wang, H., Bartlett, B. R., Aulakh, L. K., et al. (2017). Mismatch repair deficiency predicts response of solid tumors to PD-1 blockade. *Science, 357*, 409–413.

Le, D. T., Uram, J. N., Wang, H., Bartlett, B. R., Kemberling, H., Eyring, A. D., et al. (2015). PD-1 blockade in tumors with mismatch-repair deficiency. *The New England Journal of Medicine, 372*, 2509–2520.

Lee, J. M., Ledermann, J. A., & Kohn, E. C. (2014). PARP inhibitors for BRCA1/2 mutation-associated and BRCA-like malignancies. *Annals of Oncology, 25*, 32–40.

Li, M., Liu, F., Zhang, F., Zhou, W., Jiang, X., Yang, Y., et al. (2019). Genomic ERBB2/ERBB3 mutations promote PD-L1-mediated immune escape in gallbladder cancer: A whole-exome sequencing analysis. *Gut, 68*, 1024–1033.

Li, M., Zhang, Z., Li, X., Ye, J., Wu, X., Tan, Z., et al. (2014). Whole-exome and targeted gene sequencing of gallbladder carcinoma identifies recurrent mutations in the ErbB pathway. *Nature Genetics, 46*, 872–876.

Liu, X. S., & Mardis, E. R. (2017). Applications of immunogenomics to cancer. *Cell, 168*, 600–612.

Lowery, M. A., Ptashkin, R., Jordan, E., Berger, M. F., Zehir, A., Capanu, M., et al. (2018). Comprehensive molecular profiling of intrahepatic and extrahepatic Cholangiocarcinomas: Potential targets for intervention. *Clinical Cancer Research, 24*, 4154–4161.

Marcus, L., Donoghue, M., Aungst, S., Myers, C. E., Helms, W. S., Shen, G., et al. (2021). FDA approval summary: Entrectinib for the treatment of NTRK gene fusion solid tumors. *Clinical Cancer Research, 27*, 928–932.

Martin-Serrano, M. A., Kepecs, B., Torres, M., Bramel, E., Haber, P., Maeda, M., et al. (2021). A novel stroma, tumor, immune microenvironment-based classification of intrahepatic cholangiocarcinoma. In *AASLD Conference 2021, Publication Number 197.*

McNamara, M. G., Lopes, A., Wasan, H., Malka, D., Goldstein, D., Shannon, J., et al. (2020). Landmark survival analysis and impact of anatomic site of origin in prospective clinical trials of biliary tract cancer. *Journal of Hepatology, 73*, 1109–1117.

Momoi, H., Itoh, T., Nozaki, Y., Arima, Y., Okabe, H., Satoh, S., et al. (2001). Microsatellite instability and alternative genetic pathway in intrahepatic cholangiocarcinoma. *Journal of Hepatology, 35*, 235–244.

Montal, R., Sia, D., Montironi, C., Leow, W. Q., Esteban-Fabró, R., Pinyol, R., et al. (2020). Molecular classification and therapeutic targets in extrahepatic cholangiocarcinoma. *Journal of Hepatology, 73*, 315–327.

Moore, A. R., Rosenberg, S. C., McCormick, F., & Malek, S. (2020). RAS-targeted therapies: Is the undruggable drugged? *Nature Reviews. Drug Discovery, 19*, 533–552.

Nakamura, H., Arai, Y., Totoki, Y., Shirota, T., Elzawahry, A., Kato, M., et al. (2015). Genomic spectra of biliary tract cancer. *Nature Genetics, 47*, 1003–1010.

Nakamura, Y., Taniguchi, H., Ikeda, M., Bando, H., Kato, K., Morizane, C., et al. (2020). Clinical utility of circulating tumor DNA sequencing in advanced gastrointestinal cancer: SCRUM-Japan GI-SCREEN and GOZILA studies. *Nature Medicine, 26*, 1859–1864.

Nepal, C., O'Rourke, C. J., Oliveira, D., Taranta, A., Shema, S., Gautam, P., et al. (2018). Genomic perturbations reveal distinct regulatory networks in intrahepatic cholangiocarcinoma. *Hepatology, 68*, 949–963.

Nepal, C., Zhu, B., O'Rourke, C. J., Bhatt, D. K., Lee, D., Song, L., et al. (2021). Integrative molecular characterisation of gallbladder cancer reveals micro-environment-associated subtypes. *Journal of Hepatology, 74*, 1132–1144.

Oishi, N., Kumar, M. R., Roessler, S., Ji, J., Forgues, M., Budhu, A., et al. (2012). Transcriptomic profiling reveals hepatic stem-like gene signatures and interplay of miR-200c and epithelial-mesenchymal transition in intrahepatic cholangiocarcinoma. *Hepatology, 56,* 1792–1803.

Ong, C. K., Subimerb, C., Pairojkul, C., Wongkham, S., Cutcutache, I., Yu, W., et al. (2012). Exome sequencing of liver fluke-associated cholangiocarcinoma. *Nature Genetics, 44,* 690–693.

Pandey, A., Stawiski, E. W., Durinck, S., Gowda, H., Goldstein, L. D., Barbhuiya, M. A., et al. (2020). Integrated genomic analysis reveals mutated ELF3 as a potential gallbladder cancer vaccine candidate. *Nature Communications, 11,* 4225.

Piha-Paul, S. A., Oh, D. Y., Ueno, M., Malka, D., Chung, H. C., Nagrial, A., et al. (2020). Efficacy and safety of pembrolizumab for the treatment of advanced biliary cancer: Results from the KEYNOTE-158 and KEYNOTE-028 studies. *International Journal of Cancer, 147,* 2190–2198.

Rawla, P., Sunkara, T., Thandra, K. C., & Barsouk, A. (2019). Epidemiology of gallbladder cancer. *Journal of Clinical and Experimental Hepatology, 5,* 93–102.

Ribas, A., & Wolchok, J. D. (2018). Cancer immunotherapy using checkpoint blockade. *Science, 359,* 1350–1355.

Rizvi, S., & Gores, G. J. (2013). Pathogenesis, diagnosis, and management of cholangiocarcinoma. *Gastroenterology, 145,* 1215–1229.

Rolfo, C., Ruiz, R., Giovannetti, E., Gil-Bazo, I., Russo, A., Passiglia, F., et al. (2015). Entrectinib: A potent new TRK, ROS1, and ALK inhibitor. *Expert Opinion on Investigational Drugs, 24,* 1493–1500.

Ross, J. S., Wang, K., Gay, L., Al-Rohil, R., Rand, J. V., Jones, D. M., et al. (2014). New routes to targeted therapy of intrahepatic cholangiocarcinomas revealed by next-generation sequencing. *The Oncologist, 19,* 235–242.

Ross, J., Wang, K., Javle, M., Catenacci, D., Shroff, R., Ali, S., et al. (2015). (2015). Comprehensive genomic profiling of biliary tract cancers to reveal tumor-specific differences and frequency of clinically relevant genomic alterations. *Journal of Clinical Oncology: Official Journal of the American Society of Clinical Oncology, 33*(15), 4009.

Saha, S. K., Gordan, J. D., Kleinstiver, B. P., Vu, P., Najem, M. S., Yeo, J. C., et al. (2016). Isocitrate dehydrogenase mutations confer Dasatinib hypersensitivity and SRC dependence in intrahepatic cholangiocarcinoma. *Cancer Discovery, 6,* 727–739.

Saha, S. K., Parachoniak, C. A., Ghanta, K. S., Fitamant, J., Ross, K. N., Najem, M. S., et al. (2014). Mutant IDH inhibits HNF-4α to block hepatocyte differentiation and promote biliary cancer. *Nature, 513,* 110–114.

Sia, D., Hoshida, Y., Villanueva, A., Roayaie, S., Ferrer, J., Tabak, B., et al. (2013). Integrative molecular analysis of intrahepatic cholangiocarcinoma reveals 2 classes that have different outcomes. *Gastroenterology, 144,* 829–840.

Sia, D., Losic, B., Moeini, A., Cabellos, L., Hao, K., Revill, K., et al. (2015). Massive parallel sequencing uncovers actionable FGFR2–PPHLN1 fusion and ARAF mutations in intrahepatic cholangiocarcinoma. *Nature Communications, 6,* 6087.

Sia, D., Villanueva, A., Friedman, S. L., & Llovet, J. M. (2017). Liver cancer cell of origin, molecular class, and effects on patient prognosis. *Gastroenterology, 152,* 745–761.

Silverman, I. M., Hollebecque, A., Friboulet, L., Owens, S., Newton, R. C., Zhen, H., et al. (2021). Clinicogenomic analysis of FGFR2-rearranged cholangiocarcinoma identifies correlates of response and mechanisms of resistance to Pemigatinib. *Cancer Discovery, 11,* 326–339.

Skoulidis, F., Li, B. T., Dy, G. K., Price, T. J., Falchook, G. S., Wolf, J., et al. (2021). Sotorasib for lung cancers with KRAS p.G12C mutation. *The New England Journal of Medicine, 384,* 2371–2381.

Spizzo, G., Puccini, A., Xiu, J., Goldberg, R. M., Grothey, A., Shields, A. F., et al. (2020). Molecular profile of BRCA-mutated biliary tract cancers. *ESMO Open*, *5*, e000682.

Sulkowski, P. L., Corso, C. D., Robinson, N. D., Scanlon, S. E., Purshouse, K. R., Bai, H., et al. (2017). 2-hydroxyglutarate produced by neomorphic IDH mutations suppresses homologous recombination and induces PARP inhibitor sensitivity. *Science Translational Medicine*, *9*.

Tamma, R., Annese, T., Ruggieri, S., Brunetti, O., Longo, V., Cascardi, E., et al. (2019). Inflammatory cells infiltrate and angiogenesis in locally advanced and metastatic cholangiocarcinoma. *European Journal of Clinical Investigation*, *49*, e13087.

Terashima, T., Umemoto, K., Takahashi, H., Hosoi, H., Takai, E., Kondo, S., et al. (2019). Germline mutations in cancer-predisposition genes in patients with biliary tract cancer. *Oncotarget*, *10*, 5949–5957.

Tyson, G. L., & El-Serag, H. B. (2011). Risk factors for cholangiocarcinoma. *Hepatology*, *54*, 173–184.

Valle, J. W., Borbath, I., Khan, S. A., Huguet, F., Gruenberger, T., Arnold, D., et al. (2016). Biliary cancer: ESMO clinical practice guidelines for diagnosis, treatment and follow-up. *Annals of Oncology*, *27*, v28–v37.

Valle, J. W., Lamarca, A., Goyal, L., Barriuso, J., & Zhu, A. X. (2017). New horizons for precision medicine in biliary tract cancers. *Cancer Discovery*, *7*, 943–962.

Valle, J., Wasan, H., Palmer, D. H., Cunningham, D., Anthoney, A., Maraveyas, A., et al. (2010). Cisplatin plus gemcitabine versus gemcitabine for biliary tract cancer. *The New England Journal of Medicine*, *362*, 1273–1281.

Wardell, C. P., Fujita, M., Yamada, T., Simbolo, M., Fassan, M., Karlic, R., et al. (2018). Genomic characterization of biliary tract cancers identifies driver genes and predisposing mutations. *Journal of Hepatology*, *68*, 959–969.

Weinberg, B. A., Xiu, J., Lindberg, M. R., Shields, A. F., Hwang, J. J., Poorman, K., et al. (2019). Molecular profiling of biliary cancers reveals distinct molecular alterations and potential therapeutic targets. *Journal of Gastrointestinal Oncology*, *10*, 652–662.

Wu, Y. M., Su, F., Kalyana-Sundaram, S., Khazanov, N., Ateeq, B., Cao, X., et al. (2013). Identification of targetable FGFR gene fusions in diverse cancers. *Cancer Discovery*, *3*, 636–647.

Yachida, S., Wood, L. D., Suzuki, M., Takai, E., Totoki, Y., Kato, M., et al. (2016). Genomic sequencing identifies ELF3 as a driver of ampullary carcinoma. *Cancer Cell*, *29*, 229–240.

Yoshida, T., Sugai, T., Habano, W., Nakamura, S., Uesugi, N., Funato, O., et al. (2000). Microsatellite instability in gallbladder carcinoma: Two independent genetic pathways of gallbladder carcinogenesis. *Journal of Gastroenterology*, *35*, 768–774.

Zhang, M., Yang, H., Wan, L., Wang, Z., Wang, H., Ge, C., et al. (2020). Single-cell transcriptomic architecture and intercellular crosstalk of human intrahepatic cholangiocarcinoma. *Journal of Hepatology*, *73*, 1118–1130.

Zhang, Y., Zuo, C., Liu, L., Hu, Y., Yang, B., Qiu, S., et al. (2021). Single-cell RNA-sequencing atlas reveals an MDK-dependent immunosuppressive environment in ErbB pathway-mutated gallbladder cancer. *Journal of Hepatology*, *75*, 1128–1141.

Zou, S., Li, J., Zhou, H., Frech, C., Jiang, X., Chu, J. S., et al. (2014). Mutational landscape of intrahepatic cholangiocarcinoma. *Nature Communications*, *5*, 5696.

CHAPTER SEVEN

Cancer-associated fibroblasts in intrahepatic cholangiocarcinoma progression and therapeutic resistance

Aashreya Ravichandra[a], Sonakshi Bhattacharjee[a], and Silvia Affò[b],*
[a]Klinikum rechts der Isar, Technical University of Munich, Munich, Germany
[b]Institut d'Investigacions Biomèdiques August Pi i Sunyer, Barcelona, Spain
*Corresponding author: e-mail address: saffo@clinic.cat

Contents

1. Introduction	202
2. Tumor microenvironment in iCCA	203
3. Cancer-associated fibroblasts in iCCA	204
3.1 Origin and heterogeneity of cancer-associated fibroblasts	207
3.2 Plasticity of cancer-associated fibroblasts	208
4. Role of cancer-associated fibroblasts in iCCA	208
4.1 Pathways and mechanisms	209
4.2 Tumor stiffness and collagen	213
4.3 Cancer-associated fibroblasts and regulation of neo-angiogenesis	214
4.4 Modulation of inflammation by cancer-associated fibroblasts	215
5. Therapeutic relevance of cancer associated fibroblasts in iCCA	216
5.1 Cancer-associated-fibroblasts as potential therapeutic targets	216
6. Current and future perspectives	218
Acknowledgment	218
Grant support	218
Conflict of interest	218
References	219

Abstract

Cancer-associated fibroblasts (CAFs) are one of the most abundant stromal cell type in the tumor microenvironment (TME) of intrahepatic cholangiocarcinoma (iCCA), where they are actively involved in cancer progression through a complex network of interactions with other stromal cells. The majority of the studies investigating CAFs in iCCA have focused their attention on CAF tumor-promoting roles, remarking their potential as therapeutic targets. However, indiscriminate targeting of CAFs in other desmoplastic tumors has ended in failure with no effects or even accelerated cancer progression and reduced survival, indicating the urgent need to better understand the nuances and functions of CAFs to avoid deleterious effects. Indeed, recent single cell RNA sequencing studies have shown that heterogeneous CAF subpopulations coexist in the same tumor, some promoting- and other restricting- tumor growth. Moreover, recent studies have shown that in iCCA, diverse CAF subtypes interact differently with the cells of the TME, suggesting that CAFs may dynamically change their phenotypes during tumor progression, a field that remains uninvestigated. The characterization of heterogenous CAF subpopulations and their functionality, will provide a feasible and safer approach to facilitate the development of new therapeutic approaches aimed at targeting CAFs and their interactions with other stromal cells in the TME rather than solely tumor cells in iCCA. Here, we discuss the origin of CAFs, as well as their heterogeneity, plasticity, mechanisms and targeting strategies to provide a brief snapshot of the current knowledge in iCCA.

Abbreviations

CAFs cancer associated fibroblasts
iCCA intrahepatic cholangiocarcinoma
TME tumor microenvironment

1. Introduction

iCCA is the second most common primary malignancy in the liver after hepatocellular carcinoma, accounting for 10%–12% of all liver cancers (Banales et al., 2020). Classically considered to arise from the malignant transformation of cholangiocytes, recent lineage-tracing studies have demonstrated that iCCA may also arise from the trans-differentiation of hepatocytes, a phenomenon also observed in chronic liver injury (Fan et al., 2012; Sekiya & Suzuki, 2012; Tarlow et al., 2014; Yanger et al., 2013). Genetically, the mutational landscape of iCCA tumors is diverse and the most frequent genomic alterations include the *IDH1/2, FGFR2, HER2, ARID1A, TP53, KRAS, SMAD4, NTRK, PBMR1, MMR* and *BRAF* genes (Brandi, Farioli, Astolfi, Biasco, & Tavolari, 2015; Chan-On et al., 2013; Jiao et al., 2013; Lamarca, Barriuso, McNamara, & Valle, 2020; Nakamura et al., 2015; Walter et al., 2017). One of the main features of iCCA, is the presence of exacerbated and hypovascularized desmoplastic stroma with accumulation of α-smooth muscle actin (α-SMA) positive cancer-associated fibroblasts (CAFs), as well as deposition of dense extracellular matrix proteins such as collagen, hyaluronan (HA) and other proteoglycans (Razumilava & Gores, 2014; Sirica & Gores, 2014). The desmoplastic nature of iCCA is believed to contribute to poor prognosis and therapy resistance (Sirica & Gores, 2014). Indeed, clinical studies have shown a positive correlation between α-SMA-positive CAFs accumulation, increased tumor growth and poor outcomes (Banales et al., 2020; Brivio, Cadamuro, Strazzabosco, & Fabris, 2017; Zhang et al., 2017). Recent *in vivo* and *in vitro* studies highlight the presence of heterogenous CAF subpopulations that can provide both tumor-promoting and tumor-restricting signals (Affo et al., 2021; Bhattacharjee et al., 2021; LeBleu & Kalluri, 2018). Moreover, the advent of novel single cell RNA-sequencing (scRNA-Seq) studies highlights the beginning of a new *era* for the study of iCCA where the focus is not only the tumor cells, but also includes the TME. Indeed, this pioneering technique was key to characterize the heterogeneity of iCCA and its TME and to underline the complex tumor cell-CAFs crosstalk. As such, in this chapter we will summarize and provide insights on the role of CAFs in iCCA progression.

2. Tumor microenvironment in iCCA

The iCCA TME comprises a complex and dynamic stroma which includes but is not limited to: cancer cells, CAFs, tumor-associated endothelial cells (TECs), tumor-associated macrophages (TAMs), tumor-associated neutrophils (TANs), tumor infiltrating lymphocytes (TILs), and natural killer cells (NK) (Fabris, Sato, Alpini, & Strazzabosco, 2021; Sirica, Strazzabosco, & Cadamuro, 2021). All these cells form an intricate network of multi-directional interactions with overlapping or opposing functions on tumor growth, proliferation, metastasis, chemoresistance and immunosuppression (Fabris et al., 2019, 2021). In addition, the iCCA TME also includes acellular components mainly secreted by CAFs, which organize into an intricate web of ECM proteins comprised of collagens, fibronectins, proteoglycans, periostin (POSTN) and tenascin-C (TNC) within others, that create a highly desmoplastic and stiff environment associated with poor prognosis (Aishima et al., 2003; Frantz, Stewart, & Weaver, 2010; Iguchi et al., 2009; Leyva-Illades, McMillin, Quinn, & Demorrow, 2012). Another key feature of the TME in iCCA is its hypovascular nature, which may contribute to preventing the delivery of chemotherapeutic drugs by acting as a physical barrier (Leyva-Illades et al., 2012).

In the last decade, deeper sequencing techniques and advanced bioinformatic analysis tools has led to the identification and characterization of heterogeneous stromal subpopulations, thus adding more complexity to an already intricate picture. Furthermore, while many of the factors secreted by stromal cells are used to self-maintain an activated phenotype, crosstalk exists between all cell types in the TME (Kojima et al., 2010). Recent human and murine iCCA scRNA-Seq studies and characterization of ligand–receptor interactions have revealed that CAFs interact extensively with each other, with tumor cells, with endothelial cells and with macrophages, all affecting tumor growth (Affo et al., 2021). Understanding if CAFs undergo a dynamic plasticity program, and what determines shifts in CAF phenotypes, remains of interest for the development of novel therapies targeting CAFs, without inducing deleterious outcomes in patients.

3. Cancer-associated fibroblasts in iCCA

CAFs are activated mesenchymal-like cells that, as opposed to their quiescent counterparts often distributed throughout the organ, reside adjacent to tumor cells, where they crosstalk with the TME (Cadamuro, Morton, Strazzabosco, & Fabris, 2013; Vaquero, Aoudjehane, & Fouassier, 2020). CAFs, commonly identified by their expression of α-SMA, are one of the most abundant cell types in the tumor stroma of iCCA. Other described CAFs markers include vimentin (*VIM*), fibroblast specific protein-1 (*FSP1*) and fibroblast activation protein alpha (*FAP*) as well as platelet derived growth factor receptor-beta (*PDGFR-β*), decorin (*DCN*) and podoplanin (*PDPN*) (Table 1) (Augsten, 2014; Chen, McAndrews, & Kalluri, 2021). CAFs produce and secrete several ECM-related growth factors such as hepatocyte growth factor (HGF) and stromal derived growth factor-1 (SDF-1 or CXCL12), as well as ECM compounds such as collagen type I alpha I (COL1A1), POSTN and TNC, which participate in making the TME of iCCA stiff and have clinical significance as prognostic markers (Aishima et al., 2003; Campbell, Dumur, Lamour, Dewitt, & Sirica, 2012; Chuaysri et al., 2009; Midwood & Orend, 2009; Utispan et al., 2010). Clinical studies have shown a crucial role of CAFs in promoting tumor progression, where patients with high levels of α-SMA or POSTN present short-term survival (Chuaysri et al., 2009; Okabe et al., 2009). Moreover, CAFs also produce matrix metalloproteinases (MMP) including MMP-2 and MMP-9, implicated in ECM remodeling, tumor spread and growth (Prakobwong et al., 2010; Sirica et al., 2009; Terada, Okada, & Nakanuma, 1996). 3D-organotypic culture systems have additionally demonstrated enhancement of iCCA growth by α-SMA-positive CAFs and, in a syngeneic rat model of cholangiocarcinoma, the treatment with the cytotoxic drug navitoclax

Table 1 Summary of human and murine CAF subpopulations in iCCA.

CAF subpopulation	Biomarkers	References
Human PanCAF	ACTA2, COL1A1, COL1A2, COL3A1, CIR, CIS, DEC, FSP-1, FAP, PDPN PDGFR-β, SERPINF1	Affo et al. (2021), Augsten (2014), Chen, McAndrews, and Kalluri (2021), Zhang et al. (2020), Zhang et al. (2017)
Human iCAF	ADAMTS4, AGT, APOE, ARHGDIB, CCL19, CCL21, COLEC11, CPE, GEM, GJA4, GPX3, HIGD1B, IL6, ISYNA1, LHFP, MAP1B, MT1A, NDUFA4L2, PDK4, RGS5 ACTA2^{low}, CXCL1, FBLN1, IGF1, IGFBP6, SLP1, SAA1	Affo et al. (2021), Zhang et al. (2020)
Human myCAF	APOD, CCL11, COL1A1, COL1A2, COL3A1, COL5A1, COL6A3, CTGF, CTHRC1, CYP1B1, FN1, INHBA, ISLR, LUM, MMP14, POSTN, PTGDS, SERPINF1, SRFP2, SPON2, VCAN	Affo et al. (2021)
Human mesCAF	ANXA1, ANXA2, BDKRB1, C3, CALB2, CFB, CRABP2, CXCL1, CXCL6, EGFL6, EMP3, EZR, HMOX1, HP, HSPA6, IFI27, IGFBP6, ITLN1, KRT18, KRT19, KRT8, LOX, MT1E, MT1G, MT1X, MXRA5, PDPN, PRG4, PRSS23, PTGIS, S100A10, S100A16, S100A6, SAA1, SAA2, SERPINE2, SLC12A8, SLP1	Affo et al. (2021)
Human vCAF	GJA4, MHY11, MCAM, RGS5, IL-6, CCL8	Zhang et al. (2020)
Human mCAF	ACTA2^{low}, COL5A1, COL5A2, COL6A3, DCN, FN1, LUM, POSTN, VCAN	Zhang et al. (2020)
Human apCAF	CD74, HLADRA, HLA-DRB1	Zhang et al. (2020)
Human eCAF	KRT19, KRT8, SAA1	Zhang et al. (2020)
Human lCAF	APOA2, FABP1, FABP4, FRZB	Zhang et al. (2020)

Continued

Table 1 Summary of human and murine CAF subpopulations in iCCA.—cont'd

CAF subpopulation	Biomarkers	References
Mouse PanCAF	*Col1a1, Col1a2, Col3a1, C1s1, Acta2, C1ra, Serpinf1, Pdgfr-β, Col12a1*	Affo et al. (2021)
Mouse iCAF[a]	*Bmp10, Bmp2, Bmp5, C1ra, C1a1, C2, C4b, C6, Cachd1, Cadm3, Colec10, Colec11, Cxcl12, Cyr61, Dcn, Des, Fos, Fosb, Gdf10, Gdf2, Gem, Gtg5, Gm10600, H2-D1, H2-K1, H2-Q4, H2-Q6, H2-Q7, H6pd, Hand2, Hdax7, Hes1, Hgf, Hhip, Icam1, Id3, Il10rb, Il11ra1, Il6st, Itga1, Itga2b, Mertk, Myc, Ngfr, Ngf, Nkd1, Notch1, Npr3, Pdgfra, Plvap, Reln, Rgs5, Smad6, Socs3, Tnfrsf11b, Tnfrsf1b, Tnfrsf21*	Affo et al. (2021)
Mouse myCAF[a]	*Acta2, Col12a1, Col15a1, Col1a1, Col1a2, Col3a1, Cola4a5, Col5a2, Col5a3, Col6a2, Col7a1, Col8a1, Colec12, Fbln1, Fbln2, Fbn2, Fgfr1, Heyl, Igfbp5, Inhba, Lama4, Lxn, Mmp14, Mmp2, Mtch1, Nb11, Ncam1, Nduga412, Nkd2, Pdgfrl, Plat, Plaur, Ptn, Runx1, S100a16, S100a4, Serpine2, Serpinf1, Ssp1, Thy1, Tnc, Vcan, Vegfa*	Affo et al. (2021)
Mouse mesCAF[a]	*C3, Cd151, Cd200, Ddr1, Ezr, Gas6, Gpc3, Gpm6a, Il8, Krt18, Mmd, Msln, Pdpn, Rspo1, Trf, Ucp2, Upk1b, Upk3b, Wt1*	Affo et al. (2021)

[a]Denotes representative markers from Affo et al. (2021).
iCAF, inflammatory CAF; myCAF, myofibroblastic CAF; mesCAF, mesothelial CAF; vCAF, vascular CAF; mCAF, matrix CAF; apCAF, antigen-presenting CAF; lCAF, lipofibroblast CAF; eCAF, epithelial to mesenchymal (EMT)-like CAF.

triggered CAFs apoptosis and suppressed tumor outgrowth (Campbell et al., 2012; Mertens et al., 2013). The overall tumor promoting role of CAFs in iCCA, has been also recently confirmed in murine models of endogenously induced iCCA by depletion of proliferative CAFs, resulting in decreased tumor growth (Affo et al., 2021).

3.1 Origin and heterogeneity of cancer-associated fibroblasts

It is generally believed that CAFs in iCCA are largely derived from resident liver mesenchymal cells with studies pointing toward hepatic stellate cells (HSC) as a major cell source of CAF in iCCA, followed by portal fibroblasts (PF) and to a much lower extent, bone marrow derived mesenchymal cells (BMDSC) (Affo et al., 2021; Brivio et al., 2017; Zhang et al., 2020). Independently from their source, CAFs can be identified in iCCA using markers that have been included in panCAF signatures extrapolated from scRNA-Seq studies and that recapitulate some of the commonly used CAFs markers in other organs such as ACTA2, COL1A2, FAP, PDPN, DCN, COL3A1, COL6A1 (Table 1).

CAF heterogeneity is one of the most recent pieces of evidence changing a very old paradigm assigning cell heterogeneity mainly to diversity in cell source and not as a dynamic process. scRNA-Seq studies in human patients and mouse models of iCCA have described the presence of multiple CAF subtypes coexisting in the TME (Affo et al., 2021; Zhang et al., 2020). Even if there is not yet a consensus on the CAF subtype nomenclature, the presence of myofibroblastic CAF (myCAF), enriched in genes involved in ECM and collagen fiber organization; inflammatory CAF (iCAF), enriched in genes for inflammatory response regulation and complement activation and antigen-presenting CAF (apCAF) enriched in genes for major histocompatibility complex II (MHC-II) remarking the potential immunoregulatory role of CAFs in iCCA, is a common feature. Using a higher resolution analysis, in human iCCA additional subtypes such as vascular CAFs (vCAF), enriched in genes for muscle contraction and vascular development; EMT-like CAFs (eCAFs) enriched in epithelium specific markers and lipo-fibroblasts (lCAF), enriched in genes involved in lipid metabolism pathways have also been described (Zhang et al., 2020). Without a clear localization within the tumor, CAF subpopulations coexist and interact with different compounds of the TME as predicted by ligand-receptor bioinformatic tools. iCAF and vCAF, characterized by the expression of cytokines and growth factors (Table 1), seem to be the most abundant population of CAFs in both human and murine models of iCCA, followed by myCAF. Intermediate populations have been also identified, pointing toward a dynamic shift of phenotypes rather than the presence of static CAF populations. Moreover, with the new evidence that several CAF subtypes may arise from the same precursor cell, more studies will be needed to investigate such heterogeneity and plasticity. Finally, despite scRNA-Seq studies have provided a wide, new perspective on the heterogeneity of CAFs, only few have deeply investigated its significance by performing functional studies. Investigating the functionality of specific CAF subtypes will be the next step needed to evaluate the potential of each subpopulation in targeted therapy.

3.2 Plasticity of cancer-associated fibroblasts

Cell plasticity refers to the ability of cells to transform from one phenotype to another in response to external stimuli and is an important feature of CAFs, owing to their ability to transform from their cell of origin into a cell type of the desmoplastic stroma involved in tumor progression. Interestingly, the coexistence of different CAF subpopulations originating from the same cell precursor, points toward the presence of cell plasticity programs allowing for such phenotypic shifts. Indeed, while myCAFs and iCAFs have been described as two different subpopulations expressing exclusive markers (Table 1), they coexist in the same tumor and the presence of intermediate cells expressing similar levels of my- and i-CAFs markers has been noted, highlighting the potential of these cells to acquire different phenotypes, and suggesting their plasticity. However, the direction of this plasticity and how it is induced has not been described yet and will need further investigation. Supporting CAF plasticity, studies in PDAC have shown that antagonizing the iCAF JAK/STAT pathway using transforming growth factor-β (TGFβ), induced a myCAF phenotype (Biffi et al., 2019). This indicates that CAFs are fluid cells, easily susceptible to secreted factors and that may switch their phenotype.

4. Role of cancer-associated fibroblasts in iCCA

Several studies have been supporting a tumor-promoting role of CAFs in iCCA in the past years showing increased presence of α-SMA-positive CAFs in the tumor stroma correlating with larger tumor size and reduced survival (Chuaysri et al., 2009; Okabe et al., 2009); revealing increased proliferative capacity of CCA cell lines when treated with conditioned media (CM) from human α-SMA-positive CAFs (Chuaysri et al., 2009) and showing upregulation of pathways and genes involved in cellular metabolism in CAFs as compared to their quiescent counterparts, thereby "feeding" tumor expansion (Sahai et al., 2020). In CCA, a genomic signature of *TGFβ*, playing a major role in α-SMA induction and fibroblast activation, as demonstrated in a 3D organotypic culture model of iCCA, has been identified in the stromal compartment (Manzanares et al., 2017). Apart from the soluble factors released by CAFs and described to support tumor growth, such as CXCL12, interleukin (IL)-1β, PDGF-D, TGFβ, HGF, prostaglandin E2 (PGE2); extracellular vesicles have also been described to facilitate CAF-tumor interaction in iCCA by transferring proteins, lipids and nucleic acids between stromal cells, beyond the paracrine network (Brivio et al., 2017; Cadamuro et al., 2019; Leelawat, Leelawat, Narong, & Hongeng, 2007). The development of innovative 3-D organotypic systems obtained by co-culturing α-SMA-positive CAFs from orthotopic iCCA models with tumor cells in a rat type I collagen matrix to mimic the desmoplastic stroma of iCCA, has also revealed the ability of CAFs to enhance tumor cells growth, migration and invasiveness (Campbell et al., 2012). Moreover, the use of the cytotoxic drug Navitoclax triggering CAF apoptosis in a syngeneic rat model of CCA, has demonstrated reduced tumor growth and improved host survival (Mertens et al., 2013). A definitive confirmation of the tumor promoting role of CAF in iCCA is supported by a recent study where specific depletion of CAFs in endogenously arising iCCA mouse models, reduce tumor growth and proliferation (Affo et al., 2021). As such, ablation strategies to minimize the tumor-promoting role of CAFs are becoming of therapeutic interest, as it will be further discussed in the next sections.

4.1 Pathways and mechanisms

To sustain tumor proliferation, CAFs crosstalk with tumor cells and stromal cells in the TME, orchestrating a complex network of signaling pathways including inflammation, angiogenesis, ECM remodeling and regulation of the immune response (Fig. 1). A comprehensive understanding of these signaling cascades is needed to fully depict the real potential of CAFs as therapeutic target.

Among the multitude of molecules implicated in the CAFs-tumor cells crosstalk, heparin-binding epidermal growth factor (HB-EGF), has been identified to play an important role in many tumor types, including iCCA and PDAC (Clapéron et al., 2013; Miyamoto, Yagi, Yotsumoto,

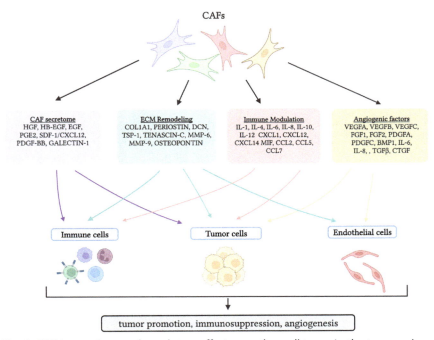

Fig. 1 CAF interactions and regulatory effects on other cell types in the tumor microenvironment in iCCA. Schematic representation of CAF-secreted cytokines and growth factors, as well as ECM-related proteins and pro-angiogenic factors and their regulatory effects on immune cells, tumor cells and endothelial cells within the tumor microenvironment in iCCA.

Kawarabayashi, & Mekada, 2006; Tarbe, Losch, Burtscher, Jarsch, & Weidle, 2002). EGF receptor (*EGFR*), often highly mutated in CCA and a marker of poor prognosis, is expressed on tumor cells and is activated by HB-EGF to induce tumor cell proliferation and migration *via* STAT-3 activation (Brivio et al., 2017). Indeed, *in vivo* transplantation studies revealed increased tumor size and metastasis upon co-transplantation of tumor cells and liver myofibroblasts, a phenotype that was reversed upon treatment with EGFR inhibitor gefitinib. Similarly, increased migratory and invasive properties of tumor cells are observed *in vitro* in iCCA, attributed to an increase in EGFR activation in tumor cells stimulated by fibroblast derived EGF (Clapéron et al., 2013; Grasset et al., 2018). Interestingly, *EGF* expression in CAFs can be enhanced by tumor cell-secreted TGF-β1, creating a continuous signaling loop (Brivio et al., 2017). However, clinical targeting of EGFR remains unsuccessful, and the IGF2/IGFR1/IR axis has been recently outlined as a signaling pathway involved in resistance to anti-EGFR therapy in CCA (Chen et al., 2015; Vaquero et al., 2018).

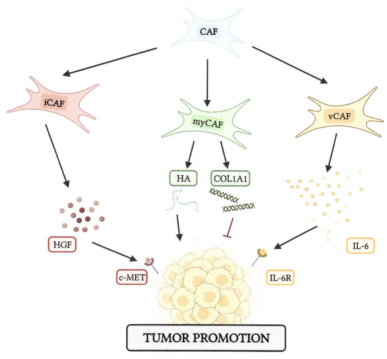

Fig. 2 CAF subpopulations and mediators in iCCA. Schematic illustration of inflammatory- (iCAF), myofibroblastic- (myCAF) and vascular- (vCAF) CAFs and their mediators interacting with tumor cells.

HGF, mainly produced by CAFs, plays a tumor promoting role *via* its receptor c-MET, mostly expressed by tumor cells in iCCA (Affo et al., 2021; Lai, Radaeva, Nakamura, & Sirica, 2000; Terada, Nakanuma, & Sirica, 1998) (Fig. 2). Upon activation of the c-MET, downstream pathways involving mitogen-activated protein kinase (MAPK), extracellular signal-related kinase (ERK), and phosphatidylinositol 3 kinase (PI3K) become activated, favoring the invasion of tumor cells (Affo et al., 2021; Goyal, Muzumdar, & Zhu, 2013). Indeed, a recent study has identified HGF as a HSC-derived CAF-secreted factor in murine models of iCCA, where it interacts with the MET receptor and induces tumor growth and proliferation mediated by ERK and AKT pathways (Affo et al., 2021). Even if MET inhibitors are not advantageous with respect to standard chemotherapy, this pathway remains of interest for CAF-driven targeting in CCA (Goyal et al., 2017; Tsimafeyeu & Temper, 2021).

CAF-secreted IL-1β, induces *CXCL5* expression in tumor cells, which is correlated with poor survival and leads to tumor cell proliferation, migration, and invasion *via* the PI3K/AKT and ERK1/2 (Okabe et al., 2012). *SDF-1*, more commonly known as *CXCL12,* is mainly expressed by CAFs in CCA where it has been associated with tumor proliferation, survival, chemotaxis and angiogenesis by binding to its G protein-coupled receptors CXCR4 and CXCR7 and activating PI3K/AKT and ERK pathways (Brivio et al., 2017; Nagarsheth, Wicha, & Zou, 2017; Ohira et al., 2006; Shi, Riese, & Shen, 2020). Blockade of CXCL12/CXCR4 signaling inhibits iCCA progression and metastasis *via* inactivation of the canonical Wnt pathway (Zhao, Wang, & Qin, 2014). Moreover, a novel scRNA-Seq study, has identified vCAF to promote iCCA tumor progression *via* the IL-6/IL-6R axis (Zhang et al., 2020), a pathway that will need further investigation (Fig. 2).

CAFs in the TME also produce and secrete ECM proteins such as POSTN and thrombospondin-1 (TSP-1), which contribute to fibrogenesis, invasion and metastasis (Kawahara et al., 1998; Riener et al., 2010; Utispan et al., 2010). POSTN has been reported to be a prognostic factor of iCCA, where it interacts with integrin $\alpha 5\beta 1$ or $\alpha 6\beta 4$ to activate and upregulate AKT and FAK signaling pathways, implicated in malignant growth. TSP-1 is an anti-angiogenic factor in iCCA, where it correlates with decreased microvessel density but with increased lymphangiogenesis (Aishima et al., 2002; Carpino et al., 2021). CAFs also produce TNC, whose expression correlates with poor prognosis (Aishima et al., 2003) and metalloproteinases, including MMP-2 and MMP-9, which are implicated in cancer progression and spread *via* ECM remodeling (Sirica et al., 2009; Terada et al., 1996).

As already mentioned, the interactions between stroma and tumor cells are bidirectional, with several studies showing a key role of tumor cells in controlling CAFs in CCA. The PDGF family of ligands have been described to play an important role in the recruitment and activation of CAFs in biliary inflammation and fibrosis and in CCA (Fingas et al., 2011). Indeed, PDGF-DD overexpressed by CCA tumor cells under hypoxia, binds to its receptor PDGFR-β expressed by CAF and induces CAF migration mediated by RHO GTPases, RAC1, CDC42 and the JNK pathways (Cadamuro et al., 2013). CAF-secreted PDGF-BB activates the hedgehog signaling (Hh) pathway by increasing Hh-mediator smoothened to the plasma membrane, resulting in glioma-associated oncogene (GLI) activation, a mechanism that improves survival of tumor cells by increasing apoptotic resistance as shown in *in vitro* and *in vivo* studies

(Fingas et al., 2011). A recent study has also shown decreased tumor growth and fibrosis in a mouse model of iCCA where *Pdgfr-β* was selectively deleted from HSC-derived CAF (Affo et al., 2021), thus confirming the relevance of this pathway. However, limited success has been observed in iCCA clinical settings employing PDGFR inhibitors, such as imatinib (Wiedmann & Mossner, 2010).

Sonic hedgehog (Shh) is highly expressed in human iCCA where it has been described to promote survival and proliferation of HSC, this suggesting that it may contribute to the activation of CAF in iCCA, in a similar mechanism as in PDAC (Olive et al., 2009; Yang et al., 2008). It has been also described that HSCs induce aberrant Hedgehog activation in CCA tumor cells, thus stimulating their proliferation, migration and invasion and suggesting that Hh signaling may be involved in the bidirectional crosstalk tumor-CAF (Kim et al., 2014; Yang et al., 2008).

Hypoxia-inducible factor 1α (*HIF-1α*) expression is a main feature of desmoplastic tumors including iCCA (Erkan et al., 2009; Sirica, Campbell, & Dumur, 2011). Indeed, hypoxia in tumor cells induces upregulation in *POSTN, HGF, SDF-1, CXCR4, CTGF, GALECTIN-1, TSP-1,* and *SPHINGOSINE KINASE-1* expression in CAFs in several cancers and has been suggested to exert similar effects in iCCA (Sirica, 2011).

4.2 Tumor stiffness and collagen

One of the main features of iCCA is its highly desmoplastic nature. Within the multiple CAF subpopulations, it seems that myCAF are the subtype more enriched in ECM and matrisome pathways and responsible for the production and secretion of ECM-proteins, including collagens. One of the main functions of the ECM, apart from providing structural support to the growing tumor, is to facilitate easy crosstalk between stromal cells and to serve as a repository of growth factors and other secreted molecules such as tumor necrosis factor (TNF), interleukins, MMPs, vascular endothelial growth factor (VEGF), HGF, EGF, HA, osteopontin and COL1A1, a fibrillar collagen and probably the most investigated in the field (Brivio et al., 2017).

COL1A1 is a fundamental component of the ECM, where it contributes to maintain the tensile strength of tissues. In the context of carcinogenesis, CAFs produce exacerbated amounts of COL1A1, which is deposited in the

ECM and contributes to increase the tissue stiffness, altering the TME organization and to enhance MMP activity (Jodele, Blavier, Yoon, & DeClerck, 2006). A recent study has shown that *Col1a1* in iCCA contributes to stiffness but not to tumor growth. Specifically, upon deleting *Col1a1* in HSC-derived CAFs in 2 models of iCCA, a reduction of tumor stiffness was observed, as well as reduction of the expression of yes-associated protein (YAP), a mechano-sensitive transcriptional coactivator known to play a role in stromal contractility (Affo et al., 2021). However, these reductions were not accompanied by changes in tumor growth, thus changing a long-standing paradigm that links collagen-mediated stiffness to desmoplastic tumor growth (Affo et al., 2021; Levental et al., 2009; Northey, Przybyla, & Weaver, 2017) (Fig. 2). Surprisingly, *in vitro* seeding of human iCCA cell line HuCCT-1 on stiff surfaces, an attempt to mimic tumoral stiffness, increased tumor cell proliferation; thus, indicating that while collagen-mediated stiffness is important for tumor proliferation, *in vivo*, additional ECM forces may play a role contributing to iCCA progression (Affo et al., 2021). In this regard the investigation of other myCAF-secreted factors rather than Col1a1 remains of interest. The study of myCAF-produced hyaluronan synthase 2 (*Has2*), the enzyme responsible for the production of HA, a viscoelastic molecule in the liver associated with poor prognosis and increased resistance to therapy has led to surprising results (McCarthy, El-Ashry, & Turley, 2018; Provenzano et al., 2012; Yang et al., 2019). Indeed, deletion of *Has2* in HSC-derived CAFs resulted in reduced HA deposition and decreased tumor growth and proliferation, without altering tumor stiffness (Affo et al., 2021) (Fig. 2). Since different molecules can contribute differently to the tumor stiffness depending on their viscosity and elasticity, further investigation will be needed to elucidate how these forces may contribute to iCCA.

4.3 Cancer-associated fibroblasts and regulation of neo-angiogenesis

The formation of new blood vessels, or neoangiogenesis, is necessary in order to supply important nutrients and oxygen to growing tumors. Increased angiogenesis in iCCA has been shown to correlate with lower survival rates (Shirabe et al., 2004; Thelen et al., 2010). While endothelial cells and TEC are the main cell type involved in tumoral neoangiogenesis, the role of CAF in this process is starting to be elucidated (Dudley, 2012). CCA is characterized by high expression of *VEGF* such as *VEGF-C* and VEGF receptors and CAFs have been recently described as the main source of VEGF in the tumor stroma. After treating fibroblasts with PDGF-D *in vitro*, CAFs secrete VEGF-A and VEGF-C, resulting in expansion of the lymphatic vasculature and tumor cell intravasation, a critical step in the early metastasis of CCA that may be blocked by blocking the PDGF-D-induced axis (Cadamuro et al., 2019). Moreover, recent data from transcriptomic analysis have revealed increased expression of pro-angiogenic factors such as neuropilin 2 (*NRP2*), platelet and endothelial cell adhesion molecule 1 (*PECAM1*), vascular cell adhesion molecule 1 (*VCAM1*) and Von Willebrand factor (*VWF*) in CAFs (Affo et al., 2021; Le Faouder et al., 2019; Zhang et al., 2020). However, it is unclear if CAFs are the main source of these factors in iCCA. A vCAF subcluster enriched in *CD146* marker and spatially distributed in the tumor core and microvascular region, has been recently identified by scRNA-Seq and bioinformatic tools predicting ligand-receptor interactions, have highlighted a high number of interactions occurring between CAF and TEC in iCCA, suggesting the relevance of this crosstalk (Affo et al., 2021; Zhang et al., 2020).

4.4 Modulation of inflammation by cancer-associated fibroblasts

To proliferate and expand, tumor cells use different strategies including immune surveillance escape and loss of antigenicity (Beatty & Gladney, 2015). However, the ability of tumors to orchestrate an immunosuppressive microenvironment also depends on interactions between the tumor cells and the TME. Indeed, CAFs have been shown to be involved in modulation of the anti-tumor response by secreting immunosuppressive factors, pro-inflammatory molecules and chemokines interacting with the immune system (Monteran & Erez, 2019; Ozdemir et al., 2014).

In iCCA, CAFs secrete IL-8, macrophage inhibitory factor (MIF), IL-13 and chemokine ligand 2 (CCL2), all identified as pro-inflammatory cytokines modulating the innate immune response (Hogdall, Lewinska, & Andersen, 2018). IL-13 has been identified in CCA to be one of the responsible factors along with IL-34 and OSTEOACTIVIN for macrophage-differentiation and invasion, as well as for *in vivo* tumor-promoting effect (Raggi et al., 2017). The FAP-STAT3-CCL2 axis has been recently shown to promote iCCA growth *via* recruitment of myeloid derived suppressor cells (MDSCs), potent immunosuppressive cells described to promote tumor progression by inhibiting T-cells and NK cell anti-tumoral functions (Lin et al., 2019; Umansky, Blattner, Gebhardt, & Utikal, 2016;

Yang et al., 2016). The expression of immune checkpoint molecules in iCCA, has been described to be expressed mainly by tumor and inflammatory cells and their expression by CAFs remains of interest (Fontugne et al., 2017).

Of note, it has been hypothesized that the CAF-secreted ECM compounds, such as HA and collagen, can also modulate immune response. Indeed, stromal accumulation of HA in patients with PDAC correlates with low immune cell infiltration, mostly due to HA acting as a physical barrier preventing immune cells from migrating to the vicinity of neoplastic cells (Tahkola et al., 2021). Moreover, a recent study has shown that depletion of Col1a1 from CAFs, results in increased MDSC accumulation and suppression of CD8-positive T cell function in PDAC models *in vivo*. (Chen et al., 2021) Since high levels of COL1A1 and HA are also detected in iCCA, such mechanisms may also be driven by CAF in iCCA and will deserve more investigation.

5. Therapeutic relevance of cancer associated fibroblasts in iCCA

For all the above mentioned pro-tumorigenic characteristics, CAFs represent an attractive therapeutic target in desmoplastic tumors. However, clinical trials targeting CAFs in other tumors have not been successful so far and in some cases even accelerated cancer progression resulting in reduced survival (Chen, McAndrews, & Kalluri, 2021). In the scRNA-Seq *era*, the discovery of CAFs heterogeneity and their multitude of interactions with stromal cells in the TME, point toward a dynamic subset of cells constantly evolving with the tumor. Understanding CAFs plasticity and how different CAFs subpopulations may be interchangeable, remains elusive and of relevance toward the development of novel therapeutic approaches.

5.1 Cancer-associated-fibroblasts as potential therapeutic targets

Based on the unsuccessful results in other organs, broad targeting of CAFs does not appear to be a feasible and achievable option for the treatment of iCCA. However, targeting CAF-secreted ECM components remains of interest. MMPs, enzymes that have the ability to remodel the ECM, are upregulated in iCCA human samples and specifically, MMP-1 and MMP-9 have been implicated in fibrosis and CCA (Prakobwong et al., 2010; Terada et al., 1996). Usage of Marimastat, an MMP inhibitor, in a small cohort of stage IV iCCA patients, resulted in improved average survival (21.5 months *vs* to >3 months) and reduction of tumor marker CA-19 suggesting that further studies in this direction will be of interest to better define the potential of CAF-secreted MMPs as targets in therapy (French, Midwinter, Bennett, Manas, & Charnley, 2005; Winer, Adams, & Mignatti, 2018). IL-6, recently identified as iCAF and vCAF mediator in CCA by scRNA-Seq, is known to play a tumor promoting role *via* various mechanisms including aberrant DNA methylation and gene expression in CCA and is of interest for targeting (Fig. 2). Indeed, IL-6 inhibitors in combination with standard chemotherapeutic regimen are currently undergoing clinical trials in pancreatic and liver cancers, including HCC and biliary cancers (Clinical trial information: NCT04258150, NCT02767557, NCT04338685) and may be further considered in iCCA. scRNA-Seq also led to the identification of iCAF and myCAF exclusively expressed markers and their manipulation in iCCA, including iCAF-*Hgf* and myCAF-*Has2* deletion in HSC-derived CAFs, shown to reduce tumor growth in mouse models of iCCA (Affo et al., 2021). Indeed, myCAF secreted HA, identified to be a prognostic marker and key component of the tumor stroma in pancreatic cancer, has been therapeutically targeted by Pegvorhyaluronidase Alfa (PEGPH20). While pre-clinical and early phase trials revealed an increased anti-tumor response and decrease in tumor growth, phase III trial of PEGPH20 with nab-paclitaxel and gemcitabine in patients with HA-high metastatic PDAC did not show any improvement in overall survival or in disease free progression, and was suspended (Hingorani et al., 2018; Thompson et al., 2010; Van Cutsem et al., 2020). While promising, targeting specific CAF subpopulations or mediators remains challenging. Current research also focuses on targeting CAF secreted factors such as TGFβ, a potential therapeutic target in iCCA based on pre-clinical models. Treatment with 1D11, a monoclonal neutralizing antibody against TGFβ, resulted in decreased fibrosis and tumor reduction in a murine model of CCA suggesting that therapeutic administration of TGFβ antagonist is a potential target for therapy in CCA (Ling et al., 2013). Clinically, a phase II trial using chimeric antibody M7824, composed of an extracellular domain of TβRII (a TGFβ antagonist) and a C-terminus PD-1 domain is underway in advanced biliary tract cancers, targeting both TGFβ pathways and immune checkpoints (Clinical Trial information: NCT03833661) and hopefully leading to promising results (Lan et al., 2018).

6. Current and future perspectives

CAFs have long been investigated as attractive therapeutic targets however, their usage in clinics has not been successful so far, most likely due to their complex network of interactions with different cells in the tumor stroma and their plasticity. With the advent of scRNA sequencing technology and sophisticated biocomputational analysis, it is becoming apparent that different CAF mediators have a dual role promoting or restricting tumor growth.

Another hindrance in the iCCA CAF field has been for long time the lack of reliable markers. In this regard, novel scRNA-Seq-derived CAF signatures have been recently developed and are useful tools to segregate different CAFs subpopulations (Affo et al., 2021; Zhang et al., 2020). However, high-quality antibodies to label CAF subtypes are scarce, requiring rigorous optimization and difficult implementation in routine pathological assessments and will need further simplification. Additionally, CAF subtypes are difficult to isolate and maintain in culture, often losing their specificity and making it hard to study specific subpopulations and their functions or to replicate the TME in 3D culture systems. While this CAF behavior highlights the plasticity and dynamicity of CAFs, it greatly inhibits pre-clinical and drug screening strategies toward successful targeting. Thanks to the implementation of genetically engineered mouse models and novel sophisticated 3D cell culture systems, the study of specific CAF subpopulations seems now possible. Concluding, CAF heterogeneity is now a reality and further studies evaluating CAF plasticity and functionality are needed toward the development of new therapies aimed to target the tumor-promoting CAF subpopulations in clinical settings.

Acknowledgment
Figures were created using BioRender.com.

Grant support
S.A. is supported by "la Caixa" Foundation (ID 100010434) and the European Union's Horizon 2020 research and innovation program under the Marie Skłodowska-Curie grant agreement N. 847648.

Conflict of interest
A.R., S.B. and S.A. have no financial or personal disclosures.

References

Affo, S., Nair, A., Brundu, F., Ravichandra, A., Bhattacharjee, S., Matsuda, M., et al. (2021). Promotion of cholangiocarcinoma growth by diverse cancer-associated fibroblast subpopulations. *Cancer Cell*, *39*(6), 866–882 e811. https://doi.org/10.1016/j.ccell.2021.03.012.

Aishima, S., Taguchi, K., Sugimachi, K., Asayama, Y., Nishi, H., Shimada, M., et al. (2002). The role of thymidine phosphorylase and thrombospondin-1 in angiogenesis and progression of intrahepatic cholangiocarcinoma. *International Journal of Surgical Pathology*, *10*(1), 47–56. https://doi.org/10.1177/106689690201000108.

Aishima, S., Taguchi, K., Terashi, T., Matsuura, S., Shimada, M., & Tsuneyoshi, M. (2003). Tenascin expression at the invasive front is associated with poor prognosis in intrahepatic cholangiocarcinoma. *Modern Pathology*, *16*(10), 1019–1027. https://doi.org/10.1097/01.MP.0000086860.65672.73.

Augsten, M. (2014). Cancer-associated fibroblasts as another polarized cell type of the tumor microenvironment. *Frontiers in Oncology*, *4*, 62. https://doi.org/10.3389/fonc.2014.00062.

Banales, J. M., Marin, J. J. G., Lamarca, A., Rodrigues, P. M., Khan, S. A., Roberts, L. R., et al. (2020). Cholangiocarcinoma 2020: The next horizon in mechanisms and management. *Nature Reviews: Gastroenterology & Hepatology*, *17*(9), 557–588. https://doi.org/10.1038/s41575-020-0310-z.

Beatty, G. L., & Gladney, W. L. (2015). Immune escape mechanisms as a guide for cancer immunotherapy. *Clinical Cancer Research*, *21*(4), 687–692. https://doi.org/10.1158/1078-0432.CCR-14-1860.

Bhattacharjee, S., Hamberger, F., Ravichandra, A., Miller, M., Nair, A., Affo, S., et al. (2021). Tumor restriction by type I collagen opposes tumor-promoting effects of cancer-associated fibroblasts. *Journal of Clinical Investigation*, *131*(11). https://doi.org/10.1172/JCI146987.

Biffi, G., Oni, T. E., Spielman, B., Hao, Y., Elyada, E., Park, Y., et al. (2019). IL1-induced JAK/STAT signaling is antagonized by TGFbeta to shape CAF heterogeneity in pancreatic ductal adenocarcinoma. *Cancer Discovery*, *9*(2), 282–301. https://doi.org/10.1158/2159-8290.CD-18-0710.

Brandi, G., Farioli, A., Astolfi, A., Biasco, G., & Tavolari, S. (2015). Genetic heterogeneity in cholangiocarcinoma: A major challenge for targeted therapies. *Oncotarget*, *6*(17), 14744–14753. https://doi.org/10.18632/oncotarget.4539.

Brivio, S., Cadamuro, M., Strazzabosco, M., & Fabris, L. (2017). Tumor reactive stroma in cholangiocarcinoma: The fuel behind cancer aggressiveness. *World Journal of Hepatology*, *9*(9), 455–468. https://doi.org/10.4254/wjh.v9.i9.455.

Cadamuro, M., Brivio, S., Mertens, J., Vismara, M., Moncsek, A., Milani, C., et al. (2019). Platelet-derived growth factor-D enables liver myofibroblasts to promote tumor lymphangiogenesis in cholangiocarcinoma. *Journal of Hepatology*, *70*(4), 700–709. https://doi.org/10.1016/j.jhep.2018.12.004.

Cadamuro, M., Morton, S. D., Strazzabosco, M., & Fabris, L. (2013). Unveiling the role of tumor reactive stroma in cholangiocarcinoma: An opportunity for new therapeutic strategies. *Translational Gastrointestinal Cancer*, *2*(3), 130–144. https://doi.org/10.3978/j.issn.2224-4778.2013.04.02.

Cadamuro, M., Nardo, G., Indraccolo, S., Dall'olmo, L., Sambado, L., Moserle, L., et al. (2013). Platelet-derived growth factor-D and Rho GTPases regulate recruitment of cancer-associated fibroblasts in cholangiocarcinoma. *Hepatology*, *58*(3), 1042–1053. https://doi.org/10.1002/hep.26384.

Campbell, D. J., Dumur, C. I., Lamour, N. F., Dewitt, J. L., & Sirica, A. E. (2012). Novel organotypic culture model of cholangiocarcinoma progression. *Hepatology Research*, *42*(11), 1119–1130. https://doi.org/10.1111/j.1872-034X.2012.01026.x.

Carpino, G., Cardinale, V., Di Giamberardino, A., Overi, D., Donsante, S., Colasanti, T., et al. (2021). Thrombospondin 1 and 2 along with PEDF inhibit angiogenesis and promote lymphangiogenesis in intrahepatic cholangiocarcinoma. *Journal of Hepatology*, 75(6), 1377–1386. https://doi.org/10.1016/j.jhep.2021.07.016.

Chan-On, W., Nairismagi, M. L., Ong, C. K., Lim, W. K., Dima, S., Pairojkul, C., et al. (2013). Exome sequencing identifies distinct mutational patterns in liver fluke-related and non-infection-related bile duct cancers. *Nature Genetics*, 45(12), 1474–1478. https://doi.org/10.1038/ng.2806.

Chen, J. S., Hsu, C., Chiang, N. J., Tsai, C. S., Tsou, H. H., Huang, S. F., et al. (2015). A KRAS mutation status-stratified randomized phase II trial of gemcitabine and oxaliplatin alone or in combination with cetuximab in advanced biliary tract cancer. *Annals of Oncology*, 26(5), 943–949. https://doi.org/10.1093/annonc/mdv035.

Chen, Y., Kim, J., Yang, S., Wang, H., Wu, C. J., Sugimoto, H., et al. (2021). Type I collagen deletion in αSMA(+) myofibroblasts augments immune suppression and accelerates progression of pancreatic cancer. *Cancer Cell*, 39(4), 548–565 e546. https://doi.org/10.1016/j.ccell.2021.02.007.

Chen, Y., McAndrews, K. M., & Kalluri, R. (2021). Clinical and therapeutic relevance of cancer-associated fibroblasts. *Nature Reviews Clinical Oncology*. https://doi.org/10.1038/s41571-021-00546-5.

Chuaysri, C., Thuwajit, P., Paupairoj, A., Chau-In, S., Suthiphongchai, T., & Thuwajit, C. (2009). Alpha-smooth muscle actin-positive fibroblasts promote biliary cell proliferation and correlate with poor survival in cholangiocarcinoma. *Oncology Reports*, 21(4), 957–969. https://doi.org/10.3892/or_00000309.

Clapéron, A., Mergey, M., Aoudjehane, L., Ho-Bouldoires, T. H., Wendum, D., Prignon, A., et al. (2013). Hepatic myofibroblasts promote the progression of human cholangiocarcinoma through activation of epidermal growth factor receptor. *Hepatology*, 58(6), 2001–2011. https://doi.org/10.1002/hep.26585.

Dudley, A. C. (2012). Tumor endothelial cells. *Cold Spring Harbor Perspectives in Medicine*, 2(3), a006536. https://doi.org/10.1101/cshperspect.a006536.

Erkan, M., Reiser-Erkan, C., Michalski, C. W., Deucker, S., Sauliunaite, D., Streit, S., et al. (2009). Cancer-stellate cell interactions perpetuate the hypoxia-fibrosis cycle in pancreatic ductal adenocarcinoma. *Neoplasia*, 11(5), 497–508. https://doi.org/10.1593/neo.81618.

Fabris, L., Perugorria, M. J., Mertens, J., Bjorkstrom, N. K., Cramer, T., Lleo, A., et al. (2019). The tumour microenvironment and immune milieu of cholangiocarcinoma. *Liver International*, 39(Suppl. 1), 63–78. https://doi.org/10.1111/liv.14098.

Fabris, L., Sato, K., Alpini, G., & Strazzabosco, M. (2021). The tumor microenvironment in cholangiocarcinoma progression. *Hepatology*, 73(Suppl. 1), 75–85. https://doi.org/10.1002/hep.31410.

Fan, B., Malato, Y., Calvisi, D. F., Naqvi, S., Razumilava, N., Ribback, S., et al. (2012). Cholangiocarcinomas can originate from hepatocytes in mice. *Journal of Clinical Investigation*, 122(8), 2911–2915. https://doi.org/10.1172/JCI63212.

Fingas, C. D., Bronk, S. F., Werneburg, N. W., Mott, J. L., Guicciardi, M. E., Cazanave, S. C., et al. (2011). Myofibroblast-derived PDGF-BB promotes Hedgehog survival signaling in cholangiocarcinoma cells. *Hepatology*, 54(6), 2076–2088. https://doi.org/10.1002/hep.24588.

Fontugne, J., Augustin, J., Pujals, A., Compagnon, P., Rousseau, B., Luciani, A., et al. (2017). PD-L1 expression in perihilar and intrahepatic cholangiocarcinoma. *Oncotarget*, 8(15), 24644–24651. https://doi.org/10.18632/oncotarget.15602.

Frantz, C., Stewart, K. M., & Weaver, V. M. (2010). The extracellular matrix at a glance. *Journal of Cell Science*, 123(Pt. 24), 4195–4200. https://doi.org/10.1242/jcs.023820.

French, J. J., Midwinter, M. J., Bennett, M. K., Manas, D. M., & Charnley, R. M. (2005). A matrix metalloproteinase inhibitor to treat unresectable cholangiocarcinoma. *HPB: The Official Journal of the International Hepato Pancreato Biliary Association*, 7(4), 289–291. https://doi.org/10.1080/13651820510042246.

Goyal, L., Muzumdar, M. D., & Zhu, A. X. (2013). Targeting the HGF/c-MET pathway in hepatocellular carcinoma. *Clinical Cancer Research*, 19(9), 2310–2318. https://doi.org/10.1158/1078-0432.CCR-12-2791.

Goyal, L., Zheng, H., Yurgelun, M. B., Abrams, T. A., Allen, J. N., Cleary, J. M., et al. (2017). A phase 2 and biomarker study of cabozantinib in patients with advanced cholangiocarcinoma. *Cancer*, 123(11), 1979–1988. https://doi.org/10.1002/cncr.30571.

Grasset, E. M., Bertero, T., Bozec, A., Friard, J., Bourget, I., Pisano, S., et al. (2018). Matrix stiffening and EGFR cooperate to promote the collective invasion of cancer cells. *Cancer Research*, 78(18), 5229–5242. https://doi.org/10.1158/0008-5472.CAN-18-0601.

Hingorani, S. R., Zheng, L., Bullock, A. J., Seery, T. E., Harris, W. P., Sigal, D. S., et al. (2018). HALO 202: Randomized phase II study of PEGPH20 plus nab-paclitaxel/gemcitabine versus nab-paclitaxel/gemcitabine in patients with untreated, metastatic pancreatic ductal adenocarcinoma. *Journal of Clinical Oncology*, 36(4), 359–366. https://doi.org/10.1200/JCO.2017.74.9564.

Hogdall, D., Lewinska, M., & Andersen, J. B. (2018). Desmoplastic tumor microenvironment and immunotherapy in cholangiocarcinoma. *Trends Cancer*, 4(3), 239–255. https://doi.org/10.1016/j.trecan.2018.01.007.

Iguchi, T., Aishima, S., Taketomi, A., Nishihara, Y., Fujita, N., Sanefuji, K., et al. (2009). Fascin overexpression is involved in carcinogenesis and prognosis of human intrahepatic cholangiocarcinoma: Immunohistochemical and molecular analysis. *Human Pathology*, 40(2), 174–180. https://doi.org/10.1016/j.humpath.2008.06.029.

Jiao, Y., Pawlik, T. M., Anders, R. A., Selaru, F. M., Streppel, M. M., Lucas, D. J., et al. (2013). Exome sequencing identifies frequent inactivating mutations in BAP1, ARID1A and PBRM1 in intrahepatic cholangiocarcinomas. *Nature Genetics*, 45(12), 1470–1473. https://doi.org/10.1038/ng.2813.

Jodele, S., Blavier, L., Yoon, J. M., & DeClerck, Y. A. (2006). Modifying the soil to affect the seed: Role of stromal-derived matrix metalloproteinases in cancer progression. *Cancer Metastasis Reviews*, 25(1), 35–43. https://doi.org/10.1007/s10555-006-7887-8.

Kawahara, N., Ono, M., Taguchi, K., Okamoto, M., Shimada, M., Takenaka, K., et al. (1998). Enhanced expression of thrombospondin-1 and hypovascularity in human cholangiocarcinoma. *Hepatology*, 28(6), 1512–1517. https://doi.org/10.1002/hep.510280610.

Kim, Y., Kim, M. O., Shin, J. S., Park, S. H., Kim, S. B., Kim, J., et al. (2014). Hedgehog signaling between cancer cells and hepatic stellate cells in promoting cholangiocarcinoma. *Annals of Surgical Oncology*, 21(8), 2684–2698. https://doi.org/10.1245/s10434-014-3531-y.

Kojima, Y., Acar, A., Eaton, E. N., Mellody, K. T., Scheel, C., Ben-Porath, I., et al. (2010). Autocrine TGF-beta and stromal cell-derived factor-1 (SDF-1) signaling drives the evolution of tumor-promoting mammary stromal myofibroblasts. *Proceedings of the National Academy of Sciences of the United States of America*, 107(46), 20009–20014. https://doi.org/10.1073/pnas.1013805107.

Lai, G. H., Radaeva, S., Nakamura, T., & Sirica, A. E. (2000). Unique epithelial cell production of hepatocyte growth factor/scatter factor by putative precancerous intestinal metaplasias and associated "intestinal-type" biliary cancer chemically induced in rat liver. *Hepatology*, 31(6), 1257–1265. https://doi.org/10.1053/jhep.2000.8108.

Lamarca, A., Barriuso, J., McNamara, M. G., & Valle, J. W. (2020). Molecular targeted therapies: Ready for "prime time" in biliary tract cancer. *Journal of Hepatology*, *73*(1), 170–185. https://doi.org/10.1016/j.jhep.2020.03.007.

Lan, Y., Zhang, D., Xu, C., Hance, K. W., Marelli, B., Qi, J., et al. (2018). Enhanced preclinical antitumor activity of M7824, a bifunctional fusion protein simultaneously targeting PD-L1 and TGF-beta. *Science Translational Medicine*, *10*(424). https://doi.org/10.1126/scitranslmed.aan5488.

Le Faouder, J., Gigante, E., Leger, T., Albuquerque, M., Beaufrere, A., Soubrane, O., et al. (2019). Proteomic landscape of cholangiocarcinomas reveals three different subgroups according to their localization and the aspect of non-tumor liver. *Proteomics Clinical Applications*, *13*(1). https://doi.org/10.1002/prca.201800128, e1800128.

LeBleu, V. S., & Kalluri, R. (2018). A peek into cancer-associated fibroblasts: Origins, functions and translational impact. *Disease Models & Mechanisms*, *11*(4). https://doi.org/10.1242/dmm.029447.

Leelawat, K., Leelawat, S., Narong, S., & Hongeng, S. (2007). Roles of the MEK1/2 and AKT pathways in CXCL12/CXCR4 induced cholangiocarcinoma cell invasion. *World Journal of Gastroenterology*, *13*(10), 1561–1568. https://doi.org/10.3748/wjg.v13.i10.1561.

Levental, K. R., Yu, H., Kass, L., Lakins, J. N., Egeblad, M., Erler, J. T., et al. (2009). Matrix crosslinking forces tumor progression by enhancing integrin signaling. *Cell*, *139*(5), 891–906. https://doi.org/10.1016/j.cell.2009.10.027.

Leyva-Illades, D., McMillin, M., Quinn, M., & Demorrow, S. (2012). Cholangiocarcinoma pathogenesis: Role of the tumor microenvironment. *Translational Gastrointestinal Cancer*, *1*(1), 71–80.

Lin, Y., Li, B., Yang, X., Cai, Q., Liu, W., Tian, M., et al. (2019). Fibroblastic FAP promotes intrahepatic cholangiocarcinoma growth via MDSCs recruitment. *Neoplasia*, *21*(12), 1133–1142. https://doi.org/10.1016/j.neo.2019.10.005.

Ling, H., Roux, E., Hempel, D., Tao, J., Smith, M., Lonning, S., et al. (2013). Transforming growth factor beta neutralization ameliorates pre-existing hepatic fibrosis and reduces cholangiocarcinoma in thioacetamide-treated rats. *PLoS One*, *8*(1), e54499. https://doi.org/10.1371/journal.pone.0054499.

Manzanares, M. A., Usui, A., Campbell, D. J., Dumur, C. I., Maldonado, G. T., Fausther, M., et al. (2017). Transforming growth factors alpha and beta are essential for modeling cholangiocarcinoma desmoplasia and progression in a three-dimensional organotypic culture model. *The American Journal of Pathology*, *187*(5), 1068–1092. https://doi.org/10.1016/j.ajpath.2017.01.013.

McCarthy, J. B., El-Ashry, D., & Turley, E. A. (2018). Hyaluronan, cancer-associated fibroblasts and the tumor microenvironment in malignant progression. *Frontiers in Cell and Developmental Biology*, *6*, 48. https://doi.org/10.3389/fcell.2018.00048.

Mertens, J. C., Fingas, C. D., Christensen, J. D., Smoot, R. L., Bronk, S. F., Werneburg, N. W., et al. (2013). Therapeutic effects of deleting cancer-associated fibroblasts in cholangiocarcinoma. *Cancer Research*, *73*(2), 897–907. https://doi.org/10.1158/0008-5472.CAN-12-2130.

Midwood, K. S., & Orend, G. (2009). The role of tenascin-C in tissue injury and tumorigenesis. *Journal of Cell Communications and Signaling*, *3*(3–4), 287–310. https://doi.org/10.1007/s12079-009-0075-1.

Miyamoto, S., Yagi, H., Yotsumoto, F., Kawarabayashi, T., & Mekada, E. (2006). Heparin-binding epidermal growth factor-like growth factor as a novel targeting molecule for cancer therapy. *Cancer Science*, *97*(5), 341–347. https://doi.org/10.1111/j.1349-7006.2006.00188.x.

Monteran, L., & Erez, N. (2019). The dark side of fibroblasts: Cancer-associated fibroblasts as mediators of immunosuppression in the tumor microenvironment. *Frontiers in Immunology*, *10*, 1835. https://doi.org/10.3389/fimmu.2019.01835.

Nagarsheth, N., Wicha, M. S., & Zou, W. (2017). Chemokines in the cancer microenvironment and their relevance in cancer immunotherapy. *Nature Reviews Immunology, 17*(9), 559–572. https://doi.org/10.1038/nri.2017.49.

Nakamura, H., Arai, Y., Totoki, Y., Shirota, T., Elzawahry, A., Kato, M., et al. (2015). Genomic spectra of biliary tract cancer. *Nature Genetics, 47*(9), 1003–1010. https://doi.org/10.1038/ng.3375.

Northey, J. J., Przybyla, L., & Weaver, V. M. (2017). Tissue force programs cell fate and tumor aggression. *Cancer Discovery, 7*(11), 1224–1237. https://doi.org/10.1158/2159-8290.CD-16-0733.

Ohira, S., Sasaki, M., Harada, K., Sato, Y., Zen, Y., Isse, K., et al. (2006). Possible regulation of migration of intrahepatic cholangiocarcinoma cells by interaction of CXCR4 expressed in carcinoma cells with tumor necrosis factor-alpha and stromal-derived factor-1 released in stroma. *American Journal of Pathology, 168*(4), 1155–1168. https://doi.org/10.2353/ajpath.2006.050204.

Okabe, H., Beppu, T., Hayashi, H., Horino, K., Masuda, T., Komori, H., et al. (2009). Hepatic stellate cells may relate to progression of intrahepatic cholangiocarcinoma. *Annals of Surgical Oncology, 16*(9), 2555–2564. https://doi.org/10.1245/s10434-009-0568-4.

Okabe, H., Beppu, T., Ueda, M., Hayashi, H., Ishiko, T., Masuda, T., et al. (2012). Identification of CXCL5/ENA-78 as a factor involved in the interaction between cholangiocarcinoma cells and cancer-associated fibroblasts. *International Journal of Cancer, 131*(10), 2234–2241. https://doi.org/10.1002/ijc.27496.

Olive, K. P., Jacobetz, M. A., Davidson, C. J., Gopinathan, A., McIntyre, D., Honess, D., et al. (2009). Inhibition of Hedgehog signaling enhances delivery of chemotherapy in a mouse model of pancreatic cancer. *Science, 324*(5933), 1457–1461. https://doi.org/10.1126/science.1171362.

Ozdemir, B. C., Pentcheva-Hoang, T., Carstens, J. L., Zheng, X., Wu, C. C., Simpson, T. R., et al. (2014). Depletion of carcinoma-associated fibroblasts and fibrosis induces immunosuppression and accelerates pancreas cancer with reduced survival. *Cancer Cell, 25*(6), 719–734. https://doi.org/10.1016/j.ccr.2014.04.005.

Prakobwong, S., Yongvanit, P., Hiraku, Y., Pairojkul, C., Sithithaworn, P., Pinlaor, P., et al. (2010). Involvement of MMP-9 in peribiliary fibrosis and cholangiocarcinogenesis via Rac1-dependent DNA damage in a hamster model. *International Journal of Cancer, 127*(11), 2576–2587. https://doi.org/10.1002/ijc.25266.

Provenzano, P. P., Cuevas, C., Chang, A. E., Goel, V. K., Von Hoff, D. D., & Hingorani, S. R. (2012). Enzymatic targeting of the stroma ablates physical barriers to treatment of pancreatic ductal adenocarcinoma. *Cancer Cell, 21*(3), 418–429. https://doi.org/10.1016/j.ccr.2012.01.007.

Raggi, C., Correnti, M., Sica, A., Andersen, J. B., Cardinale, V., Alvaro, D., et al. (2017). Cholangiocarcinoma stem-like subset shapes tumor-initiating niche by educating associated macrophages. *Journal of Hepatology, 66*(1), 102–115. https://doi.org/10.1016/j.jhep.2016.08.012.

Razumilava, N., & Gores, G. J. (2014). Cholangiocarcinoma. *Lancet, 383*(9935), 2168–2179. https://doi.org/10.1016/S0140-6736(13)61903-0.

Riener, M. O., Fritzsche, F. R., Soll, C., Pestalozzi, B. C., Probst-Hensch, N., Clavien, P. A., et al. (2010). Expression of the extracellular matrix protein periostin in liver tumours and bile duct carcinomas. *Histopathology, 56*(5), 600–606. https://doi.org/10.1111/j.1365-2559.2010.03527.x.

Sahai, E., Astsaturov, I., Cukierman, E., DeNardo, D. G., Egeblad, M., Evans, R. M., et al. (2020). A framework for advancing our understanding of cancer-associated fibroblasts. *Nature Reviews Cancer, 20*(3), 174–186. https://doi.org/10.1038/s41568-019-0238-1.

Sekiya, S., & Suzuki, A. (2012). Intrahepatic cholangiocarcinoma can arise from Notch-mediated conversion of hepatocytes. *Journal of Clinical Investigation*, *122*(11), 3914–3918. https://doi.org/10.1172/JCI63065.

Shi, Y., Riese, D. J., 2nd, & Shen, J. (2020). The role of the CXCL12/CXCR4/CXCR7 chemokine axis in cancer. *Frontiers in Pharmacology*, *11*. https://doi.org/10.3389/fphar.2020.574667, 574667.

Shirabe, K., Shimada, M., Tsujita, E., Aishima, S., Maehara, S., Tanaka, S., et al. (2004). Prognostic factors in node-negative intrahepatic cholangiocarcinoma with special reference to angiogenesis. *American Journal of Surgery*, *187*(4), 538–542. https://doi.org/10.1016/j.amjsurg.2003.12.044.

Sirica, A. E. (2011). The role of cancer-associated myofibroblasts in intrahepatic cholangiocarcinoma. *Nature Reviews Gastroenterology & Hepatology*, *9*(1), 44–54. https://doi.org/10.1038/nrgastro.2011.222.

Sirica, A. E., Campbell, D. J., & Dumur, C. I. (2011). Cancer-associated fibroblasts in intrahepatic cholangiocarcinoma. *Current Opinion in Gastroenterolgy*, *27*(3), 276–284. https://doi.org/10.1097/MOG.0b013e32834405c3.

Sirica, A. E., Dumur, C. I., Campbell, D. J., Almenara, J. A., Ogunwobi, O. O., & Dewitt, J. L. (2009). Intrahepatic cholangiocarcinoma progression: Prognostic factors and basic mechanisms. *Clinical Gastroenterology and Hepatology*, 7(11 Suppl), S68–S78. https://doi.org/10.1016/j.cgh.2009.08.023.

Sirica, A. E., & Gores, G. J. (2014). Desmoplastic stroma and cholangiocarcinoma: Clinical implications and therapeutic targeting. *Hepatology*, *59*(6), 2397–2402. https://doi.org/10.1002/hep.26762.

Sirica, A. E., Strazzabosco, M., & Cadamuro, M. (2021). Intrahepatic cholangiocarcinoma: Morpho-molecular pathology, tumor reactive microenvironment, and malignant progression. *Advances in Cancer Research*, *149*, 321–387. https://doi.org/10.1016/bs.acr.2020.10.005.

Tahkola, K., Ahtiainen, M., Mecklin, J. P., Kellokumpu, I., Laukkarinen, J., Tammi, M., et al. (2021). Stromal hyaluronan accumulation is associated with low immune response and poor prognosis in pancreatic cancer. *Scientific Reports*, *11*(1), 12216. https://doi.org/10.1038/s41598-021-91796-x.

Tarbe, N., Losch, S., Burtscher, H., Jarsch, M., & Weidle, U. H. (2002). Identification of rat pancreatic carcinoma genes associated with lymphogenous metastasis. *Anticancer Research*, *22*(4), 2015–2027.

Tarlow, B. D., Pelz, C., Naugler, W. E., Wakefield, L., Wilson, E. M., Finegold, M. J., et al. (2014). Bipotential adult liver progenitors are derived from chronically injured mature hepatocytes. *Cell Stem Cell*, *15*(5), 605–618. https://doi.org/10.1016/j.stem.2014.09.008.

Terada, T., Nakanuma, Y., & Sirica, A. E. (1998). Immunohistochemical demonstration of MET overexpression in human intrahepatic cholangiocarcinoma and in hepatolithiasis. *Human Pathology*, *29*(2), 175–180. https://doi.org/10.1016/s0046-8177(98)90229-5.

Terada, T., Okada, Y., & Nakanuma, Y. (1996). Expression of immunoreactive matrix metalloproteinases and tissue inhibitors of matrix metalloproteinases in human normal livers and primary liver tumors. *Hepatology*, *23*(6), 1341–1344. https://doi.org/10.1053/jhep.1996.v23.pm0008675149.

Thelen, A., Scholz, A., Weichert, W., Wiedenmann, B., Neuhaus, P., Gessner, R., et al. (2010). Tumor-associated angiogenesis and lymphangiogenesis correlate with progression of intrahepatic cholangiocarcinoma. *American Journal of Gastroenterology*, *105*(5), 1123–1132. https://doi.org/10.1038/ajg.2009.674.

Thompson, C. B., Shepard, H. M., O'Connor, P. M., Kadhim, S., Jiang, P., Osgood, R. J., et al. (2010). Enzymatic depletion of tumor hyaluronan induces antitumor responses in preclinical animal models. *Molecular Cancer Therapeutics*, *9*(11), 3052–3064. https://doi.org/10.1158/1535-7163.MCT-10-0470.

Tsimafeyeu, I., & Temper, M. (2021). Cholangiocarcinoma: An emerging target for molecular therapy. *Gastrointestinal Tumors*, *8*(4), 153–158. https://doi.org/10.1159/000517258.

Umansky, V., Blattner, C., Gebhardt, C., & Utikal, J. (2016). The role of myeloid-derived suppressor cells (MDSC) in cancer progression. *Vaccines (Basel)*, *4*(4). https://doi.org/10.3390/vaccines4040036.

Utispan, K., Thuwajit, P., Abiko, Y., Charngkaew, K., Paupairoj, A., Chau-in, S., et al. (2010). Gene expression profiling of cholangiocarcinoma-derived fibroblast reveals alterations related to tumor progression and indicates periostin as a poor prognostic marker. *Molecular Cancer*, *9*, 13. https://doi.org/10.1186/1476-4598-9-13.

Van Cutsem, E., Tempero, M. A., Sigal, D., Oh, D. Y., Fazio, N., Macarulla, T., et al. (2020). Randomized phase III trial of pegvorhyaluronidase alfa with nab-paclitaxel plus gemcitabine for patients with Hyaluronan-high metastatic pancreatic adenocarcinoma. *Journal of Clinical Oncology*, *38*(27), 3185–3194. https://doi.org/10.1200/JCO.20.00590.

Vaquero, J., Aoudjehane, L., & Fouassier, L. (2020). Cancer-associated fibroblasts in cholangiocarcinoma. *Current Opinion in Gastroenterolgy*, *36*(2), 63–69. https://doi.org/10.1097/MOG.0000000000000609.

Vaquero, J., Lobe, C., Tahraoui, S., Claperon, A., Mergey, M., Merabtene, F., et al. (2018). The IGF2/IR/IGF1R pathway in tumor cells and myofibroblasts mediates resistance to EGFR inhibition in cholangiocarcinoma. *Clinical Cancer Research*, *24*(17), 4282–4296. https://doi.org/10.1158/1078-0432.CCR-17-3725.

Walter, D., Doring, C., Feldhahn, M., Battke, F., Hartmann, S., Winkelmann, R., et al. (2017). Intratumoral heterogeneity of intrahepatic cholangiocarcinoma. *Oncotarget*, *8*(9), 14957–14968. https://doi.org/10.18632/oncotarget.14844.

Wiedmann, M. W., & Mossner, J. (2010). Molecular targeted therapy of biliary tract cancer—Results of the first clinical studies. *Current Drug Targets*, *11*(7), 834–850. https://doi.org/10.2174/138945010791320818.

Winer, A., Adams, S., & Mignatti, P. (2018). Matrix metalloproteinase inhibitors in cancer therapy: Turning past failures into future successes. *Molecular Cancer Therapeutics*, *17*(6), 1147–1155. https://doi.org/10.1158/1535-7163.MCT-17-0646.

Yang, X., Lin, Y., Shi, Y., Li, B., Liu, W., Yin, W., et al. (2016). FAP promotes immunosuppression by cancer-associated fibroblasts in the tumor microenvironment via STAT3-CCL2 signaling. *Cancer Research*, *76*(14), 4124–4135. https://doi.org/10.1158/0008-5472.CAN-15-2973.

Yang, Y. M., Noureddin, M., Liu, C., Ohashi, K., Kim, S. Y., Ramnath, D., et al. (2019). Hyaluronan synthase 2-mediated hyaluronan production mediates Notch1 activation and liver fibrosis. *Science Translational Medicine*, *11*(496). https://doi.org/10.1126/scitranslmed.aat9284.

Yang, L., Wang, Y., Mao, H., Fleig, S., Omenetti, A., Brown, K. D., et al. (2008). Sonic hedgehog is an autocrine viability factor for myofibroblastic hepatic stellate cells. *Journal of Hepatology*, *48*(1), 98–106. https://doi.org/10.1016/j.jhep.2007.07.032.

Yanger, K., Zong, Y., Maggs, L. R., Shapira, S. N., Maddipati, R., Aiello, N. M., et al. (2013). Robust cellular reprogramming occurs spontaneously during liver regeneration. *Genes & Development*, *27*(7), 719–724. https://doi.org/10.1101/gad.207803.112.

Zhang, X. F., Dong, M., Pan, Y. H., Chen, J. N., Huang, X. Q., Jin, Y., et al. (2017). Expression pattern of cancer-associated fibroblast and its clinical relevance in intrahepatic cholangiocarcinoma. *Human Pathology*, *65*, 92–100. https://doi.org/10.1016/j.humpath.2017.04.014.

Zhang, M., Yang, H., Wan, L., Wang, Z., Wang, H., Ge, C., et al. (2020). Single-cell transcriptomic architecture and intercellular crosstalk of human intrahepatic cholangiocarcinoma. *Journal of Hepatology*, *73*(5), 1118–1130. https://doi.org/10.1016/j.jhep.2020.05.039.

Zhao, S., Wang, J., & Qin, C. (2014). Blockade of CXCL12/CXCR4 signaling inhibits intrahepatic cholangiocarcinoma progression and metastasis via inactivation of canonical Wnt pathway. *Journal of Experimental & Clinical Cancer Research*, *33*, 103. https://doi.org/10.1186/s13046-014-0103-8.

CHAPTER EIGHT

Mechanisms and clinical significance of TGF-β in hepatocellular cancer progression

Sobia Zaidi[a,b,c], Nancy R. Gough[a,c], and Lopa Mishra[a,c,d],*

[a]The Institute for Bioelectronic Medicine, Feinstein Institutes for Medical Research, Manhasset, NY, United States
[b]Cold Spring Harbor Laboratory, Cold Spring Harbor, NY, United States
[c]Department of Medicine, Division of Gastroenterology and Hepatology, Northwell Health, New Hyde Park, NY, United States
[d]Center for Translational Medicine, Department of Surgery, The George Washington University, Washington, DC, United States
*Corresponding author: e-mail addresses: lopamishra2@gmail.com; lmishra1@northwell.edu

Contents

1. Introduction	228
1.1 A global disease with shifting risk factors	228
1.2 A heterogenous cancer	229
2. Overview of the TGF-β pathway	231
3. TGF-β signaling in liver homeostasis	233
4. Mouse models for TGF-β-Smad signaling in HCC development and progression	233
5. TGF-β pathway activity in HCC patients	236
5.1 Defining patient groups by transcriptional profiling	236
5.2 Developing treatments using TGF-β pathway insights	238
Grant support	242
Conflict of interest statement	242
References	242

Abstract

Despite progress in treating or preventing viral hepatitis, a leading cause of liver cancer, hepatocellular cancer (HCC) continues to be a major cause of cancer-related deaths globally. HCC is a highly heterogeneous cancer with many genetic alterations common within a patient's tumor and between different patients. This complicates therapeutic strategies. In this review, we highlight the critical role that the Smad-mediated transforming growth factor β (TGF-β) pathway plays both in liver homeostasis and in the development and progression of HCC. We summarize the mouse models that have enabled the exploration of the dual nature of this pathway as both a tumor suppressor and a tumor promoter. Finally, we highlight how the insights gained from evaluating pathway activity using transcriptional profiling can be used to stratify HCC patients toward rational therapeutic regimens based on the differences in patients with early or late TGF-β pathway activity or activated, normal, or inactivated profiles of this key pathway.

Abbreviations

ALD	alcohol-related liver disease
CTNNB1	β-catenin
ECM	extracellular matrix
EMT	epithelial-to-mesenchymal transition
HBV	hepatitis B virus
HCC	hepatocellular carcinoma
HCV	hepatitis C virus
HRAS	Harvey rat sarcoma viral oncogene
IGF2	insulin-like growth factor 2
ITIH4	inter-alpha-trypsin inhibitor heavy chain 4
KRAS	Kirsten rat sarcoma viral oncogene
LAP	latency-associated peptide
MAPK	mitogen-activated protein kinase
MDM2	mouse double minute 2 homolog
MTOR	mechanistic target of rapamycin
NAFLD	non-alcoholic fatty liver disease
NASH	non-alcoholic steatohepatitis
PD-1	programmed death-1 protein
PD-L1	programmed death-ligand 1 protein
PTEN	phosphatase and tensin homolog
SMAD	suppressor of mothers against decapentaplegic
TERT	telomerase reverse transcriptase
TGFBR	transforming growth factor receptor
TGF-β	transforming growth factor β
TP53	p53
TSP1	thrombospondin-1
VEGFA	vascular endothelial growth factor A
WNT	wingless/integrated
YAP1	yes1 associated transcriptional regulator (also known as yes-associated protein 1)
SPTBN1	spectrin beta, non-erythrocytic 1

1. Introduction
1.1 A global disease with shifting risk factors

Global cancer statistics from 2020 reveal that more than 900,000 new cases of liver cancer are diagnosed annually (Sung et al., 2021). Furthermore, with more than 800,000 global deaths per year, liver cancer is the third leading cause of cancer death. Hepatocellular carcinoma (HCC), resulting from malignant transformation of hepatocytes, is the most common type of liver cancer. Risk factors for HCC include viral hepatitis, alcohol-related liver disease (ALD), non-alcoholic fatty liver disease (NAFLD), and non-alcoholic steatohepatitis (NASH) (Global Burden of Disease Liver Cancer Collaboration et al., 2017; Kulik & El-Serag, 2019). All of these can lead to cirrhosis, which is a leading cause of liver-associated death (Paik, Golabi, Younossi, Mishra, & Younossi, 2020; Sepanlou et al., 2020) and considered a high risk factor for HCC (Moon, Singal, & Tapper, 2020).

As antiviral therapies and vaccination have reduced the burden of viral hepatitis in many parts of the world, the contribution of non-alcoholic fatty liver disease (NAFLD) and non-alcoholic steatohepatitis (NASH) to HCC has increased and NAFLD is predicted to become the leading cause of HCC in many countries (Huang, El-Serag, & Loomba, 2021; Ioannou, 2021). In the United States, NAFLD is the fastest growing cause of HCC (Ioannou, 2021; Paik et al., 2020).

The etiology of liver disease affects the mechanisms by which HCC develops. In viral hepatitis, viral infection drives chronic inflammation and alters host cell epigenetics and microRNAs to alter transcription. In hepatitis B virus (HBV) infection, viral DNA can integrate into the cell's genome causing oncogenesis. In both HBV and hepatitis C virus (HCV) infection, viral proteins can trigger oxidative stress and dysregulate cellular pathways to promote oncogenesis. HCC in the context of HCV infection is mostly associated with cirrhosis; whereas a subset of patients with chronic HBV infection without cirrhosis develop HCC (Chayanupatkul et al., 2017).

Most HCC occurs in the context of cirrhosis; however, non-fibrotic HCC is associated with NAFLD (Alexander, Torbenson, Wu, & Yeh, 2013; Kulik & El-Serag, 2019; Mittal et al., 2016). Furthermore, many of the risk factors for NAFLD and the progression to HCC overlap with those associated with metabolic syndrome, which is also associated with an increased risk of non-fibrotic HCC (Ioannou, 2021; Mittal et al., 2016). A subset of NAFLD patients develop NASH and a subset of the NASH patients develop cirrhosis (Huang et al., 2021). Thus, both NAFLD and NASH patients are at high risk of HCC and may benefit from proactive screening for HCC (Younes & Bugianesi, 2018).

1.2 A heterogenous cancer

The context in which HCC arises influences the molecular and genetic profile of the cancer. For example, mutations in the WNT pathway gene

CTNNB1 (encoding β-catenin) are associated with alcohol-related HCC and mutations in the tumor suppressor gene *TP53* (encoding p53) are associated with HCC in the context of HBV infection (Chen et al., 2018; Schulze et al., 2015).

Despite the association of some genes with specific underlying etiologies, HCC is heterogenous at genetic, epigenetic, histopathological, and immunological levels (Ally et al., 2017; Calderaro et al., 2017; Chen et al., 2018; De Battista et al., 2021; Letouzé et al., 2017; Losic et al., 2020; Rebouissou & Nault, 2020; Schulze et al., 2015). On the basis of the heterogeneity within the various samples of each study and the approaches used for analysis, multiple classifications schemes for HCC have been developed. A recurrent theme among these schemes is the involvement of genes in inflammation, telomere maintenance, regulation of the cell cycle and apoptosis, WNT signaling, epigenetic modification, control of cell growth and angiogenesis through receptor tyrosine kinase signaling, and oxidative stress. Recurrently, altered genes among the various studies include *TERT*, encoding telomerase, *CTNNB1*, and *TP53*. However, HCC tumors within a single patient may have multiple genetic alterations, making both stratifying patients for therapy and developing tolerable targeted therapeutic combinations challenging.

Transformation occurs through a multistep process during which genetic and epigenetic modifications accumulate. The liver is a highly regenerative tissue, enabling homeostatic recovery to injury or inflammation. As a regenerative tissue, liver is also susceptible to dysplasia and cancer (Tomasetti, Li, & Vogelstein, 2017). Transforming growth factor β (TGF-β) signaling, particularly that mediated by the TGF-β subgroup TGF-β1, TGF-β2, and TGF-β3, is essential for liver homeostasis and the response to injury and is a key modulator of the immune system (Batlle & Massagué, 2019; Fabregat et al., 2016). Because HCC develops in the context of inflammation that is often caused by chronic viral infection or metabolic disease and TGF-β signaling is a key pathway involved in this context, there is great interest in understanding this pathway in HCC development and progression. Such insight could lead to targeted approaches to influence the outcome of this pathway either in preventing HCC development in high-risk patients or in treating HCC and limiting its progression. Another reason TGF-β signaling is particularly relevant for HCC is that this pathway exhibits crosstalk with many of the other pathways implicated in HCC of various subtypes.

2. Overview of the TGF-β pathway

Here, the focus is on the "canonical" TGF-β pathway mediated by activation of transcriptional regulators of the Smad family. TGF-β ligands are synthesized as longer precursor proteins that are cleaved by endoproteases and the cleaved peptides form dimers associated with latency-associated peptide (LAP). This inactive LAP-bound complex is secreted and associates with the extracellular matrix (ECM). ECM proteases, thrombospondin-1 (TSP1), or integrins release the bioactive dimers (Fabregat et al., 2016; Li, Turpin, & Wang, 2017; Shi et al., 2011; Yang et al., 2007). Once released, TGF-β1, TGF-β2, or TGF-β3 bind and activate heteromeric complexes of a pairs of TGFBR1 and TGFBR2, along with the coreceptor betaglycan (David & Massagué, 2018; Derynck & Budi, 2019) (Fig. 1). Upon ligand binding, TGFBR2 phosphorylates and activates TGFBR1, which then phosphorylates serine residues on the R-Smads SMAD2 and SMAD3 (collectively referred to as SMAD2/3). Phosphorylated SMAD2 and SMAD3 form heterodimers or homodimers that then interact with SMAD4, referred to as c-Smad (common or comediator Smad), and translocate to the nucleus to regulate gene expression.

Context-dependent output arises through posttranslational modifications of Smads and the interaction of the SMAD2/3 and SMAD4 complex with other transcriptional regulators, chromatin-modifying factors, or adaptor proteins (Bertero et al., 2018; Chen et al., 2016; David & Massagué, 2018; Derynck & Budi, 2019; Hill, 2016; Oh, Swiderska-Syn, Jewell, Premont, & Diehl, 2018; Tang et al., 2003). The posttranslational modifications of Smads as well as the interactions with other transcription factors enable extensive crosstalk with other signaling pathway. The activity and functions of various components of the pathway are also controlled by positive or negative regulators or by posttranslational modifications. Indeed, one type of Smad family member is the inhibitory Smad (I-Smad). SMAD7 is the I-Smad for the pathway involving the TGF-β subfamily and activation of TGFBR1/TGFBR2 complexes (Miyazawa & Miyazono, 2017). In addition to regulation by posttranslational modifications, such as acetylation, ubiquitylation, ADP-ribosylation, cleavage, and phosphorylation, many components are also regulated at the transcript level by microRNAs or long-noncoding RNAs (Baek et al., 2016; Dahl et al., 2014; Papoutsoglou & Moustakas, 2020; Xu, Liu, & Derynck, 2012).

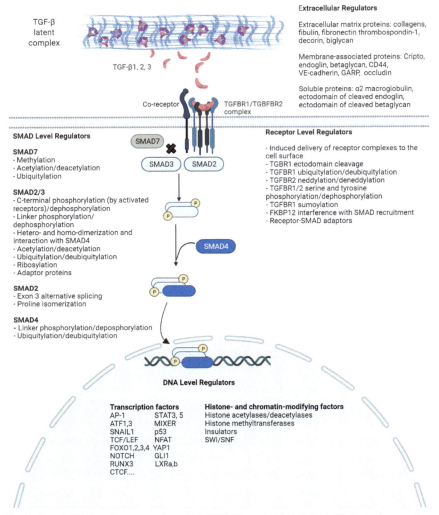

Fig. 1 Core TGF-β pathway mediated by SMAD2 and SMAD3. Not illustrated are activation of ligands by integrins or release from GARP. Created with Biorender.com.

Another mechanism by which context-dependent output is achieved is by the interaction of TGF-β ligands with the co-receptor endoglin instead of betaglycan. Endoglin alters the receptor specificity of TGF-β signaling from receptor complexes with TGFBR1 to those with ALK5 (Velasco, Alvarez-Muñoz, Pericacho, Bernabéu, et al., 2008). Endoglin-mediated redirection of the TGF-β signal is particularly relevant in vascular endothelial cells (Lebrin et al., 2004).

3. TGF-β signaling in liver homeostasis

TGF-β signaling in the inflamed or injured liver involves hepatocytes, hepatic stellate cells, and Kupffer cells, as well as infiltrating immune cells, and vascular endothelial cells. Many of the processes of liver regeneration or the response to injury are also important for HCC development or progression. TGF-β signaling has diverse, context-specific effects that require tight spatiotemporal regulation for appropriate resolution of inflammation and regeneration without initiation of malignancy, presenting potential opportunities to target TGF-β signaling in a context-specific manner in the cancer cells themselves, in the tumor microenvironment, or in the immune response.

The response of hepatocytes to TGF-β signals is complex and spatially and temporally regulated so that there is an initial proliferative phase followed by a termination phase (Fabregat et al., 2016). Inappropriate activation of the pathway impairs liver regeneration (Russell, Coffey, Ouellette, & Moses, 1988). In hepatocytes, TGF-β signaling through SMAD2 is critical for liver regeneration and involves crosstalk between this pathway and the transcriptional regulator YAP1 (Oh et al., 2018). A key role of TGF-β1 signaling in the injured liver is the activation of quiescent hepatic stellate cells, which produce pro-inflammatory signals, remodel the ECM, and deposit collagen (Hellerbrand, Stefanovic, Giordano, Burchardt, & Brenner, 1999; Tsuchida & Friedman, 2017). Injury leads to the recruitment of platelets and innate immune cells and stimulation of the resident macrophages (Kupffer cells), which are all sources of TGF-β (Ghafoory et al., 2018; Karlmark et al., 2009; Meijer et al., 2000). Additionally, many immune cells respond to TGF-β signals (Batlle & Massagué, 2019). By limiting the immune response, TGF-β signaling ensures that immune-mediated inflammation does not contribute to excessive tissue damage.

Not only is TGF-β signaling critical to liver homeostasis, but it is critical for homeostasis and repair of many tissues, as well as regulation of immune responses under many conditions. The broadly distributed function and diverse roles of TGF-β signaling complicates targeting this pathway with agents that block the pathway in any responsive cell.

4. Mouse models for TGF-β-Smad signaling in HCC development and progression

During the multistep process that is carcinogenesis and the transition to metastatic disease, TGF-β signaling plays both roles in tumor suppression and in cancer progression. Multiple reviews describe the mechanisms and dual nature of TGF-β signaling in cancer (Batlle & Massagué, 2019; Fabregat et al., 2016; Gough, Xiang, & Mishra, 2021; Morikawa, Derynck, & Miyazono, 2016). Here, the focus is on studies specific to HCC, particularly on genetically engineered mouse models that demonstrate the duality in the TGF-β pathway in HCC. Because of the importance of this pathway in development and in the immune system, most mouse models for studying the role of TGF-β signaling in HCC rely on either heterozygous strains or liver-specific mutants (Table 1).

In the liver, TGF-β activates multiple tumor-suppressing events in hepatocytes. These involve blocking transformation of hepatocytes to hepatocarcinoma by limiting hepatic cell proliferation (Bissell, Wang, Jarnagin, & Roll, 1995; Herrera et al., 2004; Russell et al., 1988), triggering hepatocyte apoptosis (Jang et al., 2002; Ramjaun, Tomlinson, Eddaoudi, & Downward, 2007; Yoo et al., 2003; Zhang et al., 2006; Zhang, Alexander, & Wang, 2017), and regulating hepatic stem cells (Chen et al., 2016; Majumdar et al., 2012; Rao et al., 2017; Tang et al., 2008). TGF-β signaling is also linked to cellular senescence through regulation of telomerase (TERT) (Chen et al., 2016). As studies with mouse models show, disruption of these functions of TGF-β signaling results in increased HCC occurrence or risk (Table 1) (Gough et al., 2021). Increased susceptibility to chemically induced HCC or spontaneous HCC occurs in mice with reduced or altered pathway activity: $Tgfb1^{+/-}$ mice, mice with liver-specific expression of dominant-negative TGFBR2 or $Tgfbr2$ heterozygosity, or $Sptbn1$ heterozygosity with or without $Smad3$ deficiency.

However, there is also ample evidence for tumor-promoting effects of TGF-β signaling in HCC (Table 1). Indeed, some of the components that function as tumor suppressors under some conditions contribute to cancer progression in other conditions. For example, mice heterozygous for $TGFB1$ have increased susceptibility to chemically induced HCC, but mice overexpressing TGF-β1 specifically in the liver are also susceptible to HCC. Thus, even the ligand itself, TGF-β1, has both tumor-suppressing and tumor-promoting activities depending on the site of activity and the amount of activity. The SMAD3 adaptor protein SPTBN1 (also known as β-spectrin, or non-erythroid spectrin) also exhibits tissue-specific anti- or pro-cancer effects that depend on genetic context. For example, compared to $Sptbn1$ heterozygous mice, knocking out $Itih4$, which is a serine protease that is stimulated by interleukin-6, in the context of $Sptbn1$ heterozygosity reduces HCC susceptibility (Tang et al., 2008). Loss of the I-Smad,

Table 1 Mouse models with altered susceptibility to HCC indicate TGF-β has both tumor-suppressing and tumor-promoting roles in HCC.

Mouse model	HCC susceptibility	Role	References
$Tgfb1^{+/-}$ Tgfb1 heterozygosity	Increased	Tumor suppressor	Tang et al. (1998)
Alb/TGF-B1 Liver-specific expression of TGF-β1	Increased	Tumor promoter	Factor et al. (1997)
myc/TGF-B1 Liver-specific expression of MYC and TGF-β1	Increased	Tumor promoter	Factor et al. (1997)
Alb-TGF-β1/ LFABP-cyclin D1 Liver-specific expression of TGF-β1 and multi-tissue expression of cyclin D1	Increased	Tumor promoter	Deane et al. (2004)
$Tgfbr2^{+/-}$ Tgfbr2 heterozygosity	Increased	Tumor suppressor	Im et al. (2001)
CRP/ΔkTβRII Liver-specific expression of dominant-negative TGFBR2	Increased	Tumor suppressor	Kanzler et al. (2001)
$Pten^{LKO}Tgfbr2^{LKO}$ Liver-specific deletion of Pten and $Tgfbr^a$	Increased	Tumor suppressor	Morris et al. (2015) and Steiger et al. (2020)
Smad7 KO Smad7 knockout	Increased	Tumor promoter	Wang et al. (2013)
TTR-Cre-SMAD7 KO Hepatocyte-specific Smad7 knockout	Increased	Tumor promoter	Feng et al. (2017)
$Smad3^{+/-}Sptbn1^{+/-}$ Heterozygous loss of function of both Smad3 and Sptbn1	Increased	Tumor suppressor	Chen et al. (2016)
$Sptbn1^{+/-}$ Sptbn1 heterozygosity	Increased	Tumor suppressor	Baek et al. (2008), Kitisin et al. (2007), and Tang et al. (2008)
$Sptbn1^{LKO}$ Liver-specific knockout of Sptbn1	Decreased	Tumor promoter	Rao et al. (2021)

[a]Primarily cholangiocarcinoma with some mixed HCC and cholangiocarcinoma.

SMAD7, is associated with increased HCC susceptibility indicating that excessive TGF-β signaling is pro-tumorigenic.

The TGF-β pathway has a well-established role in promoting full or partial epithelial-to-mesenchymal transition (EMT) (Song, 2007). EMT is associated with increased metastatic potential in cancer. In liver regeneration, TGF-β signaling is important for a partial EMT that enables hepatocytes to escape the proliferation-suppressing and apoptosis-inducing effects of this cytokine (Oh et al., 2018). However, cellular transformation could arise if these EMT-promoting effects are not inactivated as the tissue injury is resolved or if the injury remains unresolved. Consistent with these divergent effects of TGF-β1 in HCC, different HCC lines respond to TGF-β1 with increased or decreased proliferation or apoptosis (Dzieran et al., 2013) or with a full EMT or a partial EMT and markers of stem cells (Malfettone et al., 2017).

A transition from the tumor-suppressing effects to the tumor-promoting effects of this pathway appears to occur in HCC (Gough et al., 2021). This may relate to the successive accumulation of genetic alterations in the transformed cells (Marquardt et al., 2014), as indicated by the mouse genetic models, such as liver-specific expression of TGF-β1 with increased expression of MYC or cyclin D1 or decreased expression of PTEN (Deane et al., 2004; Factor et al., 1997; Morris et al., 2015; Steiger et al., 2020). Alternatively, the condition-specific effects may relate to effects on infiltrating macrophages, dendritic cells, and T cells, effects that reduce tumor elimination (Chen, Gingold, & Su, 2019; David & Massagué, 2018; Ma et al., 2016; Shen et al., 2015), or effects on hepatocyte progenitor cells that lead to cancer stem cell formation (Chen et al., 2013, 2016; Mishra, Derynck, & Mishra, 2005; Tang et al., 2008).

5. TGF-β pathway activity in HCC patients
5.1 Defining patient groups by transcriptional profiling

Multiple studies link differences in TGF-β signaling to subtypes of HCC in patients (Chen et al., 2018; Coulouarn, Factor, & Thorgeirsson, 2008; Hoshida et al., 2009). Coulouarn et al. provided the first analysis of TGF-β pathway activity in stratifying HCC patients. To develop a gene expression profile, they used primary hepatocytes from mice with conditional knockout of *TGFBR2* and from control mice and defined a biphasic TGF-β-induced gene expression signature. The "early" signature, corresponding to the response to 1–2 h of TGF-β stimulation, was associated with induction of genes involved in mediating cell cycle arrest and apoptosis. The "late" signature, corresponding to 4–24 h of stimulation, was associated with cytoskeletal reorganization, cell motility, and ECM remodeling. Thus, the early signature represented a tumor-suppressing profile and the late signature a tumor-progressing profile. By analyzing the human orthologs all of the differentially expressed genes (early and late), the 139 HCC cases were divided into a cluster positive for TGF-β signaling and cluster negative for TGF-β signaling. The group positive for TGF-β signaling were analyzed further, which revealed four subclusters. One of the TGF-β positive subcluster corresponded to the TGF-β early signature and one corresponded to the TGF-β late signature. Consistent with the late signature reflecting disease progression, metastatic HCC had this signature and patients with the late signature had significantly worse survival. This study provides clear clinical support for distinct effects of TGF-β signaling in different phases of HCC progression.

Hoshida et al. used transcriptional microarray data to define three subtypes of HCC (Hoshida et al., 2009). One of these subtypes was associated with a signature of high WNT pathway activity and high TGF-β pathway activity, suggesting oncogenic crosstalk between these pathways. Studies with HCC cell lines showed that TGF-β stimulated β-catenin-dependent gene expression through a mechanism that did not involve induction of *CTNNB1* at the transcriptional level. Since this study, crosstalk among the TGF-β, WNT, and YAP pathways in the liver, particularly in fibrosis, has been well established (Piersma, Bank, & Boersema, 2015). Much of this crosstalk occurs at gene promoters through consensus binding of multiple mediators to different sites, the formation of transcriptional regulatory complexes containing mediators from each of these pathways, or through the epigenetic chromatin modifications induced by factors recruited by Smads (Fig. 1).

Chen et al. selected a set of 18 genes related to TGF-β or BMP signaling, which they defined as the "TGF-β pathway" and used the expression of these genes to define 4 clusters of HCC, a quantitative score of activity of this 18-gene pathway, and a qualitative signature of the 18-gene pathway activity (Chen et al., 2018). Clusters A and B were classified as qualitatively "activated" and represented groups with higher pathway activity than that in normal adjacent liver tissue. Cluster A had the highest quantitative pathway activity score. Cluster C had a similar pathway activity score to that in normal adjacent tissue, receiving a qualitative classification of "normal." Cluster D had a score lower than that in normal tissue, indicating that HCC in cluster D was associated with suppression of the pathway and receiving a qualitative classification of "inactivated." Survival analysis showed that the group corresponding to cluster D with inactivated status had the shortest survival and those in cluster C with normal pathway activity had the longest survival. They evaluated the relationship between the qualitative activity groups (activated, normal, inactivated) and expression of oncogenes associated with HCC, genes implicated in changes in the tumor microenvironment, genes encoding acetyltransferases or deacetylases, and genes encoding proteins involved in DNA repair. From evaluating these relationships, specific potential interventions for patients with activated or inactivated pathway profiles could be predicted. Not surprisingly, the activated group was associated with increased expression of genes associated with fibrosis and inflammation and indicating increased TGF-β signaling in the tumor microenvironment. Surprisingly, expression of DNA repair genes was significantly less in the HCC cases with the inactivated pathway group. Thus, treatments for the group with activated TGF-β signaling could benefit from targeted therapies against the pathways altered by the oncogenes with increased expression, *MDM2* (decreased p53 signaling), *MTOR* (increased phosphoinositide-3 kinase to mechanistic target of rapamycin signaling), *IGF2* (increased insulin-like growth factor signaling), and *VEGFA* (increased angiogenic signaling), whereas the group with inactivated signaling could benefit from DNA-damaging therapies or therapies that promote oxidative stress (Marquardt, 2018). Intriguingly, *KRAS* expression was increased in the group with the activated pathway and *HRAS* was increased in the group with the inactivated pathway, suggesting that aberrant activity of the mitogen-activated protein kinase (MAPK) pathway was a therapeutic target in both groups.

Collectively, these studies indicate that a subset of HCC patients could benefit from targeting the TGF-β pathway. Furthermore, altered activity of the pathway could differentiate patients for specific therapies that do not necessarily target TGF-β signaling specifically but target downstream pro-tumor activities that result from either too much or too little TGF-β activity.

5.2 Developing treatments using TGF-β pathway insights

Multiple treatments that interfere with TGF-β signaling are undergoing or have completed clinical trials (Liu, Ren, & ten Dijke, 2021). These range from molecules that interfere with the ligands to inhibitors of the kinase activity of the receptors to antisense oligonucleotides to reduce the abundance of the ligands (Table 2). An antibody targeting endoglin is included,

Table 2 Clinical trials targeting TGF-β signaling for patients with advanced solid tumors or HCC.

Drug name	Therapeutic strategy	Trial	Notes
SAR439459	Neutralizing antibody against TGF-β1, 2, 3	NCT03192345 Advanced solid tumors	Monotherapy or combined with cemiplimab (antibody against PD-1)
NIS793	Neutralizing antibody against TGF-β1, 2, 3	NCT02947165 HCC and other solid tumors	Monotherapy or combined with spartalizumab (also known as PDR001, an antibody against PD-1)
ABBV-151	Neutralizing antibody against GARP:TGF-β1	NCT03821935 Advanced solid tumors	Monotherapy or combined with ABBV-181 (also known as budigalimab, an antibody against PD-1)
AVID200	Ligand trap for TGF-β1 and 3 (engineered subtype selective TGF-β receptor ectodomain fused to Fc)	NCT03834662 Advanced solid tumors	Monotherapy
M7824	Ligand trap for TGF-β1 and PD-L1 (engineered fusion of an antibody Fab for PD-L1 and the ectodomain of TGFBR2)	NCT03436563 Advanced solid tumors	Monotherapy
Trabedersen	Antisense oligonucleotide against *TGFB2*	NCT00844064 Advanced solid tumors with high TGF-β2 (HCC not a listed condition)	Monotherapy
IMC-TR1	Neutralizing antibody against TGFBR2	NCT01646203 Advanced solid tumors	Monotherapy

Continued

Table 2 Clinical trials targeting TGF-β signaling for patients with advanced solid tumors or HCC.—cont'd

Drug name	Therapeutic strategy	Trial	Notes
Galunisertib (also known as LY2157299)	Small molecule inhibitor of TGFBR1	NCT02906397 HCC	Combined with radiation therapy
		NCT01246986 HCC	Combined with sorafenib (multiple tyrosine kinase inhibitor) or ramucirumab (antibody against VEGFR2)
		NCT02178358 HCC	Monotherapy or combined with sorafenib
		NCT02423343 Advanced refractory solid tumors	Combined with nivolumab (antibody against PD-1)
Vactosertib (also known as TEW-7197)	Small molecule inhibitor of TGFBR1	NCT02160106 Advanced solid tumors	Monotherapy
LY3200882	Small molecule inhibitor of TGFBR1	NCT02937272 Solid tumors	Monotherapy or combined with LY3300054 (antibody against PD-L1) or gemcitabine (cell cycle inhibitor pyrimidine analog) and nab-paclitaxel (cell cycle inhibitor targeting tubulin combined with albumin) cisplatin (DNA-damaging agent) and radiotherapy
PF06952229	Small molecule inhibitor of TGFBR1	NCT03685591 Advanced solid tumors	Monotherapy
Carotixumab (also known as TRC105)	Neutralizing antibody against endoglin	NCT01306058 HCC	Combined with sorafenib

Table 2 Clinical trials targeting TGF-β signaling for patients with advanced solid tumors or HCC.—cont'd

Drug name	Therapeutic strategy	Trial	Notes
Cilengitide	αvβ3 or αvβ5 integrin inhibitor	NCT00077155 Advanced solid tumors	Monotherapy
		NCT00022113 Advanced solid tumors	Monotherapy
		NCT01276496 Advanced solid tumors	Combined with paclitaxel (cell cycle inhibitor targeting tubulin)
		NCT00004258 Advanced or metastatic cancer	Monotherapy

because endoglin plays key roles in angiogenesis through balancing TGF-β signals through distinct receptor complexes (Lebrin et al., 2004). An inhibitor of αvβ3 or αvβ5 integrins is included because these integrins not only participate in cell adhesion and signaling but also promote the release of TGF-β from latent complexes. For treating patients with HCC, these are all in either phase I or phase II. Many studies are determining safe doses and then testing combinations with other therapies. Neutralizing antibodies targeting all three members of the TGF-β subfamily (SAR439459 and NIS793), as well as the small-molecule inhibitor of the kinase activity of TGFBR1 (Galunisertib), are undergoing tests with immune checkpoint inhibitors targeting either the receptor PD-1 or the ligand PD-L1. Radiation and conventional chemotherapy agents are also being tested with TGF-β pathway inhibitors.

Results from patients with other cancer reveal the challenges of translating preclinical findings into the clinic. A TGFBR2-targeted antibody that was effective in mouse models, including the promotion of anti-tumor immune response (Zhong et al., 2010), was toxic in patients and thus was not pursued (Tolcher et al., 2017). The integrin αvβ6-inhibiting antibody 264RAD successfully reduced release of TGF-β, reduced cell proliferation and survival in vitro, and stimulated an anti-tumor immune response in either xenograft studies of human pancreatic cancer or in a mouse model of pancreatic cancer (Reader et al., 2019). However, the beneficial effect may vary, and stratification of patients may be necessary. A previous study with a different mouse model found that blocking either TGF-β or this integrin promoted invasive disease and metastasis (Hezel et al., 2012).

Evaluating patients for TGF-β pathway activity signatures could help stratify HCC patients into trials with the most likely effective combinations. For example, patients with an activated TGF-β pathway profile, which is associated with immune and inflammatory signaling (Chen et al., 2018) might be expected to respond well to the combination of inhibition of TGF-β signaling and immune checkpoint inhibitors. Furthermore, other potential combinations for patients with the activated signature include MTOR inhibitors, such as everolimus, sirolimus, or temsirolimus; MAPK inhibitors, such as trametinib or cobimetinib; or angiogenesis inhibitors, such as bevacizumab. Patients with low TGF-β activity or with reduced function due to mutation or low gene expression of specific components of the pathway, such as SMAD3 or SPTNB1, could be particularly responsive to DNA-damaging chemotherapeutics or radiation therapy (Chen et al., 2018). Thus, rather than thinking of only targeting the TGF-β pathway, using knowledge of the activity of the pathway in specific patients could serve as a guiding principle for rational therapy.

Grant support
NIH grants R01AA023146 (L. Mishra), NIH R01CA236591 (L. Mishra), NIH U01 CA230690-01 (L. Mishra).

Conflict of interest statement
The authors have no financial or personal disclosures relevant to this manuscript.

References
Alexander, J., Torbenson, M., Wu, T.-T., & Yeh, M. M. (2013). Non-alcoholic fatty liver disease contributes to hepatocarcinogenesis in non-cirrhotic liver: A clinical and pathological study. *Journal of Gastroenterology and Hepatology, 28*(5), 848–854. https://doi.org/10.1111/jgh.12116.

Ally, A., Balasundaram, M., Carlsen, R., Chuah, E., Clarke, A., Dhalla, N., et al. (2017). Comprehensive and integrative genomic characterization of hepatocellular carcinoma. *Cell, 169*(7), 1327–1341.e23. https://doi.org/10.1016/j.cell.2017.05.046.

Baek, H. J., Lee, Y. M., Kim, T. H., Kim, J.-Y., Park, E. J., Iwabuchi, K., et al. (2016). Caspase-3/7-mediated cleavage of β2-spectrin is required for acetaminophen-induced liver damage. *International Journal of Biological Sciences, 12*(2), 172–183. https://doi.org/10.7150/ijbs.13420.

Baek, H. J., Lim, S. C., Kitisin, K., Jogunoori, W., Tang, Y., Marshall, M. B., et al. (2008). Hepatocellular cancer arises from loss of transforming growth factor beta signaling adaptor protein embryonic liver fodrin through abnormal angiogenesis. *Hepatology, 48*(4), 1128–1137. https://doi.org/10.1002/hep.22460.

Batlle, E., & Massagué, J. (2019). Transforming growth Factor-β signaling in immunity and cancer. *Immunity*, *50*(4), 924–940. https://doi.org/10.1016/j.immuni.2019.03.024.

Bertero, A., Brown, S., Madrigal, P., Osnato, A., Ortmann, D., Yiangou, L., et al. (2018). The SMAD2/3 interactome reveals that TGFβ controls m6A mRNA methylation in pluripotency. *Nature*, *555*(7695), 256–259. https://doi.org/10.1038/nature25784.

Bissell, D. M., Wang, S. S., Jarnagin, W. R., & Roll, F. J. (1995). Cell-specific expression of transforming growth factor-beta in rat liver. Evidence for autocrine regulation of hepatocyte proliferation. *The Journal of Clinical Investigation*, *96*(1), 447–455. https://doi.org/10.1172/JCI118055.

Calderaro, J., Couchy, G., Imbeaud, S., Amaddeo, G., Letouzé, E., Blanc, J.-F., et al. (2017). Histological subtypes of hepatocellular carcinoma are related to gene mutations and molecular tumour classification. *Journal of Hepatology*, *67*(4), 727–738. https://doi.org/10.1016/j.jhep.2017.05.014.

Chayanupatkul, M., Omino, R., Mittal, S., Kramer, J. R., Richardson, P., Thrift, A. P., et al. (2017). Hepatocellular carcinoma in the absence of cirrhosis in patients with chronic hepatitis B virus infection. *Journal of Hepatology*, *66*(2), 355–362. https://doi.org/10.1016/j.jhep.2016.09.013.

Chen, J., Gingold, J. A., & Su, X. (2019). Immunomodulatory TGF-β signaling in hepatocellular carcinoma. *Trends in Molecular Medicine*. https://doi.org/10.1016/j.molmed.2019.06.007.

Chen, C.-L., Tsukamoto, H., Liu, J.-C., Kashiwabara, C., Feldman, D., Sher, L., et al. (2013). Reciprocal regulation by TLR4 and TGF-β in tumor-initiating stem-like cells. *The Journal of Clinical Investigation*, *123*(7), 2832–2849. https://doi.org/10.1172/JCI65859.

Chen, J., Yao, Z.-X., Chen, J.-S., Gi, Y. J., Muñoz, N. M., Kundra, S., et al. (2016). TGF-β/β2-spectrin/CTCF-regulated tumor suppression in human stem cell disorder Beckwith-Wiedemann syndrome. *The Journal of Clinical Investigation*, *126*(2), 527–542. https://doi.org/10.1172/JCI80937.

Chen, J., Zaidi, S., Rao, S., Chen, J.-S., Phan, L., Farci, P., et al. (2018). Analysis of genomes and transcriptomes of hepatocellular carcinomas identifies mutations and gene expression changes in the transforming growth Factor-β pathway. *Gastroenterology*, *154*(1), 195–210. https://doi.org/10.1053/j.gastro.2017.09.007.

Coulouarn, C., Factor, V. M., & Thorgeirsson, S. S. (2008). Transforming growth factor-β gene expression signature in mouse hepatocytes predicts clinical outcome in human cancer. *Hepatology*, *47*(6), 2059–2067. https://doi.org/10.1002/hep.22283.

Dahl, M., Maturi, V., Lönn, P., Papoutsoglou, P., Zieba, A., Vanlandewijck, M., et al. (2014). Fine-tuning of Smad protein function by poly(ADP-ribose) polymerases and poly(ADP-ribose) Glycohydrolase during transforming growth Factor β signaling. *PLoS One*, *9*(8), e103651. https://doi.org/10.1371/journal.pone.0103651.

David, C. J., & Massagué, J. (2018). Contextual determinants of TGFβ action in development, immunity and cancer. *Nature Reviews Molecular Cell Biology*, *19*(7), 419–435. https://doi.org/10.1038/s41580-018-0007-0.

De Battista, D., Zamboni, F., Gerstein, H., Sato, S., Markowitz, T. E., Lack, J., et al. (2021). Molecular signature and immune landscape of HCV-associated hepatocellular carcinoma (HCC): Differences and similarities with HBV-HCC. *Journal of Hepatocellular Carcinoma*, *8*, 1399–1413. https://doi.org/10.2147/JHC.S325959.

Deane, N. G., Lee, H., Hamaamen, J., Ruley, A., Washington, M. K., LaFleur, B., et al. (2004). Enhanced tumor formation in cyclin D1 × transforming growth Factor β1 double transgenic mice with characterization by magnetic resonance imaging. *Cancer Research*, *64*(4), 1315–1322. https://doi.org/10.1158/0008-5472.CAN-03-1772.

Derynck, R., & Budi, E. H. (2019). Specificity, versatility, and control of TGF-β family signaling. *Science Signaling*, *12*(570), eaav5183. https://doi.org/10.1126/scisignal.aav5183.

Dzieran, J., Fabian, J., Feng, T., Coulouarn, C., Ilkavets, I., Kyselova, A., et al. (2013). Comparative analysis of TGF-β/Smad signaling dependent Cytostasis in human hepatocellular carcinoma cell lines. *PLoS One*, *8*(8), e72252. https://doi.org/10.1371/journal.pone.0072252.

Fabregat, I., Moreno-Càceres, J., Sánchez, A., Dooley, S., Dewidar, B., Giannelli, G., et al. (2016). TGF-β signalling and liver disease. *The FEBS Journal*, *283*(12), 2219–2232. https://doi.org/10.1111/febs.13665.

Factor, V. M., Kao, C.-Y., Santoni-Rugiu, E., Woitach, J. T., Jensen, M. R., & Thorgeirsson, S. S. (1997). Constitutive expression of mature transforming growth Factor β1 in the liver accelerates hepatocarcinogenesis in transgenic mice. *Cancer Research*, *57*(11), 2089–2095. https://cancerres.aacrjournals.org/content/57/11/2089.

Feng, T., Dzieran, J., Yuan, X., Dropmann, A., Maass, T., Teufel, A., et al. (2017). Hepatocyte-specific Smad7 deletion accelerates DEN-induced HCC via activation of STAT3 signaling in mice. *Oncogene*, *6*(1), e294. https://doi.org/10.1038/oncsis.2016.85.

Ghafoory, S., Varshney, R., Robison, T., Kouzbari, K., Woolington, S., Murphy, B., et al. (2018). Platelet TGF-β1 deficiency decreases liver fibrosis in a mouse model of liver injury. *Blood Advances*, *2*(5), 470–480. https://doi.org/10.1182/bloodadvances.2017010868.

Global Burden of Disease Liver Cancer Collaboration, Akinyemiju, T., Abera, S., Ahmed, M., Alam, N., Alemayohu, M. A., et al. (2017). The burden of primary liver cancer and underlying etiologies from 1990 to 2015 at the global, regional, and National Level: Results from the global burden of disease study 2015. *JAMA Oncology*, *3*(12), 1683–1691. https://doi.org/10.1001/jamaoncol.2017.3055.

Gough, N. R., Xiang, X., & Mishra, L. (2021). TGF-β signaling in liver, pancreas, and gastrointestinal diseases and cancer. *Gastroenterology*, *161*(2), 434–452. e15 https://doi.org/10.1053/j.gastro.2021.04.064.

Hellerbrand, C., Stefanovic, B., Giordano, F., Burchardt, E. R., & Brenner, D. A. (1999). The role of TGFβ1 in initiating hepatic stellate cell activation in vivo. *Journal of Hepatology*, *30*(1), 77–87. https://doi.org/10.1016/S0168-8278(99)80010-5.

Herrera, B., Álvarez, A. M., Beltrán, J., Valdés, F., Fabregat, I., & Fernández, M. (2004). Resistance to TGF-β-induced apoptosis in regenerating hepatocytes. *Journal of Cellular Physiology*, *201*(3), 385–392. https://doi.org/10.1002/jcp.20078.

Hezel, A. F., Deshpande, V., Zimmerman, S. M., Contino, G., Alagesan, B., O'Dell, M. R., et al. (2012). TGF-β and αvβ6 integrin act in a common pathway to suppress pancreatic cancer progression. *Cancer Research*, *72*(18), 4840–4845. https://doi.org/10.1158/0008-5472.CAN-12-0634.

Hill, C. S. (2016). Transcriptional control by the SMADs. *Cold Spring Harbor Perspectives in Biology*, *8*(10), a022079. https://doi.org/10.1101/cshperspect.a022079.

Hoshida, Y., Nijman, S. M. B., Kobayashi, M., Chan, J. A., Brunet, J.-P., Chiang, D. Y., et al. (2009). Integrative transcriptome analysis reveals common molecular subclasses of human hepatocellular carcinoma. *Cancer Research*, *69*(18), 7385–7392. https://doi.org/10.1158/0008-5472.CAN-09-1089.

Huang, D. Q., El-Serag, H. B., & Loomba, R. (2021). Global epidemiology of NAFLD-related HCC: Trends, predictions, risk factors and prevention. *Nature Reviews. Gastroenterology & Hepatology*, *18*(4), 223–238. https://doi.org/10.1038/s41575-020-00381-6.

Im, Y. H., Kim, H. T., Kim, I. Y., Factor, V. M., Hahm, K. B., Anzano, M., et al. (2001). Heterozygous mice for the transforming growth factor-beta type II receptor gene have increased susceptibility to hepatocellular carcinogenesis. *Cancer Research*, *61*(18), 6665–6668.

Ioannou, G. N. (2021). Epidemiology and risk-stratification of NAFLD-associated HCC. *Journal of Hepatology*. S0168827821020079 https://doi.org/10.1016/j.jhep.2021.08.012.

Jang, C.-W., Chen, C.-H., Chen, C.-C., Chen, J., Su, Y.-H., & Chen, R.-H. (2002). TGF-beta induces apoptosis through Smad-mediated expression of DAP-kinase. *Nature Cell Biology*, *4*(1), 51–58. https://doi.org/10.1038/ncb731.

Kanzler, S., Meyer, E., Lohse, A. W., Schirmacher, P., Henninger, J., Galle, P. R., et al. (2001). Hepatocellular expression of a dominant-negative mutant TGF-β type II receptor accelerates chemically induced hepatocarcinogenesis. *Oncogene*, *20*(36), 5015–5024. https://doi.org/10.1038/sj.onc.1204544.

Karlmark, K. R., Weiskirchen, R., Zimmermann, H. W., Gassler, N., Ginhoux, F., Weber, C., et al. (2009). Hepatic recruitment of the inflammatory Gr1 + monocyte subset upon liver injury promotes hepatic fibrosis. *Hepatology*, *50*(1), 261–274. https://doi.org/10.1002/hep.22950.

Kitisin, K., Ganesan, N., Tang, Y., Jogunoori, W., Volpe, E. A., Kim, S. S., et al. (2007). Disruption of transforming growth factor-β signaling through β-spectrin ELF leads to hepatocellular cancer through cyclin D1 activation. *Oncogene*, *26*(50), 7103–7110. https://doi.org/10.1038/sj.onc.1210513.

Kulik, L., & El-Serag, H. B. (2019). Epidemiology and Management of Hepatocellular Carcinoma. *Gastroenterology*, *156*(2), 477–491. https://doi.org/10.1053/j.gastro.2018.08.065.

Lebrin, F., Goumans, M.-J., Jonker, L., Carvalho, R. L. C., Valdimarsdottir, G., Thorikay, M., et al. (2004). Endoglin promotes endothelial cell proliferation and TGF-beta/ALK1 signal transduction. *The EMBO Journal*, *23*(20), 4018–4028. https://doi.org/10.1038/sj.emboj.7600386.

Letouzé, E., Shinde, J., Renault, V., Couchy, G., Blanc, J.-F., Tubacher, E., et al. (2017). Mutational signatures reveal the dynamic interplay of risk factors and cellular processes during liver tumorigenesis. *Nature Communications*, *8*(1), 1315. https://doi.org/10.1038/s41467-017-01358-x.

Li, Y., Turpin, C. P., & Wang, S. (2017). Role of thrombospondin 1 in liver diseases. *Hepatology Research: The Official Journal of the Japan Society of Hepatology*, *47*(2), 186–193. https://doi.org/10.1111/hepr.12787.

Liu, S., Ren, J., & ten Dijke, P. (2021). Targeting TGFβ signal transduction for cancer therapy. *Signal Transduction and Targeted Therapy*, *6*(1), 1–20. https://doi.org/10.1038/s41392-020-00436-9.

Losic, B., Craig, A. J., Villacorta-Martin, C., Martins-Filho, S. N., Akers, N., Chen, X., et al. (2020). Intratumoral heterogeneity and clonal evolution in liver cancer. *Nature Communications*, *11*(1), 291. https://doi.org/10.1038/s41467-019-14050-z.

Ma, C., Kesarwala, A. H., Eggert, T., Medina-Echeverz, J., Kleiner, D. E., Jin, P., et al. (2016). NAFLD causes selective CD4 + T lymphocyte loss and promotes hepatocarcinogenesis. *Nature*, *531*(7593), 253–257. https://doi.org/10.1038/nature16969.

Majumdar, A., Curley, S. A., Wu, X., Brown, P., Hwang, J. P., Shetty, K., et al. (2012). Hepatic stem cells and transforming growth factor β in hepatocellular carcinoma. *Nature Reviews. Gastroenterology & Hepatology*, *9*(9), 530–538. https://doi.org/10.1038/nrgastro.2012.114.

Malfettone, A., Soukupova, J., Bertran, E., Crosas-Molist, E., Lastra, R., Fernando, J., et al. (2017). Transforming growth factor-β-induced plasticity causes a migratory stemness phenotype in hepatocellular carcinoma. *Cancer Letters*, *392*, 39–50. https://doi.org/10.1016/j.canlet.2017.01.037.

Marquardt, J. U. (2018). The role of transforming growth Factor-β in human hepatocarcinogenesis: Mechanistic and therapeutic implications from an integrative multiomics approach. *Gastroenterology*, *154*(1), 17–20. https://doi.org/10.1053/j.gastro.2017.11.015.

Marquardt, J. U., Seo, D., Andersen, J. B., Gillen, M. C., Kim, M. S., Conner, E. A., et al. (2014). Sequential transcriptome analysis of human liver cancer indicates late stage acquisition of malignant traits. *Journal of Hepatology*, *60*(2), 346–353. https://doi.org/10.1016/j.jhep.2013.10.014.

Meijer, C., Wiezer, M. J., Diehl, A. M., Yang, S.-Q., Schouten, H. J., Meijer, S., et al. (2000). Kupffer cell depletion by CI2MDP-liposomes alters hepatic cytokine expression and delays liver regeneration after partial hepatectomy. *Liver*, *20*(1), 66–77. https://doi.org/10.1034/j.1600-0676.2000.020001066.x.

Mishra, L., Derynck, R., & Mishra, B. (2005). Transforming growth Factor-ß signaling in stem cells and cancer. *Science*, *310*(5745), 68–71. https://doi.org/10.1126/science.1118389.

Mittal, S., El-Serag, H. B., Sada, Y. H., Kanwal, F., Duan, Z., Temple, S., et al. (2016). Hepatocellular carcinoma in the absence of cirrhosis in United States veterans is associated with nonalcoholic fatty liver disease. *Clinical Gastroenterology and Hepatology: The Official Clinical Practice Journal of the American Gastroenterological Association*, *14*(1), 124–131. https://doi.org/10.1016/j.cgh.2015.07.019.

Miyazawa, K., & Miyazono, K. (2017). Regulation of TGF-β family signaling by inhibitory Smads. *Cold Spring Harbor Perspectives in Biology*, *9*(3), a022095. https://doi.org/10.1101/cshperspect.a022095.

Moon, A. M., Singal, A. G., & Tapper, E. B. (2020). Contemporary epidemiology of chronic liver disease and cirrhosis. *Clinical Gastroenterology and Hepatology*, *18*(12), 2650–2666. https://doi.org/10.1016/j.cgh.2019.07.060.

Morikawa, M., Derynck, R., & Miyazono, K. (2016). TGF-β and the TGF-β family: Context-dependent roles in cell and tissue physiology. *Cold Spring Harbor Perspectives in Biology*, *8*(5), a021873. https://doi.org/10.1101/cshperspect.a021873.

Morris, S. M., Carter, K. T., Baek, J. Y., Koszarek, A., Yeh, M. M., Knoblaugh, S. E., et al. (2015). TGF-β signaling alters the pattern of liver tumorigenesis induced by Pten inactivation. *Oncogene*, *34*(25), 3273–3282. https://doi.org/10.1038/onc.2014.258.

Oh, S.-H., Swiderska-Syn, M., Jewell, M. L., Premont, R. T., & Diehl, A. M. (2018). Liver regeneration requires Yap1-TGFβ-dependent epithelial-mesenchymal transition in hepatocytes. *Journal of Hepatology*, *69*(2), 359–367. https://doi.org/10.1016/j.jhep.2018.05.008.

Paik, J. M., Golabi, P., Younossi, Y., Mishra, A., & Younossi, Z. M. (2020). Changes in the global burden of chronic liver diseases from 2012 to 2017: The growing impact of NAFLD. *Hepatology*, *72*(5), 1605–1616. https://doi.org/10.1002/hep.31173.

Papoutsoglou, P., & Moustakas, A. (2020). Long non-coding RNAs and TGF-β signaling in cancer. *Cancer Science*, *111*(8), 2672–2681. https://doi.org/10.1111/cas.14509.

Piersma, B., Bank, R. A., & Boersema, M. (2015). Signaling in fibrosis: TGF-β, WNT, and YAP/TAZ converge. *Frontiers in Medicine*, *2*, 59. https://doi.org/10.3389/fmed.2015.00059.

Ramjaun, A. R., Tomlinson, S., Eddaoudi, A., & Downward, J. (2007). Upregulation of two BH3-only proteins, Bmf and Bim, during TGFβ-induced apoptosis. *Oncogene*, *26*(7), 970–981. https://doi.org/10.1038/sj.onc.1209852.

Rao, S., Yang, X., Ohshiro, K., Zaidi, S., Wang, Z., Shetty, K., et al. (2021). B2-spectrin (SPTBN1) as a therapeutic target for diet-induced liver disease and preventing cancer development. *Science Translational Medicine*, *13*(624), eabk2267. https://doi.org/10.1126/scitranslmed.abk2267.

Rao, S., Zaidi, S., Banerjee, J., Jogunoori, W., Sebastian, R., Mishra, B., et al. (2017). Transforming growth factor-β in liver cancer stem cells and regeneration. *Hepatology Communications*, *1*(6), 477–493. https://doi.org/10.1002/hep4.1062.

Reader, C. S., Vallath, S., Steele, C. W., Haider, S., Brentnall, A., Desai, A., et al. (2019). The integrin αvβ6 drives pancreatic cancer through diverse mechanisms and represents an effective target for therapy. *The Journal of Pathology*, *249*(3), 332–342. https://doi.org/10.1002/path.5320.

Rebouissou, S., & Nault, J.-C. (2020). Advances in molecular classification and precision oncology in hepatocellular carcinoma. *Journal of Hepatology, 72*(2), 215–229. https://doi.org/10.1016/j.jhep.2019.08.017.

Russell, W. E., Coffey, R. J., Ouellette, A. J., & Moses, H. L. (1988). Type beta transforming growth factor reversibly inhibits the early proliferative response to partial hepatectomy in the rat. *Proceedings of the National Academy of Sciences of the United States of America, 85*(14), 5126–5130. https://doi.org/10.1073/pnas.85.14.5126.

Schulze, K., Imbeaud, S., Letouzé, E., Alexandrov, L. B., Calderaro, J., Rebouissou, S., et al. (2015). Exome sequencing of hepatocellular carcinomas identifies new mutational signatures and potential therapeutic targets. *Nature Genetics, 47*(5), 505–511. https://doi.org/10.1038/ng.3252.

Sepanlou, S. G., Safiri, S., Bisignano, C., Ikuta, K. S., Merat, S., Saberifiroozi, M., et al. (2020). The global, regional, and national burden of cirrhosis by cause in 195 countries and territories, 1990–2017: A systematic analysis for the global burden of disease study 2017. *The Lancet Gastroenterology & Hepatology, 5*(3), 245–266. https://doi.org/10.1016/S2468-1253(19)30349-8.

Shen, Y., Wei, Y., Wang, Z., Jing, Y., He, H., Yuan, J., et al. (2015). TGF-β regulates hepatocellular carcinoma progression by inducing Treg cell polarization. *Cellular Physiology and Biochemistry, 35*(4), 1623–1632. https://doi.org/10.1159/000373976.

Shi, M., Zhu, J., Wang, R., Chen, X., Mi, L., Walz, T., et al. (2011). Latent TGF-β structure and activation. *Nature, 474*(7351), 343–349. https://doi.org/10.1038/nature10152.

Song, J. (2007). EMT or apoptosis: A decision for TGF-β. *Cell Research, 17*(4), 289–290. https://doi.org/10.1038/cr.2007.25.

Steiger, K., Gross, N., Widholz, S. A., Rad, R., Weichert, W., & Mogler, C. (2020). Genetically engineered mouse models of liver tumorigenesis reveal a wide histological Spectrum of neoplastic and non-neoplastic liver lesions. *Cancers, 12*(8), E2265. https://doi.org/10.3390/cancers12082265.

Sung, H., Ferlay, J., Siegel, R. L., Laversanne, M., Soerjomataram, I., Jemal, A., et al. (2021). Global cancer statistics 2020: GLOBOCAN estimates of incidence and mortality worldwide for 36 cancers in 185 countries. *CA: A Cancer Journal for Clinicians, 71*(3), 209–249. https://doi.org/10.3322/caac.21660.

Tang, B., Böttinger, E. P., Jakowlew, S. B., Bagnall, K. M., Mariano, J., Anver, M. R., et al. (1998). Transforming growth factor-β1 is a new form of tumor suppressor with true haploid insufficiency. *Nature Medicine, 4*(7), 802–807. https://doi.org/10.1038/nm0798-802.

Tang, Y., Katuri, V., Dillner, A., Mishra, B., Deng, C.-X., & Mishra, L. (2003). Disruption of transforming growth factor-beta signaling in ELF beta-spectrin-deficient mice. *Science (New York, N.Y.), 299*(5606), 574–577. https://doi.org/10.1126/science.1075994.

Tang, Y., Kitisin, K., Jogunoori, W., Li, C., Deng, C.-X., Mueller, S. C., et al. (2008). Progenitor/stem cells give rise to liver cancer due to aberrant TGF-β and IL-6 signaling. *Proceedings of the National Academy of Sciences, 105*(7), 2445–2450. https://doi.org/10.1073/pnas.0705395105.

Tolcher, A. W., Berlin, J. D., Cosaert, J., Kauh, J., Chan, E., Piha-Paul, S. A., et al. (2017). A phase 1 study of anti-TGFβ receptor type-II monoclonal antibody LY3022859 in patients with advanced solid tumors. *Cancer Chemotherapy and Pharmacology, 79*(4), 673–680. https://doi.org/10.1007/s00280-017-3245-5.

Tomasetti, C., Li, L., & Vogelstein, B. (2017). Stem cell divisions, somatic mutations, cancer etiology, and cancer prevention. *Science (New York, N.Y.), 355*(6331), 1330–1334. https://doi.org/10.1126/science.aaf9011.

Tsuchida, T., & Friedman, S. L. (2017). Mechanisms of hepatic stellate cell activation. *Nature Reviews Gastroenterology & Hepatology, 14*(7), 397–411. https://doi.org/10.1038/nrgastro.2017.38.

Velasco, S., Alvarez-Muñoz, P., Pericacho, M., ten Dijke, P., Bernabéu, C., Lopez-Novoa, J. M., et al. (2008). L- and S-endoglin differentially modulate TGFβ1 signaling mediated by ALK1 and ALK5 in L6E9 myoblasts. *Journal of Cell Science*, *121*(6), 913–919. https://doi.org/10.1242/jcs.023283.

Wang, J., Zhao, J., Chu, E. S., Mok, M. T., Go, M. Y., Man, K., et al. (2013). Inhibitory role of Smad7 in hepatocarcinogenesis in mice and in vitro. *The Journal of Pathology*, *230*(4), 441–452. https://doi.org/10.1002/path.4206.

Xu, P., Liu, J., & Derynck, R. (2012). Post-translational regulation of TGF-β receptor and Smad signaling. *FEBS Letters*, *586*(14), 1871–1884. https://doi.org/10.1016/j.febslet.2012.05.010.

Yang, Z., Mu, Z., Dabovic, B., Jurukovski, V., Yu, D., Sung, J., et al. (2007). Absence of integrin-mediated TGFbeta1 activation in vivo recapitulates the phenotype of TGFbeta1-null mice. *The Journal of Cell Biology*, *176*(6), 787–793. https://doi.org/10.1083/jcb.200611044.

Yoo, J., Ghiassi, M., Jirmanova, L., Balliet, A. G., Hoffman, B., Fornace, A. J., et al. (2003). Transforming growth Factor-β-induced apoptosis is mediated by Smad-dependent expression of GADD45b through p38 activation *. *Journal of Biological Chemistry*, *278*(44), 43001–43007. https://doi.org/10.1074/jbc.M307869200.

Younes, R., & Bugianesi, E. (2018). Should we undertake surveillance for HCC in patients with NAFLD? *Journal of Hepatology*, *68*(2), 326–334. https://doi.org/10.1016/j.jhep.2017.10.006.

Zhang, Y., Alexander, P. B., & Wang, X.-F. (2017). TGF-β family signaling in the control of cell proliferation and survival. *Cold Spring Harbor Perspectives in Biology*, *9*(4), a022145. https://doi.org/10.1101/cshperspect.a022145.

Zhang, H., Ozaki, I., Mizuta, T., Hamajima, H., Yasutake, T., Eguchi, Y., et al. (2006). Involvement of programmed cell death 4 in transforming growth factor-beta1-induced apoptosis in human hepatocellular carcinoma. *Oncogene*, *25*(45), 6101–6112. https://doi.org/10.1038/sj.onc.1209634.

Zhong, Z., Carroll, K. D., Policarpio, D., Osborn, C., Gregory, M., Bassi, R., et al. (2010). Anti-transforming growth factor beta receptor II antibody has therapeutic efficacy against primary tumor growth and metastasis through multieffects on cancer, stroma, and immune cells. *Clinical Cancer Research: An Official Journal of the American Association for Cancer Research*, *16*(4), 1191–1205. https://doi.org/10.1158/1078-0432.CCR-09-1634.

CHAPTER NINE

Matricellular proteins in intrahepatic cholangiocarcinoma

Alphonse E. Sirica*

Emeritus Professor of Pathology, Virginia Commonwealth University School of Medicine, Richmond, VA, United States
*Corresponding author: e-mail address: alphonse.sirica@vcuhealth.org

Contents

1. Introduction 250
2. Overview of classes, structures, and complexities of matricellular proteins aberrantly expressed in iCCA 251
3. Matricellular proteins as modulators of iCCA microenvironment and malignant progression 254
   3.1 Postn 254
4. Clinical relevance of matricellular proteins in iCCA 260
   4.1 Matricellular proteins as prognostic biomarkers 260
   4.2 Potential as therapeutic targets 271
5. Perspectives and conclusions 272
Grant support 273
Conflict of interest statement 273
References 273

Abstract

Intrahepatic cholangiocarcinoma (iCCA) is typically characterized by a prominent desmoplastic stroma that is often the most dominant feature of the tumor. This tumor reactive stroma is comprised of a dense fibro-collagenous-enriched extracellular matrix (ECM) surrounding the cancer cells, together with other ECM proteins/peptides, specifically secreted matricellular glycoproteins and proteolytic enzymes, growth factors, and cytokines. Moreover, as enjoined by cholangiocarcinoma cells, this enriched tumor microenvironment is populated by various stromal cell types, most prominently, cancer-associated myofibroblasts (CAFs), along with variable numbers of tumor-associated macrophages (TAMs), inflammatory and vascular cell types. While it is now well appreciated that the interplay between cholangiocarcinoma cells, CAFs, and TAMs in particular play a critical role in promoting cholangiocarcinoma progression, therapeutic resistance, and immune evasion, it is also becoming increasingly evident that over-expression and secretion into the tumor microenvironment of functionally overlapping matricellular glycoproteins, including periostin, osteopontin, tenascin-C, thrombospondin-1, mesothelin and others have an important role to play in regulating or modulating a variety of pro-oncogenic cellular functions, including cholangiocarcinoma cell proliferation, invasion, and metastasis, epithelial-mesenchymal transition, ECM remodeling, and immune evasion. Matricellular proteins have also shown promise as potential prognostic factors for iCCA and may provide unique therapeutic opportunities particularly in relation to targeting iCCA pre-metastatic and metastatic niches, tumor cell dormancy, and immune evasion. This review will highlight timely research and its translational implications for salient matricellular proteins in terms of their structure-function relationships, as modulators of intrahepatic cholangiocarcinoma microenvironment and progression, and potential clinical value for iCCA prognosis and therapy.

Abbreviations

BMP-1	bone morphogenetic protein-1
CAFs	cancer-associated fibroblasts
DDC	3,5-diethoxycarbonyl-1,4-dihydrocollidine
ECM	extracellular matrix
EMT	epithelial-mesenchymal transition
iCCA	intrahepatic cholangiocarcinoma
Msln	mesothelin
Postn	periostin
VER	very early recurrence
α-SMA	α-smooth muscle actin

1. Introduction

Anatomically, iCCAs exhibit three major macroscopic growth patterns, classified either as mass-forming, periductular infiltrative, or intraductal growth type (Brindley et al., 2021; Kendall et al., 2019; Nakanuma et al., 2010; Sirica, Strazzabosco, & Cadamuro, 2021; Vijgen, Terris, & Rubbia-Brandt., 2017). Of these, the mass-forming type of iCCA is the most common type, accounting for about 65% of all iCCAs (Kendall et al., 2019; Nakanuma et al., 2010; Vijgen et al., 2017). Mixed patterns combining features of both the periductular infiltrating type and mass-forming type represent the second most common macroscopic growth pattern, comprising about 25% of all iCCAs.

Ninety to ninety-five percent of iCCAs are histologically classified as adenocarcinomas, which may be well-, moderately, or poorly differentiated. They may be either of the small bile duct or the large bile duct type, with small duct type iCCAs displaying no or minimum mucin production, while large duct type iCCAs are mucin-producing adenocarcinomas (Brindley et al., 2021; Kendall et al., 2019; Nakanuma et al., 2010; Sirica et al., 2021). The small duct type is almost exclusively of the peripheral mass-forming type, whereas the large bile duct type typically characterizes the intraductal growth, periductular infiltrating and mixed periductular infiltrating and mass-forming types (Kendall et al., 2019; Nakanuma et al., 2010; Vijgen et al., 2017).

Both small duct and large duct types of iCCAs frequently exhibit a prominent desmoplastic and hypovascularized stroma characterized by a dense fibro-collagenous enriched matrix along with other extracellular matrix (ECM) proteins, and containing an abundance of cancer-associated fibroblasts (CAFs), the vast majority of which are positive for α-smooth muscle actin (a biomarker of myofibroblast differentiation), together with variable numbers of inflammatory cell and vascular cell types (Fabris, Sato, Alpini, & Strazzabosco, 2021; Sirica et al., 2021; Sirica & Gores, 2014). It is now appreciated that this evolving and complex microenvironment together with the deleterious interplay between cholangiocarcinoma cells and stromal cellular and ECM components plays a critical role in promoting iCCA progression, therapeutic resistance, and immunosuppression (Brivio, Cadamuro, Strazzabosco, & Fabris, 2017; Cadamuro et al., 2018; Fabris et al., 2019; Sirica, 2012; Sirica et al., 2021; Vaquero, Aoudjehane, & Fouassier, 2020).

Whereas much of the focus to date has been on the interactive relationships between CAFs, tumor associated macrophages, and cholangiocarcinoma cells in relation to enhanced malignant behavior of iCCA, it is also now becoming increasingly recognized that ECM proteins also have an important role to play in regulating/modulating iCCA functions and malignant aggressiveness, as well as to have potential clinical value as prognostic biomarkers and molecular targets (Carpino et al., 2019; Fabris, Cadamuro, Cagnin, Strazzabosco, & Gores, 2020). In iCCA, as in other cancers, matricellular proteins have emerged as a salient group of non-structural functional glycoproteins whose aberrant expression and secretion into the tumor extracellular environment support multiple hallmarks of cancer. This review will highlight what is currently know about pertinent matricellular proteins, in terms of their (1) structure-function relationships, (2) the roles they play as modulators/regulators of the iCCA microenvironment and of malignant progression, and (3) their potential clinical relevance as biomarkers and molecular targets for iCCA therapy.

2. Overview of classes, structures, and complexities of matricellular proteins aberrantly expressed in iCCA

The concept of matricellular proteins was first proposed by Paul Bornstein more than a quarter of century ago to define a unique group of dynamically expressed modular glycoproteins secreted into the ECM, which unlike structural proteins (i.e., collagen fibrils; basement membrane components), do not play a primary structural role within the extracellular microenvironment (Bornstein, 1995; Bornstein & Sage, 2002; Sage & Bornstein, 1991). Rather, through their distinct evolutionary conserved functional domains, they function by binding to other ECM proteins, cell surface receptors, secreted extracellular signaling molecules, including growth factors, chemokines, or cytokines, as well as proteases to modulate or regulate a variety of critical cellular functions, including cell matrix interactions, cellular differentiation, proliferation, adhesion, motility, angiogenesis, and ECM remodeling in a complex and content-specific manner (Bornstein, 2009; Bornstein & Sage, 2002; Chong, Tan, Huang, & Tan, 2012; Feng & Gerarduzzi, 2020; Gerarduzzi, Hartmann, Leask, & Drobetsky, 2020; Murphy-Ullrich & Sage, 2014; Wong & Rustgi, 2013).

Matricellular proteins are dynamically expressed at high levels during early development, subsiding in normal adult tissues, but transiently becoming overexpressed during injury repair, especially in fibroinflammatory diseases (Gerarduzzi et al., 2020; Murphy-Ullrich & Sage, 2014). Targeted gene knockout of specific matricellular protein genes in mice, (e.g., periostin (Postn), osteopontin, tenascin-C) have been observed to produce either a grossly normal or a subtle phenotypes that become exacerbated upon injury (Bornstein, 2009; Bornstein & Sage, 2002; Wong & Rustgi, 2013) Such observations that mice with targeted inactivation of matricellular protein genes were born alive and on first glance without apparent gross developmental defects, suggested a limited role for these secreted ECM proteins in development, which is now known not to be the case (Gremlich et al., 2020; Murphy-Ullrich & Sage, 2014; Norris et al., 2008; Wen et al., 2015). There is also now ample evidence to support matricellular proteins as having broad importance in chronic diseases, such as various forms of cardiovascular and renal diseases (Landry, Cohen, & Dixon, 2017; Vianello et al., 2020; Wallace, 2019). Moreover, it is increasingly apparent that in human solid cancers types, including iCCA, overexpression of different matricellular proteins not only contribute to the evolving complexity of the tumor microenvironment, but also to influencing molecular signaling, biomechanical, and metastatic properties supporting aggressive malignant behavior (Fabris et al., 2020; Gerarduzzi et al., 2020; González-González & Alonso, 2018; Vincent & Postovit, 2018; Wong & Rustgi, 2013).

Since the initial pioneering studies of Bornstein and his colleagues, the growing discovery of additional matricellular proteins has led to their classification into families grouped on the basis of shared domains that reflect functional diversity and complexity (Gerarduzzi et al., 2020). Table 1 summarizes select ECM matricellular protein families and relevant key family

Table 1 Classification and structural-functional domains of select key ECM matricellular proteins in intrahepatic cholangiocarcinoma.

Family	Representative member expressed in iCCA	Modular domains	Number of identified human splice variants	References
Gla-protein family	Postn[a] (~90 kDa glycoprotein); also known as OSF2	Single peptide (SP) region, EMILIN (EMI) domain, tandem repeats of 4 fasciclin-like 1 (FAS1) domains, C-terminal domain	8 plus full length wild type Postn	Sirica et al. (2014); Viloria and Hill (2016); González-González and Alonso (2018); Kudo (2019); Gerarduzzi et al. (2020)
TNC Family	Tenascin-C (180–330 kDa glycoprotein)	TA (cysteine-rich tenascin assembly domain), 14.5 epidermal growth factor (EGF)-like tandem repeats, 17 fibronectin type III tandem repeats, C-terminal fibrinogen-like globe (FBG) domain	20	Giblin and Midwood (2015); Viloria and Hill (2016); Gerarduzzi et al. (2020)
SIBLING Family	Osteopontin (35–75 kDa phosphoglycoprotein); also known as SPP1	Several conserved domains including RGD domain, SVVYGLR domain, heparin binding domain, calcium binding domain, CD44 receptor, MMP and thrombin cleavage sites	5	Gimba et al. (2019); Moorman et al. (2020); Song et al., 2021)
TSP Family	Thrombospondin-1 (~450 kDa glycoprotein)	N-terminal globular pentraxin-like domain, von Willebrand factor C domain, TSP-1 type 1 (TSR) repeats, type 2 EGF-like repeats, and type 3 calcium binding repeats, carboxy terminal globular domain	8	Viloria and Hill (2016); Isenberg & Roberts, 2020; Kaur et al. (2021)

[a]Postn, periostin; TNC, tenascin; SIBLING, small integrin-binding ligand N-linked glycoprotein; MMP, metalloproteinase; TSP, thrombospondin; EGF, epidermal growth factor.

members, which will be discussed below in the context of iCCA microenvironment and malignant progression. It should also be noted that complexity of matricellular proteins and the contextual specificity of their interactions and functions is further complicated by the recognition of distinct splice variants (Fausther, Lavole, & Dranoff, 2017; Gimba, Brum, & DeMorales, 2019; Hagiwara et al., 2020; Viloria & Hill, 2016), single nucleotide polymorphisms (Zhao, Ma, Su, & He, 2014), as well as post-translational modifications that include specialized forms of glycosylation, gammacarboxylation, and phosphorylation (González-González & Alonso, 2018; Murphy-Ullrich & Sage, 2014). Because our current understanding of the functional relevance of specific matricellular protein isoforms to iCCA progression remains largely unstudied, further discussion of splice variants will only be selectively addressed in a limited fashion in the remaining sections of this review.

3. Matricellular proteins as modulators of iCCA microenvironment and malignant progression

Although still evolving, there is now a mounting body of evidence to convincingly implicate matricellular proteins, as those listed in Table 1, as playing multifaceted roles in regulating or modulating diverse cellular receptor-mediated and ECM-biomechanical functions in iCCA that cumulatively support many of cancer's hallmarks, as defined by Hanahan and Weinberg (2011). Table 2 calls attention to various matricellular proteins aberrantly expressed in iCCA in terms of their pro-tumorigenic functions, receptors, and proposed signaling pathways in promoting malignant progression, advancing a tumor supportive microenvironment, and facilitating invasion and metastasis. Here, one can appreciate the complex and overlapping interactions and redundant intracellular pathways activated by various matricellular proteins overexpressed in iCCA modulating malignant behavior, including ECM composition, remodeling, and stiffness, cell proliferation and survival, epithelial-mesenchymal transition (EMT), invasion and metastasis, tumor hypovascularity and hypoxia, immunosuppression, as well as priming pre-metastatic and metastatic niches.

3.1 Postn

Postn is is representative of a key multifunctional matricellular protein gaining increasing attention in recent years as having functional and

Table 2 Receptors, signaling pathways, and pro-tumorigenic functions of select ECM matricellular proteins aberrantly expressed in intrahepatic cholangiocarcinoma.

Matricellular protein	Key receptors	Signaling pathways	Functions	References
Postn[a]	Integrins $\alpha v\beta 1$, $\alpha v\beta 3$, $\alpha v\beta 5$ $\alpha 6\beta 4$	Activation PI3K/Akt/FAK and NF-κB; modulates Notch1 and WNT/β-catenin signaling; SMAD 2/3 activation	Scaffold for assembly of extracellular matrix proteins; promotes Collagen type 1 fibrillogenesis and collagen crosslinking by activating hepatic myofibroblast lysyl oxidase, leading to desmoplasia and ECM stiffness; induces cholangiocarcinoma cell proliferation, survival, EMT, migration, and invasion; promotes angiogenesis; likely plays crucial role in conditioning pre-metastatic and metastatic niches	Utispan et al. (2010); Thuwajit et al. (2017); Kumar et al. (2018); González-González and Alonso (2018); Kii (2019); Sonongbua et al. (2020); Fabris et al. (2020)
Tenascin-C	Integrins $\alpha v\beta 1$, $\alpha v\beta 3$; weak binding to EGFR	MAPK, Wnt, TGF-β, EGFR, HB-EGF, Notch1, HGF-Met, PDGF	Induces cholangiocarcinoma cell proliferation, migration and invasion; activates TGF-β signaling linked to stromal fibrogenesis, stiffness, and EMT; promotes angiogenesis; can interacts with Postn to promote pre-metastatic and metastatic niches	Chong et al. (2012); Yoshida et al. (2015); Lowy and Oskarsson (2015); Katoh et al. (2020); Fabris et al. (2020); Aubert et al. (2021)

Continued

Table 2 Receptors, signaling pathways, and pro-tumorigenic functions of select ECM matricellular proteins aberrantly expressed in intrahepatic cholangiocarcinoma.—cont'd

Matricellular protein	Key receptors	Signaling pathways	Functions	References
Osteopontin	Integrins $\alpha v \beta 1$, $\beta 3, \beta 5, \beta 6$ and $\beta 8$; ($\alpha 5$, $\alpha 8)\beta 1$; CD44 Varient	PI3K/Akt, MAPK, NFκB, Wnt/β-catenin, TGF-β1, VEGF	Collagen fibrillogenesis, cell proliferation/migration, survival, MMP activities, myofibroblast differentiation, angiogenesis, immunomodulation, metastasis	Lenga et al., 2008, Wong and Rustgi (2013); Chong et al. (2012); Murphy-Ullrich and Sage (2014); Weber et al. (2015); Zeng et al. (2018); Moorman et al. (2020); McQuitty et al. (2020); Song et al. (2021); Jing et al., 2021
TSP-1	Latent TGF-β, syndecans, Integrins (i.e., $\alpha v \beta 3$, $\alpha 3 \beta 1$ $\alpha 4 \beta 1$, $\alpha 6 \beta 1$, $\alpha 9 \beta 1$), CD36, CD47	VEGF signaling pathway antagonist	Latent TGF-β activation, regulator of collagen matrix organization, anti-angiogenesis and promoter of lymphangiogenesis in intrahepatic cholangiocarcinoma	Chong et al. (2012); Murphy-Ullrich and Sage (2014); Murphy-Ullrich (2019); Carpino et al. (2021)

[a] Postn, periostin; TSP-1, thrombospondin-1, ECM, extracellular matrix, EMT, epithelial mesenchymal transition; MMP, metalloproteinases; TGF-β transforming growth factor-β; EGFR, epidermal growth factor; HB-EGF, heparin binding-epidermal growth factor; PDGF, platelet derived growth factor.

clinicopathological significance for iCCA. Independent studies have shown Postn to be frequently overexpressed in the desmoplastic stroma of both human (Carpino et al., 2019; Darby et al., 2010; Fujimoto et al., 2011; Riener et al., 2010; Thuwajit et al., 2017; Utispan et al., 2010) and rat iCCAs (Manzanares, Campbell, Maldonado, & Sirica, 2018; Sirica et al., 2009), with high levels of expression correlating with increased malignant progression, as well as shorter overall survival times in patients who had undergone surgical resection (see below Section 4.1). On the other hand, Postn has been mostly shown not to be detected or at best only weakly expressed in normal human and rat adult livers and paired non-cancerous liver tissues (Darby et al., 2010; Fujimoto et al., 2011; Manzanares et al., 2017, 2018; Riener et al., 2010; Sirica, Almenara, & Li, 2014; Utispan et al., 2010). Low to no Postn expression had also been reported for select benign liver diseases and hepatocellular carcinoma (Utispan et al., 2010). Furthermore, Postn was not detected by Western blotting at 21 days after bile duct ligation in cholestatic rat liver with a periportal bile ductular reaction (Sirica et al., 2009). Sugiyama et al. (2016), however, demonstrated positive α-SMA expression together with Postn enhancement in the fine fibrotic septa around proliferating bile ducts/ductules emanating from the hepatic portal areas of wild-type mice induced by cholestatic injury produced by dietary administered 3,5-diethoxycarbonyl-1,4-dihydrocollidine (DDC). This study further showed the ductular reaction induced by DDC to be dramatically reduced in Postn$^{-/-}$ mice, and further, that Postn$^{-/-}$ mice also developed less noticeable hepatic fibrosis induced by hepatotoxic treatments with CCL4 or thioacetamine. In the human, the desmoplastic stroma of iCCA commonly presents a more prominent α-SMA+ and Postn+ immunoreactivity than that of non-malignant cholestatic cholangiopathies and HCC (Fujimoto et al., 2011; Sirica & Gores, 2014; Utispan et al., 2010).

CAFs within the desmoplastic stroma of iCCAs represent the primary if not sole source of Postn expressed in human and rat iCCA, with cholangiocarcinoma cells in each case being essentially negative for Postn mRNA and protein expression (Campbell, Dumur, Lamour, DeWitt, & Sirica, 2012; Dumur, Campbell, DeWitt, Oyesanya, & Sirica, 2010; Manzanares et al., 2017; Sirica et al., 2021; Utispan et al., 2010). Utispan et al. (2010) also did not detected Postn expression in infiltrating immune cells in their analyzed samples of human iCCA. However, cholangiocarcinoma cells positive for the cancer stem cell marker CD44^{+} isolated by cell sorting from human primary iCCA were reported by Zeng et al. (2018) to also express Postn protein, which was associated with the recruitment of $CD206^+$ tumor-associated macrophages into parts of the iCCA tumor niche.

With respect to cell origin of Postn in iCCA, the findings of Zhang et al. (2020) are particularly compelling. Employing single-cell transcriptomic analysis, these authors defined six distinct fibroblastic subsets from 8 human iCCAs and adjacent tissues, of which the subtype designated mCAFs-c1-Postn expressed high levels of Postn. Interestingly, this subtype localized to the invasive front of tumor nests, suggesting a close association with iCCA invasion.

The ECM of the desmoplastic stroma of iCCA commonly is characterized by high levels of collagen fibers, organized into thick fibrous bundles that are largely comprised of collagen type 1, together with high levels of Postn (Carpino et al., 2019; Manzanares et al., 2017; Sirica et al., 2021). TGF-β functions as a key molecular driver of the desmoplastic reaction in iCCA by activating and stimulating the proliferation of CAFs, as well as being a well established inducer of Postn (González-González & Alonso, 2018; Kudo, 2019; Sirica et al., 2014). Down-regulation of Postn was shown to suppress TGFβ1-induced hepatic stellate cell proliferation and significantly inhibited TGF-β1-induced expression of α-SMA, and collagen type 1, as well as attenuated TGF-β1-induced Smad 2/3 in hepatic stellate cells, [a notable source together with activated portal fibroblasts of CAFs in iCCA] (Hong, Shejiao, Fenrong, Gang, & Lei, 2015). Postn has more recently been demonstrated to promote extracellular collagen-type 1 production and fibrogenesis by activating lysyl oxidase in hepatic stellate cells through Smad 2/3 activation independent of TGF-β receptors (Kumar et al., 2018), possibly mediated by integrins $α_vβ_5$ and $α_vβ_3$ (Sugiyama et al., 2016).

Postn can directly interact with collagen type 1 and regulates collagen cross-linking and fibrillogenesis (González-González & Alonso, 2018; Kii, 2019; Kii & Ito, 2017; Norris et al., 2007; Sirica et al., 2014) by acting as a scaffold for the assembly of ECM proteins (i.e., collagen 1, fibronectin, tenascin-C). Moreover, Postn can recruit bone morphogenetic protein-1 (BMP-1), a pro-collagen C-proteinase, onto the fibronectin matrix to activate lysyl oxidase for collagen type 1 crosslinking (Cui, Huang, Liu, & Ouyang, 2017; Kii, 2019; Kii & Ito, 2017; Kudo & Kii, 2018). Taken together with the findings described above, it is reasonable to infer that prominent deposition of Postn into the iCCA microenvironment fosters the construction of a dense fibro-collagenous-enriched desmoplastic matrix associated with increased tumor stiffness, which has been linked to mechano-activation of YAP/TAZ in fibroblastic cells, leading them to acquire a CAF phenotype, and also to a hypovascularized tumor stroma resulting from the collapse of micro-blood vessels within the tumor (Sirica et al., 2021).

A majority of the matricellular proteins now known to be overexpressed in iCCA bind to diverse integrins to activate downstream signaling pathways implicated in malignant cell growth and progression (Table 2, Fabris et al., 2020). Postn interacts through its fasciclin (FAS)1 domains to bind to and activate multiple integrins, including $\alpha_v\beta_1$, $\alpha_v\beta_3$, $\alpha_v\beta_5$, and $\alpha_6\beta_4$ (McQuitty, Williams, Chokshi, & Urbani, 2020; Murphy-Ullrich & Sage, 2014; Sirica et al., 2014). Notably, integrins $\alpha_5\beta_1$ and $\alpha_6\beta_4$ were shown to be highly expressed in non-tumorigenic cholangiocytes and various human cholangiocarcinoma cell lines (Utispan et al., 2012). Furthermore, β_4 and β_6 integrin expression was found to be more strongly expressed in specific subtypes of iCCA, with higher expressions being observed in in periductular infiltrating type and intraductal growth sub-types than in the peripheral mass-forming type (Soejima, Takeuchi, Akashi, Sawabe, & Fukusato, 2018).

Postn has been demonstrated to bind to the integrin $\alpha_5\beta_1$ receptor on cholangiocarcinoma cells, activating the PI3K/AKT signaling pathway to promote cell proliferation, migration, and invasion (Utispan et al., 2010, 2012). More recently, Postn has also been shown to induce EMT in iCCA (Mino et al., 2017; Sonongbua et al., 2020), which was further linked to integrin $\alpha_5\beta_1$- mediated activation of TWIST-2 (Sonongbua et al., 2020).

Like integrin $\alpha_5\beta_1$, integrin $\alpha_6\beta_4$ also functions as a favorable receptor for Postn in cholangiocarcinoma cells (Utispan et al., 2012). High β_4 and β_6 integrin expressions were found to be related to infiltrative growth and bile duct invasion of iCCA (Soejima et al., 2018). These authors further reported TGF-β1 expression to correlate with β_6 expression and poor overall survival of iCCA patients. Of added relevance, Baril et al. (2007) had previously shown that that integrin $\alpha_6\beta_4$ acts as a receptor for Postn in human pancreatic adenocarcinoma cells, resulting in activation of the PI3K/AKT and FAK pathways to promote invasiveness and resistance of the cancer cells to hypoxia-induced cell death.

Postn also binds to tenascin-C through its FAS1 domain, supporting the incorporation of this matricellular protein into the desmoplastic ECM (Kudo, 2019; Kudo & Kii, 2018; Sirica et al., 2014). Tenascin-C, like Postn, is a multifunctional matricellular protein overexpressed in the desmoplastic stroma around cholangiocarcinoma cells of both human and rat mass-forming iCCA (Mertens et al., 2013; Sirica et al., 2009; Soejima et al., 2018). As has been the case for other solid cancers (Lowy & Oskarsson, 2015), strong stromal tenascin immunoreactivity has been observed at the invasive front of human iCCA, with tenascin-positive cholangiocarcinoma cells also being detected at the cancer-stromal interface (Aishima et al., 2003).

The structure of tenascin-C includes epidermal growth factor-like repeats which can act as a low affinity ligand for the epidermal growth factor receptor whose activation induces mitogen-activated protein kinase (MAPK) signaling (Giblin & Midwood, 2015). Tenascin-C also interacts with a variety of integrins, including $\alpha_v\beta_1$ and $\alpha_v\beta_3$ (Yoshida, Akatsuka, & Imanaka-Yoshida, 2015), as well as a wide range of growth factors, such as TGF-β, FGF, VEGF, and PDGF (Giblin & Midwood, 2015; McQuitty et al., 2020) known to be secreted into the ECM of iCCA and contributing to malignant progression and an immunosuppressive microenvironment.

Other critical signaling pathways in iCCA modulated by their interactions with Postn are the Notch 1 and Wnt/β-catenin pathways. The EMI domain of Postn binds to Notch 1 and Postn expression modulates Notch 1 activity in tumors (Sirica et al., 2014; Sriram et al., 2015). Notch 1 is overexpressed in human iCCA and has been shown to be associated with cholangiocarcinoma cell proliferation, invasiveness, and survival (Guo et al., 2019; Wu et al., 2014). Malanchi et al. (2011) have also shown that Postn recruits Wnt ligands and thereby increases Wnt signaling in cancer stem cells, and further, that blocking its function prevents metastatic colonization. Ghajar et al. (2013) have further demonstrated that high Postn and TGF-β1 expression surrounding neo-vascular tip cells comprising a micrometastatic niche enabled disseminated tumor cells to exit from dormancy, promoting metastatic relapse. Several other matricellular proteins, including thrombospondin-1, tenascin-C, and osteopontin have also been implicated as extracellular regulators in tumor dormancy (Wu & Ouyang, 2014).

4. Clinical relevance of matricellular proteins in iCCA
4.1 Matricellular proteins as prognostic biomarkers

Over the past decade or so, there have been an increasing number of non-randomized observational cohort studies aimed at establishing the utility of various aberrantly expressed matricellular proteins as potential prognostic biomarkers for iCCA. Table 3 describes independent findings for select matricellular proteins as being predictive of poor prognosis in iCCA patients

Table 3 Select matricellular proteins as prognostic biomarkers for resectable intrahepatic cholangiocarcinoma.

Single center study site	Matricellular protein	Number of cases analyzed	Specimen analyzed	Results	References
Thailand	Postn[a]	51	iCCA Tumor Tissue	Significant shorter overall survival in patients with high Postn expression in CAFs	Utispan et al. (2010)
Thailand	Postn	66	iCCA Tumor Tissue	Overall survival of resected patients with high tissue Postn significantly reduced from those with low tissue Postn levels	Thuwajit et al. (2017)
Thailand	Postn	68	Serum from iCCA patients	Patients with high preoperative serum levels of Postn had significantly shorter overall survival times than those with low Postn levels	Thuwajit et al. (2017)
Germany	Osteopontin	107	Serum from biliary tract cancer patients that included iCCA patients	High pre-and postoperative serum levels of osteopontin significantly correlated with reduced long-term survival	Loosen et al. (2017)
P.R. China	Osteopontin	121 (resected tumor); 60 (preoperative plasma)	iCCA tumor tissue tumor and plasma	High tumor tissue and plasma levels of osteopontin closely related to a shorter overall survival and high probability of relapse after curative resection. Patients with high tumor levels of osteopontin and β-catenin had the worst prognosis	Zheng et al. (2018)

Continued

Table 3 Select matricellular proteins as prognostic biomarkers for resectable intrahepatic cholangiocarcinoma.—cont'd

Single center study site	Matricellular protein	Number of cases analyzed	Specimen analyzed	Results	References
P.R. China	Osteopontin		iCCA tumor tissue	Low expression of cancer cell osteopontin significantly associated with shorter overall survival	Zhou et al. (2019)
P.R. China	Osteopontin	124	Serum from iCCA patients	Patients with a low level of circulating osteopontin per tumor volume exhibited significantly reduced overall survival and disease free survival than those with high circulating osteopontin per tumor volume	Zhou et al. (2019)
Japan	Osteopontin	73	iCCA tumor tissue	Overall survival was significantly lower among patients with a negative cancer cell expression of osteopontin than those with positive osteopontin expression	Terashi et al. (2004)
France	Osteopontin	40	iCCA tumor tissue	Stromal overexpression of osteopontin significantly correlated with reduced overall survival and disease free survival	Sulpice et al. (2013)
Thailand	Osteopontin	354	iCCA tumor tissue	Osteopontin expression in either cancer cells or in stroma showed no influence on patients' survival	Laohaviroj et al. (2016)

Japan	Tenascin-C	48	iCCA tumor tissue	Tenascin-C expression was not correlated with overall survival	Soejima et al. (2018)
Japan	Tenascin	78	iCCA tumor tissue	Positive tenascin expression at the invasive front significantly correlated with shorter overall survival, but no significant difference in survival rates was demonstrated between positive versus negative tenascin expression in the intra-tumoral stroma	Aishima et al. (2003)
Italy	CTGF	49	iCCA tumor tissue	Overall survival was significantly greater in patients whose tumors expressed high CTGF expression over negative expression	Gardini et al. (2005)
Japan	WISP1v (splice variant)	39	iCCA tumor tissue	Positive expression of WISP1v significantly associated with poor prognosis and reduced overall survival	Tanaka et al. (2003)
Taiwan	SPARC	78	iCCA tumor tissue	SPARC cancer positive and stroma negative immune-staining and curative-intent resection predicted favorable overall survival in patients with mass-forming iCCA	Cheng et al. (2015)

Continued

Table 3 Select matricellular proteins as prognostic biomarkers for resectable intrahepatic cholangiocarcinoma.—cont'd

Single center study site	Matricellular protein	Number of cases analyzed	Specimen analyzed	Results	References
Japan	SFRP1	50	iCCA tumor tissue	Negative SFRP1 expression significantly correlated with poorer overall survival and disease free survival	Davaadorj et al. (2017)
Japan	TSP-1	67	iCCA tumor tissue	Patients with positive TSP-1 expression had a tendency for shorter survival than those with negative expression, but those with positive TSP-1 as well as positive TP (an angiogenic factor) expression had significantly shorter overall survival than those with negative TP and positive TSP-1 expression	Aishima et al. (2002)

[a]Postn, Periostin; CTGF, Connective tissue growth factor; WISP1v, WNT1-Inducible Secreted Protein 1 variant; SPARC, Secreted Protein Acidic and Rich in Cysteine; SFRP1, Secreted Frizzled-Related Protein-1; TSP-1, Thrombospondin-1; TP, Thymidine Phosphorylase.

following surgical resection. Briefly, iCCA patients who had undergone curative-intent surgery at Khon Kaen University (Thailand) and whose tumors exhibited high tumor stomal and circulating levels of Postn were demonstrated to have significantly lower overall survival outcomes than those with low expression (Thuwajit et al., 2017; Utispan et al., 2010). As further summarized in Table 3, studies to date relating osteopontin expression to survival outcomes in iCCA patients after surgical resection have yielded contradictory results in terms of whether high or low levels of tumor and circulating levels of this matricellular protein may serve as independent prognostic factors for poor patient survival rates (Loosen et al., 2017; Terashi et al., 2004; Zhou et al., 2019). Iguchi et al. (2009) divided iCCA patients diagnosed at Kyushu University (Japan) into four subgroups based on positive immunoreactivity for osteopontin and Ki 67 in cholangiocarcinoma cells and stromal immunoreactivity for tenascin-C. Among these groups, iCCA patients whose tumors were negative for cancer cell osteopontin, but positive for stromal tenascin-C, exhibited the poorest overall survival rates.

Differences in cytoplasmic *versus* apical location and immunoreactivity of osteopontin in cholangiocarcinoma cells, reflecting cellular differentiation grade (Demarez, Hubert, Sempoux, & Lemaigre, 2016), as well as levels of osteopontin expression in the tumor stroma may possibly account in part for some of the opposing results. Sulpice et al. (2013) using an unsupervised gene expression analysis of the stroma of the mass-forming type of iCCA resected from patients at Rennes-University hospital (France), combined with Kaplan-Meier method and multivariate analysis, identified increased expression of stromal osteopontin as an independent prognostic marker for overall and disease-free survival. Laohaviroj, Chamgramol, Pairojkul, Mulvenna, and Sripa (2016) also observed stromal osteopontin to be significantly associated with tumor size, tumor direct invasion into the liver parenchyma, regional lymph node metastasis, and higher staging. However, in this study, Kaplan-Meier survival analysis demonstrated both stromal and cholangiocarcinoma cell osteopontin expression not to influence survival outcomes of the iCCA patients, and further, that in contrast to the findings of Sulpice et al., multivariant analysis revealed stromal osteopontin not to be an independent prognostic factor for iCCA following surgical resection.

Osteopontin promoter polymorphisms at locus-443 was further shown to associate with metastasis and poor prognosis in a population of Chinese iCCA patients (Zhao et al., 2014). Of relevance, in a pilot study, osteopontin b and c splicing isoforms were shown to be expressed exclusively in mRNA samples from non-long surviving (NLS) patients with pancreatic adenocarcinoma, when compared with samples from long-surviving patients (LS), whereas both the NSL and LS samples showed the same type of periostin and tenascin-C isoforms (Fiorino et al., 2020). The overall survival rates of iCCA patients whose resected tumors were positive for stromal tenascin immunoreactivity were reported by Aishima et al. (2003) not to differ significantly from those with negative tenascin immunostaining. However, in this same study, tenascin immunostaining at the invasive front was demonstrated to associate with poor prognosis in iCCA. Tenascin-C expression levels were also reported by Soejima et al. (2018) to be significantly higher in infiltrative-growth types of iCCA than in expansive growth types, but in this study was also found not to correlate with overall survival. Conversely, in pancreatic adenocarcinoma, strong stromal expression of the large splice variant of tenascin-C in the tumor, as well as of annexin A2, a cancer cell surface receptor for tenascin C, has recently been reported to be associated with poor prognosis (Hagiwara et al., 2020). Here, it is reasonable to consider that the expression of different isoforms of matricellular proteins, such as periostin, osteopontin, and tenascin-C, need to be taken into account when evaluating their potential as prognostic biomarkers for iCCA.

Mass-forming iCCA patients with cholangiocarcinoma cell-positive and stromal negative immunostaining for SPARC were reported to have a favorable overall survival after curative resection (Cheng et al., 2015). Overexpression of WISP1v, a CCN family gene product formed by alternative splicing and shown to be localized to CAF-enriched tumor stromal but not expressed in cholangiocarcinoma cells of iCCA was also demonstrated to significantly associated with poor patient survival rates after surgical resection when compared with WISP1v-negative cases (Tanaka et al., 2003). Moreover, Gardini et al. (2005) had reported that high expression of connective tissue growth factor (CTGF), another CCN protein family member localized to both cholangiocarcinoma cells and stromal CAFs to be an independent prognostic indicator of both tumor recurrence and overall survival for resected iCCA patients.

Nomura et al. (2013) demonstrated mesothelin (Msln) to be an independent predictor of shorter postoperative survival in iCCA patients. Overall survival analysis had also previously shown that patients whose resected iCCA were immunoreactive for thrombospondin-1 expression has a tendency for shorter survival than those with negative thrombospondin immunostaining (Aishima et al., 2002). On the other hand, loss of Secreted Frizzled-Related Protein-1 was determined to be an independent poor prognostic factor for the overall survival of resected iCCA patients (Davaadorj et al., 2017).

While matricellular proteins such as Postn, osteopontin, and tenascin-C show promise as prognostic biomarkers for iCCA, they are unlikely at this time to be useful as diagnostic serum biomarkers for iCCA, since elevated serum levels of these matricellular proteins are also detected in a variety of different non-neoplastic fibro-inflammatory diseases, including those of liver, as well as in other gastrointestinal cancers (Matsumoto & Aoki, 2020; Song et al., 2021; Takahashi et al., 2019; Wu et al., 2018; Zhu et al., 2016). Thuwajit et al. (2017) further found that using serum Postn to distinguish iCCA from some benign liver diseases was not statistically significant.

4.1.1 Postn and Msln predict malignant progression in a rat iCCA model

A synergistic, immunologically competent rat model of orthotopic intrahepatic cholangiocarcinoma (Sirica et al., 2008) has been used to investigate the potential of Postn and Msln to predict malignant progression in iCCA (Manzanares et al., 2018). This model, which has been well characterized, is now considered to faithfully reproduce clinicopathological, cellular, and molecular features of progressive human iCCA, including recapitulating the desmoplastic reaction typical of the mass-forming type, as well as peritoneal metastasis (Cadamuro et al., 2018a; Loeuillard, Fischbach, Gores, & Rizvi, 2019; Manzanares et al., 2017; Mohr et al., 2020; Sirica et al., 2008). The model was established by comparatively transplanting via bile duct inoculation three distinct tumorigenic rat cholangiocyte/cholangiocarcinoma cell lines/strains (BDEneu, BDEsp, TDE$_{CC}$) having a common cell parentage (Manzanares et al., 2017, 2018; Sirica et al., 2008), but different malignant cell growth rates and metastatic potential into the left hepatic lobes of syngeneic young adult male Fischer 344 rats. Using this model, Manzanares et al. (2018) carried out a systematic investigation demonstrating a strong positive correlation between tumor and serum Postn and Msln and increasing liver tumor mass and associated peritoneal metastases, that also reflected differences in cholangiocarcinoma aggressiveness and malignant grade. As is the case for human mass-forming iCCA, Postn expression was exclusively localized to the desmoplastic stroma and observed to be most prominent in the more aggressive liver tumors and associated peritoneal metastases (Fig. 1). Also consistent with human findings (Nomura et al., 2013; Yu et al., 2010), Msln was most highly expressed in the cytoplasm and cell surface membrane of the cholangiocarcinoma cells of more aggressive rat tumors.

In this syngeneic rat model of iCCA progression, two distinct molecular weight forms of Msln were identified, a more heavily glycosylated 50 kDa form overexpressed predominately at the apical luminal surface, which was

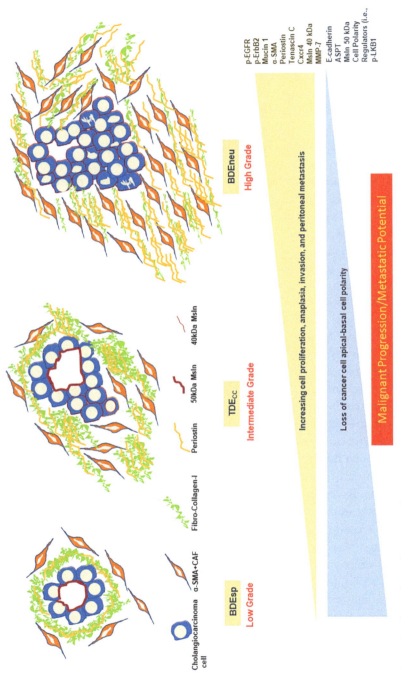

Fig. 1 See figure legend on opposite page.

predictive of a more differentiated, less aggressive iCCA phenotype, and a 40 kD cytoplasmic/diffuse cell membrane form which was associated with increased malignant progression (Fig. 1). Of particular interest, it was further determined that co-culturing rat CAFs with TDE_{CC} cholangiocarcinoma cells in 3-D culture resulted in a mean 2.8-fold increase in their protein expression of the 40 kD form over that expressed by TDE_{CC} cells cultured alone, which correlated with increased anaplasia, loss of ductal-like polarity, and enhanced invasiveness *in vitro* (Manzanares et al., 2018; Sirica et al., 2021). These results support both Postn and Msln as being predictive of tumor progression in a rat model of human iCCA and encourage the need for future comprehensive mechanistic and clinical studies using human samples to validate combined use of CAF Postn and cholangiocarcinoma cell Msln to predict iCCA progression (Mandrekar & Cardinale, 2018).

Fig. 1 Schematic reflecting differences in immunohistochemically detected expression levels of stromal periostin and cholangiocarcinoma cell mesothelin (Msln) characterizing slow growing well differentiated BDEsp cholangiocarcinoma, intermediate growing moderately differentiated TDE_{CC} cholangiocarcinoma, and rapidly growing high grade malignant BDEneu cholangiocarcinoma formed in a syngeneic rat model of orthotopic intrahepatic cholangiocarcinoma progression. Postn immunoreactivity was localized solely to the tumor stroma and not expressed in the cholangiocarcinoma cells. Further, Postn immunostaining was observed to be most prominently expressed in the desmoplastic stroma of the larger sized more highly aggressive liver tumor types, most particularly the BDEneu cholangiocarcinoma, than in the less malignant low grade BDEsp cholangiocarcinoma. In comparison, Msln immunoreactivity was more strongly expressed in the cholangiocarcinoma cells than in the tumor stroma, although in the case of BDEneu cholangiocarcinoma, stromal Msln immunostaining became more intense with increasing BDEneu liver tumor size together with associated peritoneal metastases. Peritoneal metastases were detected early in rats bearing BDEneu liver tumors and later in those with TDE_{CC} liver tumors, but not detected in BDEsp tumor-bearing rats. Apical luminal cell surface Msln immunoreactivity corresponded to a 50 kDa form was demonstrated to be predictive of a less aggressive iCCA phenotype, whereas diffuse cell surface Msln corresponding to a 40 kDa form was characteristic of a higher grade invasive and metastatic iCCA. p-EGFR, phosphorylated (activated) epidermal growth factor receptor; pErbB2/neu, activated ErbB2/neu; MMP-7, metalloproteinase-7; ASBT, apical sodium-dependent bile acid transporter. *See Manzanares, M. Á, Campbell, D.J.W. Maldonado, G.T., & Sirica, A.E et al. (2018). Overexpression of periostin and distinct mesothelin forms predict malignant progression in a rat cholangiocarcinoma model. Hepatology Communications, 2: 155–172 and Yao, W., Qian, Y., Campbell, D., Wang, J., and Sirica, A.E. (2018) Underexpression of phospho-LKB1, atypical PKCι, and Hes1 associated with disruption of polarized morphogenesis in a 3-dimensional culture model of rat cholangiocarcinoma progression mediated by activated neu receptor. Gastroenterology, 154, Supplement 1: S-1152 (Abstract Su1455) for experimental details.*

4.1.2 Interplay between TGF-β and matricellular proteins

The TGF-β signaling pathway is prominently upregulated in iCCA, and is crucially involved in generating the fibro-collagenous desmoplastic tumor microenvironment, inducing CAF myofibroblast differentiation and proliferation, mediating EMT and immune evasion, and promoting iCCA progression, invasion, and metastasis (Chen et al., 2015; Huang et al., 2016; Manzanares et al., 2017; Papoutsoglou, Louis, & Coulouarn, 2019; Ghahremanifard, Chanda, Bonni, & Bose, 2020; Lustri et al., 2017; Chung et al., 2021). TGF-β1 is overexpressed in most iCCAs, being highly expressed in cholangiocarcinoma cells, but also detected in tumor stromal cells (Clapéron et al., 2013; Manzanares et al., 2017; Sirica et al., 2019). Clinically, high expression of TGF-β1 in resected iCCA tissue samples significantly correlated with lymph node metastasis, and shorter survival time (Chen et al., 2015; Kimawaha et al., 2020). Moreover, TGF-β1 was shown to be elevated in the serum of iCCA patients compared to the normal healthy control group, with high serum TGF-β1 levels significantly correlating with metastasis status.

ECM matricellular proteins are not only regulated by TGF-β, but also regulate latent TGF-β activation. For example, in addition to being an inducer of Postn expression, TGF-β has been shown to upregulate the expression of SPARC (Shibata & Ishiyama, 2013) and tenascin-C (Chiovaro, Chiquet-Ehrismann, & Chiquet, 2015). Thrombospondin-1 and tenascin-C have further been demonstrated to regulate latent TGF-β activation (Aubert et al., 2021; Murphy-Ullrich & Suto, 2018). Moreover, the interactions between ECM matricellular proteins such as Postn, osteopontin, and tenascin-C with the TGF-β1 pathway have been shown to be essentially linked to myofibroblast differentiation characteristic of activated CAFs (Hong et al., 2015; Katoh et al., 2020; Lenga et al., 2008; Weber et al., 2015; Yue et al., 2021). Msln has also been found to regulate TGF-β 1-inducible activation of portal fibroblasts by disrupting the formation of an inhibitory Thy-1-TGFβR1 complex (Koyama et al., 2017). Thy-1 and TGFβR1 were also shown to be strongly expressed in CAFs expressing portal fibroblast biomarkers that enriched the desmoplastic stroma of the rat iCCA model described above.

While there have been promising results with different TGF-β pharmacological and antibody inhibitors in cell culture and animal cancer models, including cholangiocarcinoma (Ling et al., 2013; Huang et al., 2016; Manzanares et al., 2017), patient clinical trials against different types of cancers with on-target anti-TGF-β therapies have either not been or only marginally successful in improving survival outcomes, and even associated with adverse effects [i.e., cardiovascular toxicity] (Teixeira, ten Dijke, & Zhu, 2020; Suto et al., 2020). Targeting of specific matricellular proteins regulating/modulating TGF-β signaling might offer a more selective approach towards the development of anti-TGF-β therapies for desmoplastic cancers like iCCA.

4.2 Potential as therapeutic targets

Currently, curative-intent (R0) surgical resection remains as the only potentially effective treatment option for obtaining 5-year survival outcomes for iCCA patients with localized resectable iCCA. However, most patients present at an advanced stage at the time of diagnosis and thus are considered ineligible for R0 resection. Furthermore, even for those select patients that have undergone R0 surgical resection, the recurrence rates are unacceptably high, generally ranging between 50% and 80% (Chan et al., 2018; Sirica et al., 2019; Tsilimigras et al., 2020; Tsukamoto et al., 2017). In a recent study conducted at Ohio State University, 880 patients identified in an international multi-institutional database that had undergone curative-intent resection for iCCA, approximately one-fourth were shown to develop very early recurrence (VER; within 6 months) after resection, resulting in 5-year overall survival rate of 8.9% *versus* 49.8% for those without VER (Tsilimigras et al., 2020).

Extrahepatic metastases from iCCA most commonly develop in lung, peritoneum, bone, and lymph nodes (Archer, Paro, Elfadaly, Tsilimigras, & Pawlik, 2021; Cheng, Du, Ye, Wang, & Chen, 2019; Hahn et al., 2020) reflecting distant sites of premetastatic and metastatic niches in which matricellular proteins, such as Postn, tenascin-C, osteopontin, thrombospondin-1, and Msln have been proposed to function to prime the microenvironment to prepare for metastatic cell seeding, for providing a permissive environment for distant metastatic cell outgrowth, and for regulating metastatic dormancy and immune evasion (Celià-Terrassa & Kang, 2018; Cui et al., 2017; Ghajar et al., 2013; González-González & Alonso, 2018; Lowy & Oskarsson, 2015; Shurin, 2018; Wang et al., 2016; Wong & Rustgi, 2013; Wu et al., 2014).

Targeting select matricellular proteins could potentially offer a unique therapeutic opportunity to prevent or treat iCCA-derived extrahepatic metastases, including intraperitoneal and lung tumors. Postn and osteopontin have been shown to be promising targets to overcome chemoresistance and/or immune evasion in different tumor models

(González-González & Alonso, 2018; Moorman et al., 2020; Nakazawa et al., 2018). The use of thrombospondin-1/TGF-β antagonists have also been proposed to potentially represent a more selective approach to targetingTGF-β activity in metastatic progression, fibrotic diseases, and immune suppression (Murphy-Ullrich & Suto, 2018). A neutralizing monoclonal antibody (MZ-1) to Postn was further shown to inhibit human ovarian tumor growth and peritoneal metastasis in a mouse xenograft model (Zhu et al., 2011). In addition, anti-Msln monoclonal antibody amatuximab in combination with gemcitabine was demonstrated to exhibited synergistic killing of Msln-high expressing pancreatic cancer cells in a peritoneal metastasis mouse model (Mizukami et al., 2018). In a recent Phase 1 study, anetumab ravtansine, an antibody-drug conjugant of anti-Msln linked to maytansininoid DM4, a tubulin inhibitor, was demonstrated to be well tolerated in patients with advanced, metastatic, and recurrent Msln-expressing solid cancers, with preliminary anti-tumor activity being observed in patients with mesothelioma and ovarian cancer (Hassan et al., 2020). Based on these findings, several clinical trials across multiple tumor types overexpressing Msln, including cholangiocarcinoma (NCT03102320) are now under way. Current and developing strategies for therapeutic targeting of Msln in iCCA are discussed in chapter "Immunotherapy for hepatobiliary cancers: Emerging targets and translational advances" by Li et al. The potential use of various other matricellular proteins in drug delivery targeted to the tumor microenvironment, including CAFs, is also being explored (Baghban et al., 2020; Chen et al., 2016; Sawyer & Kyriakides, 2016).

5. Perspectives and conclusions

Matricellular proteins have emerged as an important class of oncogenic modulators playing in mediating iCCA ECM remodeling, promoting malignant progression, priming of pre-and metastatic niches, and controlling immunosuppression in the tumor microenvironment. It is also increasingly apparent that multiple matricellular proteins secreted into the iCCA microenvironment can mechanistically converge to induce a range of pro-oncogenic phenomena, including cell proliferation, EMT, tumor ECM stiffness, invasion and metastasis via activation of an array of downstream signaling pathways. Elucidating the functional and temporal relationships posed the convergence of diverse matricellular proteins produced by tumor stomal cells and cholangiocarcinoma cells functioning to promote aberrant ECM remodeling and cholangiocarcinoma progression through redundant signaling pathways will be key to translating these relationships into novel strategies for iCCA therapy based on multitargeting approaches. Such strategies may prove to be particularly useful in providing innovative therapeutic approaches towards targeting iCCA pre-metastatic and metastatic niches, tumor dormancy, and immune evasion.

Furthermore, the exploratory studies highlighted in Table 3 demonstrating high circulating levels of ECM matricellular proteins such as Postn to be predictive of iCCA recurrence and patient survival outcomes following curative-intent resection represent a promising first step towards the establishment of relatively non-invasive and simple serum-based matricellular protein biomarker bioassays for predicting progression and survival outcomes following surgical resection. However, for the development of such matricellular-based bioassays to be fully realized, large scale and rigorously controlled international validation studies now need to be performed. Future studies that address the clinical significance of distinct patterns matricellular protein splice variants for their prognostic value in iCCA are clearly warranted, along with rigorous assessments of specificity and sensitivity. Moreover, efforts to establish novel serum-based assays based on multiplexing of matricellular proteins (i.e., Postn, osteopontin, and Msln, together with TGF-β) for reliably and accurately predicting iCCA progression and therapeutic responses also seem to be well justified. Lastly, increasing our understanding of how matricellular proteins such as Postn, osteopontin, thrombospondin-1, tenascin-C, and Msln mechanistically interact to prepare pre-metastatic and metastatic niches, regulate disseminated cholangiocarcinoma cell dormancy, and foster immunosuppression is now needed to support the development and testing of novel combinatorial therapeutic approaches designed to prevent and treat metastatic iCCA.

Grant support
None.

Conflict of interest statement
A.E.S. has no financial or personal disclosures relevant to the contents of this manuscript.

References
Aishima, S., Taguchi, K., Sugimachi, K., Asayama, Y., Nishi, H., Shimada, M., et al. (2002). The role of thymidine phosphorylase and thrombospondin-1 in angiogenesis and progression of intrahepatic cgolangiocarcinoma. *International Journal of Surgical Pathology, 10,* 47–56.

Aishima, S., Taguchi, K., Terashi, T., Matsuura, S., Shimada, M., & Tsuneyoshi, M. (2003). Tenascin expression at the invasive front is associated with poor prognosis in intrahepatic cholangiocarcinoma. *Modern Pathology, 16,* 1019–1027.

Archer, A. W., Paro, A., Elfadaly, A., Tsilimigras, D., & Pawlik, T. M. (2021). Intrahepatic cholangiocarcinoma: A summative review of biomarkers and targeted therapies. *Cancers, 13,* 5169.

Aubert, A., Mercier-Gouy, P., Aguero, S., Berthier, L., Liot, S., Prigent, L., et al. (2021). Latent TGF-β activation is a hallmark of the tenascin family. *Frontiers in Immunology, 12,* 613438.

Baghban, R., Roshangar, L., Jahanban-Esfahlan, R., Seidi, K., Ebrahimi-Kalan, A., Jaymand, M., et al. (2020). Tumor microenvironment complexity and therapeutic implications at a glance. *Cell Communication and Signaling: CCS, 18,* 59.

Baril, P., Gangeswaran, R., Mahon, P. C., Caulee, K., Kocher, H. M., Harada, T., et al. (2007). Periostin promotes invasiveness and resistance of pancreatic cancer cells to hypoxia-induced cell death: Role of the beta 4 integrin and PI3k pathway. *Oncogene, 26,* 2082–2094.

Bornstein, P. (1995). Diversity of function is inherent in matricellular proteins: An appraisal of thrombospondin 1. *Journal of Cell Biology, 130,* 503–506.

Bornstein, P. (2009). Matricellular proteins: An overview. *Journal of Cell Communication and Signaling, 3,* 163–165.

Bornstein, P., & Sage, E. H. (2002). Matricellular proteins: Extracellular modulators of cell function. *Current Opinion in Cell Biology, 14,* 608–616.

Brindley, P., Bachini, M., Ilyas, S., Khan, S., Loukas, A., Sirica, A. E., et al. (2021). Cholangiocarcinoma. *Nature Reviews. Disease Primers, 7,* 65.

Brivio, S., Cadamuro, M., Strazzabosco, M., & Fabris, L. (2017). Tumor reactive stroma in cholangiocarcinoma: The fuel behind cancer aggressiveness. *World Journal of Hepatology, 9,* 455–468.

Cadamuro, M., Brivio, S., Stecca, T., Kaffe, E., Mariotti, V., Milani, C., et al. (2018). Animal models of cholangiocarcinoma: What they teach us about the human disease. *Clinics and Research in Hepatology and Gastroenterology, 42,* 403–415.

Cadamuro, M., Stecca, T., Brivio, S., Mariotti, V., Fiorotto, R., Spirli, C., et al. (2018). The deleterious interplay between tumor epithelia and stroma in cholangiocarcinoma. *Biochimica et Biophysica Acta-Molecular Basis of Disease, 1864*(4PtB), 1435–1443.

Campbell, D. J. W., Dumur, C. I., Lamour, N. F., DeWitt, J. L., & Sirica, A. E. (2012). Novel organotypic culture model of cholangiocarcinoma progression. *Hepatology Research, 42,* 1119–1130.

Carpino, G., Cardinale, V., Di Giamberardino, A., Overi, D., Donsante, S., Colasanti, T., et al. (2021). Thrombospondin 1 and 2 along with PEDF inhibit angiogenesis and promote lymphangiogenesis in intrahepatic cholangiocarcinoma. *Journal of Hepatology, 75,* 1377–1386.

Carpino, G., Overi, D., Melandro, F., Grimaldi, A., Cardinale, V., Matteo, S., et al. (2019). Matrisome analysis of intrahepatic cholangiocarcinoma unveils a peculiar cancer-associated extracellular matrix structure. *Clinical Proteomics, 16,* 37.

Celià-Terrassa, T., & Kang, Y. (2018). Metastatic niche functions and therapeutic opportunities. *Nature Cell Biology, 20,* 868–877.

Chan, K.-M., Tsai, C.-Y., Yeh, C.-N., Yea, T.-S., Lee, W.-C., Jan, Y.-Y., et al. (2018). Characterization of intrahepatic cholangiocarcinoma after curative resection: Outcome, prognostic factor, and recurrence. *BMC Gastroenterology, 18,* 180.

Chen, B., Wang, Z., Sun, J., Song, Q., He, B., et al. (2016). A tenascin C targeted nanoliposome with navitoclax for specifically eradicating of cancer-associated fibroblasts. *Nanomedicine, 12,* 131–141.

Chen, Y., Ma, L., He, Q., Zhang, S., Zhang, C., & Wei, J. (2015). TGF-β expression is associated with invasion and metastasis of intrahepatic cholangiocarcinoma. *Biological Research, 48*, 26.
Cheng, C.-T., Chu, Y.-Y., Yeh, C.-N., Huang, S.-C., Chen, M.-H., Wang, S.-Y., et al. (2015). Peritumoral SPARC expression and patient outcome with resectable intrahepatic cholangiocarcinoma. *Oncotargets and Therapy, 8*, 1899–1907.
Cheng, R., Du, Q., Ye, J., Wang, B., & Chen, Y. (2019). Prognostic value of site-specific metastases for patients with advanced intrahepatic cholangiocarcinoma. A SEER data base analysis. *Medicine, 98*(49), e18191.
Chiovaro, F., Chiquet-Ehrismann, R., & Chiquet, M. (2015). Transcriptional regulation of tenascin genes. *Cell Adhesion & Migration, 9*(1–2), 34–47.
Chong, H. C., Tan, C. K., Huang, R.-L., & Tan, N. S. (2012). Matricellular proteins: A sticky affair with cancer. *Journal of Oncology, 351089*.
Chung, J. Y.-F., Chan, M. K.-K., Li, J. S.-F., Chan, A. S.-W., Tang, P. C.-T., Leung, K.-T., et al. (2021). TGF-β signaling: From tissue fibrosis to tumor microenvironment. *International Journal of Molecular Sciences, 22*, 7575.
Clapéron, A., Mergey, M., Aoudjehane, L., Ho-Bouldoires, T. H. N., Wendum, D., Prignon, A., et al. (2013). Hepatic myofibroblasts promote the progression of human cholangiocarcinoma through activation of epidermal growth factor receptor. *Hepatology, 58*, 2001–2011.
Cui, D., Huang, Z., Liu, Y., & Ouyang, G. (2017). The multifaceted role of periostin in priming the tumor microenvironments for tumor progression. *Cellular and Molecular Life Sciences, 74*, 4287–4291.
Darby, I. A., Vuillier-Devillers, K., Pinault, É., Sarrazy, V., Lepreux, S., Balabaud, C., et al. (2010). Proteomic analysis of differentially expressed proteins in peripheral cholangiocarcinoma. *Cancer Microenvironment, 4*, 73–91.
Davaadorj, M., Saito, Y., Morine, Y., Ikemoto, T., Imura, S., Takasu, C., et al. (2017). Loss of secreted frizzled-related protein-1 expression is associated with poor prognosis in intrahepatic cholangiocarcinoma. *European Journal of Surgical Oncology, 43*, 344–350.
Demarez, C., Hubert, C., Sempoux, C., & Lemaigre, F. (2016). Expression of molecular differentiation markers does not correlate with histological differentiation grade in intrahepatic cholangiocarcinoma. *PLoS One, 11*, e0157140.
Dumur, C. I., Campbell, D. J. W., DeWitt, J. W., Oyesanya, R. A., & Sirica, A. E. (2010). Differential gene expression profiling of cultured *neu*-transformed *versus* spontaneously-transformed rat cholangiocytes and of corresponding cholangiocarcinomas. *Experimental and Molecular Pathology, 89*, 227–235.
Fabris, L., Cadamuro, M., Cagnin, S., Strazzabosco, M., & Gores, G. J. (2020). Liver matrix in benign and malignant biliary tract disease. *Seminars in Liver Disease, 40*(3), 282–297.
Fabris, L., Perugorria, M. J., Mertens, J., Björkström, N. K., Cramer, T., Lleo, A., et al. (2019). The tumour microenvironment and immune milieu of cholangiocarcinoma. *Liver International, 39*, 63–78.
Fabris, L., Sato, K., Alpini, G., & Strazzabosco, M. (2021). The tumor microenvironment in cholangiocarcinoma progression. *Hepatology, 73*, 75–85.
Fausther, M., Lavole, E. G., & Dranoff, J. A. (2017). Liver myofibroblasts of murine origins express mesothelin: Identification of novel rat mesothelin splice variants. *PLoS One, 12*, e0184499.
Feng, D., & Gerarduzzi, C. (2020). Emerging role of matricellular proteins in systemic sclerosis. *International Journal of Molecular Sciences, 21*, 4776.
Fiorino, S., Visani, M., Masetti, M., Acquaviva, G., Tallini, G., De Leo, A., et al. (2020). Periostin, tenascin, osteopontin isoforms in long- and non-long survival patients with pancreatic cancer: a pilot study. *Molecular Biology Reports, 47*, 8235–8241.

Fujimoto, K., Kawaguchi, T., Nakashima, O., Ono, J., Ohta, S., Kawaguchi, A., et al. (2011). Periostin, a matrix protein, has potential as a novel serodiagnostic marker for cholangiocarcinoma. *Oncology Reports, 25*, 1211–1216.

Gardini, A., Corti, B., Fiorentino, M., Altimari, A., Ercolani, G., Grazi, G. L., et al. (2005). Expression of connective tissue growth factor is a prognostic marker for patients with intrahepatic cholangiocarcinoma. *Digestive and Liver Disease, 37*, 269–274.

Gerarduzzi, C., Hartmann, U., Leask, A., & Drobetsky, E. (2020). The matrix revolution: Matricellular proteins and restructuring of the cancer microenvironment. *Cancer Research, 80*, 2705–2717.

Ghahremanifard, P., Chanda, A., Bonni, S., & Bose, P. (2020). TGF-β mediated immune evasion in cancer-spotlight on cancer-associated fibroblasts. *Cancers, 12*, 3650.

Ghajar, C. M., Peinado, H., Mori, H., Matei, I. R., Evason, K. J., Brazier, H., et al. (2013). The perivascular niche regulates breast cancer dormancy. *Nature Cell Biology, 15*, 807–817.

Giblin, S. P., & Midwood, K. S. (2015). Tenascin-C: Form versus function. *Cell Adhesion & Migration, 9*(1–2), 48–82.

Gimba, E. R. P., Brum, M. C. M., & DeMorales, G. N. (2019). Full-length osteopontin and its splice variants as modulators of chemoresistance and radioresistance (review). *International Journal of Oncology, 54*, 420–430.

González-González, L., & Alonso, J. (2018). Periostin: A matricellular protein with multiple functions in cancer development and progression. *Frontiers in Oncology, 8*, 225.

Gremlich, S., Roth-Kleiner, M., Equey, L., Fytianos, K., Schittny, J., & Cremona, T. P. (2020). Tenascin-C inactivation impacts lung structure and function beyond lung development. *Scientific Reports, 10*, 5118.

Guo, J., Fu, W., Xiang, M., Zhang, Y., Zhou, K., Xu, C.-R., et al. (2019). Notch 1 drives the formation of intrahepatic cholangiocarcinoma. *Current Medical Science, 39*, 929–937.

Hagiwara, K., Harimoto, N., Yokobori, T., Muranushi, R., Hoshino, K., Gantumur, D., et al. (2020). High co-expression of large tenascin C splice variants in stromal tissue and annexin A2 in cancer cell membranes is associated with poor prognosis in pancreatic cancer. *Annals of Surgical Oncology, 27*, 924–930.

Hahn, F., Müller, L., Mähringer-Kunz, A., Tanyildizi, Y., dos Santos, D. P., Düber, C., et al. (2020). Distant metastases in patients with intrahepatic cholangiocarcinoma: Does location matter? A retrospective analysis of 370 patients. *Journal of Oncology.* Article ID 7195373.

Hanahan, D., & Weinberg, R. A. (2011). Hallmarks of cancer: The next generation. *Cell, 144*, 646–674.

Hassan, R., Blumenschein, G. R., Moore, K. N., Santin, A. D., Kindler, H. L., Nemunaitis, J. J., et al. (2020). First-in-human, multicenter, phase I dose-escalation and expansion study of anti-mesothelin antibody-drug conjugate anetumab ravtansine in advanced or metastatic tumors. *Journal of Clinical Oncology, 38*, 1824–1835.

Hong, L., Shejiao, D., Fenrong, C., Gang, Z., & Lei, D. (2015). Periostin down-regulation attenuates the pro-fibrogenic response of hepatic stellate cells induced by TGF-β1. *Journal of Cellular and Molecular Medicine, 19*, 2462–2468.

Huang, C.-K., Aihara, A., Iwagami, Y., Yu, T., Carlson, R., Koga, H., et al. (2016). Expression of transforming growth factor β1 promotes cholangiocarcinoma development and progression. *Cancer Letters, 380*, 153–162.

Iguchi, T., Yamashita, N., Aishima, S., Kuroda, Y., Terashi, T., Sugimachi, K., et al. (2009). A comprehensive analysis of immunohistochemical studies in intrahepatic cholangiocarcinoma using the survival tree model. *Oncology, 76*, 293–300.

Isenberg, J. S., & Roberts, D. D. (2020). THBS1 (thrombospondin 1). *Atlas of Genetics and Cytogenetics in Oncology and Hematology, 24*, 291–299.

Jing, C.-Y., Fu, Y.-P., Zhou, C., Zhang, M.-X., Huang, J.-L., Gan, W., et al. (2021). Hepatic stellate cells promote intrahepatic cholangiocarcinoma progression via NR4A2/osteopontin/Wnt signaling axis. *Oncogene, 40*, 2910–2922.

Katoh, D., Kozuka, Y., Noro, A., Ogawa, T., Imanaka-Yoshida, K., & Yoshida, T. (2020). Tenascin-C induces phenotypic changes in fibroblasts to myofibroblasts with high contractility through the integrin αvβ1/transforming growth factorβ/SMAD signaling axis in human breast cancer. *American Journal of Pathology, 190*, 2123–2135.

Kaur, S., Bronson, S. M., Pal-Nath, D., Miller, T. W., Soto-Pantoja, D. R., & Roberts, D. D. (2021). Functions of Thrombospondin-1 in the tumor microenvironment. *International Journal of Molecular Sciences, 22*, 4570.

Kendall, T., Verheij, J., Gaudio, E., Evert, M., Guido, M., Goeppert, B., et al. (2019). Anatomical, histomorphological and molecular classification of cholangiocarcinoma. *Liver International, 39*, 7–18.

Kii, I. (2019). Periostin functions as a scaffold for assembly of extracellular proteins. *Advances in Experimental Medicine and Biology, 1132*, 23–32.

Kii, I., & Ito, H. (2017). Periostin and its interacting proteins in the construction of extracellular architectures. *Cellular and Molecular Life Sciences, 74*, 4269–4277.

Kimawaha, P., Jusakul, A., Junsawang, P., Loilome, W., Khuntikeo, N., & Techasen, A. (2020). Circulating TGF-β1 as the potential epithelial mesenchymal transition-biomarker for diagnosis of cholangiocarcinoma. *Journal of Gastrointestinal Oncology, 11*, 304–318.

Koyama, Y., Wang, P., Liang, S., Iwaisako, K., Liu, X., Xu, J., et al. (2017). Mesothelin/mucin 16 signaling in activated portal fibroblasts regulates cholestatic liver fibrosis. *The Journal of Clinical Investigation, 127*, 1254–1270.

Kudo, A. (2019). The structure of the periostin gene, its transcriptional control and alternative splicing and protein expression. In A. Kudo (Ed.), *Periostin, advances in experimental medicine and biology* (pp. 7–20). Springer Nature Singapore Pte Lte.

Kudo, A., & Kii, I. (2018). Periostin functions in communication with extracellular matrices. *Journal of Cell Communication and Signaling, 12*, 301–308.

Kumar, P., Smith, T., Raeman, R., Chopyk, D. M., Brink, H., Liu, Y., et al. (2018). Periostin promotes liver fibrogenesis by activating lysyl oxidase in hepatic stellate cells. *Journal of Biological Chemistry, 293*, 12781–12792.

Landry, N., Cohen, S., & Dixon, I. M. C. (2017). Periostin in cardiovascular disease and development: A tale of two distinct roles. *Basic Research in Cardiology, 113*, 1.

Laohaviroj, M., Chamgramol, Y., Pairojkul, C., Mulvenna, J., & Sripa, B. (2016). Clinicopathological significance of osteopontin in cholangiocarcinoma cases. *Asian Pacific Journal of Cancer Prevention, 17*, 201–205.

Lenga, Y., Koh, A., Perera, A. S., McCulloch, C. A., Sodek, J., & Zohar, R. (2008). Osteopontin expression is required for myofibroblast differentiation. *Circulation Research, 102*, 319–327.

Ling, H., Roux, E., Hempel, D., Tao, J., Smith, M., Lonning, S., et al. (2013). Transforming growth factor β neutralization ameliorates pre-existing hepatic fibrosis and reduces cholangiocarcinoma in thioacetamide-treated rats. *PLoS One, 8*, E54499.

Loeuillard, E., Fischbach, S. R., Gores, G. J., & Rizvi, S. (2019). Animal models of cholangiocarcinoma. *Biochimica et Biophysica Acta-Molecular Basis of Disease, 1865*, 982–992.

Loosen, S., Roderburg, C., Kauertz, K. L., Pombeiro, I., Leyh, C., Benz, F., et al. (2017). Elevated levels of circulating osteopontin are associated with a poor survival after resection of cholangiocarcinoma. *Journal of Hepatology, 67*, 749–757.

Lowy, C. M., & Oskarsson, T. (2015). Tenascin C in metastasis: A view from the invasive front. *Cell Adhesion & Migration, 9*(1–2), 112–124.

Lustri, A. M., Di Matteo, S., Fraveto, A., Costantini, D., Cantafora, A., Napoletano, C., et al. (2017). TGF-β signaling is an effective target to impair survival and induce apoptosis of human cholangiocarcinoma cells: a study on human primary cell cultures. *PLoS One, 12*, e0183932.

Malanchi, I., Santamaria-Martínez, A., Susanto, E., Peng, H., Lehr, H.-A., Delaloye, J.-F., et al. (2011). Interactions between cancer stem cells and their niche govern metastatic colonization. *Nature, 481*, 85–89.

Mandrekar, P., & Cardinale, V. (2018). Periostin and Mesothelin: Potential predictors of malignant progression in intrahepatic cholangiocarcinoma. *Hepatology Communications, 2*, 481–483.

Manzanares, M.Á., Campbell, D. J. W., Maldonado, G. T., & Sirica, A. E. (2018). Overexpression of periostin and distinct mesothelin forms predict malignant progression in a rat cholangiocarcinoma model. *Hepatology Communications, 2*, 155–172.

Manzanares, M.Á., M Usui, A., Campbell, D. J., Dumur, C. I., Maldonado, G. T., Fausther, M., et al. (2017). Transforming growth factor α and β are essential for modeling cholangiocarcinoma desmoplasia and progression in a three-dimensional organotypic culture model. *American Journal of Pathology, 187*, 1068–1092.

Matsumoto, K., & Aoki, H. (2020). The roles of tenascins in cardiovascular, inflammatory, and heritable connective tissue diseases. *Frontiers in Immunology, 11*, 609752.

McQuitty, C. E., Williams, R., Chokshi, S., & Urbani, L. (2020). Immunoregulatory role of the extracellular matrix within the liver disease microenvironment. *Frontiers in Immunology, 11*, 574276.

Mertens, J. C., Fingas, C. D., Christensen, J. D., Smoot, R. L., Bronk, S. F., Werneburg, N. W., et al. (2013). Therapeutic effects of deleting cancer-associated fibroblasts in cholangiocarcinoma. *Cancer Research, 73*, 897–907.

Mino, M., Kanno, K., Okimoto, K., Sugiyama, A., Kishikawa, N., Kobayashi, T., et al. (2017). Periostin promotes malignant potential by induction of epithelial-mesenchymal transition in intrahepatic cholangiocarcinoma. *Hepatology Communications, 1*, 1099–1109.

Mizukami, T., Kamachi, H., Fujii, Y., Matsuzawa, F., Einama, T., Kawamata, F., et al. (2018). The anti-mesothelin monoclonal antibody amatuximab enhances the anti-tumor effect of gemcitabine against mesothelin-high expressing pancreatic cancer cells in a peritoneal metastasis mouse model. *Oncotarget, 9*, 33844–33852.

Mohr, R., Özdirik, B., Knorr, J., Wree, A., Demir, M., Tacke, F., et al. (2020). In vivo models for cholangiocarcinoma-what can we learn for human disease. *International Journal of Molecular Sciences, 21*, 4993.

Moorman, H. R., Poschel, D., Klement, J. D., Lu, C., Redd, P. S., & Liu, K. (2020). Osteopontin: A key regulator of tumor progression and immunomodulation. *Cancers, 12*, 3379.

Murphy-Ullrich, J. E. (2019). Thrombospondin 1 and its diverse roles as a regulator of extracellular matrix in fibrotic disease. *Journal of Histochemistry and Cytochemistry, 67*, 683–699.

Murphy-Ullrich, J. E., & Sage, E. H. (2014). Revisiting the matricellular concept. *Matrix Biology, 37*, 1–14.

Murphy-Ullrich, J. E., & Suto, M. J. (2018). Thrombospondin-1 regulation of latent TGF-β activation: A therapeutic target for fibrotic disease. *Matrix Biology, 68-69*, 28–43.

Nakanuma, Y., Sato, Y., Harada, K., Sasaki, M., Xu, J., & Ikeda, H. (2010). Pathological classification of intrahepatic cholangiocarcinoma based on a new concept. *World Journal of Hepatology, 2*, 419–427.

Nakazawa, Y., Taniyama, Y., Sanada, F., Morishita, R., Nakamori, S., Morimoto, K., et al. (2018). Periostin blockade overcomes chemoresistance via restricting the expansion of mesenchymal tumor subpopulations in breast cancer. *Scientific Reports, 8*, 4013.

Nomura, R., Fujii, H., Abe, M., Sugo, H., Ishizaki, Y., Kawasaki, S., et al. (2013). Mesothelin expression is a prognostic factor in cholangiocellular carcinoma. *International Surgery, 98*, 164–169.

Norris, R. A., Damon, B., Mironov, V., Kasyanov, V., Ramamurthi, A., Moreno-Rodriguez, R., et al. (2007). Periostin regulates collagen fibrillogenesis and the biomechanical properties of connective tissue. *Journal of Cellular Biochemistry, 101*, 695–711.

Norris, R. A., Moreno-Rodriguez, R., Sugi, Y., Hoffman, S., Amos, J., Hart, M. M., et al. (2008). Periostin regulates atrioventricular valve maturation. *Developmental Biology, 316*, 200–213.

Papoutsoglou, P., Louis, C., & Coulouarn, C. (2019). Transforming growth factor -beta (TGFβ) signaling pathway in cholangiocarcinoma. *Cell, 8*, 960.

Riener, M.-O., Fritzsche, F. R., Soll, C., Pestalozzi, B. C., Probst-Hensch, N., Clavien, P.-A., et al. (2010). Expression of the extracellular matrix protein periostin in liver tumours and bile duct carcinomas. *Histopathology, 56*, 600–606.

Sage, E. H., & Bornstein, P. (1991). Extracellular proteins that modulate cell-matrix interactions. SPARC, tenascin, and thrombospondin. *Journal of Biological Chemistry, 266*, 14831–14834.

Sawyer, A. J., & Kyriakides, T. R. (2016). Matricellular proteins in drug delivery: Therapeutic targets, active agents, and therapeutic location. *Advanced Drug Delivery Reviews, 97*, 56–68.

Shibata, S., & Ishiyama, J. (2013). Secreted protein acidic and rich in cysteine (SPARC) is upregulated by transforming growth factor (TGF)-β and is required for TGF-β-induced hydrogen peroxide production in fibroblasts. *Fibrogenesis & Tissue Repair, 6*, 6.

Shurin, M. R. (2018). Osteopontin controls immunosuppression in the tumor microenvironment. *Journal of Clinical Investigation, 128*, 5209–5212.

Sirica, A. E. (2012). The role of cancer-associated myofibroblasts in intrahepatic cholangiocarcinoma. *Nature Reviews Gastroenterology & Hepatology, 9*, 44–54.

Sirica, A. E., Almenara, J. A., & Li, C. (2014). Periostin in intrahepatic cholangiocarcinoma: Pathobiological insights and clinical implications. *Experimental and Molecular Pathology, 97*, 515–524.

Sirica, A. E., Dumur, C. I., Campbell, D. J. W., Almenara, J. A., Ogunwobi, O. O., & DeWitt, J. L. (2009). Intrahepatic cholangiocarcinoma progression: Prognostic factors and basic mechanisms. *Clinical Gastroenterology and Hepatology, 7*, S68–S78.

Sirica, A. E., & Gores, G. J. (2014). Desmoplastic stroma and cholangiocarcinoma: Clinical implications and therapeutic targeting. *Hepatology, 59*, 2397–2402.

Sirica, A. E., Gores, G. J., Groopman, J. D., Selaru, F. M., Strazzabosco, M., Wang, X. W., et al. (2019). Intrahepatic cholangiocarcinoma: Continuing challenges and translational advances. *Hepatology, 69*, 1803–1815.

Sirica, A. E., Strazzabosco, M., & Cadamuro, M. (2021). Intrahepatic cholangiocarcinoma: Morpho-molecular pathology, tumor reactive microenvironment, and malignant progression. *Advances in Cancer Research, 149*, 321–387.

Sirica, A. E., Zhang, Z., Lai, G.-H., Asano, T., Shen, X.-N., Ward, D. J., et al. (2008). A novel "patient-like" model of cholangiocarcinoma progression based on bile duct inoculation of tumorigenic rat cholangiocyte cell lines. *Hepatology, 47*, 1178–1190.

Soejima, Y., Takeuchi, M., Akashi, T., Sawabe, M., & Fukusato, T. (2018). β4 and β6 integrin expression is associated with the subclassification and clinicopathological features of intrahepatic cholangiocarcinoma. *International Journal of Molecular Sciences, 19*, 1004.

Song, Z., Chen, W., Athavale, D., Ge, X., Desert, R., Das, S., et al. (2021). Osteopontin takes center stage in chronic liver disease. *Hepatology, 73*, 1594–1608.

Sonongbua, J., Siritungyong, S., Thongchot, S., Kamolhan, T., Utispan, K., Thuwajit, P., et al. (2020). Periostin induces epithelial-to-mesenchymal transition via the integrin α5β1/TWIST axis in cholangiocarcinoma. *Oncology Reports, 43*, 1147–1158.

Sriram, R., Lo, V., Pryce, B., Antonova, L., Mears, A. J., Daneshmand, M., et al. (2015). Loss of periostin/OSF-2 in ErbB2/Neu-driven tumors results in androgen receptor-positive molecular apocrine-like tumors with reduced Notch1 activity. *Breast Cancer Research, 17*, 7.

Sugiyama, A., Kanno, K., Nishimichi, N., Ohta, S., Ono, J., Conway, S. J., et al. (2016). Periostin promotes hepatic fibrosis in mice by modulating hepatic stellate cell activation via α_v integrin interaction. *Journal of Gastroenterology, 51,* 1161–1174.

Sulpice, L., Rayar, M., Desille, M., Turlin, B., Fautrel, A., Boucher, E., et al. (2013). Molecular profiling of stroma identifies osteopontin as an independent predictor of poor prognosis in intrahepatic cholangiocarcinoma. *Hepatology, 58,* 1992–2000.

Suto, M. J., Gupta, V., Mathew, B., Zhang, W., Pallero, M. A., & Murphy-Ullrich, J. E. (2020). Identification of inhibitors of thrombospondin 1 activation of TGF-β. *ACS Medicinal Chemistry Letters, 11,* 1130–1136.

Takahashi, K., Meguro, K., Kawashima, H., Kashiwakuma, D., Kagami, S., Ohta, S., et al. (2019). Serum periostin levels serve as a biomarker for both eosinophilic airway inflammation and fixed airflow limitation in well-controlled asthmatics. *Journal of Asthma, 56,* 236–243.

Tanaka, S., Sugimachi, K., Kameyama, T., Maehara, S., Shirabe, K., Shimada, M., et al. (2003). Human Wisp1v, a member of the CCN family, is associated with invasive cholangiocarcinoma. *Hepatology, 37,* 1122–1129.

Teixeira, A. F., ten Dijke, P., & Zhu, H.-J. (2020). On-target anti-TGF-β therapies are not succeeding in clinical cancer treatments: What are remaining challenges? *Frontiers in Cell and Development Biology, 8,* 605.

Terashi, T., Aishima, S., Taguchi, K., Asayama, Y., Sugimachi, K., Matsuura, S., et al. (2004). Decreased expression of osteopontin is related to tumor aggressiveness and clinical outcome of intrahepatic cholangiocarcinoma. *Liver International, 24,* 38–45.

Thuwajit, C., Thuwajit, P., Jamjantra, P., Pairojkul, C., Wongkham, S., Bhudhisawasdi, V., et al. (2017). Clustering of patients with intrahepatic cholangiocarcinoma based on serum periostin may be predictive of prognosis. *Oncology Letters, 14,* 623–634.

Tsilimigras, D. I., Sahara, K., Wu, L., Moris, D., Bagante, F., Guglielmi, A., et al. (2020). Very early recurrence after liver resection for intrahepatic cholangiocarcinoma. *JAMA Surgery, 155,* 1–9.

Tsukamoto, M., Yamashita, Y., Imai, K., Umezaki, N., Yamao, T., Okabe, H., et al. (2017). Predictors of cure of intrahepatic cholangiocarcinoma after hepatic resection. *Anticancer Research, 37,* 6971–6975.

Utispan, K., Sonongbua, J., Thuwajit, P., Chau-In, S., Pairojkul, C., Wongkham, S., et al. (2012). Periostin activates integrin $\alpha 5\beta 1$ through a PI3K/AKT-dependent pathway in invasion of cholangiocarcinoma. *International Journal of Oncology, 41,* 1110–1118.

Utispan, K., Thuwajit, P., Abiko, Y., Charngkaew, K., Paupairoj, A., Chau-in, S., et al. (2010). Gene expression profiling of cholangiocarcinoma-derived fibroblasts reveals alterations related to tumor progression and indicates periostin as a poor prognostic marker. *Molecular Cancer, 9,* 13.

Vaquero, J., Aoudjehane, L., & Fouassier, L. (2020). Cancer-associated fibroblasts in cholangiocarcinoma. *Current Opinion in Gastroenterology, 36,* 63–69.

Vianello, E., Kalousová, M., Dozio, E., Tacchini, L., Zima, T., & Romanelli, M. M. C. (2020). Osreopontin: The molecular bridge between fat and cardiac-renal disorders. *International Journal of Molecular Sciences, 21,* 5568.

Vijgen, S., Terris, B., & Rubbia-Brandt. (2017). Pathology of intrahepatic cholangiocarcinoma. *HepatoBiliary Surgery and Nutrition, 6,* 22–34.

Viloria, K., & Hill, N. J. (2016). Embracing the complexity of matricellular proteins: The functional and clinical significance of splice variation. *Biomolecular Concepts, 7,* 117–132.

Vincent, K. M., & Postovit, L.-M. (2018). Matricellular proteins in cancer: A focus on secreted frizzled-related proteins. *Journal of Cell Communication and Signaling, 12,* 103–112.

Wallace, D. P. (2019). Periostin in the kidney. *Advances in Experimental Medicine and Biology, 1132,* 99–112.

Wang, Z., Xiong, S., Mao, Y., Chen, M., Ma, X., Zhou, X., et al. (2016). Periostin promotes immunosuppressive premetastatic niche formation to facilitate breast tumour metastasis. *Journal of Pathology, 239*, 484–495.

Weber, C. E., Kothari, A. N., Wai, P. Y., Li, N. Y., Driver, J., Zapf, M. A. C., et al. (2015). Osteopontin mediates an MZF1-TGF-β1-dependent transformation of mesenchymal stem cells into cancer-associated fibroblasts in breast cancer. *Oncogene, 34*, 4821–4833.

Wen, Y., Feng, D., Wu, H., Liu, W., Li, H., Wang, F., et al. (2015). Defective initiation of liver regeneration in osteopontin-deficient mice after partial hepatectomy due to insufficient activation of IL-6/Stat3 pathway. *International Journal of Biological Sciences, 11*, 1236–1247.

Wong, G. S., & Rustgi, A. K. (2013). Matricellular proteins: Priming the tumour microenvironment for cancer development and metastasis. *British Journal of Cancer, 108*, 755–761.

Wu, T., & Ouyang, G. (2014). Matricellular proteins: Multifaceted extracellular regulators in tumor dormancy. *Protein & Cell, 5*, 249–252.

Wu, H., Zhang, H., Hu, L.-Y., Zhang, T.-Y., Zheng, Y.-J., Shen, F., et al. (2018). Is osteopontin a promising biomarker for cholangiocarcinoma? *Journal of Hepatology, 68*, 199–220.

Wu, W.-R., Zhang, R., Shi, X.-D., Zhu, M.-S., Xu, L.-B., Zeng, H., et al. (2014). Notch1 is overexpressed in human intrahepatic cholangiocarcinoma and is associated with its proliferation, invasiveness, and sensitivity to 5-fluorouracil in vitro. *Oncology Reports, 31*, 2515–2524.

Yoshida, T., Akatsuka, T., & Imanaka-Yoshida, K. (2015). Tenascin-C and integrins in cancer. *Cell Adhesion & Migration, 9*(1–2), 96–104.

Yu, L., Feng, M., Kim, H., Phung, Y., Kleiner, D. E., Gores, G. J., et al. (2010). Mesothelin as a potential therapeutic target in human cholangiocarcinoma. *Journal of Cancer, 1*, 141–149.

Yue, H., Li, W., Chen, R., Wang, J., Lu, X., & Li, J. (2021). Stromal eriostin induced by TGF-β1 facilitates the migration and invasion of ovarian cancer. *Gyenecologic Oncology, 160*, 530–538.

Zeng, J., Liu, Z., Sun, S., Xie, J., Cao, L., Lv, P., et al. (2018). Tumor-associated macrophages recruited by periostin in intrahepatic cholangiocarcinoma stem cells. *Oncology Letters, 15*, 8681–8686.

Zhang, M., Yang, H., Wan, L., Wang, Z., Wang, H., Ge, C., et al. (2020). Single-cell transcriptomic architecture and intercellular crosstalk of human intrahepatic cholangiocarcinoma. *Journal of Hepatology, 73*, 1118–1130.

Zhao, X.-Q., Ma, H.-X., Su, M.-S., & He, L. (2014). Osteopontin promoter polymorphisms at locus −443 are associated with metastasis and poor prognosis of human intrahepatic cholangiocarcinoma in Chinese population. *International Journal of Clinical and Experimental Pathology, 7*, 6914–6921.

Zheng, Y., Zhou, C., Yu, X.-X., Wu, C., Jia, H.-L., Gao, X.-M., et al. (2018). Osteopontin promotes metastasis of intrahepatic cholangiocarcinoma through recruiting MAPK1 and mediating Ser675 phosphorylation of β-catenin. *Cell Death and Disease, 9*, 179. https://doi.org/10.1038/s41419-017-0226-x.

Zhou, K.-Q., Liu, W.-F., Yang, L.-X., Sun, Y.-F., Hu, J., Chen, F.-Y., et al. (2019). Circulating osteopontin per tumor volume as a prognostic biomarker for resectable intrahepatic cholangiocarcinoma. *Hepatobiliary Surgery and Nutrition, 8*, 582–596.

Zhu, M., Saxton, R. E., Ramos, L., Chang, D. D., Karlan, B. Y., Gasson, J. C., et al. (2011). Neutralizing monoclonal antibody to periostin inhibits ovarian tumor growth and metastasis. *Molecular Cancer Therapeutics, 10*, 1500–1508.

Zhu, J.-Z., Zhu, H.-T., Dai, Y.-N., Li, C.-X., Fang, Z.-Y., Zhao, D.-J., et al. (2016). Serum periostin is a potential biomarker for non-alcoholic fatty liver disease: A case-control study. *Endocrine, 51*, 91–100.

CHAPTER TEN

YAP1 activation and Hippo pathway signaling in the pathogenesis and treatment of intrahepatic cholangiocarcinoma

Sungjin Ko[a,b,]*, Minwook Kim[c], Laura Molina[a,b], Alphonse E. Sirica[d], and Satdarshan P. Monga[a,b,e,]*

[a]Division of Experimental Pathology, Department of Pathology, University of Pittsburgh School of Medicine, Pittsburgh, PA, United States
[b]Pittsburgh Liver Research Center, Pittsburgh, PA, United States
[c]Department of Developmental Biology, University of Pittsburgh School of Medicine, Pittsburgh, PA, United States
[d]Department of Pathology, Virginia Commonwealth University School of Medicine, Richmond, VA, United States
[e]Division of Gastroenterology, Hepatology, and Nutrition, University of Pittsburgh and UPMC, Pittsburgh, PA, United States
*Corresponding authors: e-mail address: sungjin@pitt.edu; smonga@pitt.edu

Contents

1. Introduction	286
2. Overview of Hippo-YAP1 signaling pathway	287
3. Biologic function and pathologic impact of Hippo-YAP1 pathway in ICCA	293
3.1 Liver cell fate conversion and transformation	293
3.2 iCCA metabolism	297
3.3 iCCA immunology	301
4. Clinical relevance and therapeutic potential of modulation of Hippo-YAP1 pathway in ICCA	302
5. Concluding remarks	308
Acknowledgments	308
Conflict of interest statement	308
References	308

Abstract

Intrahepatic cholangiocarcinoma (iCCA), the second most common primary liver cancer, is a highly lethal epithelial cell malignancy exhibiting features of cholangiocyte differentiation. iCCAs can potentially develop from multiple cell types of origin within liver, including immature or mature cholangiocytes, hepatic stem cells/progenitor cells, and from transdifferentiation of hepatocytes. Understanding the molecular mechanisms and genetic drivers that diversely drive specific cell lineage pathways leading to iCCA has important biological and clinical implications. In this context, activation of the YAP1-TEAD dependent transcription, driven by Hippo-dependent or -independent diverse mechanisms that lead to the stabilization of YAP1 is crucially important to biliary fate commitment in hepatobiliary cancer. In preclinical models, YAP1 activation in hepatocytes or cholangiocytes is sufficient to drive their malignant transformation into iCCA. Moreover, nuclear YAP1/TAZ is highly prevalent in human iCCA irrespective of the varied etiology, and significantly correlates with poor prognosis in iCCA patients. Based on the ubiquitous expression and diverse physiologic roles for YAP1/TAZ in the liver, recent studies have further revealed distinct functions of active YAP1/TAZ in regulating tumor metabolism, as well as the tumor immune microenvironment. In the current review, we discuss our current understanding of the various roles of the Hippo-YAP1 signaling in iCCA pathogenesis, with a specific focus on the roles played by the Hippo-YAP1 pathway in modulating biliary commitment and oncogenicity, iCCA metabolism, and immune microenvironment. We also discuss the therapeutic potential of targeting the YAP1/TAZ-TEAD transcriptional machinery in iCCA, its current limitations, and what future studies are needed to facilitate clinical translation.

Abbreviation

AAV	adeno-associated virus
ACTN4	actinin alpha 4
Alb	albumin
α-KG	α-ketoglutarate
AMPK	AMP-activates protein kinase
ANKRD1	ankyrin repeat domain 1
ATP	adenosine triphosphate
β-TRCP	beta-transducin repeat containing
CAFs	cancer-associated fibroblasts
CCL2	C-C motif chemokine ligand 2
CCND1	cyclin D1
CDK	cyclin Dependent Kinase
cHCC-iCCA	combined hepatocellular carcinoma-cholangiocarcinoma
circ	circular RNA
CK1	casein kinase 1
CMV	*Cytomegalovirus*
CTGF/CCN2	cellular communication network factor 2
CYR61/CCN1	cellular communication network factor 1
DDC	3,5-diethoxycarbonyl-1,4-dihydrocollidine
D-2HG	D-2-hydroxyglutarate
ECM	extracellular matrix
FAK	focal adhesion kinase
Fbxw7	F-Box and WD repeat domain containing 7
FGFR	fibroblast growth factor receptor
Foxm1	forkhead box M1
FZD	frizzled class receptor

GLUT1	glucose transporter 1
G9a	G9A histone methyltransferase
HCC	hepatocellular carcinoma
HDAC1	histone deacetylase1
HDTVI	hydrodynamic tail vein injection
iCCA	intrahepatic cholangiocarcinoma
IDH	isocitrate dehydrogenase
IG	intraductal growing
IGF-1	insulin like growth factor 1
IL-6	interleukin 6
KO	knockout
LATS1/2	large tumor suppressor kinase 1 and 2
LGG	low-grade glioma
lnc	*long non-coding*
LPCs	liver progenitor cells
mAkt	myristoylated Akt
MFAP5	microfibrillar-associated protein 5
MAP4K	mitogen-activated protein kinase 4
MCL1	MCL1 apoptosis regulator, BCL2 family member
MED1	mediator complex subunit 1
MF	mass-forming
miR	microRNA
MOB1	MOB kinase activator 1
MST1/2	mammalian Ste20-like serine/threonine protein kinase 1 and 2
mRNA	messenger RNA
mTOR	mammalian target of rapamycin
MYPT1	myosin phosphatase-targeting subunit 1
NICD	notch intracellular domain
NF2	neurofibromin 2
NRas	NRAS proto-oncogene, GTPase
nTAZ	TAZ nuclear localization
NUAK2	NUAK family kinase 2
nYAP1	YAP1 nuclear localization
O-GlcNAcylation	O-linked-N-acetylglucosaminylation
OCT4	octamer-binding protein 4
PDGFR	platelet derived growth factor receptor
PDX	patient-derived xenograft
PES1	pescadillo ribosomal biogenesis factor 1
PI	periductular infiltrating
PIEZO1	piezo type mechanosensitive ion channel component 1
PIK3CA	phosphatidylinositol-4,5-bisphosphate 3-kinase catalytic subunit alpha
RASSF1A	Ras association domain family member 1A
RhoA	Ras homology family member A
rtTA	reverse tetracycline-controlled transactivator
SAV1	Salvador 1
SCF	SKP-Cullin-Fbxw
sh	short hairpin

SHP2	protein tyrosine phosphatase non-receptor type 11
SOX9	SRY-box transcription factor 9
SRC	SRC proto-oncogene, non-receptor tyrosine kinase
STAT3	signal transducer and activator of transcription 3
SULF2	sulfatase 2
TAOK1-3	Tao kinases 1-3
TAZ	PDZ-binding motif
TBX5	T-box transcription factor 5
TEAD1–4	TEA domain transcription factors 1–4
TetO	tetracycline operator
TGF-β	transforming growth factor-β
TBG	thyroxine-binding globulin
TME	tumor microenvironment
YAP1	Yes1 associated transcriptional regulator

1. Introduction

Next to hepatocellular carcinoma (HCC), intrahepatic cholangiocarcinoma (iCCA) is the second most common primary cancer of liver, comprising around 10–25% of hepatobiliary cancers. Histologically, 90–95% of iCCAs are classified as adenocarcinomas, which depending on anatomic location may be subclassified into either small-duct type or large bile duct type tumors. Based on macroscopic growth patterns, iCCAs have been grossly classifies as mass-forming (MF), periductular infiltrating (PI), intraductal growing (IG) and mixed MF-PI types (Brindley et al., 2021; Kendall et al., 2019; Sirica, Strazzabosco, & Cadamuro, 2021). The MF type is the most common form, accounting for about 65% of all iCCAs, followed by the MF-PI type, comprising about 25% of cholangiocarcinomas formed in liver. MF-iCCA almost exclusively is of the small duct type and is often associated with chronic non-bile duct liver diseases, such as viral hepatitis and cirrhosis (Cardinale et al., 2018; Kendall et al., 2019). In contrast, the large bile duct type typically characterized the PI and PI + MF types of iCCA, such as those associated with high risk fibroinflammatory cholangiopathies, such as primary sclerosing cholangitis, hepatolithiasis, liver fluke infestation, and Caroli's disease (Brindley et al., 2021; Sirica et al., 2021). Unique rare variants include combined cHCC-iCCA, intestinal-type iCCA, and cholangiolocellular carcinoma (reviewed in Sirica et al., 2021).

Current experimental findings based largely on preclinical studies performed on genetic mouse models; cell lineage tracing, and immunophenotyping, indicate that iCCAs can potentially develop from multiple liver cells of origin, including cholangiocytes, liver progenitor cells (LPCs), and from transdifferentiation of hepatocytes (Sirica et al., 2021). Notably both HCCs and iCCAs have been shown to share common risk factors associated with chronic fibroinflammatory and cirrhotic liver diseases (Brindley et al., 2021; Clements, Eliahoo, Kim, Taylor-Robinson, & Khan, 2020; Liu et al., 2019; Moeini, Haber, & Sia, 2021), further suggesting that specific molecular signatures determine the cancer types that may originate from distinct cell types in the chronic liver/cirrhosis stage. The well-proven cell plasticity of hepatocytes and cholangiocytes in chronically diseased liver (Ko, Russell, Molina, & Monga, 2019) and clinical presentation of cHCC-iCCA (0.4–14.2% of primary liver cancers) (Gera, Ettel, Acosta-Gonzalez, & Xu, 2017; Stavraka, Rush, & Ross, 2019) also support this hypothesis (Moeini et al., 2021).

The Hippo/Yes1 Associated Transcriptional Regulator (YAP1) signaling pathway is a master regulator of cell viability, fate commitment and tissue homoeostasis, and plays critical roles in tumorigenesis and regeneration (Ma, Meng, Chen, & Guan, 2019) and has been intensively studied in bile duct pathophysiology, including iCCA, where it acts as important modulator of biliary fate and oncogenesis. In this chapter, we will comprehensively review essential pathway components of the Hippo/YAP1 cascade, highlight current preclinical iCCA models generated by Hippo/YAP1 modulation and its role in biliary fate conversion and malignant transformation, and discuss the role of Hippo/YAP1 signaling as a regulator of iCCA metabolism and immune microenvironment. Lastly, we will review studies supporting the clinical relevance of the Hippo/YAP1 signaling pathway for human iCCA, and discuss the potential and limitations of this pathway as a therapeutic target in advanced iCCA.

2. Overview of Hippo-YAP1 signaling pathway

Mechanistically, the Hippo/YAP1 signaling pathway comprises a large and complex kinase-based protein network (Fig. 1A and B). Essential signaling pathway components are allocated into three phases of the cascade: (1) the various physical extracellular stimuli which activate

Hippo kinases; (2) regulatory Hippo kinase modules in the cytoplasm; and (3) transcriptional modulation machinery in the nucleus. Although diverse physical and biochemical stimuli, including mechano–sensors, extracellular stress, and disturbances of cell polarity are defined as upstream Hippo

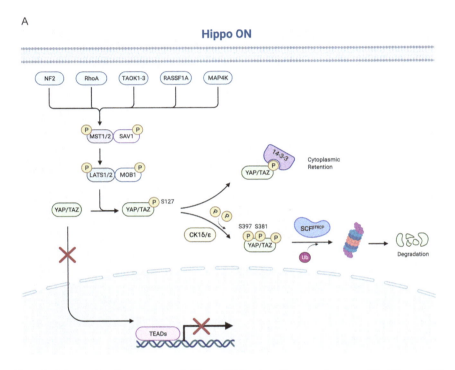

Fig. 1 Schematic illustration of Hippo/YAP signaling pathways. (A) Hippo ON: diverse upstream signals (i.e., NF2, RhoA, TAOK1-3, RASSF1A and MAP4K) phosphorylate MST1/2 to phosphorylate and activate LATS1/2 which become active with MOB1 and in turn phosphorylate YAP1 at Serine 127 site and TAZ, to induce their cytoplasmic sequestration by 14-3-3 Protein. Phosphorylation of YAP1 and TAZ leads their ubiquitination by SCF-β-TRCP E3 ubiquitin ligase, and subsequent degradation by the proteolytic proteasome. (B) Hippo OFF: Non-phosphorylated YAP1 and TAZ trans-localized into the nucleus to form a transcription complex with their definite binding partner TEAD to transcribe down-stream target genes. (C) Non-canonical, Hippo-independent YAP1 activation: PDGFR signaling pathway increases nuclear YAP1 via phosphorylation of Tyr357 by Src Family Kinases, thereby inducing MCL-1 transcription. In contrast, SHP2 dephosphorylates YAP1 Y357, thus inhibiting the nuclear translocation of YAP1. YAP1/TAZ and TBX5 form a transcription complex to transcribe FGFR1,2 and 4, in turn, FGFR signaling pathway inhibits phosphorylation of YAP1 S127. MFAP5 or PES1 can directly target YAP1 nuclear translocation. Besides, PIEZO-1 can inhibits LATS1 activity thus result in YAP1/TAZ translocation. *Adapted from BioRender (2022). Hippo Pathway in Mammals. Retrieved from https://app.biorender.com/biorender-templates/t-5ee8e905305f3a00 ae6a51b7-hippo-pathway-in-mammals.*

B

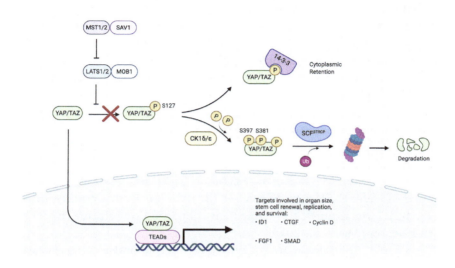

C

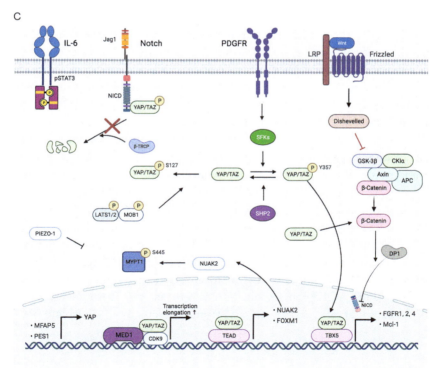

Fig. 1—Cont'd

regulators, detailed mechanisms for respective factors are largely unknown and no single, specific ligand switching Hippo activity on or off is yet to be determined in the liver (Ma et al., 2019).

In the cytoplasm, mammalian Ste20-like serine/threonine protein kinase 1 and 2 (MST1/2) and the large tumor suppressor kinase 1 and 2 (LATS1/2) axis are the core kinases of the Hippo pathway, which function to block the downstream transcription module. Once MST1/2 is primed by binding with Salvador (SAV1), it can either be auto-phosphorylated or phosphorylated (activated) by the action of diverse upstream kinases (i.e., Mitogen-activated protein kinase 4 (MAP4K) family, Ras association domain family member 1A (RASSF1A), Tao kinases 1-3 (TAOK1-3), Ras homology family member A (RhoA), Neurofibromin 2 (NF2)). Subsequently, active MST1/2 phosphorylates downstream LATS1/2 in cooperation with MOB Kinase Activator 1 (MOB1) (Ma et al., 2019; Nguyen-Lefebvre, Selzner, Wrana, & Bhat, 2021). In this process, SAV1 and MOB1 are required to recruit their canonical downstream targets for kinase action (Ma et al., 2019).

Activated (phosphorylated) LATS1/2 phosphorylates the crucial transcription regulator components of Hippo kinase cascade (Ma et al., 2019): YAP1 and transcriptional coactivator with the PDZ-binding motif (TAZ). TAZ, with 47% amino acid sequence identity to YAP1 and encoded by a paralogue gene, is known for its redundancy as a transcriptional co-activator together with YAP1 (Reggiani, Gobbi, Ciarrocchi, & Sancisi, 2021). Phosphorylation of several sites on YAP1 determines its stability, balancing sequestration, or degradation in the cytoplasm with translocation into the nucleus for transcriptional regulation (Ma et al., 2019; Nguyen-Lefebvre et al., 2021). In other cases, YAP1/TAZ also binds with cytoskeletal/junctional proteins Tight Junction Protein 2 or Claudin, responding to aberrant cellular density, which eventually inhibits their transcription of downstream genes (Driskill & Pan, 2021; Ma et al., 2019).

Noncanonically, Notch, Wnt/β-Catenin, SRC Proto-Oncogene, Non-Receptor Tyrosine Kinase (SRC)-Focal Adhesion Kinase (FAK), Transforming Growth Factor-β (TGF-β), Peroxisome Proliferator Activated Receptor α and mammalian target of rapamycin (mTOR) signaling also can regulate YAP1 stability and nuclear localization in a Hippo-independent manner (Fig. 1C) (Fan et al., 2022; Hu et al., 2017; Kim et al., 2017; Szeto et al., 2016). In addition, Protein 14-3-3-driven YAP1 phospho-S127 sequestration is most critical for cytoplasmic retention and degradation, preventing nuclear translocation. Thus, the *YAP1 S127A* mutant has been widely used as constitutionally-active form of YAP1. Phosphorylation on Ser 397 or Ser 381 by Casein Kinase 1 (CK1)δ/ε provides phosphodegron substrates for the SKP-Cullin-Fbxw (SCF)-Beta-Transducin Repeat Containing E3 Ubiquitin Protein Ligase (β-TRCP E3 ubiquitin ligase) complex, forcing ubiquitination of YAP1 and subsequent proteasomal degradation. Once non-phosphorylated YAP1/TAZ is translocated into the nucleus by escaping from diverse kinase action, these activators bind to canonical transcription factor TEA domain transcription factors 1–4 (TEAD1–4) (Ma et al., 2019) or non-canonical transcription factors including RUNX family transcription factor, SMAD family member 2/3, p73 or T-Box transcription factor 5 (TBX5) (Kim, Jang, & Bae, 2018).

Given the ubiquitous expression of YAP1/TAZ and tight regulation of transcriptional activity by quick turnover, the Hippo/YAP1 pathway is critical for numerous essential biologic processes, such as cell viability, proliferation, cell cycle regulation, differentiation and transformation (Ma et al., 2019). In the liver, reflecting its abundant physiologic role, Hippo/YAP1 signaling is crucially involved in pathophysiology including development, homeostasis, regeneration, and malignancy in diverse cell types, functions which have been extensively reviewed elsewhere (Driskill & Pan, 2021; Nguyen-Lefebvre et al., 2021).

Notably, modulation of Hippo/YAP1 signaling determines the quantity and quality of bile duct development in mammals, as well as the survival of mature bile ducts in homeostasis (Molina et al., 2021). Manipulation of the Hippo/YAP1 cascade induces a striking effect on liver size (Driskill & Pan, 2021). Genetic inactivation of Hippo kinases in liver parenchyma provokes massive hepatomegaly by substantially increasing hepatocyte proliferation, which could be rapidly reverted by interruption of YAP1 stabilization (Yimlamai, Fowl, & Camargo, 2015). Importantly, extended YAP1 activation in these diverse models of Hippo inactivation (Genetic deletion of *Lats1/2, Mst1/2, Sav1, Mob1a/b, Nf2,* or *Kibra*) eventually provokes liver cancer development, including iCCA formation (Table 1). In this process, TEADs have been identified as crucial transcription factors in the absence of competition from Vestigial Like Family Member 4 for hepatomegaly and subsequent tumor formation (Driskill & Pan, 2021).

Table 1 Currently available iCCA models generated by Hippo/YAP1 modulation.

Methodology	Background	Allele	Gene delivery	Treatment	Liver cancer type	Refs
Genetic mutation	Alb-Cre	$Sav1^{fl/fl}$		DDC diet	Mixed iCCA/HCC	Lee et al. (2010)
	Alb-Cre	$Nf2^{fl/fl};Wwc1^{-/-}$			iCCA	Park et al. (2020)
	Alb-Cre	$Nf2^{fl/fl}$			Mixed iCCA/HCC	Zhang et al. (2010)
	Sox9-CreERT2	$Lats1^{fl/fl};Lats2^{fl/fl}$			iCCA	Park et al. (2020)
	Alb-Cre	$Mob1a^{fl/fl};Mob1b^{-/-}$			mixed iCCA/HCC	Nishio et al. (2016)
	CAGGCre-ER	$Mst1^{-/-};Mst2^{fl/-}$			mixed iCCA/HCC	Song et al. (2010)
	Alb-Cre	$Nf2^{fl/fl};Sav1^{fl/fl}$			biliary hyperplasia, iCCA like	Yin et al. (2013)
	Rosa26-rtTa	$TetO:YAP1^{S127A}$			iCCA	Liu et al. (2021)
AAV8: pan HC transduction	C57BL/6	WT	$AAV8\text{-}TBG\text{-}YAP1^{S127A}$		iCCA	Liu, Zhuo, et al. (2021)
HDTVI: mosaic HC transfection	FVB/N	WT	$mAKT + YAP1^{S127A}$		Mixed iCCA/HCC (iCCA dominant)	Wang et al. (2018)
	FVB/N	WT	$PIK3CA^{H1047} + YAP1^{S127A}$		Mixed iCCA/HCC	Li et al. (2015)
	FVB/N	WT	$sh\text{-}p53 + YAP1^{S127A} \pm NICD$		iCCA	Tschaharganeh et al. (2014)
Cholangiocyte transfection	C57BL/6	WT	$mAKT + YAP1^{S127A}$	Bile duct ligation ± IL-33 administration	iCCA	Yamada et al. (2015)

3. Biologic function and pathologic impact of Hippo-YAP1 pathway in ICCA

3.1 Liver cell fate conversion and transformation

Despite the ongoing debate about its clinical relevance, bi-directional cell plasticity of hepatocyte-cholangiocytes and of LPCs is well described in mammalian model systems using diverse cell lineage tracing methods (Gadd, Aleksieva, & Forbes, 2020; Ko et al., 2019). Cholangiocytes are readily activated in response to diverse hepatic injury insults and give rise to LPCs or ductular reaction, although the differentiation of activated cholangiocytes/LPCs into functional hepatocytes occurs only when the primary liver regeneration capability (i.e. proliferation of hepatocytes) is completely abrogated, such as in acute liver injury combined with genetically defective proliferation or senescence of all hepatocytes or in the setting of long-term chronic liver injury (Ko et al., 2019). Several regulators/signaling pathways important for cholangiocyte-to-hepatocyte conversion have been identified in zebrafish and in mice (Jung et al., 2021; Ko et al., 2019), but the master regulators or drivers and relevant mechanisms remain largely unknown, especially in mammals (Gadd et al., 2020; Ko et al., 2019).

Hippo/YAP1 signaling is required for the ductular reaction (specifically in cholangiocyte reactivation) (Bai et al., 2012; Planas-Paz et al., 2019), but there is no direct evidence for an effect of Hippo/YAP1 modulation in cholangiocyte/LPC differentiation into terminal hepatocytes. Given the crucial role for Hippo/YAP1 in cholangiocyte-to-hepatocyte conversion (Bai et al., 2012; Yimlamai et al., 2014) and recently developed YAP1-TEAD inhibitors (Holden et al., 2020; Li et al., 2020), it would be important to address the effect of YAP1 inhibition in cholangiocyte-to-hepatocyte conversion using diverse injury models, since it may be relevant for end-stage cirrhosis patients who desperately need restoration of functional hepatocytes.

Likewise, hepatocyte-to-cholangiocyte transdifferentiation has been well proven in mammals using diverse hepatocyte lineage-tracing systems such as *Adeno-associated virus (AAV)8-Thyroxine-binding globulin (TBG)-Cre* or *AAV8-TBG-Flp* in a variety of chemical and genetic liver injury models (Han et al., 2019; Ko et al., 2019). Importantly, the Stanger laboratory showed human hepatocyte transdifferentiation into cholangiocyte *in vivo* using 3,5-Diethoxycarbonyl-1,4-Dihydrocollidine (DDC)-fed *Fah;Rag2; Il2rg* KO background humanized liver mouse model, indicating the clinical relevance of this fate switch (Merrell et al., 2021). Detailed molecular mechanisms and genetic drivers of hepatocyte-to-cholangiocyte conversion and/or metaplasia are a subject of intense investigation and also remain to be fully elucidated, although biliary fate commitment pathways such as Notch, Hippo/YAP1, and *TGF-β* signaling have all been implicated in this process.

iCCA is traditionally considered to originate from cholangiocytes, as tumor cells exhibit luminal structures and express cholangiocyte-specific markers such as SRY-Box Transcription Factor 9 (SOX9), Cytokeratin-7, and MUCIN1. However, given the frequent detection of human iCCA in the pericentral area of the liver lobule, which is anatomically distinct from the native biliary structure, and the well-proven occurrence of hepatocyte-to-cholangiocyte differentiation in various chronic cholestasis models, hepatocytes have also been theorized as the cell of origin of a subset of human iCCA. Interestingly, several groups provided evidence for this theory by provoking hepatocyte-derived iCCA using hepatocyte-specific co-expression of proto-oncogenes and biliary lineage commitment genes such as *myristoylated Akt (mAkt)* along with either *Notch intracellular domain (NICD)* or a constitutionally active form of YAP1, the *YAP1 S127A* mutant (Fan et al., 2012; Wang et al., 2018). Overexpression of these two oncogenes initially induce transdifferentiation of transduced hepatocytes into cholangiocyte-like cells, producing lesions with morphologies similar to that observed in human cholestatic conditions (Fan et al., 2012; Wang et al., 2018). While delivery of m*Akt* alone induces HCC and a small number of benign iCCA lesions in long-term studies (28 weeks), the combination of *Akt* with *NICD* or *YAP1* rapidly provokes severe iCCA within 4 weeks (Calvisi et al., 2011; Fan et al., 2012). Moreover, long-term administration of thioacetamide in mice also induced hepatocyte-driven iCCA development, again supporting the relevance of hepatocyte-originated iCCA generated by chronic toxin exposure. These observations suggest that hepatocyte-to-cholangiocyte differentiation may be pathologically contributing to a subset of iCCA and that key drivers of cell fate conversion may cooperate with second-hit oncogenes for hepatocyte-derived cholangiocarcinogenesis. A better understanding of these mechanisms can help identify potential therapeutic targets especially for chemoprevention and, if this process turns out to be dynamic and important in tumor maintenance, then also for the treatment of iCCA.

As described above, forced expression of *YAP1 S127A* in hepatocytes has been shown in the mouse to be sufficient to induce hepatocyte dedifferentiation/transdifferentiation into the cholangiocyte lineage through direct regulation of *Notch2*. *Notch2* deletion reduces YAP1-mediated hepatocyte adoption of cholangiocyte characteristics (Wang et al., 2018; Yimlamai et al., 2014). Similarly, diverse models of genetic inhibition of Hippo kinases, which result in YAP1 hyperactivation in hepatocytes, induces hepatocyte dedifferentiation into LPCs and/or fate conversion into cholangiocytes (Sugihara, Isomoto, Gores, & Smoot, 2019). Currently, there is no valid marker to distinguish dedifferentiated LPCs and cholangiocytes, since most markers overlapped in these two cell types. Moreover, cell fate conversion is a dynamic process in diseased liver. Thus, it is difficult to clearly address the effect of hepatocyte-specific YAP1 activation as transdifferentiation into cholangiocyte *versus* dedifferentiation into bipotent LPC-like cell. Given that long-term YAP1 activation in hepatoblasts or hepatocytes actually induces diverse liver cancers, including hepatoblastoma, HCC and iCCA (Driskill & Pan, 2021; Moeini et al., 2021), theoretically it is adequate to consider that YAP1 alone is sufficient to revert hepatocytes back to bi-potent LPC status (dedifferentiation) and thus stimulate malignant transformation irrespective of lineage. This means it is critical to identify signaling partners, regulators and downstream targets of YAP1 specifically in iCCA context to distinguish its roles in biliary type cancer formation from hepatocytes.

Although hydrodynamic delivery of *YAP1 S127A* alone into hepatocytes is sufficient to differentiate the hepatocytes into LPC-like cells in the mouse, *YAP1 S127A* expression together with *mAkt* also induced hepatocyte-derived iCCA in a *Notch2*-dependent manner (Wang et al., 2018), supporting the roles for a YAP1-Notch2 axis in hepatocyte dedifferentiation (Yimlamai et al., 2014). Interestingly, co-expression of *mAkt-YAP1 S127A* in hepatocytes using the hydrodynamic tail vein injection (HDTVI) technique results in mixed HCC/iCCA formation, suggesting the importance of *YAP1* activity on liver tumor fate in the presence of *mAkt* (Hu et al., 2020). Singular transduction of *mAKT* does not induce expression of biliary markers in transduced hepatocytes, but instead remarkably enhances the number of *YAP1 S127A* transduced cells compared to *YAP1 S127A* singular delivery, suggesting a permissive and supportive role for *mAkt* in hepatocyte-to-iCCA conversion. Indeed, the fate of *YAP1* active hepatocytes is completely distinct based on hepatocyte status; in case of HDTVI, a majority of transduced hepatocytes are eliminated by the immune system (Miyamura et al., 2017). In contrast, hepatocyte-specific YAP1 activation using non-invasive genetic methods results in YAP1-dependent proliferation, also synergizing with *mAkt* in development of liver cancer (Miyamura et al., 2017). Importantly, Liu et al. recently showed that long-term activation of *YAP1* alone, using non-invasive Tet-ON system is sufficient to provokes iCCA, again indicating the proficiency of YAP1 activation in hepatocyte-to-iCCA transformation (Liu, Zhuo, et al., 2021).

In the case of *mAkt-YAP1 S127A*-driven iCCA model, *mTOR-1/-2* signaling strongly activates while genetic *mTOR-1/-2* inhibition significantly represses iCCA development (Song et al., 2019; Zhang et al., 2017). A recent publication also showed that hepatocyte-specific *YAP1 S127A* expression using a Tet-ON system, along with *Sox9* deletion, induces a molecular phenotype switch of iCCA to an aggressive HCC, addressing a role for *Sox9* in commitment to *YAP1 S127A*-driven iCCA-phenotype, despite its apparent lack of role in Notch driven iCCA as described above (Liu, Zhuo, et al., 2021). Given that YAP1 can directly regulate *Notch2* and *Notch2-Sox9* signaling cascade can facilitate hepatocyte-to-cholangiocyte/iCCA conversion (Wang et al., 2018; Yimlamai et al., 2014; Zong et al., 2009), future identification of YAP1-Notch functional downstream target genes besides *Sox9* will be important.

Interestingly, YAP1 or TEAD inhibition prevents hepatocytes from adopting cholangiocyte characteristics and re-expression of DNMT1 rescues this impaired hepatocyte-to-cholangiocyte transition, suggesting YAP1's involvement in methylome remodeling during fate-switch. Several *Dnmt1* downstream target genes, including *Sox17*, *Foxa2*, and microRNAs that silence tumor suppressor genes, have been found in iCCA, although understanding the upstream *Dnmt1* transcription regulators is sparse. Previously, IL-6 control of *DNMT1* transcription via *microRNA (miR)148-1/152* was identified in human iCCA, however diverse HDTVI-mediated iCCA models are robustly DNMT1-positive without any significant inflammation or *Il-6* up-regulation, implying numerous alternative DNMT1 expression regulators in iCCA. Methylome remodeling is an initial phase event in oncogenic transformation in several cancers, including iCCA. Along these lines, inhibiting the *Notch-YAP1/Tead-Dnmt1* axis completely eliminates iCCA formation by preventing HC conversion to biliary fate, while showing no therapeutic effect on established iCCA. Remarkably, pharmacologic DNMT1 inhibition selectively prevent iCCA nodules, but had no effect on overall HCC tumor burden in YAP1 S127A-driven mixed HCC/iCCA indicating its specific roles in hepatocyte-to-ICC fate conversion (Hu et al., 2020). Given the strong positive roles for YAP1 and DNMT1 in various hepatocyte-originated iCCA models driven by disparate drivers, it would be important to validate function of YAP1 and DNMT1 in human cholangiocarcinogenesis.

In the case of chronic cholestasis, we may also need to carefully observe the cholestatic liver microenvironment in which hepatocyte-to-cholangiocyte conversion is actually happening: specifically that severe peri-portal inflammation, stromal expansion, cell death together with active conversion of SOX9 positive hepatocytes into cholangiocytes can possibly potentiate oncogenic transformation (Clements et al., 2020). Indeed, chromatin structure and the epigenome are largely remodeled during cell fate changes, which potentiates a higher risk for malignant transformation of the cell (Nebbioso, Tambaro, Dell'Aversana, & Altucci, 2018), supporting the theory of hepatocyte transformation into iCCA in cholestasis. It has also been shown that modulation of tumor microenvironment (TME) is sufficient to direct the fate of hepatocyte-originated liver cancer (*c-Myc*-mediated) into HCC *versus* iCCA, with *Tbx3* and *Prdm5* identified as major microenvironment-dependent epigenetic regulator directing lineage-commitment (Seehawer et al., 2018). This study strongly claims the sufficiency of TMEs (i.e., inflammation) determines the lineage commitment of liver cancer. In fact, the Halder laboratory demonstrated that peri-tumoral YAP1 activity is critical regulator for YAP1-mediated iCCA formation in mouse (Moya et al., 2019). Altogether, these observations suggest that peri-tumoral YAP1 activation is critical for hepatocyte dedifferentiation-following transformation and oncogenic microenvironment may be important to determine tumor cell fate and lineage.

3.2 iCCA metabolism

Like other cancer cells, cholangiocarcinoma cells also undergo metabolic reprogramming to sustain proliferation under the energy stress condition (Pant, Richard, Peixoto, & Gradilone, 2020). Malignant iCCA cells upregulate the expression of the *glucose transporter 1 (GLUT1)* (Zimmerman, Fogt, Burke, & Murakata, 2002), which maximizes glucose absorption across plasma membranes and activates the cellular energy sensor AMP-activates protein kinase (AMPK) pathways to replenish Adenosine triphosphate (ATP) by reducing cell autonomous ATP-consuming and enhancing cell-autonomous ATP production. Previously, GLUT1 has been suggested as a specific marker for distinguishing iCCA from native bile duct or HCC (Roh, Jeong, Kim, Kim, & Hong, 2004; Zimmerman et al., 2002), and multiple papers have described its critical functions in iCCA pathophysiology, including carcinogenesis, metastasis, and chemoresistance (Thamrongwaranggoon et al., 2021; Tiemin et al., 2020). Importantly, the *YAP1/TEAD-GLUT1* axis has been well established in a variety of cancers, including HCC (Li et al., 2019; Lin & Xu, 2017). However, evidence of *YAP1-Glut1* regulation in iCCA is limited. Interestingly, AMPK activation strongly represses YAP1 activity through phosphorylation and stabilization of the Hippo component AMOTL1 (DeRan et al., 2014) and direct phosphorylation of YAP1 (Wang et al., 2015). AMPK-YAP1 crosstalk has also been implicated in the proliferation of colorectal cancer cells (Chen et al., 2021) and hepatoblastoma cells (Chen, Yen, Lin, Lai, & Lee, 2020). Given that publications address the beneficial effect of Metformin (FDA-approved AMPK activator) treatment in clinical iCCA, particularly with metabolic disturbances via the anti-Warburg effect (Di Matteo et al., 2021; Tang et al., 2018), the role of Hippo/YAP1 regulation by AMPK in the metabolic imbalance condition needs to be further elucidated in the iCCA setting.

Additionally, elevated blood glucose levels activate the Hexosamine biosynthesis pathway-mediated O-linked-N-acetylglucosaminylation (O-GlcNAcylation), which is frequently perturbed during cancer metabolic reprogramming (Ciraku, Esquea, & Reginato, 2022; Phoomak et al., 2017). Increased O-GlcNAcylation augments Nuclear Factor kappa B and AKT-extracellular signal-regulated kinase activity in iCCA, ultimately worsening the prognosis (Phoomak et al., 2016; Phoomak et al., 2018). Notably, Hippo/YAP1 has been regarded as a key target for O-GlcNAcylation in a variety of cancer types, including thyroid and breast cancer (Kim et al., 2020; Li et al., 2021; Peng et al., 2017) and O-GlcNAcylation of YAP1 was reported to be essential for high-glucose-induced HCC carcinogenesis in mice (Zhang et al., 2017). Given the high frequency of YAP1 nuclear localization (nYAP1) activity regardless of the etiology, it would be interesting to examine the function of O-GlcNAcylation in maintaining YAP1 activity in iCCA patients with metabolic disturbances.

In tumor cell proliferation, the activation of purine and pyrimidine biosynthesis is crucial to supply substrates for nucleotide replication. Glutamate metabolism is fundamental for nucleotide biosynthesis. YAP1 contributes to the human iCCA development by regulating the expression of glutamine transporters, *Slc1a5*, *Slc7a5*, and Slc38a1 (Lu et al., 2021). In particular, mTORC1 activity potentiates the expression of amino acid transporters in the *mAkt-YAP1 S127A* iCCA model (Lu et al., 2021) and numerous studies have demonstrated a crosstalk between PI3K/AKT/mTORC1 and Hippo/YAP1 pathways to be crucial for cell proliferation in other cancer types (Liu et al., 2017; Xu et al., 2021). These observations suggest pathologic contributions of Hippo/YAP1-dependent glutamate metabolism in human iCCA.

Lipid metabolism is also essential to generating an energy source and membrane components for iCCA cells (Liu et al., 2019; Yang et al., 2015). Importantly, several studies have shown a beneficial effect of lipid-lowering in iCCA (Kitagawa et al., 2020; Yang et al., 2015). Liver-specific deletion of Hippo components (*Mst1/2* or *Lats1/2*) results in excessive lipid accumulation partly by YAP1-dependent *de novo* lipogenesis (Nguyen-Lefebvre et al., 2021), which eventually leads to HCC and/or iCCA, suggesting that Hippo/YAP1 may contribute to liver cancer development through modulation of hepatic lipid metabolism. There is more evidence of role of lipid metabolism-mediated YAP1 signaling in cancer as well. The inhibition of Stearoyl-CoA-desaturase 1, the enzyme involved in monounsaturated fatty acids synthesis, decreases expression, stabilization, nuclear translocation, and transcriptional activity of YAP1/TAZ, thereby regulating the stemness of lung cancer (Noto et al., 2017). Further, lymph node metastasis was also shown to require YAP1-dependent induction of fatty acid oxidation. Moreover, mevalonate pathway was required for the activation of Rho GTPase, which activates YAP1/TAZ by inhibiting its phosphorylation and inducing its nuclear translocation (Sorrentino et al., 2014). Interestingly, in iCCA cell, FDA-approved statins which are inhibitors of the rate-limiting enzyme β-Hydroxy β-methylglutaryl-CoA reductase in the mevalonate pathway, suppressed cell growth by inhibiting YAP1 nuclear translocation and TEAD transcriptional activity (Kitagawa et al., 2020). Considering the significance of distinct liver functions of YAP1 in lipid metabolism such as in ketogenesis, cholesterol biosynthesis, or bile acid metabolism, it may impact iCCA pathology via regulating any of these processes.

Tumor cells also secrete metabolites responsible for signal transduction or epigenetic regulation in a cell-based intrinsic manner. D-2-hydroxyglutarate (D-2HG) is one of the most prevalent oncometabolites with limited quantity in normal tissues but increased in the sera or tissue of patients harboring diverse cancers (Dang et al., 2010; DiNardo et al., 2013; Yan et al., 2009). The aberrant accumulation of this D-2HG results from conversion of α-ketoglutarate (α-KG) to D-2HG by mutated *Isocitrate Dehydrogenase (IDH)1* or *IDH2*. The increase in level of D-2HG can alter epigenome through prolylhydroxylation of 5-methyl cytosine in DNA and inhibition of histone lysine demethylases (Figueroa et al., 2010). In iCCA, 15–20% of patients harbor *IDH1/2* missense mutations, hence IDH1/2 specific inhibitor has been approved as first line therapy in selected patients (Razumilava & Gores, 2014). In contrast, the *IDH1/2* mutation has not been identified in clinical HCC, suggesting tumor-specificity. It was shown that the aberrant accumulation of D-2HG not only promotes tumorigenesis in cholangiocyte but also inhibits Hepatocyte Nuclear Factor 4 Alpha to block hepatocyte differentiation (Saha et al., 2014). Despite numerous studies demonstrating important roles for Hippo/YAP1 in *IDH1/2* mutation dependent pathologic events in iCCA, the detailed underlying molecular mechanism of YAP1 in iCCA tumor harboring *IDH1/2* mutation is yet to be fully understood.

Studies regarding YAP1 signaling in *IDH*-mutated glioma were recently reported. In low-grade glioma (LGG) patients with *IDH* mutation, the expression of *TEAD4* was higher due to high copy number variation frequency of its genomic locus. Moreover, higher *TEAD4* expression level significantly correlated with poor prognosis in LGG patient cohort (Yuan et al., 2021). The promoter of *LATS2* is hypermethylated in almost all *IDH*-mutated LGG clinical samples and may suggest *LATS2* downregulation by hypermethylation can activate YAP1 signaling (Gu et al., 2020). In addition, YAP1-regulation and its relationship with telomerase were shown in human glioma cell line to have an impact on cellular physiology, such as reactive oxygen species stress, mitochondrial fragmentation, or apoptosis, (Patrick, Gowda, Lathoria, Suri, & Sen, 2021).

Under hypoxic conditions in solid tumors, including iCCA (Vanichapol, Leelawat, & Hongeng, 2015), glutamine is a major source for intermediates of the tricarboxylic acid cycle to support cancer cell growth. Glutamine is transported into the cell by SLC1A5 which is up-regulated by YAP1/TAZ (Edwards et al., 2017) and converted into α-KG by Glutaminase 1, Glutamic-Oxaloacetic Transaminase 1, and Phosphoserine Aminotransferase 1, whose expressions are also governed by the YAP1-TEAD complex (Edwards et al., 2017; Yang et al., 2018). In turn, α-KG is converted into D-2HG by mutated *IDH1/2*, thereby inducing the epigenetic alterations in tumor cells. Taken together, the evidence supports that YAP1 can contribute to the production of α-KG in iCCA with *IDH1/2* mutation, which leads to an increase in D-2HG and triggers epigenetic alteration. Given these roles of YAP1 signaling in other malignancies with *IDH* mutations, the regulation of YAP1 signaling in iCCA with *IDH1/2* mutation should be substantially investigated in terms of metabolic alterations, biliary maturity, and epigenetic consequences.

3.3 iCCA immunology

Several studies have reported Hippo/YAP1 signaling to regulate immune microenvironment in hepatic diseases. It was shown that *Mst1*-deleted mice were more susceptible to induce a phenotype of autoimmune disease in the liver compared to wild-types (Du et al., 2014). Moreover, the expression level of *Interferon beta 1* was significantly higher in the liver of $YAP1^{+/-}$ mice than $YAP1^{+/+}$ mice. The liver of $YAP1^{+/-}$ mice was also shown to potentiate antiviral responses (Wang et al., 2017). In addition, a relationship between YAP1 signaling pathway and liver inflammation has been identified. It was revealed that YAP1 was activated by phosphorylation at Y357 in primary hepatocytes that were stimulated with IL-6, and in the regenerating livers after undergoing a partial hepatectomy (Taniguchi et al., 2015).

In HCC tumorigenesis, Hippo signaling in hepatocytes suppresses macrophage infiltration during the establishment of a protumoral milieu during HCC tumorigenesis. This occurs via inhibition of YAP1-dependent *monocyte chemoattractant protein 1* and *C-C motif chemokine ligand 2 (Ccl2)* expression (Kim et al., 2018). In addition, *Ccl2* was identified as a direct target of YAP1-TEAD in a tumor-initiating cell of HCC using ChIP-seq analysis. YAP1-dependent *Ccl2* upregulation attracts tumor-associated macrophages, which are essential for tumorigenesis (Guo et al., 2017). There was correlation between YAP1 activation or overexpression and the increase in regulatory T cells in HCCs. It was also assessed *in vitro* that increased YAP1 signaling promotes regulatory T cell differentiation via transcriptional enhancement of *TGFBR2*, thereby resulting in immunosuppression in HCC (Fan et al., 2017).

One of pivotal factors that mediates immunosuppression via TME is extracellular matrix (ECM) stiffening. Cancer-associated fibroblasts (CAFs) in the tumor stroma are in charge of synthesis of collagen and other ECM proteins, as well as the enzymes essential for their remodeling, thereby contributing to the generation of a mechanically aberrant TME (Zanconato, Cordenonsi, & Piccolo, 2016). YAP1 is an active inducer of CAF function, while ECM stiffening activates YAP1 via Src, thereby restructuring the ECM via a positive feedback loop (Calvo et al., 2013). Similarly, YAP1 signaling in hepatocytes contributes to HCC tumorigenesis through its mechanosensing role. Increased stiffness of ECM inhibits the interaction of YAP1/TAZ with SWI/SNF complex, resulting in binding of YAP1/TAZ to TEAD. Hence, it induces transcriptional reprogramming, cell proliferation, cancer stem cell traits, and plasticity (Chang et al., 2018).

Even though there is evidence of the direct relationship between Hippo/YAP1 signaling and TME, its role in the modulation of immunity in iCCA remains to be a matter of further and careful investigation.

4. Clinical relevance and therapeutic potential of modulation of Hippo-YAP1 pathway in ICCA

Based on the many important roles of Hippo/YAP1 signaling in biliary commitment and oncogenicity, multiple studies have investigated the role of this signaling pathway in clinical HCC and CCA. Except a few publications, majority of studies have shown a high prevalence of nYAP1 ranging from 64% to 98% in CCA patients, and demonstrated a significant association between nYAP1/TAZ and poor prognosis or shorter patient survival times (Marti et al., 2015; Sugihara et al., 2019; Van Haele et al., 2019). For example, the Tchorz group reported 85% of nYAP1 positivity in 107 CCA cases and identified proangiogenic microfibrillar-associated protein 5 (MFAP5) as a direct YAP1/TEAD target gene which contributes to vasculogenesis (Marti et al., 2015). Similarly, another study reported 94% of nYAP1 in 90 CCA patients, significantly correlated with worse prognosis (Pei et al., 2015). Mimori's laboratory detected nYAP1 in only 31.8% in a Japanese iCCA patient cohort. Interestingly, they discovered that 47.7% of iCCA patients displayed reduced expression of the Hippo kinase MOB1 in the population with nYAP1, suggesting the roles for MOB1 in YAP1 activation in this iCCA cohort (Sugimachi et al., 2017).

Given the recent studies describing the distinct roles for YAP1 and TAZ in the liver, it would be interesting to examine TAZ nuclear localization (nTAZ) expression from same cohort to clarify the observed discrepancy between the prevalence of MOB1 and nYAP1. Indeed, studies addressed the heterogeneous expression pattern of YAP1 and TAZ in liver cancer (Liu et al., 2021). Toth et al. reported both nYAP1 and nTAZ to be detectable only in 13%, whereas nYAP1 or nTAZ alone were positive in 60% and 48% of 152 iCCA patients respectively, implying both overlapping and distinct roles in human iCCA (Toth et al., 2021). Roskam's laboratory also showed that both nYAP1 and nTAZ were detected in all of iCCA and mixed HCC/iCCA tumor samples, while 65.9% and 23.4% of HCC patients were found to be positive for nYAP1 and nTAZ, respectively, suggesting the dominance of YAP1/TAZ activation in CCA compared to HCC (Van Haele et al., 2019). Similarly, Wu et al. (2016) also reported higher YAP1 expression in CCA (N=122) compared to HCC (N=137), which was associated with tumor size, vascular invasion and intrahepatic metastasis (Wu et al., 2016). Overall, these studies support the clinical relevance of studying Hippo/YAP1 signaling in human iCCA, it is also clear that additional studies are likely needed to carefully examine the limitations raised below.

Specifically, several groups have raised questions about discrepancy between nYAP1/TAZ and their transcriptional activity in iCCA cells as assessed by routine IHC staining (Sugihara et al., 2019). This possibility suggests that nYAP1 staining alone may be of limited value as a surrogate of YAP1 activity and requires co-staining of valid YAP1 target genes including Cellular Communication Network Factor 2 (CTGF/CCN2) and Cellular Communication Network Factor 1 (CYR61/CCN1) in human iCCA. There is a further need to determine cognate markers to reflect true YAP1 activity, especially in clinical iCCA samples. In this regard, although aberrant Hippo/YAP1 signaling has been defined as iCCA driver in animal studies (Liu, Zhuo, et al., 2021; Yimlamai et al., 2015), its role in fully developed clinical iCCA remain elusive. In particular, several studies have demonstrated a synergistic effect of YAP1 inhibition in combination with Gemcitabine/Cisplatin treatment, while the effect of YAP1 singular inhibition was not significant (Sugihara et al., 2019). These observations appear to suggest a unique role of YAP1 in chemically-injured cholangiocarcinoma cells and the existence of a distinct driver modulating the Hippo/YAP1 cascade in chemical injury-induced iCCA. Indeed, the mutation rate of Hippo kinase is extremely low in iCCA (around 5%) (Sugihara et al., 2019), thus suggesting transient and context dependent upstream regulation of the Hippo/YAP1 pathway at diverse stages of clinical iCCA pathogenesis.

As already noted, *TNF superfamilt member 10 (TRAIL)* and *MFAP5* have been identified as downstream targets of active YAP1 in human iCCA (Marti et al., 2015). However, evidence for upstream regulators for Hippo/YAP1 is limited, likely due to the difficulty of mimicking a diverse pathophysiologic environment to which Hippo responds (Nguyen-Lefebvre et al., 2021). A heterogeneous and complex iCCA microenvironment is extremely difficult to recapitulate *in vivo* especially in xenograft models using immunocompromised mice. Employing an *in vivo* system, Gores' laboratory identified Fibroblast Growth Factor (FGF), Platelet derived growth factor receptor (PDGFR), and LCK-SRC family kinase signaling (Sugihara et al., 2019) as Hippo-independent YAP1 activators in human iCCA, implying diverse mechanisms of YAP1 activation in human iCCA pathogenesis.

Additional studies using *in vivo* systems closely reflecting the human disease are essential to translate these finding to the bedside.

Given that active Hippo kinases degrade YAP1 and the low mutational rates of Hippo components in human iCCA (Sugihara et al., 2019; Yimlamai et al., 2015), therapeutic approaches have focused on the inhibition of transcriptional activity of YAP1-TEAD complex. Table 2 summarizes studies, describing the effect of Hippo/YAP modulation in diverse CCA models. These trials are based on the hypothesis that YAP1-TEAD transcription is critical for survival of iCCA cells and blockade of YAP1-TEAD is sufficient to reduce YAP1$^+$ iCCA burden *in vivo*. However, the therapeutic benefit of chemical/genetic YAP1 inhibition in iCCA has to date only been shown to be modest and often leads to tumor relapse *in vivo*. For now, efficacy of blockade of YAP1-TEAD transcription in fully developed iCCA cell viability remains questionable. Classical YAP1-TEAD binding inhibitor Vertepofrin has been tested in several studies with debatable results possibly due to adequate efficacy and/or *in vivo* toxicity of this drug (Sugiura et al., 2019). To this end, based on their validated potency, newly developed diverse set of TEAD inhibitors are expected to be tested in various iCCA models (Holden et al., 2020; Li et al., 2020).

Additionally, tumor intrinsic adaptive resistance mechanism replenishing YAP1-TEAD target gene transcription need to also be comprehensively investigated in an *in vivo* setting. A fundamental question is whether permanent blockade of TEAD transcription (independent of YAP1) is sufficient to reduce human iCCA burden thereby prolonging the lifespan of patients with advanced iCCA. In various malignancies, TAZ, the paralogue of YAP1, has been shown to play a redundant role in the absence YAP1. *YAP1* or *TAZ* singular deletion mildly delays HCC formation, whereas simultaneous co-deletion of *YAP1* and *TAZ* completely eliminates *mAkt-NRAS proto-oncogene, GTPase (NRas)*-driven HCC development in mouse (Wang et al., 2021). In addition, our group revealed that TEAD inhibition using dominant-negative *TEAD* prevents Notch-dependent hepatocyte-originated iCCA development, supporting the concept of YAP1-TAZ compensation via forming a complex with TEAD (Hu et al., 2020). This observation suggests the potential of testing the newly developed TEAD palmitoylation inhibitor MGH-CP1, which acts independent of YAP1 as described by Li et al. (2020) against iCCA. Intriguingly, we also found that co-deletion of *Sox9* and *YAP1* completely prevents *mAkt-NICD*-iCCA formation, indicating SOX9-YAP1 transcriptional compensation under Notch during hepatocyte-driven cholangiocarcinogenesis (Hu et al., 2020).

Table 2 Summary of the effect for direct/indirect Hippo/YAP1 modulation in diverse iCCA models.

	CCA models	Upstream Hippo/YAP regulator	Hippo/YAP modulation	Downstream effector of YAP	Effect on iCCA	Refs.
In vitro	HuCCT1	NA	YAP1 S127 over expression	MFAP5 and CD31	Angiogenesis	Marti et al. (2015)
	HCCC-9810 and RBE	PES1	YAP1 mRNA expression	NA	Tumor cell growth	Xu et al. (2021)
	RBE and FRH0201	circACTN4	YAP1 mRNA & protein ↓	miR-424-5p & FZD7 transcription → β-Catenin & YAP1 interaction ↑	Tumor cell growth	Chen et al. (2022)
	SNU1196	HDAC inhibitor, CG200745 → miR-509-3p	YAP1 protein ↓	NA	Tumor cell growth	Jung, Park, Kim, Kim, and Song (2017)
	HUCCT1	MED1 and CDK9		NA	Tumor cell growth	Galli et al. (2015)
	CCLP1 and SG231	G9a	LATS2↓→ p-YAP1^{S127} ↑	p-MYPT^{S445}	Growth of CCA cells	Ma et al. (2020)
In vivo, xenograft model	HuCCT-1 and TFK-1, tail vein injection	PIEZO-1	p-LATS 1 ↓→ p-YAP1^{S127} ↑	E-CADHEREIN, N-CADHEREIN and VIMENTIN	Metastasis	Zhu, Qian, Han, Bai, and Hou (2021)
	KMCH xenograft	SHP2 KO	p-YAP1^{Y357} ↑	MCL1	Chemoresistance (gemcitabine and cisplatin)	Buckarma et al. (2020)
	HCCC9810 and QBC939 xenografts	Gankyrin	YAP1 overexpression	miR-29c, IGF-1, AKT and GANKRYN	Tumorigenesis and metastasis (EMT)	Pei et al. (2015)
	KMCH, KMBC & PDX	FGFR signaling (FGFR1,2,4)	Feed-forward loop: FGFR signaling inhibition → p-YAP1^{S127} ↑)	MCL1	Tumor cell growth and survival	Rizvi et al. (2016)

Continued

Table 2 Summary of the effect for direct/indirect Hippo/YAP1 modulation in diverse iCCA models.—cont'd

CCA models	Upstream Hippo/YAP regulator	Hippo/YAP modulation	Downstream effector of YAP	Effect on iCCA	Refs.
HUCCT-1 and SB-1, xenografts	PDGFR signaling → SFKs	p-YAP1^{Y357} ↑	MCL1	Tumor cell growth and survival	Smoot et al. (2018)
HuCCT-1, xenograft	Verteporfin	YAP1-TEAD inhibition	IL-6-STAT3 Signaling and OCT4	Tumor cell growth and stemness	Sugiura et al. (2019)
TetO-YAP1^{S127A}	NA	p-YAP1^{S127} ↑	NUAK2	Tumor cell growth and survival	Yuan et al. (2018)
mAkt-NICD, HDTVI	G9a inhibitor, UNC0642	LATS2 ↑ → p-YAP1^{S127} ↓	P-MYPT1^{S445}	Tumor cell growth and survival	Ma et al. (2020)
HuCCT-1 and QBC-939, S.C injection	lncRNA-HPR	nYAP1 ↑	NA	Tumor cell growth and survival	Zhang et al. (2021)
HuCCT-1, xenograft	lncRNA-PAICC-miR-141-3p/27a-3p	nYAP1 ↓	NA	Tumor cell growth and survival	Xia et al. (2020)
HuCCT-1, xenograft	SULF2 → PDGFR	p-YAP1Y357 ↑	MCL1 and CCND1	Tumor cell growth and survival	Luo et al. (2021)
HuCCT-1, xenograft	Gemcitabine and atorvastatin	nYAP1 ↓	CTGF, CYR61, ANKRD1 and MFAP5	Tumor cell growth and survival	Kitagawa et al. (2020)
SNU-1196, xenograft	HDAC inhibitor, CG200745 → miR-509-3p	YAP1 and TEAD4 protein ↓	NA	Tumor cell growth and survival	Jung et al. (2017)

In vivo, genetic model	NICD-YAP1^{S127A}, HDTVI	CMV-Cre (HDTVI)-Raptor$^{fl/fl}$, conditional KO	NA	NA	Tumorigenesis	Lu et al. (2021)
	mAkt-NRas^{V12}, HDTVI	DN-TEAD, HDTVI	YAP1-TEAD inhibition	NA	Tumorigenesis and reduce iCCA region	Zhang et al. (2018)
		Lats2, HDTVI	p-YAP1^{S127} ↓	NA	Tumorigenesis and reduce iCCA region	
	mAkt-YAP1^{S127A} (biliary transfection) + BDL + IL-33	NA	p-YAP1^{S127} ↑	Chromosome instability↑ → Foxm1↑	Tumorigenesis and proliferation	Rizvi et al. (2018)
	NICD-YAP1^{S127A}, HDTVI	CMV-Cre (HDTVI)-Fak(fl/fl), conditional KO	p-YAP1^{Y357} ↓	NA	Tumorigenesis	Song et al. (2021)
		CreERT2 (HDTVI)-Fak(fl/fl), conditional and inducible KO	p-YAP1^{Y357} ↓	NA	Tumor cell growth and survival	
	mAkt-Fbxw7^{ΔF}, HDTVI	CMV-Cre (HDTVI)-YAP1(fl/fl), conditional KO	YAP1 KO	NA	Tumorigenesis	Wang et al. (2019)
	NICD-YAP1^{S127A}, HDTVI	pan-mTOR inhibitor, MLN0128	NA	NA	Tumor cell survival	Zhang, Song, et al. (2017)

Again, these findings would need further and careful validation. In summary, the role of YAP1 and TAZ as therapeutic targets in advanced iCCA requires further studies which may allow eventual translation into the clinic.

5. Concluding remarks

YAP1 activation is higly prevalent in clinical iCCA, and aberrant Hippo-YAP1 signaling in the liver is sufficient to lead to iCCA development in mouse models, suggesting its essential role in iCCA pathophysiology. Despite extensive research exploring Hippo-YAP1 signaling in iCCA, therapeutic intervention targeting this important pathway remains far from reaching its full potential for treatment of this lethal tumor. This signaling cascade is complicated as it interacts with a variety of regulatory signaling networks, while responding to mechano-sensory stimuli, epigenetic factors, and microenvironmental cues, to in turn regulating malignant transformation, TME and other aspects critical for tumor maintenance. Furthermore, while the mutational rate in the components of canonical Hippo-YAP1 signaling are low in human iCCA, several non-canonical regulators of this signaling cascade have been identified, making it cumbersome to target specific and most relevant upstream regulators. Thus, substantial understanding of context-dependent and truly functional activation of this pathway, identification of the missing nuclear co-regulators, and binding partners that are crucial for crosstalk with other pathways will be essential for effectively targeting and treating YAP1-activated human iCCA and pivotal for translating the most relevant preclinical observations to the bedside.

Acknowledgments

Funding was provided by NIH grant 1R01CA258449 and PLRC Pilot & Feasibility grant PF 2019-05 to S.K through 1P30DK120531, and by NIH grants R01CA204586, R01CA251155, R01CA250227, and Endowed Chair for Experimental Pathology to S.P.M. and NIH grant 1P30DK120531 to the Pittsburgh Liver Research Center.

Conflict of interest statement

None of the authors have any interests to declare related to this study.

References

Bai, H., Zhang, N., Xu, Y., Chen, Q., Khan, M., Potter, J. J., et al. (2012). Yes-associated protein regulates the hepatic response after bile duct ligation. *Hepatology*, 56(3), 1097–1107. https://doi.org/10.1002/hep.25769.

Brindley, P. J., Bachini, M., Ilyas, S. I., Khan, S. A., Loukas, A., Sirica, A. E., et al. (2021). Cholangiocarcinoma. *Nature Reviews Disease Primers*, 7(1), 65. https://doi.org/10.1038/s41572-021-00300-2.

Buckarma, E. H., Werneburg, N. W., Conboy, C. B., Kabashima, A., O'Brien, D. R., Wang, C., et al. (2020). The YAP-interacting phosphatase SHP2 can regulate transcriptional coactivity and modulate sensitivity to chemotherapy in cholangiocarcinoma. *Molecular Cancer Research*, *18*(10), 1574–1588. https://doi.org/10.1158/1541-7786.MCR-20-0165.

Calvisi, D. F., Wang, C., Ho, C., Ladu, S., Lee, S. A., Mattu, S., et al. (2011). Increased lipogenesis, induced by AKT-mTORC1-RPS6 signaling, promotes development of human hepatocellular carcinoma. *Gastroenterology*, *140*(3), 1071–1083. https://doi.org/10.1053/j.gastro.2010.12.006.

Calvo, F., Ege, N., Grande-Garcia, A., Hooper, S., Jenkins, R. P., Chaudhry, S. I., et al. (2013). Mechanotransduction and YAP-dependent matrix remodelling is required for the generation and maintenance of cancer-associated fibroblasts. *Nature Cell Biology*, *15*(6), 637–646. https://doi.org/10.1038/ncb2756.

Cardinale, V., Bragazzi, M. C., Carpino, G., Di Matteo, S., Overi, D., Nevi, L., et al. (2018). Intrahepatic cholangiocarcinoma: review and update. *Hepatoma Research*, *4*(6). https://doi.org/10.20517/2394-5079.2018.46.

Chang, L., Azzolin, L., Di Biagio, D., Zanconato, F., Battilana, G., Lucon Xiccato, R., et al. (2018). The SWI/SNF complex is a mechanoregulated inhibitor of YAP and TAZ. *Nature*, *563*(7730), 265–269. https://doi.org/10.1038/s41586-018-0658-1.

Chen, Y. C., Chien, C. Y., Hsu, C. C., Lee, C. H., Chou, Y. T., Shiah, S. G., et al. (2021). Obesity-associated leptin promotes chemoresistance in colorectal cancer through YAP-dependent AXL upregulation. *American Journal of Cancer Research*, *11*(9), 4220–4240. https://www.ncbi.nlm.nih.gov/pubmed/34659884.

Chen, Q., Wang, H., Li, Z., Li, F., Liang, L., Zou, Y., et al. (2022). Circular RNA ACTN4 promotes intrahepatic cholangiocarcinoma progression by recruiting YBX1 to initiate FZD7 transcription. *Journal of Hepatology*, *76*(1), 135–147. https://doi.org/10.1016/j.jhep.2021.08.027.

Chen, Y. L., Yen, I. C., Lin, K. T., Lai, F. Y., & Lee, S. Y. (2020). 4-Acetylantrocamol LT3, a new ubiquinone from antrodia cinnamomea, inhibits hepatocellular carcinoma HepG2 cell growth by targeting YAP/TAZ, mTOR, and WNT/beta-catenin signaling. *The American Journal of Chinese Medicine*, *48*(5), 1243–1261. https://doi.org/10.1142/S0192415X20500615.

Ciraku, L., Esquea, E. M., & Reginato, M. J. (2022). O-GlcNAcylation regulation of cellular signaling in cancer. *Cellular Signaling*, *90*, 110201. https://doi.org/10.1016/j.cellsig.2021.110201.

Clements, O., Eliahoo, J., Kim, J. U., Taylor-Robinson, S. D., & Khan, S. A. (2020). Risk factors for intrahepatic and extrahepatic cholangiocarcinoma: A systematic review and meta-analysis. *Journal of Hepatology*, *72*(1), 95–103. https://doi.org/10.1016/j.jhep.2019.09.007.

Dang, L., White, D. W., Gross, S., Bennett, B. D., Bittinger, M. A., Driggers, E. M., et al. (2010). Cancer-associated IDH1 mutations produce 2-hydroxyglutarate. *Nature*, *465*(7300), 966. https://doi.org/10.1038/nature09132.

DeRan, M., Yang, J., Shen, C. H., Peters, E. C., Fitamant, J., Chan, P., et al. (2014). Energy stress regulates hippo-YAP signaling involving AMPK-mediated regulation of angiomotin-like 1 protein. *Cell Reports*, *9*(2), 495–503. https://doi.org/10.1016/j.celrep.2014.09.036.

Di Matteo, S., Nevi, L., Overi, D., Landolina, N., Faccioli, J., Giulitti, F., et al. (2021). Metformin exerts anti-cancerogenic effects and reverses epithelial-to-mesenchymal transition trait in primary human intrahepatic cholangiocarcinoma cells. *Scientific Reports*, *11*(1), 2557. https://doi.org/10.1038/s41598-021-81172-0.

DiNardo, C. D., Propert, K. J., Loren, A. W., Paietta, E., Sun, Z., Levine, R. L., et al. (2013). Serum 2-hydroxyglutarate levels predict isocitrate dehydrogenase mutations and clinical outcome in acute myeloid leukemia. *Blood*, *121*(24), 4917–4924. https://doi.org/10.1182/blood-2013-03-493197.

Driskill, J. H., & Pan, D. (2021). The Hippo pathway in liver homeostasis and pathophysiology. *Annual Review of Pathology: Mechanisms of Disease, 16*, 299–322. https://doi.org/10.1146/annurev-pathol-030420-105050.

Du, X., Shi, H., Li, J., Dong, Y., Liang, J., Ye, J., et al. (2014). Mst1/Mst2 regulate development and function of regulatory T cells through modulation of Foxo1/Foxo3 stability in autoimmune disease. *Journal of Immunology, 192*(4), 1525–1535. https://doi.org/10.4049/jimmunol.1301060.

Edwards, D. N., Ngwa, V. M., Wang, S., Shiuan, E., Brantley-Sieders, D. M., Kim, L. C., et al. (2017). The receptor tyrosine kinase EphA2 promotes glutamine metabolism in tumors by activating the transcriptional coactivators YAP and TAZ. *Science Signaling, 10*(508). https://doi.org/10.1126/scisignal.aan4667.

Fan, S., Gao, Y., Qu, A., Jiang, Y., Li, H., Xie, G., et al. (2022). YAP-TEAD mediates PPAR alpha-induced hepatomegaly and liver regeneration in mice. *Hepatology, 75*(1), 74–88. https://doi.org/10.1002/hep.32105.

Fan, Y., Gao, Y., Rao, J., Wang, K., Zhang, F., & Zhang, C. (2017). YAP-1 promotes tregs differentiation in hepatocellular carcinoma by enhancing TGFBR2 transcription. *Cellular Physiology and Biochemistry, 41*(3), 1189–1198. https://doi.org/10.1159/000464380.

Fan, B., Malato, Y., Calvisi, D. F., Naqvi, S., Razumilava, N., Ribback, S., et al. (2012). Cholangiocarcinomas can originate from hepatocytes in mice. *Journal of Clinical Investigation, 122*(8), 2911–2915. https://doi.org/10.1172/JCI63212.

Figueroa, M. E., Abdel-Wahab, O., Lu, C., Ward, P. S., Patel, J., Shih, A., et al. (2010). Leukemic IDH1 and IDH2 mutations result in a hypermethylation phenotype, disrupt TET2 function, and impair hematopoietic differentiation. *Cancer Cell, 18*(6), 553–567. https://doi.org/10.1016/j.ccr.2010.11.015.

Gadd, V. L., Aleksieva, N., & Forbes, S. J. (2020). Epithelial plasticity during liver injury and regeneration. *Cell Stem Cell, 27*(4), 557–573. https://doi.org/10.1016/j.stem.2020.08.016.

Galli, G. G., Carrara, M., Yuan, W. C., Valdes-Quezada, C., Gurung, B., Pepe-Mooney, B., et al. (2015). YAP drives growth by controlling transcriptional pause release from dynamic enhancers. *Molecular Cell, 60*(2), 328–337. https://doi.org/10.1016/j.molcel.2015.09.001.

Gera, S., Ettel, M., Acosta-Gonzalez, G., & Xu, R. (2017). Clinical features, histology, and histogenesis of combined hepatocellular-cholangiocarcinoma. *World Journal of Hepatology, 9*(6), 300–309. https://doi.org/10.4254/wjh.v9.i6.300.

Gu, Y., Wang, Y., Wang, Y., Luo, J., Wang, X., Ma, M., et al. (2020). Hypermethylation of LATS2 promoter and its prognostic value in IDH-mutated low-grade gliomas. *Frontiers in Cell and Developmental Biology, 8*, 586581. https://doi.org/10.3389/fcell.2020.586581.

Guo, X., Zhao, Y., Yan, H., Yang, Y., Shen, S., Dai, X., et al. (2017). Single tumor-initiating cells evade immune clearance by recruiting type II macrophages. *Genes and Development, 31*(3), 247–259. https://doi.org/10.1101/gad.294348.116.

Han, X., Wang, Y., Pu, W., Huang, X., Qiu, L., Li, Y., et al. (2019). Lineage tracing reveals the bipotency of SOX9(+) hepatocytes during liver regeneration. *Stem Cell Reports, 12*(3), 624–638. https://doi.org/10.1016/j.stemcr.2019.01.010.

Holden, J. K., Crawford, J. J., Noland, C. L., Schmidt, S., Zbieg, J. R., Lacap, J. A., et al. (2020). Small molecule dysregulation of TEAD lipidation induces a dominant-negative inhibition of Hippo pathway signaling. *Cell Reports, 31*(12), 107809. https://doi.org/10.1016/j.celrep.2020.107809.

Hu, J. K., Du, W., Shelton, S. J., Oldham, M. C., DiPersio, C. M., & Klein, O. D. (2017). An FAK-YAP-mTOR signaling axis regulates stem cell-based tissue renewal in mice. *Cell Stem Cell, 21*(1), 91–106. e106 https://doi.org/10.1016/j.stem.2017.03.023.

Hu, S., Molina, L., Tao, J., Liu, S., Hassan, M., Singh, S., et al. (2020). NOTCH-YAP1/TEAD-DNMT1 axis regulates hepatocyte reprogramming into intrahepatic cholangiocarcinoma. *bioRxiv*. https://doi.org/10.1101/2020.12.03.410993.

Jung, K., Kim, M., So, J., Lee, S. H., Ko, S., & Shin, D. (2021). Farnesoid X receptor activation impairs liver progenitor cell-mediated liver regeneration via the PTEN-PI3K-AKT-mTOR axis in zebrafish. *Hepatology*, 74(1), 397–410. https://doi.org/10.1002/hep.31679.

Jung, D. E., Park, S. B., Kim, K., Kim, C., & Song, S. Y. (2017). CG200745, an HDAC inhibitor, induces anti-tumour effects in cholangiocarcinoma cell lines via miRNAs targeting the Hippo pathway. *Scientific Reports*, 7(1), 10921. https://doi.org/10.1038/s41598-017-11094-3.

Kendall, T., Verheij, J., Gaudio, E., Evert, M., Guido, M., Goeppert, B., et al. (2019). Anatomical, histomorphological and molecular classification of cholangiocarcinoma. *Liver International*, 39(Suppl. 1), 7–18. https://doi.org/10.1111/liv.14093.

Kim, M. K., Jang, J. W., & Bae, S. C. (2018). DNA binding partners of YAP/TAZ. *BMB Reports*, 51(3), 126–133. https://doi.org/10.5483/bmbrep.2018.51.3.015.

Kim, E., Kang, J. G., Kang, M. J., Park, J. H., Kim, Y. J., Kweon, T. H., et al. (2020). O-GlcNAcylation on LATS2 disrupts the Hippo pathway by inhibiting its activity. *Proceedings of the National Academy of Sciences of the United States of America*, 117(25), 14259–14269. https://doi.org/10.1073/pnas.1913469117.

Kim, W., Khan, S. K., Gvozdenovic-Jeremic, J., Kim, Y., Dahlman, J., Kim, H., et al. (2017). Hippo signaling interactions with Wnt/beta-catenin and Notch signaling repress liver tumorigenesis. *Journal of Clinical Investigation*, 127(1), 137–152. https://doi.org/10.1172/JCI88486.

Kim, W., Khan, S. K., Liu, Y., Xu, R., Park, O., He, Y., et al. (2018). Hepatic Hippo signaling inhibits protumoural microenvironment to suppress hepatocellular carcinoma. *Gut*, 67(9), 1692–1703. https://doi.org/10.1136/gutjnl-2017-314061.

Kitagawa, K., Moriya, K., Kaji, K., Saikawa, S., Sato, S., Nishimura, N., et al. (2020). Atorvastatin augments gemcitabine-mediated anti-cancer effects by inhibiting yes-associated protein in human cholangiocarcinoma cells. *International Journal of Molecular Sciences*, 21(20). https://doi.org/10.3390/ijms21207588.

Ko, S., Russell, J. O., Molina, L. M., & Monga, S. P. (2019). Liver progenitors and adult cell plasticity in hepatic injury and repair: Knowns and unknowns. *Annual Review of Pathology: Mechanisms of Disease*, 15, 23–50. https://doi.org/10.1146/annurev-pathmechdis-012419-032824.

Lee, K. P., Lee, J. H., Kim, T. S., Kim, T. H., Park, H. D., Byun, J. S., et al. (2010). The Hippo-Salvador pathway restrains hepatic oval cell proliferation, liver size, and liver tumorigenesis. *Proceedings of the National Academy of Sciences of the United States of America*, 107(18), 8248–8253. https://doi.org/10.1073/pnas.0912203107.

Li, H., Fu, L., Liu, B., Lin, X., Dong, Q., & Wang, E. (2019). Ajuba overexpression regulates mitochondrial potential and glucose uptake through YAP/Bcl-xL/GLUT1 in human gastric cancer. *Gene*, 693, 16–24. https://doi.org/10.1016/j.gene.2019.01.018.

Li, Q., Sun, Y., Jarugumilli, G. K., Liu, S., Dang, K., Cotton, J. L., et al. (2020). Lats1/2 sustain intestinal stem cells and Wnt Activation through TEAD-dependent and independent transcription. *Cell Stem Cell*, 26(5), 675–692 e678. https://doi.org/10.1016/j.stem.2020.03.002.

Li, X., Tao, J., Cigliano, A., Sini, M., Calderaro, J., Azoulay, D., et al. (2015). Co-activation of PIK3CA and Yap promotes development of hepatocellular and cholangiocellular tumors in mouse and human liver. *Oncotarget*, 6(12), 10102–10115. https://doi.org/10.18632/oncotarget.3546.

Li, X., Wu, Z., He, J., Jin, Y., Chu, C., Cao, Y., et al. (2021). OGT regulated O-GlcNAcylation promotes papillary thyroid cancer malignancy via activating YAP. *Oncogene*, *40*(30), 4859–4871. https://doi.org/10.1038/s41388-021-01901-7.

Lin, C., & Xu, X. (2017). YAP1-TEAD1-Glut1 axis dictates the oncogenic phenotypes of breast cancer cells by modulating glycolysis. *Biomedicine and Pharmacotherapy*, *95*, 789–794. https://doi.org/10.1016/j.biopha.2017.08.091.

Liu, Z., Alsaggaf, R., McGlynn, K. A., Anderson, L. A., Tsai, H. T., Zhu, B., et al. (2019). Statin use and reduced risk of biliary tract cancers in the UK Clinical Practice Research Datalink. *Gut*, *68*(8), 1458–1464. https://doi.org/10.1136/gutjnl-2018-317504.

Liu, M., Lin, Y., Zhang, X. C., Tan, Y. H., Yao, Y. L., Tan, J., et al. (2017). Phosphorylated mTOR and YAP serve as prognostic markers and therapeutic targets in gliomas. *Laboratory Investigation*, *97*(11), 1354–1363. https://doi.org/10.1038/labinvest.2017.70.

Liu, Y., Wu, Y., Ouyang, Y. S., He, J. J., Shen, L. L., Qi, H., et al. (2021). The combination of YAP/TAZ predicts the clinical prognosis in patients with cholangiocarcinoma after radical resection. *Zhonghua Nei Ke Za Zhi*, *60*(7), 637–643. https://doi.org/10.3760/cma.j.cn112138-20210201-00091.

Liu, Y., Zhuo, S., Zhou, Y., Ma, L., Sun, Z., Wu, X., et al. (2021). Yap-Sox9 signaling determines hepatocyte plasticity and lineage-specific hepatocarcinogenesis. *Journal of Hepatology*. https://doi.org/10.1016/j.jhep.2021.11.010.

Lu, X., Peng, B., Chen, G., Pes, M. G., Ribback, S., Ament, C., et al. (2021). YAP accelerates notch-driven cholangiocarcinogenesis via mTORC1 in mice. *American Journal of Pathology*, *191*(9), 1651–1667. https://doi.org/10.1016/j.ajpath.2021.05.017.

Luo, X., Campbell, N. A., He, L., O'Brien, D. R., Singer, M. S., Lemjabbar-Alaoui, H., et al. (2021). Sulfatase 2 (SULF2) monoclonal antibody 5D5 suppresses human cholangiocarcinoma xenograft growth through regulation of a SULF2-platelet-derived growth factor receptor beta-yes-associated protein signaling axis. *Hepatology*, *74*(3), 1411–1428. https://doi.org/10.1002/hep.31817.

Ma, W., Han, C., Zhang, J., Song, K., Chen, W., Kwon, H., et al. (2020). The histone methyltransferase G9a promotes cholangiocarcinogenesis through regulation of the Hippo pathway kinase LATS2 and YAP signaling pathway. *Hepatology*, *72*(4), 1283–1297. https://doi.org/10.1002/hep.31141.

Ma, S., Meng, Z., Chen, R., & Guan, K. L. (2019). The Hippo pathway: Biology and pathophysiology. *Annual Review of Biochemistry*, *88*, 577–604. https://doi.org/10.1146/annurev-biochem-013118-111829.

Marti, P., Stein, C., Blumer, T., Abraham, Y., Dill, M. T., Pikiolek, M., et al. (2015). YAP promotes proliferation, chemoresistance, and angiogenesis in human cholangiocarcinoma through TEAD transcription factors. *Hepatology*, *62*(5), 1497–1510. https://doi.org/10.1002/hep.27992.

Merrell, A. J., Peng, T., Li, J., Sun, K., Li, B., Katsuda, T., et al. (2021). Dynamic transcriptional and epigenetic changes drive cellular plasticity in the liver. *Hepatology*, *74*(1), 444–457. https://doi.org/10.1002/hep.31704.

Miyamura, N., Hata, S., Itoh, T., Tanaka, M., Nishio, M., Itoh, M., et al. (2017). YAP determines the cell fate of injured mouse hepatocytes in vivo. *Nature Communications*, *8*, 16017. https://doi.org/10.1038/ncomms16017.

Moeini, A., Haber, P. K., & Sia, D. (2021). Cell of origin in biliary tract cancers and clinical implications. *JHEP Reports*, *3*(2), 100226. https://doi.org/10.1016/j.jhepr.2021.100226.

Molina, L. M., Zhu, J., Li, Q., Pradhan-Sundd, T., Krutsenko, Y., Sayed, K., et al. (2021). Compensatory hepatic adaptation accompanies permanent absence of intrahepatic biliary network due to YAP1 loss in liver progenitors. *Cell Reports*, *36*(1), 109310. https://doi.org/10.1016/j.celrep.2021.109310.

Moya, I. M., Castaldo, S. A., Van den Mooter, L., Soheily, S., Sansores-Garcia, L., Jacobs, J., et al. (2019). Peritumoral activation of the Hippo pathway effectors YAP and TAZ suppresses liver cancer in mice. *Science, 366*(6468), 1029–1034. https://doi.org/10.1126/science.aaw9886.

Nebbioso, A., Tambaro, F. P., Dell'Aversana, C., & Altucci, L. (2018). Cancer epigenetics: Moving forward. *PLoS Genetics, 14*(6), e1007362. https://doi.org/10.1371/journal.pgen.1007362.

Nguyen-Lefebvre, A. T., Selzner, N., Wrana, J. L., & Bhat, M. (2021). The hippo pathway: A master regulator of liver metabolism, regeneration, and disease. *The FASEB Journal, 35*(5), e21570. https://doi.org/10.1096/fj.202002284RR.

Nishio, M., Sugimachi, K., Goto, H., Wang, J., Morikawa, T., Miyachi, Y., et al. (2016). Dysregulated YAP1/TAZ and TGF-beta signaling mediate hepatocarcinogenesis in Mob1a/1b-deficient mice. *Proceedings of the National Academy of Sciences of the United States of America, 113*(1), E71–E80. https://doi.org/10.1073/pnas.1517188113.

Noto, A., De Vitis, C., Pisanu, M. E., Roscilli, G., Ricci, G., Catizone, A., et al. (2017). Stearoyl-CoA-desaturase 1 regulates lung cancer stemness via stabilization and nuclear localization of YAP/TAZ. *Oncogene, 36*(32), 4573–4584. https://doi.org/10.1038/onc.2017.75.

Pant, K., Richard, S., Peixoto, E., & Gradilone, S. A. (2020). Role of glucose metabolism reprogramming in the pathogenesis of cholangiocarcinoma. *Frontiers in Medicine (Lausanne), 7*, 113. https://doi.org/10.3389/fmed.2020.00113.

Park, J., Kim, J. S., Nahm, J. H., Kim, S. K., Lee, D. H., & Lim, D. S. (2020). WWC1 and NF2 prevent the development of intrahepatic cholangiocarcinoma by regulating YAP/TAZ activity through LATS in mice. *Molecules and Cells, 43*(5), 491–499. https://doi.org/10.14348/molcells.2020.0093.

Patrick, S., Gowda, P., Lathoria, K., Suri, V., & Sen, E. (2021). YAP1-mediated regulation of mitochondrial dynamics in IDH1 mutant gliomas. *Journal of Cell Science, 134*-(22). https://doi.org/10.1242/jcs.259188.

Pei, T., Li, Y., Wang, J., Wang, H., Liang, Y., Shi, H., et al. (2015). YAP is a critical oncogene in human cholangiocarcinoma. *Oncotarget, 6*(19), 17206–17220. https://doi.org/10.18632/oncotarget.4043.

Peng, C., Zhu, Y., Zhang, W., Liao, Q., Chen, Y., Zhao, X., et al. (2017). Regulation of the Hippo-YAP pathway by glucose sensor O-GlcNAcylation. *Molecular Cell, 68*(3), 591–604. e595 https://doi.org/10.1016/j.molcel.2017.10.010.

Phoomak, C., Silsirivanit, A., Park, D., Sawanyawisuth, K., Vaeteewoottacharn, K., Wongkham, C., et al. (2018). O-GlcNAcylation mediates metastasis of cholangiocarcinoma through FOXO3 and MAN1A1. *Oncogene, 37*(42), 5648–5665. https://doi.org/10.1038/s41388-018-0366-1.

Phoomak, C., Vaeteewoottacharn, K., Sawanyawisuth, K., Seubwai, W., Wongkham, C., Silsirivanit, A., et al. (2016). Mechanistic insights of O-GlcNAcylation that promote progression of cholangiocarcinoma cells via nuclear translocation of NF-kappaB. *Scientific Reports, 6*, 27853. https://doi.org/10.1038/srep27853.

Phoomak, C., Vaeteewoottacharn, K., Silsirivanit, A., Saengboonmee, C., Seubwai, W., Sawanyawisuth, K., et al. (2017). High glucose levels boost the aggressiveness of highly metastatic cholangiocarcinoma cells via O-GlcNAcylation. *Scientific Reports, 7*, 43842. https://doi.org/10.1038/srep43842.

Planas-Paz, L., Sun, T., Pikiolek, M., Cochran, N. R., Bergling, S., Orsini, V., et al. (2019). YAP, but not RSPO-LGR4/5, signaling in biliary epithelial cells promotes a ductular reaction in response to liver injury. *Cell Stem Cell, 25*(1), 39–53. e10 https://doi.org/10.1016/j.stem.2019.04.005.

Razumilava, N., & Gores, G. J. (2014). Cholangiocarcinoma. *Lancet*, *383*(9935), 2168–2179. https://doi.org/10.1016/S0140-6736(13)61903-0.
Reggiani, F., Gobbi, G., Ciarrocchi, A., & Sancisi, V. (2021). YAP and TAZ are not identical twins. *Trends in Biochemical Sciences*, *46*(2), 154–168. https://doi.org/10.1016/j.tibs.2020.08.012.
Rizvi, S., Fischbach, S. R., Bronk, S. F., Hirsova, P., Krishnan, A., Dhanasekaran, R., et al. (2018). YAP-associated chromosomal instability and cholangiocarcinoma in mice. *Oncotarget*, *9*(5), 5892–5905. https://doi.org/10.18632/oncotarget.23638.
Rizvi, S., Yamada, D., Hirsova, P., Bronk, S. F., Werneburg, N. W., Krishnan, A., et al. (2016). A Hippo and fibroblast growth factor receptor autocrine pathway in cholangiocarcinoma. *Journal of Biological Chemistry*, *291*(15), 8031–8047. https://doi.org/10.1074/jbc.M115.698472.
Roh, M. S., Jeong, J. S., Kim, Y. H., Kim, M. C., & Hong, S. H. (2004). Diagnostic utility of GLUT1 in the differential diagnosis of liver carcinomas. *Hepatogastroenterology*, *51*(59), 1315–1318. https://www.ncbi.nlm.nih.gov/pubmed/15362741.
Saha, S. K., Parachoniak, C. A., Ghanta, K. S., Fitamant, J., Ross, K. N., Najem, M. S., et al. (2014). Mutant IDH inhibits HNF-4alpha to block hepatocyte differentiation and promote biliary cancer. *Nature*, *513*(7516), 110–114. https://doi.org/10.1038/nature13441.
Seehawer, M., Heinzmann, F., D'Artista, L., Harbig, J., Roux, P. F., Hoenicke, L., et al. (2018). Necroptosis microenvironment directs lineage commitment in liver cancer. *Nature*, *562*(7725), 69–75. https://doi.org/10.1038/s41586-018-0519-y.
Sirica, A. E., Strazzabosco, M., & Cadamuro, M. (2021). Intrahepatic cholangiocarcinoma: Morpho-molecular pathology, tumor reactive microenvironment, and malignant progression. *Advances in Cancer Research*, *149*, 321–387. https://doi.org/10.1016/bs.acr.2020.10.005.
Smoot, R. L., Werneburg, N. W., Sugihara, T., Hernandez, M. C., Yang, L., Mehner, C., et al. (2018). Platelet-derived growth factor regulates YAP transcriptional activity via Src family kinase dependent tyrosine phosphorylation. *Journal of Cellular Biochemistry*, *119*(1), 824–836. https://doi.org/10.1002/jcb.26246.
Song, X., Liu, X., Wang, H., Wang, J., Qiao, Y., Cigliano, A., et al. (2019). Combined CDK4/6 and Pan-mTOR inhibition is synergistic against intrahepatic cholangiocarcinoma. *Clinical Cancer Research*, *25*(1), 403–413. https://doi.org/10.1158/1078-0432.CCR-18-0284.
Song, H., Mak, K. K., Topol, L., Yun, K., Hu, J., Garrett, L., et al. (2010). Mammalian Mst1 and Mst2 kinases play essential roles in organ size control and tumor suppression. *Proceedings of the National Academy of Sciences of the United States of America*, *107*(4), 1431–1436. https://doi.org/10.1073/pnas.0911409107.
Song, X., Xu, H., Wang, P., Wang, J., Affo, S., Wang, H., et al. (2021). Focal adhesion kinase (FAK) promotes cholangiocarcinoma development and progression via YAP activation. *Journal of Hepatology*, *75*(4), 888–899. https://doi.org/10.1016/j.jhep.2021.05.018.
Sorrentino, G., Ruggeri, N., Specchia, V., Cordenonsi, M., Mano, M., Dupont, S., et al. (2014). Metabolic control of YAP and TAZ by the mevalonate pathway. *Nature Cell Biology*, *16*(4), 357–366. https://doi.org/10.1038/ncb2936.
Stavraka, C., Rush, H., & Ross, P. (2019). Combined hepatocellular cholangiocarcinoma (cHCC-CC): An update of genetics, molecular biology, and therapeutic interventions. *Journal of Hepatocellular Carcinoma*, *6*, 11–21. https://doi.org/10.2147/JHC.S159805.
Sugihara, T., Isomoto, H., Gores, G., & Smoot, R. (2019). YAP and the Hippo pathway in cholangiocarcinoma. *Journal of Gastroenterology*, *54*(6), 485–491. https://doi.org/10.1007/s00535-019-01563-z.
Sugimachi, K., Nishio, M., Aishima, S., Kuroda, Y., Iguchi, T., Komatsu, H., et al. (2017). Altered expression of Hippo signaling pathway molecules in intrahepatic cholangiocarcinoma. *Oncology*, *93*(1), 67–74. https://doi.org/10.1159/000463390.

Sugiura, K., Mishima, T., Takano, S., Yoshitomi, H., Furukawa, K., Takayashiki, T., et al. (2019). The expression of Yes-associated protein (YAP) maintains putative cancer stemness and is associated with poor prognosis in intrahepatic cholangiocarcinoma. *American Journal of Pathology*, *189*(9), 1863–1877. https://doi.org/10.1016/j.ajpath.2019.05.014.

Szeto, S. G., Narimatsu, M., Lu, M., He, X., Sidiqi, A. M., Tolosa, M. F., et al. (2016). YAP/TAZ are mechanoregulators of TGF-beta-Smad signaling and renal fibrogenesis. *Journal of the American Society of Nephrology*, *27*(10), 3117–3128. https://doi.org/10.1681/ASN.2015050499.

Tang, D., Xu, L., Zhang, M., Dorfman, R. G., Pan, Y., Zhou, Q., et al. (2018). Metformin facilitates BG45induced apoptosis via an antiWarburg effect in cholangiocarcinoma cells. *Oncology Reports*, *39*(4), 1957–1965. https://doi.org/10.3892/or.2018.6275.

Taniguchi, K., Wu, L. W., Grivennikov, S. I., de Jong, P. R., Lian, I., Yu, F. X., et al. (2015). A gp130-Src-YAP module links inflammation to epithelial regeneration. *Nature*, *519*(7541), 57–62. https://doi.org/10.1038/nature14228.

Thamrongwaranggoon, U., Sangkhamanon, S., Seubwai, W., Saranaruk, P., Cha'on, U., & Wongkham, S. (2021). Aberrant GLUT1 expression is associated with carcinogenesis and progression of liver fluke-associated cholangiocarcinoma. *In Vivo*, *35*(1), 267–274. https://doi.org/10.21873/invivo.12255.

Tiemin, P., Peng, X., Qingfu, L., Yan, W., Junlin, X., Zhefeng, H., et al. (2020). Dysregulation of the miR-148a-GLUT1 axis promotes the progression and chemoresistance of human intrahepatic cholangiocarcinoma. *Oncogenesis*, *9*(2), 19. https://doi.org/10.1038/s41389-020-0207-2.

Toth, M., Wehling, L., Thiess, L., Rose, F., Schmitt, J., Weiler, S. M. E., et al. (2021). Co-expression of YAP and TAZ associates with chromosomal instability in human cholangiocarcinoma. *BMC Cancer*, *21*(1), 1079. https://doi.org/10.1186/s12885-021-08794-5.

Tschaharganeh, D. F., Xue, W., Calvisi, D. F., Evert, M., Michurina, T. V., Dow, L. E., et al. (2014). p53-dependent Nestin regulation links tumor suppression to cellular plasticity in liver cancer. *Cell*, *158*(3), 579–592. https://doi.org/10.1016/j.cell.2014.05.051.

Van Haele, M., Moya, I. M., Karaman, R., Rens, G., Snoeck, J., Govaere, O., et al. (2019). YAP and TAZ heterogeneity in primary liver cancer: An analysis of its prognostic and diagnostic role. *International Journal of Molecular Sciences*, *20*(3). https://doi.org/10.3390/ijms20030638.

Vanichapol, T., Leelawat, K., & Hongeng, S. (2015). Hypoxia enhances cholangiocarcinoma invasion through activation of hepatocyte growth factor receptor and the extracellular signalregulated kinase signaling pathway. *Molecular Medicine Reports*, *12*(3), 3265–3272. https://doi.org/10.3892/mmr.2015.3865.

Wang, J., Dong, M., Xu, Z., Song, X., Zhang, S., Qiao, Y., et al. (2018). Notch2 controls hepatocyte-derived cholangiocarcinoma formation in mice. *Oncogene*, *37*(24), 3229–3242. https://doi.org/10.1038/s41388-018-0188-1.

Wang, J., Wang, H., Peters, M., Ding, N., Ribback, S., Utpatel, K., et al. (2019). Loss of Fbxw7 synergizes with activated Akt signaling to promote c-Myc dependent cholangiocarcinogenesis. *Journal of Hepatology*, *71*(4), 742–752. https://doi.org/10.1016/j.jhep.2019.05.027.

Wang, H., Wang, J., Zhang, S., Jia, J., Liu, X., Zhang, J., et al. (2021). Distinct and overlapping roles of Hippo effectors YAP and TAZ during human and mouse hepatocarcinogenesis. *Cellular and Molecular Gastroenterology and Hepatology*, *11*(4), 1095–1117. https://doi.org/10.1016/j.jcmgh.2020.11.008.

Wang, W., Xiao, Z. D., Li, X., Aziz, K. E., Gan, B., Johnson, R. L., et al. (2015). AMPK modulates Hippo pathway activity to regulate energy homeostasis. *Nature Cell Biology*, *17*(4), 490–499. https://doi.org/10.1038/ncb3113.

Wang, S., Xie, F., Chu, F., Zhang, Z., Yang, B., Dai, T., et al. (2017). YAP antagonizes innate antiviral immunity and is targeted for lysosomal degradation through IKKvarepsilon-mediated phosphorylation. *Nature Immunology*, *18*(7), 733–743. https://doi.org/10.1038/ni.3744.

Wu, H., Liu, Y., Jiang, X. W., Li, W. F., Guo, G., Gong, J. P., et al. (2016). Clinicopathological and prognostic significance of Yes-associated protein expression in hepatocellular carcinoma and hepatic cholangiocarcinoma. *Tumor Biology*, *37*(10), 13499–13508. https://doi.org/10.1007/s13277-016-5211-y.

Xia, L., Chen, X., Yang, J., Zhu, S., Zhang, L., Yin, Q., et al. (2020). Long non-coding RNA-PAICC promotes the tumorigenesis of human intrahepatic cholangiocarcinoma by increasing YAP1 transcription. *Frontiers in Oncology*, *10*, 595533. https://doi.org/10.3389/fonc.2020.595533.

Xu, K. D., Miao, Y., Li, P., Li, P. P., Liu, J., Li, J., et al. (2021). Licochalcone A inhibits cell growth through the downregulation of the Hippo pathway via PES1 in cholangiocarcinoma cells. *Environmental Toxicology*. https://doi.org/10.1002/tox.23422.

Xu, W., Zhang, M., Li, Y., Wang, Y., Wang, K., Chen, Q., et al. (2021). YAP manipulates proliferation via PTEN/AKT/mTOR-mediated autophagy in lung adenocarcinomas. *Cancer Cell International*, *21*(1), 30. https://doi.org/10.1186/s12935-020-01688-9.

Yamada, D., Rizvi, S., Razumilava, N., Bronk, S. F., Davila, J. I., Champion, M. D., et al. (2015). IL-33 facilitates oncogene-induced cholangiocarcinoma in mice by an interleukin-6-sensitive mechanism. *Hepatology*, *61*(5), 1627–1642. https://doi.org/10.1002/hep.27687.

Yan, H., Parsons, D. W., Jin, G., McLendon, R., Rasheed, B. A., Yuan, W., et al. (2009). IDH1 and IDH2 mutations in gliomas. *The New England Journal of Medicine*, *360*(8), 765–773. https://doi.org/10.1056/NEJMoa0808710.

Yang, S. H., Lin, H. Y., Chang, V. H., Chen, C. C., Liu, Y. R., Wang, J., et al. (2015). Lovastatin overcomes gefitinib resistance through TNF-alpha signaling in human cholangiocarcinomas with different LKB1 statuses in vitro and in vivo. *Oncotarget*, *6*(27), 23857–23873. https://doi.org/10.18632/oncotarget.4408.

Yang, C. S., Stampouloglou, E., Kingston, N. M., Zhang, L., Monti, S., & Varelas, X. (2018). Glutamine-utilizing transaminases are a metabolic vulnerability of TAZ/YAP-activated cancer cells. *EMBO Reports*, *19*(6). https://doi.org/10.15252/embr.201643577.

Yimlamai, D., Christodoulou, C., Galli, G. G., Yanger, K., Pepe-Mooney, B., Gurung, B., et al. (2014). Hippo pathway activity influences liver cell fate. *Cell*, *157*(6), 1324–1338. https://doi.org/10.1016/j.cell.2014.03.060.

Yimlamai, D., Fowl, B. H., & Camargo, F. D. (2015). Emerging evidence on the role of the Hippo/YAP pathway in liver physiology and cancer. *Journal of Hepatology*, *63*(6), 1491–1501. https://doi.org/10.1016/j.jhep.2015.07.008.

Yin, F., Yu, J., Zheng, Y., Chen, Q., Zhang, N., & Pan, D. (2013). Spatial organization of Hippo signaling at the plasma membrane mediated by the tumor suppressor Merlin/NF2. *Cell*, *154*(6), 1342–1355. https://doi.org/10.1016/j.cell.2013.08.025.

Yuan, H. Y., Lv, Y. J., Chen, Y., Li, D., Li, X., Qu, J., et al. (2021). TEAD4 is a novel independent predictor of prognosis in LGG patients with IDH mutation. *Open Life Science*, *16*(1), 323–335. https://doi.org/10.1515/biol-2021-0039.

Yuan, W. C., Pepe-Mooney, B., Galli, G. G., Dill, M. T., Huang, H. T., Hao, M., et al. (2018). NUAK2 is a critical YAP target in liver cancer. *Nature Communications*, *9*(1), 4834. https://doi.org/10.1038/s41467-018-07394-5.

Zanconato, F., Cordenonsi, M., & Piccolo, S. (2016). YAP/TAZ at the roots of cancer. *Cancer Cell*, *29*(6), 783–803. https://doi.org/10.1016/j.ccell.2016.05.005.

Zhang, N., Bai, H., David, K. K., Dong, J., Zheng, Y., Cai, J., et al. (2010). The Merlin/NF2 tumor suppressor functions through the YAP oncoprotein to regulate tissue homeostasis in mammals. *Developmental Cell, 19*(1), 27–38. https://doi.org/10.1016/j.devcel.2010.06.015.

Zhang, S., Liang, S., Wu, D., Guo, H., Ma, K., & Liu, L. (2021). LncRNA coordinates Hippo and mTORC1 pathway activation in cancer. *Cell Death and Disease, 12*(9), 822. https://doi.org/10.1038/s41419-021-04112-w.

Zhang, X., Qiao, Y., Wu, Q., Chen, Y., Zou, S., Liu, X., et al. (2017). The essential role of YAP O-GlcNAcylation in high-glucose-stimulated liver tumorigenesis. *Nature Communications, 8*, 15280. https://doi.org/10.1038/ncomms15280.

Zhang, S., Song, X., Cao, D., Xu, Z., Fan, B., Che, L., et al. (2017). Pan-mTOR inhibitor MLN0128 is effective against intrahepatic cholangiocarcinoma in mice. *Journal of Hepatology, 67*(6), 1194–1203. https://doi.org/10.1016/j.jhep.2017.07.006.

Zhang, S., Wang, J., Wang, H., Fan, L., Fan, B., Zeng, B., et al. (2018). Hippo cascade controls lineage commitment of liver tumors in mice and humans. *American Journal of Pathology, 188*(4), 995–1006. https://doi.org/10.1016/j.ajpath.2017.12.017.

Zhu, B., Qian, W., Han, C., Bai, T., & Hou, X. (2021). Piezo 1 activation facilitates cholangiocarcinoma metastasis via Hippo/YAP signaling axis. *Molecular Therapy Nucleic Acids, 24*, 241–252. https://doi.org/10.1016/j.omtn.2021.02.026.

Zimmerman, R. L., Fogt, F., Burke, M., & Murakata, L. A. (2002). Assessment of Glut-1 expression in cholangiocarcinoma, benign biliary lesions and hepatocellular carcinoma. *Oncology Reports, 9*(4), 689–692. https://www.ncbi.nlm.nih.gov/pubmed/12066193.

Zong, Y., Panikkar, A., Xu, J., Antoniou, A., Raynaud, P., Lemaigre, F., et al. (2009). Notch signaling controls liver development by regulating biliary differentiation. *Development, 136*(10), 1727–1739. https://doi.org/10.1242/dev.029140.

CHAPTER ELEVEN

Patient-derived functional organoids as a personalized approach for drug screening against hepatobiliary cancers

Ling Li and Florin M. Selaru*

Division of Gastroenterology and Hepatology, School of Medicine, The Johns Hopkins University, Baltimore, MD, United States
*Corresponding author: e-mail address: fselaru1@jhmi.edu

Contents

1. Introduction   320
2. Hepatobiliary cancer treatment: Challenges   321
3. Tumor organoids for personalized drug screening   322
       3.1 Hepatobiliary cancer organoids establishment and culture   322
       3.2 High-throughput drug screening platform to incorporate cancer organoids   326
       3.3 Potential of personalized drug screening for therapy guidance   330
       3.4 Complex organoids reflect stromal and immune components   332
4. Conclusion and future perspectives   334
Grant support   335
Conflict of interest statement   335
References   335

Abstract

Patient-derived organoids (PDOs) established from hepatobiliary cancers are seen as valuable models of the cancer of origin. More precisely, PDOs have the ability to retain the original cancer genetic, epigenetic and phenotypic features. By extension, hepatobiliary cancer PDOs have the potential to (1) increase our understanding of cancer biology; (2) allow high-throughput drug screening for more efficient identification and testing of small molecule therapeutics, and (3) permit the design of personalized drug choice approaches for patients with liver cancer. Here, we review general principles for PDO establishment from hepatocellular carcinoma and cholangiocarcinoma, their utilization in drug screening strategies, and last, the establishment of complex PDOs to include tumor stroma. We conclude that PDOs represent a promising and important development in investigating interaction between liver cancer cell types and their microenvironment, as well as for positioning PDOs for high throughput drug screening for hepatobiliary cancers, and that further work is now needed to fully realize their potential.

Abbreviations

2D	two dimensional
3D	three dimensional
CAFs	cancer-associated fibroblasts
CCA	cholangiocarcinoma
cPDOs	complex PDOs
CTLA-4	cytotoxic T-lymphocyte-associated protein 4
ECM	extracellular matrix
ePDOs	epithelial PDOs
HCC	hepatocellular carcinoma
OS	overall survival
PD-1	programmed cell death-1
PDMCs	patient-derived models of cancer
PDOs	patient-derived organoids
PDX	patient-derived xenografts
PFS	progression-free survival
TAMs	tumor associated macrophages
TIL	tumor-infiltrating T lymphocyte
TME	tumor microenvironment

1. Introduction

The promise of precision oncology has been brought forth by the genomic era and the improved understanding of driver mutations in cancer (Johnson, 2017; Konnick, 2020; Yap, Johnson, & Meric-Bernstam, 2021). The paradigm is attractive: identify an actionable mutation, then identify the small molecule drug that specifically targets the mutation (Johnson, 2017; Konnick, 2020; Yap et al., 2021). One of the caveats, however, is that only up to 6% of cancer patients are successfully paired with a drug (Davis et al., 2019; Harding et al., 2019). Therefore, there is an urgent need to develop direct, accurate, reproducible, and relatively inexpensive functional drug testing methodologies using patient-derived cancer samples. Patient-derived models of cancer (PDMCs) offer the opportunity to elucidate fundamental mechanisms involved in human cancer, as well as to predict responses to targeted agent and immune-based cancer therapies.

To date, several systems have been employed to elucidate cancer mechanisms, including cell lines (*in vitro* as well as xenotransplanted in immunocompromised rodents), genetic animal models, patient-derived xenografts (PDX) and, more recently, patient-derived organoids (PDO).

Arguably, PDX models represent the gold standard in recapitulating cancer biology (Aparicio, Hidalgo, & Kung, 2015; Zhang et al., 2013). Importantly, PDX were also shown to predict patient response to chemotherapy and further found to have in excess of 80% drug response positive predictive value in patients from whom they were derived (Aparicio et al., 2015). There are, unfortunately, practical limitations to using PDX routinely to inform drug choice in oncology practice. PDX can be slow to establish and it may take up to 1 year to establish some of these models. They are, therefore, difficult to integrate into either drug discovery pipelines or the direct clinical care of cancer patients (Shroyer, 2016). Thus, there is a critical need for *in vitro* human cancer models that (1) are quicker and relatively inexpensive to establish, (2) are able to recapitulate the biology of the cancer, and (3) can accurately predict clinical drug responses in the cancer of origin.

2. Hepatobiliary cancer treatment: Challenges

Primary liver cancer represents the sixth most diagnosed cancer and the third leading cause of cancer-related mortality worldwide (Sung et al., 2021). In 2020, there were approximately 905,677 new diagnoses of primary liver cancers and 830,180 liver cancer death globally (Sung et al., 2021). Hepatocellular carcinoma (HCC) is the most common hepatobiliary cancer, accounting for 75–85% of all cases of primary liver cancers (Sung et al., 2021), and cholangiocarcinoma (CCA) is the second most prevalent hepatobiliary cancer, accounting for 10–15% of primary liver cancers. HCC has a median survival of only 6–20 months (Golabi et al., 2017), making it one of the most lethal cancers (Villanueva, 2019). The only curative approach to HCC is surgical: surgical resection, ablation or liver transplantation (Llovet et al., 2021). Nonetheless, 50–60% of patients are diagnosed at advanced stages, when curative surgery is not an option (Llovet, Montal, Sia, & Finn, 2018).

For advanced HCC, there are a number of FDA approved systemic therapies, including sorafenib, lenvantinib, regorafenib, cabozantinib and ramucirumab (Abou-Alfa et al., 2018; Bruix et al., 2017; Kudo et al., 2018; Llovet et al., 2008), as well as the recently approved first-line combination therapy of an immune check inhibitor with a VEGF-A monoclonal antibody (atezolizumab+bevacizumab) (Casak et al., 2021; Hack et al., 2020). In earlier trials, sorafenib extended the survival of HCC patients to a median survival of 10.7 month compared with 7.9 month in the placebo arm (Llovet et al., 2008). In another trial, lenvatinib demonstrated noninferior efficacy when compared to sorafenib, which allowed for its approval (Kudo et al., 2018). The combination atezolizumab plus bevacizumab was shown to achieve better overall survival (OS) and progression-free survival (PFS) compared with sorafenib (Finn et al., 2020). This combination resulted in an OS for advanced HCC of 10–19 months (Llovet et al., 2021).

CCA, like HCC, also has a dismal median survival measured in months (Valle et al., 2010). In spite of ongoing efforts to elucidate molecular pathways in CCA (Jiao et al., 2013), the first line and mainstay of CCA chemotherapy continues to be the combination of gemcitabine and cisplatin—with a poor response rate.

Overall, the biggest challenge in hepatobiliary cancers is the poor survival in spite of years of research into developing novel therapies. This poor survival is due, at least in part, to the lack of predictive preclinical models to allow efficient drug discovery. The American Society of Clinical Oncology found that first generation functional diagnostics—Chemotherapy Sensitivity and Resistance Assays—have not been successful in surpassing published reports of clinical trials (Burstein et al., 2011). These efforts, however, had a number of limitations, including lack of standardization in regards to the study design, type of assay, reproducibility, and accuracy of such assays used (typically low), or type of drugs used (typically conventional chemotherapy) (Friedman, Letai, Fisher, & Flaherty, 2015). Issues such as assay validation, reproducibility, and more importantly, clinical translatability were incompletely and variably described. However, next generation functional diagnostics, including patient-derived organoids (PDOs), have been recently developed, which could address these concerns (Gao et al., 2014; van de Wetering et al., 2015). Here, we present a viewpoint on the potential translational application of PDOs for precise drug screening modeling in hepatobiliary cancer.

3. Tumor organoids for personalized drug screening

3.1 Hepatobiliary cancer organoids establishment and culture

The successful culture of Lgr5+ colonic epithelial stem cells was a watershed for understanding normal as well as cancer epithelial cell biology (Sato et al., 2009). The methodologies and culture medium described in the original manuscript led to adaptations tailored to epithelial cells obtained from a variety of organs (Sato et al., 2009). For example, human normal liver organoids were established in 2015 based on the original intestinal culture media

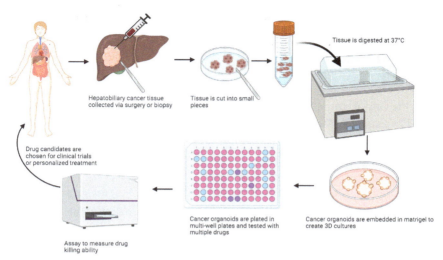

Fig. 1 Schematic representation of human hepatobiliary cancer organoid establishment and drug screening (created with Biorender.com). Tumor tissue is obtained at surgery, or biopsy and is processed to created cancer organoids. These organoids are then utilized in high throughput drug screening pipelines.

modified for hepatocytes (Huch et al., 2015). The establishment of hepatobiliary cancer organoids was then reported in 2017 (Broutier et al., 2017; Driehuis, Kretzschmar, & Clevers, 2020; Lee, Kenny, Lee, & Bissell, 2007). We and others have further adapted and refined culture conditions to establish and maintain PDOs from liver cancers (Li et al., 2019; Nuciforo et al., 2018; Saito et al., 2019). Fig. 1 represents a schematic of the establishment of PDOs as well as their utilization in drug screening pipelines. In brief, the culture of human liver cancer organoids begins with sterile fresh liver cancer tissue that is obtained from surgery or biopsy. The piece of tissue is then cut into small fragments, and digested with collagenase supplemented with DNAase. In our experience, the required digestion time is longer for cancer than for normal tissue, and even longer for HCC tumors in a background of cirrhosis. Similarly, intrahepatic CCA with a high stromal component also require a longer digestion time. Tumor pieces are digested into single cells or small clumps of cells, then mixed with an extracellular matrix hydrogel such as growth factor reduced Matrigel, basement membrane extract or Geltrex (Lee et al., 2007). Next, cells suspended in extracellular matrix are seeded as a drop in the center of a well in 24- or 48-well plates. After the extracellular matrix hydrogel solidified into a dome shape, liver cancer specific medium is added, and the cultures are grown at 37 °C. Interestingly, PDOs demonstrate similar hematoxylin and eosin and molecular staining to the original tumors (Broutier et al., 2017; Driehuis et al., 2020; Huch et al., 2013; Karthaus et al., 2014; Lee et al., 2007; Li et al., 2019; Mullenders et al., 2019; Nuciforo et al., 2018; Sachs et al., 2018; Sato et al., 2009).

One important stumbling block to developing drug screening pipelines for HCC and CCA has been the difficulty in easily establishing PDOs from hepatobiliary cancers. Studies to date, unfortunately, showed that this success rate is relatively low when compared with other epithelial cancer cell types (Nuciforo et al., 2018; Saito et al., 2019; Veninga & Voest, 2021). For example, Saito et al. (2019) reported that the PDO establishment success rate was 50% (3/6) for intrahepatic CCA, whereas Nuciforo et al. (2018) reported a hepatobiliary PDO establishment rate from ultrasound-guided needle biopsy to be 26%. Histologic differentiation of the original tumor, as well as percent necrosis, composition of growth media, time the specimen has spent in transit from operating room to the lab, size of the tissue piece retrieved, percent contribution cancer cells vs other cell types (immune cells, vascular cells, fibroblasts), and processing techniques can all influence the PDO establishment success rate (Driehuis et al., 2020; Sachs et al., 2018).

In terms of morphology, HCC PDOs tend to grow as irregular, round, solid structures (Broutier et al., 2017; Nuciforo et al., 2018). HCC PDOs can also form pseudo-glands but rarely if ever form lumens (Broutier et al., 2017; Nuciforo et al., 2018). Most CCA PDOs form irregular cystic duct shapes in contrast to homogenous cyst-like hollow shapes of liver PDOs that originated from normal cholangiocytes (Broutier et al., 2017; Nuciforo et al., 2018). HCC PDOs generally express heppar-1 and/or alpha fetoprotein, and albumin (Broutier et al., 2017; Nuciforo et al., 2018). CCA PDOs were found to express keratin 7 and 9 (Broutier et al., 2017; Nuciforo et al., 2018). Fig. 2 highlights differences in morphology in HCC, CCA as well as normal liver PDOs. Although the growth rate of liver cancer PDOs can vary significantly from patient to patient, overall, PDOs can be passaged 2–3 weeks after their establishment (Broutier et al., 2017) and can be utilized for drug screening after 4–12 weeks (Broutier et al., 2017). Long-term, liver cancer PDOs were reportedly passaged in culture for more than 1 year with no obvious change in growth ability or other characteristics such as histological morphology, genomic and transcriptome profiles of the original tumor (Broutier et al., 2017; Nuciforo et al., 2018; Saito et al., 2019), and were able to undergo freeze-thaw cycles (Saito et al., 2019). In our lab, we maintained hepatobiliary cancer PDOs for more than 3 years.

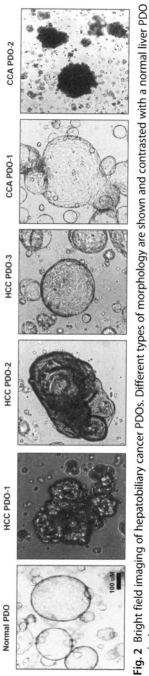

Fig. 2 Bright field imaging of hepatobiliary cancer PDOs. Different types of morphology are shown and contrasted with a normal liver PDO morphology.

3.2 High-throughput drug screening platform to incorporate cancer organoids

Two dimensional (2D) monolayer cell cultures, either cell lines or primary cancer cell cultures, have historically been the mainstay of high-throughput drug screening strategies (Cox, Reese, Bickford, & Verbridge, 2015; Ferreira, Gaspar, & Mano, 2018; Jensen & Teng, 2020; Stock et al., 2016). These culture systems are relatively inexpensive and technically easy to implement. While they allow dissection of fundamental cancer mechanisms, their ability to predict drug response *in vivo* is limited (Chwalek, Bray, & Werner, 2014; Cox et al., 2015; Ferreira et al., 2018; Fong, Harrington, Farach-Carson, & Yu, 2016; Jensen & Teng, 2020; Lang, Yeow, Nichols, & Scheer, 2006; Stock et al., 2016). Specifically, 2D drug screening strategies rarely lead to successful clinical trials (Brancato, Oliveira, Correlo, Reis, & Kundu, 2020; Chwalek et al., 2014; Cox et al., 2015; Ferreira et al., 2018; Fong et al., 2016; Jensen & Teng, 2020; Lang et al., 2006; Radhakrishnan, Varadaraj, Dash, Sharma, & Verma, 2020; Stock et al., 2016). Cell-based drug screening strategies have been used as preclinical steps to identify effective drug candidates for future drug development (Lang et al., 2006). However, it should also be noted that current cancer drug development pipelines typically have a relatively low overall success rate (Mak, Evaniew, & Ghert, 2014). For example, the success rate of drugs progressing from animal models to human clinical trials is less than 8% (Mak et al., 2014). In addition, among drugs advanced to Phase 3 clinical trials, the success rate is approximate 55% (De Martini, 2020; Thomas et al., 2016; Wong, Siah, & Lo, 2019b). A study looking at all oncology trials (Phase I, II and III) from 2000 to 2015 and found that there was only a 3.4% overall probability of success for any given small molecule entering a Phase I study (Wong, Siah, & Lo, 2019a). Moreover, the cost for any successful drug an oncology is a staggering $3 billion per approved drug, due largely in part to the high failure rates (Mak et al., 2014). It appears that one of the major limitations of traditional drug discovery pipelines is insufficient ability to model the cancer response to therapeutics.

Compared with conventional 2D culture systems, 3-dimensional (3D) cultures appear more physiologically relevant to the original tumor (Dunne et al., 2014; Jensen & Teng, 2020). In addition, these 3D culture models better emulate tumor morphology, as well as cell–cell interactions

(Chwalek et al., 2014; Fong et al., 2016). Lastly, 3D culture models can incorporate an extracellular matrix component for added complexity and better representation of *in vivo* conditions (Radhakrishnan et al., 2020). Due to these considerations, the advent of PDOs established from hepatobiliary cancers offer renewed hope for better correlation between PDO response and patient response to a novel drug, as well as for high-throughput drug screening.

Earlier work focused precisely on testing the ability of PDOs to accurately reflect original tumor biology. Indeed, these studies have demonstrated that hepatobiliary cancer PDOs maintain individual genetic and epigenetic features of the original cancer (Broutier et al., 2017; Nuciforo et al., 2018; Saito et al., 2019). Patient-derived xenografts have been shown to better predict tumor response to drugs in patients (Zhang et al., 2013). Hepatobiliary cancer patient-derived xenografts (PDX) further confirmed the ability of PDOs to predict which drugs will work *in vivo* (Broutier et al., 2017; Li et al., 2019; Li et al., 2021; Saito et al., 2019). A direct comparison of drug response in PDOs, PDO-derived PDX, and patient cancer response has been performed, although not in hepatobiliary cancer (Vlachogiannis et al., 2018). This study provided with proof of concept evidence of PDO ability to predict therapeutic drug responses in patients (Vlachogiannis et al., 2018), with 88% positive predictive value and 100% negative predictive value.

Once the PDO platform design had been outlined, the next challenge was its adaptation for high-throughput drug screening. High-throughput drug screening protocols were reported to be performed in different types of cancer PDOs (Broutier et al., 2017; Huch et al., 2013, 2015; Karthaus et al., 2014; Lancaster et al., 2013; Li et al., 2019; Mullenders et al., 2019; Roerink et al., 2018; Sachs et al., 2018; Sato et al., 2009). A recently published protocol has provided general guidelines to standardize the methodology of establishing cancer PDOs from different cancer types, as well as high-throughput drug screening in a semi-automated fashion in 384 well plates(Driehuis et al., 2020). With respect to hepatobiliary PDO drug testing, we and others have reported successful adaptation of PDO platforms for 96- and 384-well plates (Broutier et al., 2017; Li et al., 2019, 2021; Saito et al., 2019). As shown in Table 1, a number of protocols were subsequently published specifically for hepatobiliary cancers (Broutier et al., 2017; Li et al., 2019; Nuciforo et al., 2018; Saito et al., 2019, 2020).

Table 1 Details of culture conditions for hepatobiliary cancer organoids from different groups.

Cancer type	Source	Digestion	Digestion time	Medium composition	Year of publication	References
HCC and CCA	Surgery and Biopsy	Collagenase D (2.5 mg/mL)	2 h—overnight	Advanced DMEM/F12 supplemented with 1% penicillin/streptomycin, 1% Glutamax, 10-mM HEPES, 1:50 B27 supplement (without vitamin A), 1:100 N2 supplement, 1.25-mM N-acetyl-L-cysteine, 10% (vol/vol) Rspo-1 conditioned medium, 30% (vol/vol) Wnt3a-conditioned medium, 10-mM nicotinamide, 10-nM recombinant human [Leu15]-gastrin I, 50 ng/mL recombinant human EGF, 100 ng/mL recombinant human FGF10, 25 ng/mL recombinant human HGF, 10 μM forskolin, 5 μM A8301, 25 ng/mL Noggin and 10–μM Y2763 +/− 3 nM dexamethasone	2017	Broutier et al. (2017)
HCC and CCA	Biopsy	Collagenase IV + Dnase (2.5 mg/mL + 0.1 mg/mL)	Varies (goal is to avoid complete digestion into single cell)	Advanced DMEM/F-12 (GIBCO) supplemented with 1:50 B-27 (GIBCO), 1:100 N-2 (GIBCO), 10 mM nicotinamide (Sigma), 1.25 mM N-acetyl-L-cysteine (Sigma), 10 nM [Leu15]-gastrin (Sigma), 10 μM forskolin (Tocris), 5 μM A83–01 (Tocris), 50 ng/mL EGF (PeproTech), 100 ng/mL FGF10 (PeproTech), 25 ng/mL HGF (PeproTech), 10% RSpo1-conditioned medium, (homemade), and 30% Wnt3a-conditioned medium	2018	Nuciforo et al. (2018)

HCC and CCA	Surgery and Biopsy	Collagenase IV + Dnase (2.5 mg/mL + 0.1 mg/mL)	Varies (goal is to avoid complete digestion into single cell)	Advanced DMEM/F-12 (GIBCO) supplemented with 1% penicillin/streptomycin, 1% glutamax, 10-mM HEPES, 1:50 B-27 (GIBCO), 1:100 N-2 (GIBCO), 10mM nicotinamide (Sigma), 1.25mM N-acetyl-L-cysteine (Sigma), 10nM [Leu15]-gastrin (Sigma), 10 μM forskolin (Tocris), 5 μM A83–01 (Tocris), 50ng/mL EGF (PeproTech), 100ng/mL FGF10 (PeproTech), FGF-2 20ng/mL (peroTech), PGE 2(10nM)(TOCRIS), 10% RSpo1-conditioned medium, (homemade), 30–50% Wnt3a-conditioned medium (homemade), 10% noggin-conditioned medium	2019	Li et al. (2019)
CCA and gallbladder cancer	Surgery	Dispase type II (0.0125%) + Collagenase type XI (0.0125%)	1 h	Advanced DMEM/F12 (Thermo Fisher Scientific) supplemented with Glutamax 10mM HEPES, penicillin/streptomycin 1 × N2 supplement 1 × B27 supplement 50ng/mL EGF 1.25mM N-acetylcysteine 10nM gastrin 10mM nicotinamide R-spondin 1, 5 μM A83–01 10 μM forskolin 10 μM Y-27632	2020	Saito, Muramatsu, and Saito (2020)

3.3 Potential of personalized drug screening for therapy guidance

Tumor sequencing followed by identification of actionable mutations represents the current paradigm of personalized cancer treatment (Veninga & Voest, 2021). These approaches have matured and, in many hospitals and cancer centers, they represent standard of care. However, the majority of cancer patients who undergo such testing in a clinical setting are never matched with an effective drug, nor do they have a measurable increase in survival (Cobain et al., 2021). There are multiple potential explanations for the unrealized potential of tumor sequencing in impacting the personalized care and drug choice for cancer patients. The focus on exome sequencing and driver mutations excludes potentially important layers of cancer complexity originating in the non-coding regions (Cobain et al., 2021). In addition, DNA sequencing does not provide an appreciation for epigenetic modulators of cancer behavior, including response to treatment (Veninga & Voest, 2021). Also, importantly, intra-tumor heterogeneity leads to sampling biases and subsequent poor positive prediction value (Veninga & Voest, 2021).

Multiple studies have now shown PDOs to maintain several key features of the parent tumor, including histopathologic characteristics, differentiation state, as well as mutational profile, and general genomic landscape (Broutier et al., 2017; Driehuis et al., 2020; Huch et al., 2013; Karthaus et al., 2014; Lee et al., 2007; Li et al., 2019; Mullenders et al., 2019; Nuciforo et al., 2018; Sachs et al., 2018; Sato et al., 2009). In addition, some studies in breast and lung cancer indicated that the PDO response to drugs correlated with somatic mutations from the parent cancer tissue (Kim et al., 2019; Sachs et al., 2018). Furthermore, studies in breast cancer, pancreatic cancer, colorectal cancer, as well as metastatic gastrointestinal cancers have shown that PDOs predict the patient response to drugs (Ooft et al., 2019; Sachs et al., 2018; Vlachogiannis et al., 2018; Yao et al., 2020). These data, along with the refinement of PDO platforms for high-throughput drug screening, have brought forth an intriguing question: can PDOs be used in real time for patient-specific, personalized drug choice? In other words, would it be technically and logistically possible to obtain a tumor sample from a patient, establish a PDO culture, perform high-throughput drug screening, choose a drug, and then treat the patient in a duration of time that is acceptable and clinically relevant?

A recent paper from our group reported the generation of multiple PDO lines from geographically distinct areas from within hepatobiliary cancers

(Li et al., 2019). In brief, 2 HCC and 3 CCA tumors were each used to generate between 3 and 7 PDO lines per tumor. Each of these 27 PDO lines underwent (1) exome sequencing; (2) RNA sequencing and (3) drug testing. This study found no correlation between DNA or RNA alterations with response to drug testing. In addition, direct drug testing in these PDOs allowed for precise determinations of effective drug concentrations, which would not be possible based on DNA or RNA analyses. This study, while limited in number of samples, highlights the renewed hope for PDO drug testing and its role in the direct, real time drug choice for cancer patients (Fig. 3).

For a personalized drug testing pipeline to be clinically relevant, the timeframe from tumor sampling to obtaining results from drug testing needs to be relatively short. Studies from our group and others have shown that this timeframe is 3 months or shorter, which may be clinically acceptable (Driehuis et al., 2020). In contrast, PDX models, which are the only alternative with a comparable positive predictive value, may take months to years to establish and propagate and also come at a cost that can be prohibitive

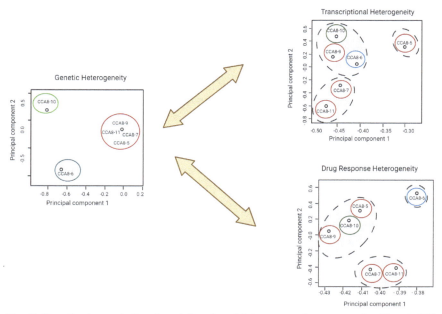

Fig. 3 Genetic, transcriptional and functional intra-tumor heterogeneity. The 6 CCA PDO lines from the same patient were grouped in an unsupervised analysis (principal components analysis). The grouping of samples based on genetic features does not translate into transcriptional groupings or into drug response groupings.

(Veninga & Voest, 2021). In spite of the promise of integrating PDOs into clinical grade protocols for informing drug choice, there are also some limitations. As noted above, the success rate for establishing PDOs is much lower than 100%, which would exclude a significant number of patients from these approaches. In addition, the success rate is far less with biopsy specimens than with resected tumor specimens. In this context, it should be noted that current clinical grade diagnostic protocols are based on needle biopsy and not on surgically resected specimens. In addition, epithelial PDOs (ePDOs) do not convey the complexity and interactions between epithelial cancer cells and mesenchymal components of tumors *in vivo*. For hepatobiliary cancers in particular (i.e., HCC that typically arises in a cirrhotic liver and CCA, which are very desmoplastic tumors) the lack of stroma in ePDOs can be a significant factor in reducing the positive predictive value of PDOs. Thus, while PDOs represent a compelling approach to personalizing HCC and CCA drug therapies, further development is necessary before their seamless integration into clinical grade protocols for informing drug choice for hepatobiliary cancer patients.

3.4 Complex organoids reflect stromal and immune components

Tumor cells grow in a complex tumor microenvironment (TME) composed of a diversity of non-epithelial cancer cell types as well as non-cellular elements, such as the extracellular matrix (ECM) (see chapter "Cancer-associated fibroblasts in intrahepatic cholangiocarcinoma progression and therapeutic resistance" by Ravichandra et al. and chapter "Matricellular proteins in intrahepatic cholangiocarcinoma" by Sirica. The TME is known to play pivotal roles in tumor formation, progression, invasion, metastases and drug resistance (Manzanares et al., 2017; Manzanares, Campbell, Maldonado, & Sirica, 2018; Quail & Joyce, 2013; Sirica, 2012; Sirica, Almenara, & Li, 2014). TME includes cancer-associated fibroblasts (CAFs), macrophages, vascular endothelial cells, and tumor-infiltrating T lymphocytes chapter "Cancer-associated fibroblasts in intrahepatic cholangiocarcinoma progression and therapeutic resistance" by Ravichandra et al. and chapter "Matricellular proteins in intrahepatic cholangiocarcinoma" by Sirica. Of importance to drug treatment as well as to *in vitro* modeling of cancer, these components are frequently reported to be active participants in the tumor behavior, including drug resistance (Joyce & Fearon, 2015; Manzanares et al., 2017, 2018; Quail & Joyce, 2013; Seidlitz & Stange, 2021; Sirica, 2012; Sirica et al., 2014; Vitale, Manic, Coussens, Kroemer, & Galluzzi, 2019).

Historically, interactions between epithelial cancer cells and TME have been explored in 2D and more recently in 3-D co-cultures systems(Manzanares et al., 2017, 2018), as well as in PDX models(Manzanares et al., 2017, 2018). Of note, in PDX models, the stroma is of rodent origin. With the emergence of PDOs, significant efforts have been spent to develop complex PDOs (cPDOs) to capture epithelial cancer cell to TME interaction (Koledova & Lu, 2017; Ling Li, Bader, & Selaru, 2019; Naruse et al., 2021; Ohlund et al., 2017; Tsai et al., 2018).

Several studies have highlighted the impact of stroma on the behavior of the culture system. For example, an early study found that gene expression in epithelial cancer cells from colorectal cancer PDOs was altered vs the original tumor, but restored closer to that of the *in situ* tumor upon co-culturing of cancer PDO cells with matched CAFs (Naruse et al., 2021). These results indicate that CAFs were able to re-create, to a certain extent, a TME similar to that *in vivo* (Naruse et al., 2021). A separate study identified several subgroups of CAFs, which can be activated upon co-culture with pancreatic PDOs and subsequently cooperate with neighboring cancer cells (Ohlund et al., 2017). Another group further demonstrated that pancreatic cancer PDOs co-cultured with CAFs showed increased resistance to gemcitabine, presumably gained secondary to the CAFs present in the co-culture (Tsai et al., 2018). Additional studies of cPDO to include cancer cells and CAFs in several other cancers, such as prostate adenocarcinomas, esophageal carcinomas, as well as breast cancer also demonstrated that CAFs can introduce elements of TME and participate in cancer progression and in anti-cancer drug resistance (Adisetiyo et al., 2014; Ebbing et al., 2019; Marusyk et al., 2016). As it was reported in other cancers, and by us in hepatobiliary cancers, drug response can be impacted by the presence of CAFs (Ling Li et al., 2019; Liu et al., 2021). For example, we reported that while ePDOs are sensitive to a number of drugs such as bortezomib, panbinostat, and ixazomib, the introduction of CAFs in co-culture resulted in relative resistance (Ling Li et al., 2019).

An area of high excitement in novel cancer therapeutics is represented by immune checkpoint inhibitors, such as anti-programmed cell death-1 (PD-1) or anti PD-L1 (programmed cell death-1 ligand) monoclonal antibodies, as well as cytotoxic T-lymphocyte-associated protein 4 (CTLA-4) antibodies(Larkin et al., 2019; Pardoll, 2012; Socinski et al., 2018). The strategy is based on stimulating the effector function of cytotoxic T cells and can be utilized in combination with conventional chemotherapy (Larkin et al., 2019; Pardoll, 2012; Socinski et al., 2018). These early clinical successes increased the interest in studies of immune cell modulation in cancer and by extension, in the modeling of systemic immune surveillance and/or local immune microenvironments. Therefore, there is an urgent need for robust experimental models and platforms to allow the investigation of such tumor-T cell interactions (Chen & Mellman, 2017; Sharma & Allison, 2015a, 2015b).

A study based on co-culturing of ePDOs and matched peripheral blood lymphocytes found that activated T cells were able to specifically recognize and kill the cancer cells but not matched healthy organoids or tissues (Dijkstra et al., 2018). Kong et al. (2018) further reported a cPDO approach that included colorectal cancer PDOs co-cultured with matched tumor-infiltrating T lymphocyte (TIL) from the same patient. Of note, this study was able to predict the response of patients to neoadjuvant therapy based on the response of matched cPDOs to same drugs (Kong et al., 2018). Furthermore, they were able to demonstrate that while TILs have low or no killing ability for patient derived PDOs, they can be reactivated by treatment with a PD1 inhibitor in co-culture (Kong et al., 2018). By extension, this TIL-ePDO co-culture platform can potentially be utilized to predict sensitivity to neoadjuvant chemoradiotherapy as well as the effect of immune check inhibitors (anti PD1). Finally, another cPDO co-culture system to include an air liquid interface PDO culture and microfluidic PDO culture platform has been reported (Veninga & Voest, 2021). This cPDO platform was able to preserve endogenous TEM components including immune cells (TIL, tumor-associated macrophages (TAMs), and others), as well as to model immune check inhibitor anti PD-1 effects alone and in combination with chemotherapy (Yuki, Cheng, Nakano, & Kuo, 2020). TAMs represent one of the major cell types in cancer stroma, and were reported to be involved in tumor related inflammation and phenotypic behavior, such as growth promotion (Koledova & Lu, 2017; Mantovani, Marchesi, Malesci, Laghi, & Allavena, 2017). TAMs also represent a potential treatment target for many cancers, including hepatobiliary cancers. Early efforts to simulate TAM-cancer cell inter-actions found that, in a transwell PDO-macrophage co-culture system, TAMs were able to promote ePDO differentiation (Noel et al., 2017).

4. Conclusion and future perspectives

The urgent requirement for effective therapeutics for hepatobiliary cancers is highlighted by the dismal overall survival rates of HCC and CCA patients, despite decades of basic and clinical research. The relatively slow rate of progress in therapeutics for hepatobiliary cancers may be traced, at least in part, to the lack of adequate predictive *in vitro* human cancer models. To be meaningful, these models need to closely reproduce of biology of the primary tumor, as well as accurately predict their response to drug therapies. The recent progress in positioning PDOs for high-throughput drug screening for hepatobiliary cancers is encouraging. Further work is now needed, to better model cancer behavior, associated stromal interactions, as well as to improve the success rate of establishing hepatobiliary cancer PDOs capable of being integrated into clinically relevant drug screening protocols.

Grant support
None.

Conflict of interest statement
L.L. and F.M.S. have no financial or personal disclosures relevant to the contents of this manuscript.

References

Abou-Alfa, G. K., Meyer, T., Cheng, A. L., El-Khoueiry, A. B., Rimassa, L., Ryoo, B. Y., et al. (2018). Cabozantinib in patients with advanced and progressing hepatocellular carcinoma. *New England Journal of Medicine, 379*(1), 54–63. https://doi.org/10.1056/NEJMoa1717002.

Adisetiyo, H., Liang, M. M., Liao, C. P., Jeong, J. H., Cohen, M. B., Roy-Burman, P., et al. (2014). Dependence of castration-resistant prostate cancer (CRPC) stem cells on CRPC-associated fibroblasts. *Journal of Cellular Physiology, 229*(9), 1170–1176. https://doi.org/10.1002/jcp.24546.

Aparicio, S., Hidalgo, M., & Kung, A. L. (2015). Examining the utility of patient-derived xenograft mouse models. *Nature Reviews Cancer, 15*(5), 311–316. https://doi.org/10.1038/nrc3944.

Brancato, V., Oliveira, J. M., Correlo, V. M., Reis, R. L., & Kundu, S. C. (2020). Could 3D models of cancer enhance drug screening? *Biomaterials, 232*. https://doi.org/10.1016/j.biomaterials.2019.119744, 119744.

Broutier, L., Mastrogiovanni, G., Verstegen, M. M. A., Francies, H. E., Gavarro, L. M., Bradshaw, C. R., et al. (2017). Human primary liver cancer-derived organoid cultures for disease modeling and drug screening. *Nature Medicine, 23*(12), 1424-+. https://doi.org/10.1038/nm.4438.

Bruix, J., Qin, S., Merle, P., Granito, A., Huang, Y. H., Bodoky, G., et al. (2017). Regorafenib for patients with hepatocellular carcinoma who progressed on sorafenib treatment (RESORCE): A randomised, double-blind, placebo-controlled, phase 3 trial. *Lancet, 389*(10064), 56–66. https://doi.org/10.1016/S0140-6736(16)32453-9.

Burstein, H. J., Mangu, P. B., Somerfield, M. R., Schrag, D., Samson, D., Holt, L., et al. (2011). American Society of Clinical Oncology clinical practice guideline update on the use of chemotherapy sensitivity and resistance assays. *Journal of Clinical Oncology, 29*(24), 3328–3330. https://doi.org/10.1200/JCO.2011.36.0354.

Casak, S. J., Donoghue, M., Fashoyin-Aje, L., Jiang, X., Rodriguez, L., Shen, Y. L., et al. (2021). FDA approval summary: Atezolizumab plus bevacizumab for the treatment of patients with advanced unresectable or metastatic hepatocellular carcinoma. *Clinical Cancer Research*, *27*(7), 1836–1841. https://doi.org/10.1158/1078-0432.CCR-20-3407.

Chen, D. S., & Mellman, I. (2017). Elements of cancer immunity and the cancer-immune set point. *Nature*, *541*(7637), 321–330. https://doi.org/10.1038/nature21349.

Chwalek, K., Bray, L. J., & Werner, C. (2014). Tissue-engineered 3D tumor angiogenesis models: Potential technologies for anti-cancer drug discovery. *Advanced Drug Delivery Reviews*, *79-80*, 30–39. https://doi.org/10.1016/j.addr.2014.05.006.

Cobain, E. F., Wu, Y. M., Chugh, R., Worden, F., Smith, D. C., Schuetze, S. M., et al. (2021). Assessment of clinical benefit of integrative genomic profiling in advanced solid tumors. *JAMA Oncology*, *7*(4), 525–533. https://doi.org/10.1001/jamaoncol.2020.7987.

Cox, M. C., Reese, L. M., Bickford, L. R., & Verbridge, S. S. (2015). Toward the broad adoption of 3D tumor models in the cancer drug pipeline. *ACS Biomaterials Science & Engineering*, *1*(10), 877–894. https://doi.org/10.1021/acsbiomaterials.5b00172.

Davis, W., Makar, G., Mehta, P., Zhu, G. G., Somer, R., Morrison, J., et al. (2019). Next-generation sequencing in 305 consecutive patients: Clinical outcomes and management changes. *JCO Oncology Practice*, *15*(12), e1028–e1034. https://doi.org/10.1200/JOP.19.00269.

De Martini, D. (2020). Empowering phase II clinical trials to reduce phase III failures. *Pharmaceutical Statistics*, *19*(3), 178–186. https://doi.org/10.1002/pst.1980.

Dijkstra, K. K., Cattaneo, C. M., Weeber, F., Chalabi, M., van de Haar, J., Fanchi, L. F., et al. (2018). Generation of tumor-reactive T cells by co-culture of peripheral blood lymphocytes and tumor organoids. *Cell*, *174*(6), 1586-+. https://doi.org/10.1016/j.cell.2018.07.009.

Driehuis, E., Kretzschmar, K., & Clevers, H. (2020). Establishment of patient-derived cancer organoids for drug-screening applications. *Nature Protocols*, *15*(10), 3380–3409. https://doi.org/10.1038/s41596-020-0379-4.

Dunne, L. W., Huang, Z., Meng, W. X., Fan, X. J., Zhang, N. Y., Zhang, Q. X., et al. (2014). Human decellularized adipose tissue scaffold as a model for breast cancer cell growth and drug treatments. *Biomaterials*, *35*(18), 4940–4949. https://doi.org/10.1016/j.biomaterials.2014.03.003.

Ebbing, E. A., van der Zalm, A. P., Steins, A., Creemers, A., Hermsen, S., Rentenaar, R., et al. (2019). Stromal-derived interleukin 6 drives epithelial-to-mesenchymal transition and therapy resistance in esophageal adenocarcinoma. *Proceedings of the National Academy of Sciences of the United States of America*, *116*(6), 2237–2242. https://doi.org/10.1073/pnas.1820459116.

Ferreira, L. P., Gaspar, V. M., & Mano, J. F. (2018). Design of spherically structured 3D in vitro tumor models -advances and prospects. *Acta Biomaterialia*, *75*, 11–34. https://doi.org/10.1016/j.actbio.2018.05.034.

Finn, R. S., Qin, S., Ikeda, M., Galle, P. R., Ducreux, M., Kim, T. Y., et al. (2020). Atezolizumab plus Bevacizumab in Unresectable Hepatocellular Carcinoma. *New England Journal of Medicine*, *382*(20), 1894–1905. https://doi.org/10.1056/NEJMoa1915745.

Fong, E. L. S., Harrington, D. A., Farach-Carson, M. C., & Yu, H. (2016). Heralding a new paradigm in 3D tumor modeling. *Biomaterials*, *108*, 197–213. https://doi.org/10.1016/j.biomaterials.2016.08.052.

Friedman, A. A., Letai, A., Fisher, D. E., & Flaherty, K. T. (2015). Precision medicine for cancer with next-generation functional diagnostics. *Nature Reviews Cancer*, *15*(12), 747–756. https://doi.org/10.1038/nrc4015.

Gao, D., Vela, I., Sboner, A., Iaquinta, P. J., Karthaus, W. R., Gopalan, A., et al. (2014). Organoid cultures derived from patients with advanced prostate cancer. *Cell*, *159*(1), 176–187. https://doi.org/10.1016/j.cell.2014.08.016.

Golabi, P., Fazel, S., Otgonsuren, M., Sayiner, M., Locklear, C. T., & Younossi, Z. M. (2017). Mortality assessment of patients with hepatocellular carcinoma according to underlying disease and treatment modalities. *Medicine (Baltimore)*, 96(9), e5904. https://doi.org/10.1097/MD.0000000000005904.

Hack, S. P., Spahn, J., Chen, M., Cheng, A. L., Kaseb, A., Kudo, M., et al. (2020). IMbrave 050: A phase III trial of atezolizumab plus bevacizumab in high-risk hepatocellular carcinoma after curative resection or ablation. *Future Oncology*, 16(15), 975–989. https://doi.org/10.2217/fon-2020-0162.

Harding, J. J., Nandakumar, S., Armenia, J., Khalil, D. N., Albano, M., Ly, M., et al. (2019). Prospective genotyping of hepatocellular carcinoma: Clinical implications of next-generation sequencing for matching patients to targeted and immune therapies. *Clinical Cancer Research*, 25(7), 2116–2126. https://doi.org/10.1158/1078-0432.CCR-18-2293.

Huch, M., Bonfanti, P., Boj, S. F., Sato, T., Loomans, C. J. M., van de Wetering, M., et al. (2013). Unlimited in vitro expansion of adult bi-potent pancreas progenitors through the Lgr5/R-spondin axis. *EMBO Journal*, 32(20), 2708–2721. https://doi.org/10.1038/emboj.2013.204.

Huch, M., Gehart, H., van Boxtel, R., Hamer, K., Blokzijl, F., Verstegen, M. M. A., et al. (2015). Long-term culture of genome-stable bipotent stem cells from adult human liver. *Cell*, 160(1–2), 299–312. https://doi.org/10.1016/j.cell.2014.11.050.

Jensen, C., & Teng, Y. (2020). Is it time to start transitioning from 2D to 3D cell culture? *Frontiers in Molecular Biosciences*, 7. https://doi.org/10.3389/fmolb.2020.00033. ARTN 33.

Jiao, Y., Pawlik, T. M., Anders, R. A., Selaru, F. M., Streppel, M. M., Lucas, D. J., et al. (2013). Exome sequencing identifies frequent inactivating mutations in BAP1, ARID1A and PBRM1 in intrahepatic cholangiocarcinomas. *Nature Genetics*, 45(12), 1470–1473. https://doi.org/10.1038/ng.2813.

Johnson, T. M. (2017). Perspective on precision medicine in oncology. *Pharmacotherapy*, 37(9), 988–989. https://doi.org/10.1002/phar.1975.

Joyce, J. A., & Fearon, D. T. (2015). T cell exclusion, immune privilege, and the tumor microenvironment. *Science*, 348(6230), 74–80. https://doi.org/10.1126/science.aaa6204.

Karthaus, W. R., Iaquinta, P. J., Drost, J., Gracanin, A., Van Boxtel, R., Wongvipat, J., et al. (2014). Identification of multipotent luminal progenitor cells in human prostate organoid cultures. *Cell*, 159(1), 163–175. https://doi.org/10.1016/j.cell.2014.08.017.

Kim, M., Mun, H., Sung, C. O., Cho, E. J., Jeon, H. J., Chun, S. M., et al. (2019). Patient-derived lung cancer organoids as in vitro cancer models for therapeutic screening. *Nature Communications*, 10. https://doi.org/10.1038/s41467-019-11867-6. ARTN 3991.

Koledova, Z., & Lu, P. F. (2017). A 3D fibroblast-epithelium co-culture model for understanding microenvironmental role in branching morphogenesis of the mammary gland. *Mammary Gland Development: Methods and Protocols*, 1501, 217–231. https://doi.org/10.1007/978-1-4939-6475-8_10.

Kong, J. C. H., Guerra, G. R., Millen, R. M., Roth, S., Xu, H. L., Neeson, P. J., et al. (2018). Tumor-infiltrating lymphocyte function predicts response to neoadjuvant chemoradiotherapy in locally advanced rectal cancer. *JCO Precision Oncology*, 2, 1–15. https://doi.org/10.1200/Po.18.00075.

Konnick, E. Q. (2020). The regulatory landscape of precision oncology laboratory medicine in the United States - perspective on the past 5 years and considerations for future regulation. *Practical Laboratory Medicine*, 21, ARTN e00172. https://doi.org/10.1016/j.plabm.2020.e00172.

Kudo, M., Finn, R. S., Qin, S., Han, K. H., Ikeda, K., Piscaglia, F., et al. (2018). Lenvatinib versus sorafenib in first-line treatment of patients with unresectable hepatocellular carcinoma: A randomised phase 3 non-inferiority trial. *Lancet*, 391(10126), 1163–1173. https://doi.org/10.1016/S0140-6736(18)30207-1.

Lancaster, M. A., Renner, M., Martin, C. A., Wenzel, D., Bicknell, L. S., Hurles, M. E., et al. (2013). Cerebral organoids model human brain development and microcephaly. *Nature, 501*(7467), 373-+. https://doi.org/10.1038/nature12517.

Lang, P., Yeow, K., Nichols, A., & Scheer, A. (2006). Cellular imaging in drug discovery. *Nature Reviews Drug Discovery, 5*(4), 343–356. https://doi.org/10.1038/nrd2008.

Larkin, J., Chiarion-Sileni, V., Gonzalez, R., Grob, J. J., Rutkowski, P., Lao, C. D., et al. (2019). Five-year survival with combined nivolumab and ipilimumab in advanced melanoma. *New England Journal of Medicine, 381*(16), 1535–1546. https://doi.org/10.1056/NEJMoa1910836.

Lee, G. Y., Kenny, P. A., Lee, E. H., & Bissell, M. J. (2007). Three-dimensional culture models of normal and malignant breast epithelial cells. *Nature Methods, 4*(4), 359–365. https://doi.org/10.1038/Nmeth1015.

Li, L., Halpert, G., Lerner, M. G., Hu, H., Dimitrion, P., Weiss, M. J., et al. (2021). Protein synthesis inhibitor omacetaxine is effective against hepatocellular carcinoma. *Jci. Insight, 6*(12). https://doi.org/10.1172/jci.insight.138197.

Li, L., Knutsdottir, H., Hui, K., Weiss, M. J., He, J., Philosophe, B., et al. (2019). Human primary liver cancer organoids reveal intratumor and interpatient drug response heterogeneity. *Jci. Insight, 4*(2), ARTN e121490. https://doi.org/10.1172/jci.insight.121490.

Ling Li, H. K., Bader, J., & Selaru, F. (2019). Complext patient derived organoids(CPDOS) modeling for anti-cancer drug screening. *Gastroenterology, 156*(6).

Liu, J. Y., Li, P. F., Wang, L., Li, M., Ge, Z. H., Noordam, L., et al. (2021). Cancer-associated fibroblasts provide a stromal niche for liver cancer organoids that confers trophic effects and therapy resistance. *Cellular and Molecular Gastroenterology and Hepatology, 11*(2), 407–431. https://doi.org/10.1016/j.jcmgh.2020.09.003.

Llovet, J. M., De Baere, T., Kulik, L., Haber, P. K., Greten, T. F., Meyer, T., et al. (2021). Locoregional therapies in the era of molecular and immune treatments for hepatocellular carcinoma. *Nature Reviews Gastroenterology & Hepatology, 18*(5), 293–313. https://doi.org/10.1038/s41575-020-00395-0.

Llovet, J. M., Kelley, R. K., Villanueva, A., Singal, A. G., Pikarsky, E., Roayaie, S., et al. (2021). Hepatocellular carcinoma. *Nature Reviewes Disease Primers, 7*(1), 6. https://doi.org/10.1038/s41572-020-00240-3.

Llovet, J. M., Montal, R., Sia, D., & Finn, R. S. (2018). Molecular therapies and precision medicine for hepatocellular carcinoma. *Nature Reviews. Clinical Oncology, 15*(10), 599–616. https://doi.org/10.1038/s41571-018-0073-4.

Llovet, J. M., Ricci, S., Mazzaferro, V., Hilgard, P., Gane, E., Blanc, J. F., et al. (2008). Sorafenib in advanced hepatocellular carcinoma. *New England Journal of Medicine, 359*(4), 378–390. https://doi.org/10.1056/NEJMoa0708857.

Mak, I. W., Evaniew, N., & Ghert, M. (2014). Lost in translation: Animal models and clinical trials in cancer treatment. *American Journal of Translational Research, 6*(2), 114–118.

Mantovani, A., Marchesi, F., Malesci, A., Laghi, L., & Allavena, P. (2017). Tumour-associated macrophages as treatment targets in oncology. *Nature Reviews. Clinical Oncology, 14*(7), 399–416. https://doi.org/10.1038/nrclinonc.2016.217.

Manzanares, M. A., Campbell, D. J. W., Maldonado, G. T., & Sirica, A. E. (2018). Overexpression of periostin and distinct mesothelin forms predict malignant progression in a rat cholangiocarcinoma model. *Hepatology Communications, 2*(2), 155–172. https://doi.org/10.1002/hep4.1131.

Manzanares, M. A., Usui, A., Campbell, D. J., Dumur, C. I., Maldonado, G. T., Fausther, M., et al. (2017). Transforming growth factors alpha and beta are essential for modeling cholangiocarcinoma desmoplasia and progression in a three-dimensional organotypic culture model. *American Journal of Pathology, 187*(5), 1068–1092. https://doi.org/10.1016/j.ajpath.2017.01.013.

Marusyk, A., Tabassum, D. P., Janiszewska, M., Place, A. E., Trinh, A., Rozhok, A. I., et al. (2016). Spatial proximity to fibroblasts impacts molecular features and therapeutic sensitivity of breast cancer cells influencing clinical outcomes. *Cancer Research*, *76*(22), 6495–6506. https://doi.org/10.1158/0008-5472.Can-16-1457.

Mullenders, J., de Jongh, E., Brousali, A., Roosen, M., Blom, J. P. A., Begthel, H., et al. (2019). Mouse and human urothelial cancer organoids: A tool for bladder cancer research. *Proceedings of the National Academy of Sciences of the United States of America*, *116*(10), 4567–4574. https://doi.org/10.1073/pnas.1803595116.

Naruse, M., Ochiai, H., Sekine, S., Taniguchi, H., Yoshida, T., Ichikawa, H., et al. (2021). Re-expression of REG family and DUOXs genes in CRC organoids by co-culturing with CAFs. *Scientific Reports*, *11*(1), ARTN 2077. https://doi.org/10.1038/s41598-021-81475-2.

Noel, G., Baetz, N. W., Staab, J. F., Donowitz, M., Kovbasnjuk, O., Pasetti, M. F., et al. (2017). A primary human macrophage-enteroid co-culture model to investigate mucosal gut physiology and host-pathogen interactions. *Scientific Reports*, *7*. https://doi.org/10.1038/srep45270. ARTN 45270.

Nuciforo, S., Fofana, I., Matter, M. S., Blumer, T., Calabrese, D., Boldanova, T., et al. (2018). Organoid models of human liver cancers derived from tumor needle biopsies. *Cell Reports*, *24*(5), 1363–1376. https://doi.org/10.1016/j.celrep.2018.07.001.

Ohlund, D., Handly-Santana, A., Biffi, G., Elyada, E., Almeida, A. S., Ponz-Sarvise, M., et al. (2017). Distinct populations of inflammatory fibroblasts and myofibroblasts in pancreatic cancer. *Journal of Experimental Medicine*, *214*(3), 579–596. https://doi.org/10.1084/jem.20162024.

Ooft, S. N., Weeber, F., Dijkstra, K. K., McLean, C. M., Kaing, S., van Werkhoven, E., et al. (2019). Patient-derived organoids can predict response to chemotherapy in metastatic colorectal cancer patients. *Science Translational Medicine*, *11*(513), ARTN eaay2574. https://doi.org/10.1126/scitranslmed.aay2574.

Pardoll, D. M. (2012). The blockade of immune checkpoints in cancer immunotherapy. *Nature Reviews Cancer*, *12*(4), 252–264. https://doi.org/10.1038/nrc3239.

Quail, D. F., & Joyce, J. A. (2013). Microenvironmental regulation of tumor progression and metastasis. *Nature Medicine*, *19*(11), 1423–1437. https://doi.org/10.1038/nm.3394.

Radhakrishnan, J., Varadaraj, S., Dash, S. K., Sharma, A., & Verma, R. S. (2020). Organotypic cancer tissue models for drug screening: 3D constructs, bioprinting and microfluidic chips. *Drug Discovery Today*, *25*(5), 879–890. https://doi.org/10.1016/j.drudis.2020.03.002.

Roerink, S. F., Sasaki, N., Lee-Six, H., Young, M. D., Alexandrov, L. B., Behjati, S., et al. (2018). Intra-tumour diversification in colorectal cancer at the single-cell level. *Nature*, *556*(7702), 457. https://doi.org/10.1038/s41586-018-0024-3.

Sachs, N., de Ligt, J., Kopper, O., Gogola, E., Bounova, G., Weeber, F., et al. (2018). A living biobank of breast cancer organoids captures disease heterogeneity. *Cell*, *172*(1–2), 373-+. https://doi.org/10.1016/j.cell.2017.11.010.

Saito, Y., Muramatsu, T., Kanai, Y., Ojima, H., Sukeda, A., Hiraoka, N., et al. (2019). Establishment of patient-derived organoids and drug screening for biliary tract carcinoma. *Cell Reports*, *27*(4), 1265-+. https://doi.org/10.1016/j.celrep.2019.03.088.

Saito, Y., Muramatsu, T., & Saito, H. (2020). Establishment and long-term culture of organoids derived from human biliary tract carcinoma. *STAR Protocols*, *1*(1), 100009. https://doi.org/10.1016/j.xpro.2019.100009.

Sato, T., Vries, R. G., Snippert, H. J., van de Wetering, M., Barker, N., Stange, D. E., et al. (2009). Single Lgr5 stem cells build crypt-villus structures in vitro without a mesenchymal niche. *Nature*, *459*(7244), 262–U147. https://doi.org/10.1038/nature07935.

Seidlitz, T., & Stange, D. E. (2021). Gastrointestinal cancer organoids-applications in basic and translational cancer research. *Experimental & Molecular Medicine*, 53(10), 1459–1470. https://doi.org/10.1038/s12276-021-00654-3.

Sharma, P., & Allison, J. P. (2015a). The future of immune checkpoint therapy. *Science*, 348(6230), 56–61. https://doi.org/10.1126/science.aaa8172.

Sharma, P., & Allison, J. P. (2015b). Immune checkpoint targeting in cancer therapy: Toward combination strategies with curative potential. *Cell*, 161(2), 205–214. https://doi.org/10.1016/j.cell.2015.03.030.

Shroyer, N. F. (2016). Tumor organoids fill the niche. *Cell Stem Cell*, 18(6), 686–687. https://doi.org/10.1016/j.stem.2016.05.020.

Sirica, A. E. (2012). The role of cancer-associated myofibroblasts in intrahepatic cholangiocarcinoma. *Nature Reviews Gastroenterology & Hepatology*, 9(1), 44–54. https://doi.org/10.1038/nrgastro.2011.222.

Sirica, A. E., Almenara, J. A., & Li, C. (2014). Periostin in intrahepatic cholangiocarcinoma: Pathobiological insights and clinical implications. *Experimental and Molecular Pathology*, 97(3), 515–524. https://doi.org/10.1016/j.yexmp.2014.10.007.

Socinski, M. A., Jotte, R. M., Cappuzzo, F., Orlandi, F., Stroyakovskiy, D., Nogami, N., et al. (2018). Atezolizumab for first-line treatment of metastatic nonsquamous NSCLC. *New England Journal of Medicine*, 378(24), 2288–2301. https://doi.org/10.1056/NEJMoa1716948.

Stock, K., Estrada, M. F., Vidic, S., Gjerde, K., Rudisch, A., Santo, V. E., et al. (2016). Capturing tumor complexity in vitro: Comparative analysis of 2D and 3D tumor models for drug discovery. *Scientific Reports*, 6, ARTN 28951. https://doi.org/10.1038/srep28951.

Sung, H., Ferlay, J., Siegel, R. L., Laversanne, M., Soerjomataram, I., Jemal, A., et al. (2021). Global cancer statistics 2020: GLOBOCAN estimates of incidence and mortality worldwide for 36 cancers in 185 countries. *CA: a Cancer Journal for Clinicians*, 71(3), 209–249. https://doi.org/10.3322/caac.21660.

Thomas, D. W., Burns, J., Audette, J., Carroll, A., Dow-Hygelund, C., & Hay, M. (2016). *Clinical development success rates 2006–2015*. BIO Industry Analysis.

Tsai, S., McOlash, L., Palen, K., Johnson, B., Duris, C., Yang, Q. H., et al. (2018). Development of primary human pancreatic cancer organoids, matched stromal and immune cells and 3D tumor microenvironment models. *BMC Cancer*, 18. https://doi.org/10.1186/s12885-018-4238-4. ARTN 335.

Valle, J., Wasan, H., Palmer, D. H., Cunningham, D., Anthoney, A., Maraveyas, A., et al. (2010). Cisplatin plus gemcitabine versus gemcitabine for biliary tract cancer. *New England Journal of Medicine*, 362(14), 1273–1281. https://doi.org/10.1056/NEJMoa0908721.

van de Wetering, M., Francies, H. E., Francis, J. M., Bounova, G., Iorio, F., Pronk, A., et al. (2015). Prospective derivation of a living organoid biobank of colorectal cancer patients. *Cell*, 161(4), 933–945. https://doi.org/10.1016/j.cell.2015.03.053.

Veninga, V., & Voest, E. E. (2021). Tumor organoids: Opportunities and challenges to guide precision medicine. *Cancer Cell*, 39(9), 1190–1201. https://doi.org/10.1016/j.ccell.2021.07.020.

Villanueva, A. (2019). Hepatocellular Carcinoma. *New England Journal of Medicine*, 380(15), 1450–1462. https://doi.org/10.1056/NEJMra1713263.

Vitale, I., Manic, G., Coussens, L. M., Kroemer, G., & Galluzzi, L. (2019). Macrophages and metabolism in the tumor microenvironment. *Cell Metabolism*, 30(1), 36–50. https://doi.org/10.1016/j.cmet.2019.06.001.

Vlachogiannis, G., Hedayat, S., Vatsiou, A., Jamin, Y., Fernandez-Mateos, J., Khan, K., et al. (2018). Patient-derived organoids model treatment response of metastatic gastrointestinal cancers. *Science*, 359(6378), 920-+. https://doi.org/10.1126/science.aao2774.

Wong, C. H., Siah, K. W., & Lo, A. W. (2019a). Estimation of clinical trial success rates and related parameters. *Biostatistics*, *20*(2), 273–286. https://doi.org/10.1093/biostatistics/kxx069.
Wong, C. H., Siah, K. W., & Lo, A. W. (2019b). Estimation of clinical trial success rates and related parameters (vol 20, pg 273, 2019). *Biostatistics*, *20*(2), 366. https://doi.org/10.1093/biostatistics/kxy072.
Yao, Y., Xu, X. Y., Yang, L. F., Zhu, J., Wan, J. F., Shen, L. J., et al. (2020). Patient-derived organoids predict chemoradiation responses of locally advanced rectal cancer. *Cell Stem Cell*, *26*(1), 17-+. https://doi.org/10.1016/j.stem.2019.10.010.
Yap, T. A., Johnson, A., & Meric-Bernstam, F. (2021). Precision medicine in oncology-toward the integrated targeting of somatic and germline genomic aberrations. *JAMA Oncology*, *7*(4), 507–509. https://doi.org/10.1001/jamaoncol.2020.7988.
Yuki, K., Cheng, N., Nakano, M., & Kuo, C. J. (2020). Organoid models of tumor immunology. *Trends in Immunology*, *41*(8), 652–664. https://doi.org/10.1016/j.it.2020.06.010.
Zhang, X. M., Claerhout, S., Prat, A., Dobrolecki, L. E., Petrovic, I., Lai, Q., et al. (2013). A renewable tissue resource of phenotypically stable, biologically and ethnically diverse, patient-derived human breast cancer xenograft models. *Cancer Research*, *73*(15), 4885–4897. https://doi.org/10.1158/0008-5472.Can-12-4081.

CHAPTER TWELVE

Molecular therapeutic targets for cholangiocarcinoma: Present challenges and future possibilities

Dan Høgdall[a,b,†], Colm J. O'Rourke[a,†], and Jesper B. Andersen[a,*]
[a]Biotech Research and Innovation Centre (BRIC), Department of Health and Medical Sciences, University of Copenhagen, Copenhagen, Denmark
[b]Department of Oncology, Herlev and Gentofte Hospital, Herlev, Copenhagen University Hospital, Copenhagen, Denmark
[*]Corresponding author: e-mail address: jesper.andersen@bric.ku.dk

Contents

1. Introduction	345
2. Genomic alterations as therapeutic targets in CCA	348
2.1 FGFR2: A success story	348
2.2 IDH1: A success story	349
3. Non-genomic alterations as therapeutic targets in CCA	350
3.1 Receptor tyrosine kinases	350
3.2 Immunotherapy	352
4. Present challenges and future possibilities for therapeutic targets in CCA	355
4.1 Undruggable oncogenes and tumor suppressors	355
5. Predictive Biomarkers & Treatment Resistance	357
6. Current & future perspectives	360
Acknowledgment	360
Conflict of interest statement	360
References	361

Abstract

A diagnosis of cholangiocarcinoma (CCA) is implicit with poor prognosis and limited treatment options, underscoring the near equivalence of incidence and mortality rates in this disease. In less than 9 years from genomic identification to FDA-approval of the corresponding inhibitors, fibroblast growth factor receptor 2 (*FGFR2*) rearrangements and isocitrate dehydrogenase 1 (*IDH1*) mutations became exemplary successes of precision oncology in subsets of patients with CCA. However, clinical trial results from multikinase inhibitors in unselected populations have been less successful, while the impact of immunotherapies are only beginning to impact this setting. Development of future

[†] Equal contribution.

therapeutics is incumbent with new challenges. Many driver alterations occur in tumor suppressor-like genes which are not directly druggable. Therapeutically, this will require identification of ensuant "non-oncogene addiction" involving genes which are not themselves oncogenes but become tumor survival dependencies when a specific driver alteration occurs. The low recurrence frequency of genomic alterations between CCA patients will require careful evaluation of targeted agents in biomarker-enrolled trials, including basket trial settings. Systematic expansion of candidate drug targets must integrate genes affected by non-genetic alterations which incorporates the fundamental contribution of the microenvironment and immune system to treatment response, disease facets which have been traditionally overlooked by DNA-centric analyses. As treatment resistance is an inevitability in advanced disease, resistance mechanisms require characterization to guide the development of combination therapies to increase the duration of clinical benefit. Patient-focused clinical, technological and analytical synergy is needed to deliver future solutions to these present therapeutic challenges.

Abbreviations

2HG	2-hydroxyglutarate
αKG	α-ketoglutarate
ABL2	ABL proto-oncogene 2
ARID1A	AT-rich interactive domain-containing protein 1A
BAP1	BRCA1 associated protein-1
BICC1	BicC family RNA binding protein 1
BRAF	v-raf murine sarcoma viral oncogene homolog B1
BTC	biliary tract cancer
CCA	cholangiocarcinoma
CPI	checkpoint inhibitors
CR	complete response
ctDNA	circulating tumor DNA
CTLA-4	cytotoxic T-lymphocyte-associated protein 4
dCCA	distal cholangiocarcinoma
dMMR	deficient mismatch repair
EGFR	epidermal growth factor receptor
EID	extracellular domain in-frame deletions
ESCAT	European society of medical oncology scale for clinical actionability of molecular targets
FDA	food and drug administration
FGFR	fibroblast growth factor receptor
FGFRi	fibroblast growth factor receptor inhibitor
FIGHT-202	fibroblast growth factor receptor inhibitor in oncology and hematology trial
FOLFOX	folinic acid, fluorouracil and oxaliplatin
HER2	human epidermal growth factor receptor 2
HNF4A	hepatocyte nuclear factor 4 alpha
HR	hazard ratio
HUGO	human gene nomenclature committee
iCCA	intrahepatic cholangiocarcinoma
IDH	isocitrate dehydrogenase

KRAS	KRAS proto-oncogene, GTPase
LRRK2	leucine rich repeat kinase 2
MEK	mitogen-activated protein kinase
MET	met proto-oncogene, receptor tyrosine kinase
MHC	major histocompatibility complex
MSI	microsatellite instability
NCT	national clinical trial number
NSCLC	non-small cell lung cancer
ORR	objective response rate
OS	overall survival
PARPi	poly(adenosine 5′-diphosphate) ribose polymerase inhibitor
PBRM1	polybromo 1
pCCA	perihilar cholangiocarcinoma
PD-1	programmed cell death protein 1
PD-L1	programmed death-ligand 1
PFS	progression-free survival
PR	partial response
ROAR	rare oncology agnostic research trial
RTK	receptor tyrosine kinase
TET	ten-eleven-translocation
TGF-β	transforming growth factor-beta
TMB-H	high tumor mutational burden
TME	tumor microenvironment
TP53	tumor protein p53
VEGF	vascular endothelial growth factor

1. Introduction

Tumors with features of cholangiocyte differentiation arising throughout the biliary tract are collectively annotated as cholangiocarcinoma (CCA). Despite this umbrella diagnosis, these malignancies are highly heterogeneous, which significantly impacts optimal clinical management of individual patients. CCA is anatomically stratified into intrahepatic CCA (iCCA) which arises inside the liver, and extrahepatic CCA (eCCA) which arises outside the liver and is further stratified into perihilar (pCCA) and distal CCA (dCCA) (Banales et al., 2020). Approximately 70% of patients are diagnosed with advanced disease (Forner et al., 2019), a scenario attributable to the largely asymptomatic development of these tumors in addition to the lack of biomarkers for general or high-risk population screening. Even among patients with potentially operable disease, post-surgical upstaging is common and progression of micrometastatic disease rapidly occurs in most patients following surgery. Therapeutically, CCA is often considered alongside gallbladder cancer as biliary tract cancer (BTC). Since 2010, doublet cytotoxic chemotherapy with gemcitabine and cisplatin remains the standard-of-care for patients with recurrent, locally advanced and metastatic CCA (Valle et al., 2010). With this regimen, median progression-free survival (PFS) is approximately 8 months and overall survival (OS) is approximately 11.7 months. However, patient benefit from first-line chemotherapy is highly variable, with disease control achieved in some patients for lengthy durations whereas rapid progression occurs in others. With folinic acid, fluorouracil and oxaliplatin (FOLFOX) now recommended as second-line chemotherapy (Lamarca et al., 2021) and targeted therapies also available in second-line for subgroups of patients, therapeutic opportunities are at their peak for advanced, chemorefractory CCA.

Overall, CCA appears genomically unremarkable compared to other cancer types, exhibiting intermediate genome mutation and aneuploidy rates, as well as average numbers of driver events per tumor (including point mutations, copy number alterations and structural alterations in coding and non-coding regions) (Consortium, 2020; Taylor et al., 2018). However, meta-analysis of genomic data from 1424 tumors spanning 10 independent studies identified only 34 cancer-associated genes that were recurrently altered in >10% of CCA patients (Boerner et al., 2021; Chan-On et al., 2013; Farshidfar et al., 2017; Jiao et al., 2013; Jolissaint et al., 2021; Jusakul et al., 2017; Lowery et al., 2018; Ong et al., 2012; Sia et al., 2015; Zou et al., 2014) (Fig. 1). Such low recurrence of genomic alterations emphasizes the extensive molecular heterogeneity characteristic of CCA patients, mirroring that observed at biological and clinical levels. Therapeutically, 85% (29/34) of the recurrent genomic alterations in CCA patients are classified as having no evidence of actionability (tier X) as defined by the European Society of Medical Oncology Scale for Clinical Actionability of Molecular Targets (ESCAT) (Mateo et al., 2018). Only genomic alterations in fibroblast growth factor receptor 2 (*FGFR2*) and isocitrate dehydrogenase 1 (*IDH1*) are currently clinically actionable (tier I), having become the first and second FDA approved targeted therapies for iCCA in 2020 and 2021, respectively. KRAS proto-oncogene, GTPase (*KRAS*) is recurrently altered in 14% CCA and hypothetically actionable based on trials in other cancers (Skoulidis et al., 2021) (tier III), but mutant KRAS-directed therapies remain to be tested for CCA in this setting. Similarly, ABL proto-oncogene 2 (*ABL2*) is recurrently amplified in 11%

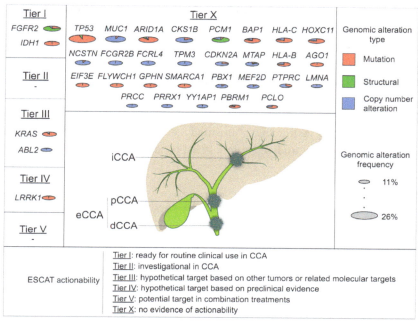

Fig. 1 Clinical actionability of recurrent genomic alterations in CCA. Diverse genomic perturbation mechanisms are operative in CCA, including mutations, structural alterations and copy number alterations. Genomic data mining (cBioPortal) of 1424 CCA (dCCA, iCCA, pCCA, unspecified CCA) patient tumors identified 34 cancer-associated genes that are altered in >10% of the population. Among these genes, *FGFR2* and *IDH1* are routinely clinically actionable (tier I); *KRAS* and ABL2 are potentially actionable based on clinical trial data in other cancers (tier III); *LRRK1* is hypothetically actionable based on pre-clinical data in other cancers (tier IV); whereas the majority of recurrently altered genes have no evidence of actionability (tier V). Target actionability with existing small molecule inhibitors or antibodies is defined by the European Society for Medical Oncology Scale for Clinical Actionability of Molecular Targets (ESCAT). Gene names are annotated according to the Human Gene Nomenclature Committee (HUGO). dCCA: distal CCA; eCCA: extrahepatic CCA; iCCA: intrahepatic CCA; pCCA: perihilar CCA.

CCA patients and is potentially actionable with multikinase inhibitors, such as imatinib, based on trial data in other cancers (Joensuu et al., 2020). Preclinical data suggest a role for leucine rich repeat kinase 2 (*LRRK2*) in sunitinib sensitivity (Sheils et al., 2021), which could be relevant in CCA with 13% of patients harboring *LRRK2* mutations. Collectively, these data present an ominous picture in which recurrently actionable targets have plateaued based on genomic analysis. Advancing new therapeutics into the clinic will require evaluation of genes that are altered at low recurrence frequencies and may be currently biologically obscure in function, as well as assessment of candidate drug targets that are altered by non-genetic mechanisms in CCA.

In this chapter, we describe the characterization of *FGFR2* and *IDH1* alterations that paved the way for the successes of the first targeted therapies in CCA. Subsequently, we critically assess the performance of drugs targeting receptor tyrosine kinases, as well as the burgeoning contribution of immunotherapy in this space. Lastly, we discuss current challenges and potential solutions facing the continued evolution of CCA therapeutics, focusing on currently undruggable targets, treatment resistance and predictive biomarkers.

2. Genomic alterations as therapeutic targets in CCA

2.1 FGFR2: A success story

Fusions involving the fibroblast growth factor receptor 2 (*FGFR2*) gene occur in approximately 15% of patients with iCCA (Nepal et al., 2018). While diverse donor genes have been reported, BicC family RNA binding protein 1 (*BICC1*) is the most common fusion partner with breakpoint regions occurring in exon 3 or exon 5 of *FGFR2* (Nepal et al., 2018). *FGFR2* fusions tend to occur mutually exclusive with mutations in *KRAS*, *IDH1* and *BRAF* (Arai et al., 2014; Jusakul et al., 2017; Nepal et al., 2018), suggesting that translocated *FGFR2* can function independently as a potent oncogene. Indeed, overexpression of human *FGFR2* fusions in mouse NIH3T3 fibroblasts (Arai et al., 2014) and tumor protein p53 ($Tp53$)$^{-/-}$ mouse liver organoids (Cristinziano et al., 2021) generated iCCA-like tumors *in vivo*. FGFR signaling is constitutively active in fusion-positive tumor organoid models and cell lines, potently stimulating downstream Ras-Erk signaling independent of *KRAS* mutations (Cristinziano et al., 2021). Kinase domain mutations impaired tumor formation *in vivo* (Arai et al., 2014) and viability of tumor models was compromised in a dose-dependent manner following treatment with the pan-FGFR inhibitor (FGFRi), infigratinib (BGJ398) (Cristinziano et al., 2021). The high recurrence of *FGFR2* fusions and experimental data supporting addiction to FGF signaling in these tumors highlighted iCCA patients as an ideal demographic in which to test pan-FGFRi that were already in development. The single-arm, phase 2 fibroblast Growth factor receptor inhibitor in oncology and Hematology Trial (FIGHT-202) trial of pemigatinib in previously treated, *FGFR2*-altered CCA achieved an objective response rate of 35.5%, including 3 complete and 35 partial responses (Abou-Alfa, Sahai, et al., 2020). Such therapeutic responses led to the accelerated FDA approval of pemigatinib, in this setting, in April 2020. Additional FGFRi, including infigratinib (Javle et al., 2021) and futibatinib (Meric-Bernstam et al., 2021), have subsequently demonstrated therapeutic potential in phase 2 and 1 trials, respectively, highlighting FGFR blockade as an already evolving mainstay in the clinical management of FGFR2-altered iCCA patients.

2.2 IDH1: A success story

Isocitrate dehydrogenase 1 (*IDH1*) is mutated in approximately 14% of patients with iCCA (Nepal et al., 2018), with these alterations typically occurring at the R132 hotpot. Such mutations generate the oncometabolite, 2-hydroxyglutarate (2HG), in place of α-ketoglutarate (αKG), resulting in impaired functionality of diverse processes which are dependent on αKG as a cofactor. Uncoupling the cancer-promoting mechanism(s) of mutant IDH1 from the wide-ranging consequences of these mutations remains challenging, unlike classic oncogenes. IDH1 mutant iCCA appears biologically distinct from wild-type iCCA, including higher copy number, expression of mitochondrial genes and lower expression of chromatin remodeling genes (Farshidfar et al., 2017). The αKG-dependent DNA demethylases, ten-eleven-translocation (TETs), become deficient in IDH mutant tumors, resulting in distinct DNA hypermethylation profiles across the genome (Farshidfar et al., 2017; Nepal et al., 2018; Wang et al., 2013). Homologous recombination deficiency arises in other IDH-mutated cancers (Pirozzi & Yan, 2021), and this also appears to be the case in iCCA based on experiments in two immortalized cell lines (Wang et al., 2020). In mice, *IDH1* mutations lead to 2HG-associated suppression of the master transcription factor, hepatocyte nuclear factor 4 alpha (HNF4A), blocking hepatic progenitor differentiation and triggering cholangiocarcinogenesis to metastatic disease when introduced alongside *Kras* mutations (Saha et al., 2014). These data indicate a dedifferentiation mechanism contributing to initiation of IDH-mutated iCCA. However, 2HG-associated TET2 dysfunction has been shown to compromise induction of interferon-γ response genes and immunoevasion, indicating an immune-epigenetic role in iCCA maintenance and progression (Wu et al., 2021). Therapeutically, IDH mutant iCCA models also appear to have distinct sensitivities to metabolic agents (Nepal et al., 2018), as well as increased SRC-dependency leading to dasatinib hypersensitivity (Saha et al., 2016). Such biological and therapeutic distinctions highlight IDH-mutated iCCA as a prime subgroup for precision medicine trials, in particular following the tumor agnostic development of the IDH mutant inhibitor, AG-120 (Dhillon, 2018). The double-blind, phase 3 ClarIDHy trial evaluated the therapeutic potential of ivosidenib (formerly AG-120) in previously treated patients with IDH1-mutated iCCA (Abou-Alfa, Macarulla, et al., 2020; Zhu et al., 2021). PFS was 2.7 months in the treatment arm ($n=124$) compared to 1.4 months in the placebo arm ($n=61$), culminating in a HR of 0.37 for ivosidenib treatment (Abou-Alfa, Macarulla, et al., 2020). Similarly, OS was 10.3 months in the treatment arm compared to 5.1 months in the placebo arm after crossover adjustment, contributing to a HR of 0.49 for ivosidenib treatment (Zhu et al., 2021). In August 2021, ivosidenib was FDA-approved as a second-line treatment for *IDH1*-mutated iCCA.

3. Non-genomic alterations as therapeutic targets in CCA

3.1 Receptor tyrosine kinases

Oncogenic addiction to receptor tyrosine kinase (RTK) signaling is a fundamental characteristic of CCA, like other cancers. RTKs are diverse transmembrane proteins that bind extracellular ligands, leading to intracellular tyrosine domain phosphorylation and downstream signaling cascades that promote cell growth and survival. While genomic alterations can result in RTK hyperactivity, such as *FGFR2* fusions, the major contributor to increased RTK signaling in CCA appears to be non-genomic alterations. A prime example is vascular endothelial growth factor (VEGF) signaling which is stimulated by multiple cell types (tumor, immune, endothelial) through tumor-intrinsic and -extrinsic mechanisms promoting CCA (Mariotti, Fiorotto, Cadamuro, Fabris, & Strazzabosco, 2021). However, the therapeutic benefit of inhibitors targeting VEGF signaling as well as other kinases has been underwhelming in clinical trials for unselected CCA patients (Fig. 2). Treatment with the multikinase inhibitor, lenvatinib, resulted in an objective response rate (ORR) of 11.5% in a phase 2 study of 26 chemorefractory BTC patients (Ueno et al., 2020). Two independent phase 2 trials of the multikinase inhibitor, regorafenib, generated an ORR of 9.1% in 33 BTC patients (Kim, Sanoff, et al., 2020) and 11% in 34 BTC patients (Sun et al., 2019), who had previously progressed on chemotherapy. Administration of the multikinase inhibitor, cabozantinib, to 19 previously treated CCA patients resulted in limited therapeutic benefit and significant toxicities in a phase 2 trial (Goyal, Zheng, et al., 2017). A phase 2 trial of 34 previously treated BTC patients with sorafenib and erlotinib, a multikinase inhibitor and epidermal growth factor receptor respectively,

Fig. 2 Performance of targeted therapies in phase 2 trials in chemotherapy-refractory biliary tract cancer (BTC) patients. Therapeutic performance is depicted as overall response rate (ORR) vs progression-free survival (PFS) and overall survival (OS) in BTC patients. BTC patients included those diagnosed with CCA and gallbladder cancer. Phase 2 trials are only considered if they investigated second-line systemic targeted therapies and have results published in PubMed. *Pemigatinib data available for PFS but not OS; [a]regorafenib reported by Sun et al. (NCT02053376); [b]regorafenib reported by Kim et al. (NCT not available); NCT: national clinical trial number.

led to early study termination due to failure in meeting the predetermined patient number alive at 4 months (El-Khoueiry et al., 2014). Collectively, receptor tyrosine kinase inhibitors have performed underwhelmingly in unselected CCA populations, though careful patient selection and appropriate treatment combinations have potential to improve trial designs in this setting.

3.2 Immunotherapy
3.2.1 First generation immune checkpoint inhibitors
CCA is associated with a dense desmoplastic stroma in a chronic inflammatory state that attenuates effective anti-tumor immune responses through cytokine and cellular modulation (Hogdall, Lewinska, & Andersen, 2018). This inflammation is prevalent in all known risk factors of CCA and is in many cases thematic of the disease impacting patient prognosis (Hogdall et al., 2020). The importance of the immunological equilibrium is biologically epitomized in patients with microsatellite instability (MSI-H), deficient mismatch repair (dMMR), and high tumor mutational burden (TMB-H). In these cases, higher mutation loads increase the abundance of immunogenic neoantigens which are presented by major histocompatibility complex (MHC) class I and promote lymphocyte infiltration (Goeppert et al., 2019). However, anti-tumor immune responses are dampened by tumor expression of immune checkpoint members, such as programmed death-ligand 1 (PD-L1), leading to failed recognition of tumor cells by the immune system. It is based in this immune checkpoint system that current checkpoint inhibitors (CPI) have been developed, achieving durable responses in patients with otherwise poor prognosis and radically altering the management of several difficult-to-treat cancers, such as melanoma and non-small cell lung cancer (NSCLC). Compared to common malignancies, clinical evidence of the benefits of immunotherapy remains immature in CCA. Pembrolizumab is a monoclonal antibody directed against the PD-L1 receptor, programmed cell death protein 1 (PD-1), that gained tumor agnostic approval for use in MSI-H (Marcus, Lemery, Keegan, & Pazdur, 2019), dMMR (Marcus et al., 2019) and TMB-H ($\geq$10 mutations per megabase) (Marcus et al., 2021) solid tumors. MSI-H is only reported in 1.3% of CCA cases (Goeppert et al., 2019), though pembrolizumab may be particularly relevant to CCA patients with concurrent Lynch syndrome as it predisposes to MSI-H and dMMR. Despite the rarity of CCA, the KEYNOTE-158 basket trial has reported data specific for MSI-H and dMMR CCA (Marabelle et al., 2020). In this phase 2 trial, 22 previously treated CCA patients with MSI-H and/or dMMR were treated with

pembrolizumab. These encouraging results reported that 9% and 32% of patients experienced complete response (CR) and partial response (PR), respectively. Further, the ORR was 40.9% with a median OS of 24.3 months with the upper confidence interval not reached (Marabelle et al., 2020). Yet, most patients did not respond, indicating that CPI indications require expansion and refinement beyond MSI-H, dMMR or TMB-H status.

CPI monotherapy has also been investigated in unselected CCA patients. A multicenter phase 2 trial investigating nivolumab (anti-PD-1) in 54 previously treated BTC patients reported an ORR of 21% and 40% for iCCA and eCCA, respectively (Kim, Chung et al., 2020). Importantly, the median duration of response was not reached with a median follow-up of 12.4 months. PD-L1 expression (>1% of tumor cells) was associated with improved PFS but not OS in this trial, and notably none of these tumors were found to be dMMR. KEYNOTE-028 (phase 1b) and KEYNOTE-158 (phase 2) reported ORRs for pembrolizumab of 13% (all PD-L1-positive) and 5.8% (mixed PD-L1-positivity) in previously treated BTC patients, respectively (Piha-Paul et al., 2020). In these studies, the median duration of response was not reached, and some patients were reported as having responses greater than 50 months (Piha-Paul et al., 2020). The ORR was slightly higher in PD-L1-positive compared to -negative patients (6.6% vs 2.9%), but no association was found between OS and PD-L1 expression status. Collectively, these data suggest that a small minority of the unselected CCA population may benefit from CPI monotherapy and that this may be independent of current molecular indications for CPI use.

3.2.2 Immune checkpoint inhibitors in combination treatments

As the majority of CCA patients show no benefit of CPI monotherapy, it is compelling to consider if combination treatments could enhance efficacy. Combined targeting of the PD-1/PD-L1 axis together with another immune checkpoint member, cytotoxic T-lymphocyte-associated protein 4 (CTLA-4), has improved response in other cancers by increasing lymphocyte priming and invasion through CTLA-4 blockade, albeit at the expense of increased toxicities. Evidence of the efficacy of PD-L1 and CTLA-4 combination treatments in CCA patients is still limited. A phase 1 study evaluating durvalumab (anti-PD-L1) with ($n=65$) or without ($n=42$) tremelimumab (anti-CTLA-4) in a pretreated Japanese patient cohort reported partial responses in 11% and 5%, respectively (Kendre et al., 2021). Median OS was 10.1 and 8.1 months in the combination and monotherapy arms, respectively. However, this indication of enhanced effect came at a price of an increased toxicity profile for the combination arm with adverse events of any grade (82% vs 64%) and grade ≥3 (23% vs 19%), as well as one fatal event with the combination. Combination toxicities have also been described in a subgroup analysis of 39 BTC patients treated with nivolumab and ipilimumab (anti-CTLA-4) in the CA209–538 (Klein et al., 2020). While an ORR of 23% was achieved, toxic events occurred in 49% of participants, including grade ≥3 in 15%. Despite the need for larger trials, these data are reflective of the increased efficacy and toxicity profiles found in doublet treatments in other malignancies. If validated, patients with potentially resectable disease requiring tumor shrinkage could under close management be considered for dual CPIs, but the increased benefits appear limited.

Several strategies are exploring combinations of CPI with other treatment modalities. The combination of chemotherapy and CPIs is standard treatment in NSCLC and may modify the tumor microenvironment (TME) by increasing neoantigen expression and upregulating checkpoint expression. One such phase 1 trial enrolled 30 patients with BTC in a treatment arm combining gemcitabine and cisplatin with nivolumab (Ueno et al., 2019). Although small, the study reported an ORR of 37% with median OS of 15.4 months and a manageable toxicity profile. Comparing these data with the 26% response rate reported for standard-of-care chemotherapy in the ABC-02 trial (Wang et al., 2020), it is encouraging to consider the addition of a CTLA-4 inhibitor. However, the IMMUNOBIL PRODIGE 57 trial in BTC patients serves as a stark warning that chemotherapy-immunotherapy combinations (durvalumab, tremelimumab and paclitaxel) can result in significant toxicities, which ultimately led to trial discontinuation (Boileve et al., 2021). Notably, further data about the efficacy of combining chemotherapy and PD-L1 inhibition will be presented with the reporting from the large randomized KEYNOTE-966 (NCT04003636) ($n=1048$) and TOPAZ-1 (NCT03875235) ($n=757$) trials investigating cisplatin and gemcitabine in combination with pembrolizumab and durvalumab, respectively.

Multikinase inhibitors are also being investigated in combination with CPIs. Combining pembrolizumab with lenvatinib resulted in an ORR of 25% and median OS of 11 months among previously treated BTC patients, with manageable toxicity profiles (Lin et al., 2020). Similarly, preliminary data on 31 previously treated BTC patients in the LEAP basket protocol reported ORR of 10% and median OS of 8.6 months, leading to expansion of the protocol (Villanueva et al., 2021). Transforming growth factor-beta

(TGF-β) may also be a viable therapeutic target in combination with PD-L1, as demonstrated in a phase 1 trial of 30 previously treated BTC patients (Yoo et al., 2020). An ORR of 20% was reported, including 83% of those responses ongoing at study cut-off, and median OS was reported as 12.7 months. Notably, the investigators observed responses irrespectively of PD-L1 expression and MSI-H status, suggesting the potential of specific targeted therapies to prompt significant immune mobilization in boosting CPI response.

4. Present challenges and future possibilities for therapeutic targets in CCA

4.1 Undruggable oncogenes and tumor suppressors

A substantial proportion of tumors harbor clear oncogenic drivers that have traditionally been considered "undruggable," exemplified by *KRAS* which is mutated in 13% of CCA and associated with short patient survival (Boerner et al., 2021) (Fig. 1). KRAS functions as a guanine diphosphate-triphosphate switch responsible for regulating signal transduction between activated receptors and intracellular proteins across diverse pathways. Single-base missense mutations at hotspots (G12, G13, Q61) render mutant KRAS constitutively active, enabling hyperactivity of downstream signaling pathways. Developing direct inhibitors of KRAS has been exceptionally challenging due to its unusually smooth protein structure and corresponding lack of grooves against which to design inhibitory peptides. Development of the selective KRAS G12C inhibitor, sotorasib, successfully overcame this issue by forming an irreversible covalent bond to the mutant-specific sulfur atom of cysteine (Canon et al., 2019), leading to its accelerated FDA-approval in KRAS G12C lung cancer (Skoulidis et al., 2021). G12C mutations are not the most common *KRAS* alterations in CCA, and sotorasib has only been evaluated in a single BTC patient in the context of a phase 1 basket trial (Hong et al., 2020). Expansion of sotorasib and new KRAS mutant therapeutics will likely impact CCA patient management in upcoming years. Nonetheless, many CCA-relevant oncogenes (e.g., *MYC* alterations) and variants (e.g., non-G12C *KRAS* mutations) remain undruggable.

Recurrent driver alterations in tumor suppressor genes pose even greater therapeutic challenges as restoring wild-type expression through gene therapy or wild-type behavior through small molecule activators remain far from oncology clinics. Most notoriously, *TP53* is mutated in 25% of CCA

(Fig. 1), these mutations are distributed heterogeneously throughout the gene and are associated with poor survival (Boerner et al., 2021). Similarly, recurrent mutations in chromatin remodelers such as AT-rich interactive domain-containing protein 1A (*ARID1A*), BRCA1 associated protein-1 (*BAP1*) and polybromo 1 (*PBRM1*) appear to function as tumor suppressor-like events in CCA (Artegiani et al., 2019; Jiao et al., 2013).

Genomics-guided evaluation of driver-associated dependencies may provide the most realistic opportunity to overcome the therapeutic challenges imposed by undruggable oncogenes and tumor suppressor genes. In this scenario, certain genes which are not functionally oncogenic themselves become necessary for tumor cell survival due to the cellular implications of driver alterations. Such "non-oncogene addiction" leads to distinct therapeutic sensitivities among subgroups of CCA patients. For example, stratification of iCCA patients according to mutation status in three genes (*IDH1/2*, *KRAS*, *TP53*) uncovered distinct mutational, structural and transcriptomic characteristics that extended beyond the immediate biological pathways in which these genes function (Nepal et al., 2018). By genotypically matching CCA cell lines to these patient subgroups, unique pharmacological sensitivities were observed corresponding to those pathways compromised in patients. These included increased sensitivity to RNA synthesis inhibitors in *IDH*-mutated tumors, microtubule inhibitors in *KRAS*-mutated tumors, topoisomerase inhibitors in *TP53*-mutated tumors, and mTOR inhibitors in tumors lacking these mutations (but including *FGFR2* fusions). Mechanistically, such non-oncogene addictions likely arise due to synthetic lethality. For example, it is known that 2HG production in IDH-mutated cells impairs homologous recombination-dependent DNA repair, resulting in poly(adenosine 5′-diphosphate) ribose polymerase inhibitor (PARPi) sensitivity that can be therapeutically exploited in models of CCA, as well as further enhanced by coadministration of localized radiation therapy (Wang et al., 2020). However, the specific interactions between drivers and resulting vulnerabilities remain mostly unclear. For example, *IDH1* mutant CCA cells are hypersensitive to dasatinib specifically due to its inhibition of SRC, but drug sensitivity did not correspond to differences in SRC expression and dasatinib sensitivity could not be conferred by overexpression of mutant IDH, establishing this definitive sensitivity as an unresolved biological question (Saha et al., 2016). Further advancing non-oncogene addiction-directed therapies in CCA will require expansive integration of "omics" in parallel with high-throughput drug screening.

5. Predictive Biomarkers & Treatment Resistance

Biomarker-guided patient selection is a key principle shared among the most successful clinical trials in CCA (Fig. 2) and will become of even greater importance as new therapeutics emerge resulting in more complex treatment decision-making algorithms (Fig. 3). It is further possible that broadening genomic inclusion criteria to encompass different perturbation mechanisms may increase the catchment of candidate responders. For example,

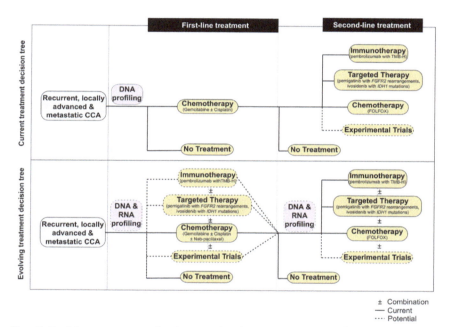

Fig. 3 Evolving treatment landscapes in the management of recurrent, locally advanced and metastatic CCA patients. Current oncological management of CCA patients involves chemotherapy in first-line followed by varied treatments in second-line, depending on patient status. DNA profiling of diagnostic biopsies may support second-line treatment with specific targeted therapies. Alternatively, second-line chemotherapy or enrolment in clinical trials may be options. Future advances may involve moving current second-line therapies into first-line in subsets of CCA patients, guided by advanced DNA and RNA profiling. Combinations of therapeutic modalities may be developed, repeat tissues biopsies may be performed following progression on first-line, and liquid biopsies may also be incorporated during patient management. *FGFR2*: fibroblast growth factor receptor 2; FOLFOX: folinic acid, fluorouracil and oxaliplatin; *IDH1*: isocitrate dehydrogenase 1; TMB-H: tumor mutational burden—high.

the FIGHT-202 trial included a small cohort B arm with other *FGF* or *FGFR* alterations among whom three out of four patients with *FGFR2* p.C382R mutations achieved disease stabilization (Silverman et al., 2021). Further, *FGFR2* extracellular domain in-frame deletions (EID) occur in approximately 2.8% iCCA and all three EID-positive patients exhibited a partial response to treatment with the FGFRi, Debio 1347 (Cleary et al., 2021). However, routine incorporation of genomic profiling into clinical practice is not without its challenges, with biopsies failing to generate molecular data in approximately 26.8% of BTC cases (mostly due to insufficient tumor cells or DNA recovery) (Lamarca et al., 2020). In contrast, the failure rate of circulating tumor DNA (ctDNA) was only 15.4%, including successful identification of *FGFR2* fusions and mutations in BTC patients (Lamarca et al., 2020).

Excluding the most commonly altered genes that are therapeutically approved (*FGFR2*, *IDH1*), likely upcoming (*KRAS*) and not actionable (*TP53*, chromatin remodelers), the remaining targets in CCA are altered at very low frequencies (<10%). This raises challenges for the design of sufficiently powered clinical trials in this rare cancer demographic. An exemplary case of a trial that directly met this challenge is the Rare Oncology Agnostic Research (ROAR) basket trial which evaluated the therapeutic potential of combining the mutant v-raf murine sarcoma viral oncogene homolog B1 (BRAF) inhibitor and the mitogen-activated protein kinase (MEK) inhibitor, dabrafenib and trametinib, in BRAF^{V600E}-mutated rare cancers. A remarkable ORR of 49% was reported in the 43 BTC patients included in this study, highlighting the potential of this treatment to significantly improve outcome of the approximately 5% of BTC cases that harbor this molecular alteration (Subbiah et al., 2020). Use of basket trial approaches has also improved outcome of CCA patients on a patient-by-patient basis, and also highlighted CCA as target-rich (albeit heterogeneous) disease that is particularly amenable to precision oncology (Horak et al., 2021; Massard et al., 2017). In the MOSCATO-01 trial, 68% (23/34) of BTC patients were found to have actionable targets compared to 24% (199/844) across all trial participants, and molecularly guided treatment resulted in increased PFS compared to non-molecularly guided treatment (Verlingue et al., 2017). Coordinated basket trial efforts focusing on specific drugs in biomarker-recruited patients will be critical to advancing the next wave of therapeutics in CCA.

Despite the successes of pemigatinib in *FGFR2*-rearranged iCCA and ivosidenib in *IDH1*-mutated iCCA, the majority of patients in these trials did not achieve an objective response even though their biopsies were molecularly confirmed as positive for the target alterations. As such, it is clear that 'single gene' inclusion criteria are insufficient to predict response to treatment and this is certainly not a curiosum specific to these two therapies. As the biological effects of a genomic alteration are dependent on upstream, downstream and parallel pathway signaling, transcriptome profiling can provide a comprehensive overview of global gene activity that also incorporates the contribution of microenvironment cells that are stable at DNA-level. Diverse transcriptomic subgroups of CCA patients have been reported with potential therapeutic implications, including: a poor survival CCA subgroup defined by high expression of epidermal growth factor receptor (EGFR), human epidermal growth factor receptor 2 (HER2) and met proto-oncogene, receptor tyrosine kinase (MET) (Andersen et al., 2012); inflammation and proliferation classes of iCCA (Sia et al., 2013); four tumor microenvironment-based subtypes of iCCA (Job et al., 2020); a CCA subgroup defined by Notch activation and predicted γ-secretase inhibitor sensitivity (O'Rourke et al., 2020); metabolic, proliferation, mesenchymal and immune subgroups of eCCA (Montal et al., 2020). While biologically compelling, such transcriptomic classifications have yet to impact patient management and must first overcome hurdles such as compatibility with biopsy material and adaptability to clinical platforms (e.g., NanoString).

Treatment resistance is an inevitability for targeted therapies and immunotherapies in advanced CCA patients. Comparison of pre-treatment and post-progression biopsies from *FGFR2* fusion-positive iCCA patients undergoing infigratinib and pemigatinib treatment identified diverse *FGFR2* kinase domain mutations exclusively present in post-progression samples (Goyal, Saha, et al., 2017; Silverman et al., 2021). Modeling these mutations *in vitro* revealed that alternative, structurally distinct FGFRi can overcome resistance, indicating a role for sequential ordering of FGFRi to counteract resistance mechanisms. This has been supported by the therapeutic activity of the FGFRi, TAS-120, in four fusion-positive patients who previously progressed on infigratinib or Debio 1347 (Goyal et al., 2019). However, TAS-120 resistance has also been documented and combination regimens to pre-emptively mitigate such mechanisms are already being suggested. Among these, preclinical data indicate that FGFRi synergizes with MEK inhibitors even against the V565F resistance mutation (Cristinziano et al., 2021), as well as synergizing with KRAS blockade (Kendre et al., 2021). Harvesting longitudinal tissue biopsies is not standard practice or feasible for every patient, warranting serial evaluation by liquid biopsies. CCA has been revealed to have the greatest frequency of fusion kinases detected in ctDNA in a pan-cancer analysis of 36,916 patient samples, emphasizing the penetrance of these genomic events in biliary tumors (Lee et al., 2021). More so, comparing matched pre-treatment and post-progression tissue and liquid biopsies, ctDNA analysis recovered a greater number of FGFR2 resistance mutations (Goyal, Saha, et al., 2017). This indicates the evolution of independent resistance mutations between tumor sites that are only captured by liquid biopsy and not by standard single tumor biopsy. Though resistance mechanism studies are only mature for FGFRi, these scenarios are pertinent to all future therapeutics for CCA.

6. Current & future perspectives

From the first report of FGFR2 fusions in iCCA in 2013 (Wu et al., 2013) to the accelerated FDA approval of pemigatinib in 2020, and the first description of IDH1 mutations in iCCA in 2012 (Borger et al., 2012) to the accelerated FDA approval of ivosidenib in 2021, genomic analyses fueled the development of therapeutics in 7–9 years that have forever changed the management of patient subgroups. While these bench-to-bedside turnaround times might become even shorter due to technological and analytical advances, the next wave of therapeutic targets will be fraught with new challenges. These include lower recurrence frequencies among patients, non-kinase functions and poorly characterized target biology. As the therapeutic landscape of CCA progressively evolves, new clinical challenges also emerge. These involve complex treatment decision-making requiring robust predictive biomarkers, risk of treatment-induced selection of aggressive tumor clones, and design of effective treatment combinations while minimizing patient toxicities. The successes to date have only been accomplished through multidisciplinary research focused on CCA patients, and continued successes will require continued synergy from diverse fields to meet tomorrow's clinical challenges today.

Acknowledgment

The laboratory of J.B.A. is funded by competitive grants from the Novo Nordisk Foundation (no. 0058419), Independent Research Fund Denmark (no. 171730), Danish Cancer Society (no. R278-A16638) and NEYE Foundation.

Conflict of interest statement

The authors have no conflict of interest.

References

Abou-Alfa, G. K., Macarulla, T., Javle, M. M., Kelley, R. K., Lubner, S. J., Adeva, J., et al. (2020). Ivosidenib in IDH1-mutant, chemotherapy-refractory cholangiocarcinoma (ClarIDHy): A multicentre, randomised, double-blind, placebo-controlled, phase 3 study. *The Lancet Oncology*, *21*(6), 796–807. https://doi.org/10.1016/S1470-2045(20)30157-1.

Abou-Alfa, G. K., Sahai, V., Hollebecque, A., Vaccaro, G., Melisi, D., Al-Rajabi, R., et al. (2020). Pemigatinib for previously treated, locally advanced or metastatic cholangiocarcinoma: A multicentre, open-label, phase 2 study. *The Lancet Oncology*, *21*(5), 671–684. https://doi.org/10.1016/S1470-2045(20)30109-1.

Andersen, J. B., Spee, B., Blechacz, B. R., Avital, I., Komuta, M., Barbour, A., et al. (2012). Genomic and genetic characterization of cholangiocarcinoma identifies therapeutic targets for tyrosine kinase inhibitors. *Gastroenterology*, *142*(4), 1021–1031 e1015. https://doi.org/10.1053/j.gastro.2011.12.005.

Arai, Y., Totoki, Y., Hosoda, F., Shirota, T., Hama, N., Nakamura, H., et al. (2014). Fibroblast growth factor receptor 2 tyrosine kinase fusions define a unique molecular subtype of cholangiocarcinoma. *Hepatology*, *59*(4), 1427–1434. https://doi.org/10.1002/hep.26890.

Artegiani, B., van Voorthuijsen, L., Lindeboom, R. G. H., Seinstra, D., Heo, I., Tapia, P., et al. (2019). Probing the tumor suppressor function of BAP1 in CRISPR-engineered human liver organoids. *Cell Stem Cell*, *24*(6), 927–943 e926. https://doi.org/10.1016/j.stem.2019.04.017.

Banales, J. M., Marin, J. J. G., Lamarca, A., Rodrigues, P. M., Khan, S. A., Roberts, L. R., et al. (2020). Cholangiocarcinoma 2020: The next horizon in mechanisms and management. *Nature Reviews. Gastroenterology & Hepatology*, *17*(9), 557–588. https://doi.org/10.1038/s41575-020-0310-z.

Boerner, T., Drill, E., Pak, L. M., Nguyen, B., Sigel, C. S., Doussot, A., et al. (2021). Genetic determinants of outcome in intrahepatic cholangiocarcinoma. *Hepatology*, *74*(3), 1429–1444. https://doi.org/10.1002/hep.31829.

Boileve, A., Hilmi, M., Gougis, P., Cohen, R., Rousseau, B., Blanc, J. F., et al. (2021). Triplet combination of durvalumab, tremelimumab, and paclitaxel in biliary tract carcinomas: Safety run-in results of the randomized IMMUNOBIL PRODIGE 57 phase II trial. *European Journal of Cancer*, *143*, 55–63. https://doi.org/10.1016/j.ejca.2020.10.027.

Borger, D. R., Tanabe, K. K., Fan, K. C., Lopez, H. U., Fantin, V. R., Straley, K. S., et al. (2012). Frequent mutation of isocitrate dehydrogenase (IDH)1 and IDH2 in cholangiocarcinoma identified through broad-based tumor genotyping. *The Oncologist*, *17*(1), 72–79. https://doi.org/10.1634/theoncologist.2011-0386.

Canon, J., Rex, K., Saiki, A. Y., Mohr, C., Cooke, K., Bagal, D., et al. (2019). The clinical KRAS(G12C) inhibitor AMG 510 drives anti-tumour immunity. *Nature*, *575*(7781), 217–223. https://doi.org/10.1038/s41586-019-1694-1.

Chan-On, W., Nairismagi, M. L., Ong, C. K., Lim, W. K., Dima, S., Pairojkul, C., et al. (2013). Exome sequencing identifies distinct mutational patterns in liver fluke-related and non-infection-related bile duct cancers. *Nature Genetics*, *45*(12), 1474–1478. https://doi.org/10.1038/ng.2806.

Cleary, J. M., Raghavan, S., Wu, Q., Li, Y. Y., Spurr, L. F., Gupta, H. V., et al. (2021). FGFR2 extracellular domain in-frame deletions are therapeutically targetable genomic alterations that function as oncogenic drivers in cholangiocarcinoma. *Cancer Discovery*, *11*(10), 2488–2505. https://doi.org/10.1158/2159-8290.CD-20-1669.

Consortium I. T. P.-C. A. o. W. G. (2020). Pan-cancer analysis of whole genomes. *Nature*, *578*(7793), 82–93. https://doi.org/10.1038/s41586-020-1969-6.

Cristinziano, G., Porru, M., Lamberti, D., Buglioni, S., Rollo, F., Amoreo, C. A., et al. (2021). FGFR2 fusion proteins drive oncogenic transformation of mouse liver organoids towards cholangiocarcinoma. *Journal of Hepatology*, *75*(2), 351–362. https://doi.org/10.1016/j.jhep.2021.02.032.

Dhillon, S. (2018). Ivosidenib: First global approval. *Drugs*, *78*(14), 1509–1516. https://doi.org/10.1007/s40265-018-0978-3.

El-Khoueiry, A. B., Rankin, C., Siegel, A. B., Iqbal, S., Gong, I. Y., Micetich, K. C., et al., & D. (2014). S0941: A phase 2 SWOG study of sorafenib and erlotinib in patients with advanced gallbladder carcinoma or cholangiocarcinoma. *British Journal of Cancer*, *110*(4), 882–887. https://doi.org/10.1038/bjc.2013.801.

Farshidfar, F., Zheng, S., Gingras, M. C., Newton, Y., Shih, J., Robertson, A. G., et al. (2017). Integrative genomic analysis of cholangiocarcinoma identifies distinct IDH-mutant molecular profiles. *Cell Reports*, *18*(11), 2780–2794. https://doi.org/10.1016/j.celrep.2017.02.033.

Forner, A., Vidili, G., Rengo, M., Bujanda, L., Ponz-Sarvise, M., & Lamarca, A. (2019). Clinical presentation, diagnosis and staging of cholangiocarcinoma. *Liver International*, *39*(Suppl. 1), 98–107. https://doi.org/10.1111/liv.14086.

Goeppert, B., Roessler, S., Renner, M., Singer, S., Mehrabi, A., Vogel, M. N., et al. (2019). Mismatch repair deficiency is a rare but putative therapeutically relevant finding in non-liver fluke associated cholangiocarcinoma. *British Journal of Cancer*, *120*(1), 109–114. https://doi.org/10.1038/s41416-018-0199-2.

Goyal, L., Saha, S. K., Liu, L. Y., Siravegna, G., Leshchiner, I., Ahronian, L. G., et al. (2017). Polyclonal secondary FGFR2 mutations drive acquired resistance to FGFR inhibition in patients with FGFR2 fusion-positive cholangiocarcinoma. *Cancer Discovery*, *7*(3), 252–263. https://doi.org/10.1158/2159-8290.CD-16-1000.

Goyal, L., Shi, L., Liu, L. Y., Fece de la Cruz, F., Lennerz, J. K., Raghavan, S., et al. (2019). TAS-120 overcomes resistance to ATP-competitive FGFR inhibitors in patients with FGFR2 fusion-positive intrahepatic cholangiocarcinoma. *Cancer Discovery*, *9*(8), 1064–1079. https://doi.org/10.1158/2159-8290.CD-19-0182.

Goyal, L., Zheng, H., Yurgelun, M. B., Abrams, T. A., Allen, J. N., Cleary, J. M., et al. (2017). A phase 2 and biomarker study of cabozantinib in patients with advanced cholangiocarcinoma. *Cancer*, *123*(11), 1979–1988. https://doi.org/10.1002/cncr.30571.

Hogdall, D., Lewinska, M., & Andersen, J. B. (2018). Desmoplastic tumor microenvironment and immunotherapy in cholangiocarcinoma. *Trends Cancer*, *4*(3), 239–255. https://doi.org/10.1016/j.trecan.2018.01.007.

Hogdall, D., O'Rourke, C. J., Dehlendorff, C., Larsen, O. F., Jensen, L. H., Johansen, A. Z., et al. (2020). Serum IL6 as a prognostic biomarker and IL6R as a therapeutic target in biliary tract cancers. *Clinical Cancer Research*, *26*(21), 5655–5667. https://doi.org/10.1158/1078-0432.CCR-19-2700.

Hong, D. S., Fakih, M. G., Strickler, J. H., Desai, J., Durm, G. A., Shapiro, G. I., et al. (2020). KRAS(G12C) inhibition with Sotorasib in advanced solid tumors. *The New England Journal of Medicine*, *383*(13), 1207–1217. https://doi.org/10.1056/NEJMoa1917239.

Horak, P., Heining, C., Kreutzfeldt, S., Hutter, B., Mock, A., Hullein, J., et al. (2021). Comprehensive genomic and transcriptomic analysis for guiding therapeutic decisions in patients with rare cancers. *Cancer Discovery*, *11*(11), 2780–2795. https://doi.org/10.1158/2159-8290.CD-21-0126.

Javle, M., Roychowdhury, S., Kelley, R. K., Sadeghi, S., Macarulla, T., Weiss, K. H., et al. (2021). Infigratinib (BGJ398) in previously treated patients with advanced or metastatic cholangiocarcinoma with FGFR2 fusions or rearrangements: Mature results from a multicentre, open-label, single-arm, phase 2 study. *The Lancet Gastroenterology & Hepatology*, *6*(10), 803–815. https://doi.org/10.1016/S2468-1253(21)00196-5.

Jiao, Y., Pawlik, T. M., Anders, R. A., Selaru, F. M., Streppel, M. M., Lucas, D. J., et al. (2013). Exome sequencing identifies frequent inactivating mutations in BAP1, ARID1A and PBRM1 in intrahepatic cholangiocarcinomas. *Nature Genetics, 45*(12), 1470–1473. https://doi.org/10.1038/ng.2813.

Job, S., Rapoud, D., Dos Santos, A., Gonzalez, P., Desterke, C., Pascal, G., et al. (2020). Identification of four immune subtypes characterized by distinct composition and functions of tumor microenvironment in intrahepatic cholangiocarcinoma. *Hepatology, 72*(3), 965–981. https://doi.org/10.1002/hep.31092.

Joensuu, H., Eriksson, M., Sundby Hall, K., Reichardt, A., Hermes, B., Schutte, J., et al. (2020). Survival outcomes associated with 3 years vs 1 year of adjuvant Imatinib for patients with high-risk gastrointestinal stromal tumors: An analysis of a randomized clinical trial after 10-year follow-up. *JAMA Oncology, 6*(8), 1241–1246. https://doi.org/10.1001/jamaoncol.2020.2091.

Jolissaint, J. S., Soares, K. C., Seier, K. P., Kundra, R., Gonen, M., Shin, P. J., et al. (2021). Intrahepatic cholangiocarcinoma with lymph node metastasis: Treatment-related outcomes and the role of tumor genomics in patient selection. *Clinical Cancer Research, 27*(14), 4101–4108. https://doi.org/10.1158/1078-0432.CCR-21-0412.

Jusakul, A., Cutcutache, I., Yong, C. H., Lim, J. Q., Huang, M. N., Padmanabhan, N., et al. (2017). Whole-genome and epigenomic landscapes of etiologically distinct subtypes of cholangiocarcinoma. *Cancer Discovery, 7*(10), 1116–1135. https://doi.org/10.1158/2159-8290.CD-17-0368.

Kendre, G., Marhenke, S., Lorz, G., Becker, D., Reineke-Plaass, T., Poth, T., et al. (2021). The co-mutational spectrum determines the therapeutic response in murine FGFR2 fusion-driven cholangiocarcinoma. *Hepatology, 74*(3), 1357–1370. https://doi.org/10.1002/hep.31799.

Kim, R. D., Chung, V., Alese, O. B., El-Rayes, B. F., Li, D., Al-Toubah, T. E., et al. (2020). A phase 2 multi-institutional study of nivolumab for patients with advanced refractory biliary tract cancer. *JAMA Oncology, 6*(6), 888–894. https://doi.org/10.1001/jamaoncol.2020.0930.

Kim, R. D., Sanoff, H. K., Poklepovic, A. S., Soares, H., Kim, J., Lyu, J., et al. (2020). A multi-institutional phase 2 trial of regorafenib in refractory advanced biliary tract cancer. *Cancer, 126*(15), 3464–3470. https://doi.org/10.1002/cncr.32964.

Klein, O., Kee, D., Nagrial, A., Markman, B., Underhill, C., Michael, M., et al. (2020). Evaluation of combination Nivolumab and Ipilimumab immunotherapy in patients with advanced biliary tract cancers: Subgroup analysis of a phase 2 nonrandomized clinical trial. *JAMA Oncology, 6*(9), 1405–1409. https://doi.org/10.1001/jamaoncol.2020.2814.

Lamarca, A., Kapacee, Z., Breeze, M., Bell, C., Belcher, D., Staiger, H., et al. (2020). Molecular profiling in daily clinical practice: Practicalities in advanced cholangiocarcinoma and other biliary tract cancers. *Journal of Clinical Medicine, 9*(9). https://doi.org/10.3390/jcm9092854.

Lamarca, A., Palmer, D. H., Wasan, H. S., Ross, P. J., Ma, Y. T., Arora, A., et al. (2021). Second-line FOLFOX chemotherapy versus active symptom control for advanced biliary tract cancer (ABC-06): A phase 3, open-label, randomised, controlled trial. *The Lancet Oncology, 22*(5), 690–701. https://doi.org/10.1016/S1470-2045(21)00027-9.

Lee, J. K., Hazar-Rethinam, M., Decker, B., Gjoerup, O., Madison, R. W., Lieber, D. S., et al. (2021). The pan-tumor landscape of targetable kinase fusions in circulating tumor DNA. *Clinical Cancer Research*. https://doi.org/10.1158/1078-0432.CCR-21-2136.

Lin, J., Yang, X., Long, J., Zhao, S., Mao, J., Wang, D., et al. (2020). Pembrolizumab combined with lenvatinib as non-first-line therapy in patients with refractory biliary tract carcinoma. *Hepatobiliary Surgery and Nutrition, 9*(4), 414–424. https://doi.org/10.21037/hbsn-20-338.

Lowery, M. A., Ptashkin, R., Jordan, E., Berger, M. F., Zehir, A., Capanu, M., et al., & K. (2018). Comprehensive molecular profiling of intrahepatic and extrahepatic cholangiocarcinomas: Potential targets for intervention. *Clinical Cancer Research*, 24(17), 4154–4161. https://doi.org/10.1158/1078-0432.CCR-18-0078.

Marabelle, A., Le, D. T., Ascierto, P. A., Di Giacomo, A. M., De Jesus-Acosta, A., Delord, J. P., et al. (2020). Efficacy of Pembrolizumab in patients with noncolorectal high microsatellite instability/mismatch repair-deficient cancer: Results from the phase II KEYNOTE-158 study. *Journal of Clinical Oncology*, 38(1), 1–10. https://doi.org/10.1200/JCO.19.02105.

Marcus, L., Fashoyin-Aje, L. A., Donoghue, M., Yuan, M., Rodriguez, L., Gallagher, P. S., et al. (2021). FDA approval summary: Pembrolizumab for the treatment of tumor mutational burdaen-high solid tumors. *Clinical Cancer Research*, 27(17), 4685–4689. https://doi.org/10.1158/1078-0432.CCR-21-0327.

Marcus, L., Lemery, S. J., Keegan, P., & Pazdur, R. (2019). FDA approval summary: Pembrolizumab for the treatment of microsatellite instability-high solid tumors. *Clinical Cancer Research*, 25(13), 3753–3758. https://doi.org/10.1158/1078-0432.CCR-18-4070.

Mariotti, V., Fiorotto, R., Cadamuro, M., Fabris, L., & Strazzabosco, M. (2021). New insights on the role of vascular endothelial growth factor in biliary pathophysiology. *JHEP Reports*, 3(3). https://doi.org/10.1016/j.jhepr.2021.100251, 100251.

Massard, C., Michiels, S., Ferte, C., Le Deley, M. C., Lacroix, L., Hollebecque, A., et al. (2017). High-throughput genomics and clinical outcome in hard-to-treat advanced cancers: Results of the MOSCATO 01 trial. *Cancer Discovery*, 7(6), 586–595. https://doi.org/10.1158/2159-8290.CD-16-1396.

Mateo, J., Chakravarty, D., Dienstmann, R., Jezdic, S., Gonzalez-Perez, A., Lopez-Bigas, N., et al. (2018). A framework to rank genomic alterations as targets for cancer precision medicine: The ESMO scale for clinical actionability of molecular targets (ESCAT). *Annals of Oncology*, 29(9), 1895–1902. https://doi.org/10.1093/annonc/mdy263.

Meric-Bernstam, F., Bahleda, R., Hierro, C., Sanson, M., Bridgewater, J., Arkenau, H. T., et al. (2021). Futibatinib, an irreversible FGFR1-4 inhibitor, in patients with advanced solid tumors harboring FGF/FGFR aberrations: A phase I dose-expansion study. *Cancer Discovery*, 12(2), 402–415. https://doi.org/10.1158/2159-8290.CD-21-0697.

Montal, R., Sia, D., Montironi, C., Leow, W. Q., Esteban-Fabro, R., Pinyol, R., et al. (2020). Molecular classification and therapeutic targets in extrahepatic cholangiocarcinoma. *Journal of Hepatology*, 73(2), 315–327. https://doi.org/10.1016/j.jhep.2020.03.008.

Nepal, C., O'Rourke, C. J., Oliveira, D., Taranta, A., Shema, S., Gautam, P., et al. (2018). Genomic perturbations reveal distinct regulatory networks in intrahepatic cholangiocarcinoma. *Hepatology*, 68(3), 949–963. https://doi.org/10.1002/hep.29764.

Ong, C. K., Subimerb, C., Pairojkul, C., Wongkham, S., Cutcutache, I., Yu, W., et al. (2012). Exome sequencing of liver fluke-associated cholangiocarcinoma. *Nature Genetics*, 44(6), 690–693. https://doi.org/10.1038/ng.2273.

O'Rourke, C. J., Matter, M. S., Nepal, C., Caetano-Oliveira, R., Ton, P. T., Factor, V. M., et al. (2020). Identification of a Pan-gamma-secretase inhibitor response signature for notch-driven cholangiocarcinoma. *Hepatology*, 71(1), 196–213. https://doi.org/10.1002/hep.30816.

Piha-Paul, S. A., Oh, D. Y., Ueno, M., Malka, D., Chung, H. C., Nagrial, A., et al. (2020). Efficacy and safety of pembrolizumab for the treatment of advanced biliary cancer: Results from the KEYNOTE-158 and KEYNOTE-028 studies. *International Journal of Cancer*, 147(8), 2190–2198. https://doi.org/10.1002/ijc.33013.

Pirozzi, C. J., & Yan, H. (2021). The implications of IDH mutations for cancer development and therapy. *Nature Reviews. Clinical Oncology*, 18(10), 645–661. https://doi.org/10.1038/s41571-021-00521-0.

Saha, S. K., Gordan, J. D., Kleinstiver, B. P., Vu, P., Najem, M. S., Yeo, J. C., et al. (2016). Isocitrate dehydrogenase mutations confer Dasatinib hypersensitivity and SRC dependence in intrahepatic cholangiocarcinoma. *Cancer Discovery*, *6*(7), 727–739. https://doi.org/10.1158/2159-8290.CD-15-1442.

Saha, S. K., Parachoniak, C. A., Ghanta, K. S., Fitamant, J., Ross, K. N., Najem, M. S., et al. (2014). Mutant IDH inhibits HNF-4alpha to block hepatocyte differentiation and promote biliary cancer. *Nature*, *513*(7516), 110–114. https://doi.org/10.1038/nature13441.

Sheils, T. K., Mathias, S. L., Kelleher, K. J., Siramshetty, V. B., Nguyen, D. T., Bologa, C. G., et al. (2021). TCRD and pharos 2021: Mining the human proteome for disease biology. *Nucleic Acids Research*, *49*(D1), D1334–D1346. https://doi.org/10.1093/nar/gkaa993.

Sia, D., Hoshida, Y., Villanueva, A., Roayaie, S., Ferrer, J., Tabak, B., et al. (2013). Integrative molecular analysis of intrahepatic cholangiocarcinoma reveals 2 classes that have different outcomes. *Gastroenterology*, *144*(4), 829–840. https://doi.org/10.1053/j.gastro.2013.01.001.

Sia, D., Losic, B., Moeini, A., Cabellos, L., Hao, K., Revill, K., et al. (2015). Massive parallel sequencing uncovers actionable FGFR2-PPHLN1 fusion and ARAF mutations in intrahepatic cholangiocarcinoma. *Nature Communications*, *6*, 6087. https://doi.org/10.1038/ncomms7087.

Silverman, I. M., Hollebecque, A., Friboulet, L., Owens, S., Newton, R. C., Zhen, H., et al., & C. (2021). Clinicogenomic analysis of FGFR2-rearranged cholangiocarcinoma identifies correlates of response and mechanisms of resistance to Pemigatinib. *Cancer Discovery*, *11*(2), 326–339. https://doi.org/10.1158/2159-8290.CD-20-0766.

Skoulidis, F., Li, B. T., Dy, G. K., Price, T. J., Falchook, G. S., Wolf, J., et al. (2021). Sotorasib for lung cancers with KRAS p.G12C mutation. *The New England Journal of Medicine*, *384*(25), 2371–2381. https://doi.org/10.1056/NEJMoa2103695.

Subbiah, V., Lassen, U., Elez, E., Italiano, A., Curigliano, G., Javle, M., et al. (2020). Dabrafenib plus trametinib in patients with BRAF(V600E)-mutated biliary tract cancer (ROAR): A phase 2, open-label, single-arm, multicentre basket trial. *The Lancet Oncology*, *21*(9), 1234–1243. https://doi.org/10.1016/S1470-2045(20)30321-1.

Sun, W., Patel, A., Normolle, D., Patel, K., Ohr, J., Lee, J. J., et al. (2019). A phase 2 trial of regorafenib as a single agent in patients with chemotherapy-refractory, advanced, and metastatic biliary tract adenocarcinoma. *Cancer*, *125*(6), 902–909. https://doi.org/10.1002/cncr.31872.

Taylor, A. M., Shih, J., Ha, G., Gao, G. F., Zhang, X., Berger, A. C., et al. (2018). Genomic and functional approaches to understanding cancer aneuploidy. *Cancer Cell*, *33*(4). https://doi.org/10.1016/j.ccell.2018.03.007. 676–689.e673.

Ueno, M., Ikeda, M., Morizane, C., Kobayashi, S., Ohno, I., Kondo, S., et al. (2019). Nivolumab alone or in combination with cisplatin plus gemcitabine in Japanese patients with unresectable or recurrent biliary tract cancer: A non-randomised, multicentre, open-label, phase 1 study. *The Lancet Gastroenterology & Hepatology*, *4*(8), 611–621. https://doi.org/10.1016/S2468-1253(19)30086-X.

Ueno, M., Ikeda, M., Sasaki, T., Nagashima, F., Mizuno, N., Shimizu, S., et al. (2020). Phase 2 study of lenvatinib monotherapy as second-line treatment in unresectable biliary tract cancer: Primary analysis results. *BMC Cancer*, *20*(1), 1105. https://doi.org/10.1186/s12885-020-07365-4.

Valle, J., Wasan, H., Palmer, D. H., Cunningham, D., Anthoney, A., Maraveyas, A., et al. (2010). Cisplatin plus gemcitabine versus gemcitabine for biliary tract cancer. *The New England Journal of Medicine*, *362*(14), 1273–1281. https://doi.org/10.1056/NEJMoa0908721.

Verlingue, L., Malka, D., Allorant, A., Massard, C., Ferte, C., Lacroix, L., et al. (2017). Precision medicine for patients with advanced biliary tract cancers: An effective strategy within the prospective MOSCATO-01 trial. *European Journal of Cancer*, *87*, 122–130. https://doi.org/10.1016/j.ejca.2017.10.013.

Villanueva, L., Lwin, Z., Chung, H. C., Gomez-Roca, C., Longo, F., Yanez, E., et al. (2021). Lenvatinib plus pembrolizumab for patients with previously treated biliary tract cancers in the multicohort phase II LEAP-005 study. *Journal of Clinical Oncology*, *39*(3 Suppl).

Wang, P., Dong, Q., Zhang, C., Kuan, P. F., Liu, Y., Jeck, W. R., et al. (2013). Mutations in isocitrate dehydrogenase 1 and 2 occur frequently in intrahepatic cholangiocarcinomas and share hypermethylation targets with glioblastomas. *Oncogene*, *32*(25), 3091–3100. https://doi.org/10.1038/onc.2012.315.

Wang, Y., Wild, A. T., Turcan, S., Wu, W. H., Sigel, C., Klimstra, D. S., et al. (2020). Targeting therapeutic vulnerabilities with PARP inhibition and radiation in IDH-mutant gliomas and cholangiocarcinomas. *Science Advances*, *6*(17), eaaz3221. https://doi.org/10.1126/sciadv.aaz3221.

Wu, M. J., Shi, L., Dubrot, J., Merritt, J., Vijay, V., Wei, T. Y., et al. (2021). Mutant-IDH inhibits interferon-TET2 signaling to promote immunoevasion and tumor maintenance in cholangiocarcinoma. *Cancer Discovery*. https://doi.org/10.1158/2159-8290.Cd-21-1077.

Wu, Y. M., Su, F., Kalyana-Sundaram, S., Khazanov, N., Ateeq, B., Cao, X., et al. (2013). Identification of targetable FGFR gene fusions in diverse cancers. *Cancer Discovery*, *3*(6), 636–647. https://doi.org/10.1158/2159-8290.CD-13-0050.

Yoo, C., Oh, D. Y., Choi, H. J., Kudo, M., Ueno, M., Kondo, S., et al. (2020). Phase I study of bintrafusp alfa, a bifunctional fusion protein targeting TGF-beta and PD-L1, in patients with pretreated biliary tract cancer. *Journal for Immunotherapy of Cancer*, *8*(1). https://doi.org/10.1136/jitc-2020-000564.

Zhu, A. X., Macarulla, T., Javle, M. M., Kelley, R. K., Lubner, S. J., Adeva, J., et al. (2021). Final overall survival efficacy results of Ivosidenib for patients with advanced cholangiocarcinoma with IDH1 mutation: The phase 3 randomized clinical ClarIDHy trial. *JAMA Oncology*, *7*(11), 1669–1677. https://doi.org/10.1001/jamaoncol.2021.3836.

Zou, S., Li, J., Zhou, H., Frech, C., Jiang, X., Chu, J. S., et al. (2014). Mutational landscape of intrahepatic cholangiocarcinoma. *Nature Communications*, *5*, 5696. https://doi.org/10.1038/ncomms6696.

CHAPTER THIRTEEN

Immunotherapies for hepatocellular carcinoma and intrahepatic cholangiocarcinoma: Current and developing strategies

Josepmaria Argemi[a,b,c], Mariano Ponz-Sarvise[a], and Bruno Sangro[a,b,c,*]

[a]HPB Oncology Area, Clinica Universidad de Navarra-CCUN, Pamplona, Spain
[b]Hepatology Program, Centro de Investigacion Medica Aplicada (CIMA), Pamplona, Spain
[c]Centro de Investigacion Biomedica en Red de Enfermedades Hepaticas y Digestivas (CIBERehd), Madrid, Spain
*Corresponding author: e-mail address: bsangro@unav.es

Contents

1. The immune biology of the liver — 369
2. The immunosuppressive environment of hepatocellular carcinoma — 370
    2.1 The role of immune and other non-parenchymal cells — 370
    2.2 The role of tumor cells — 377
    2.3 Etiology of liver cancer and mechanisms of immune evasion — 381
3. Immunotherapy of HCC — 383
    3.1 Combination therapies — 384
    3.2 Single agents — 390
    3.3 Biomarkers of response or resistance — 391
    3.4 Beyond immune checkpoint inhibitors — 393
4. Immunotherapy of intrahepatic cholangiocarcinoma — 395
    4.1 Single agents — 396
    4.2 Combination therapies — 397
5. Concluding remarks — 398
Grant support — 398
Conflict of interest — 398
References — 398

Abstract

Liver cancer including hepatocellular carcinoma (HCC) and intrahepatic cholangiocarcinoma (iCCA) is the third leading cause of cancer-related deaths worldwide. HCC arises from hepatocyte or hepatic stem cells, while iCCA originates from biliary epithelial cells, and the respective biological context are very different. Despite screening programs, the diagnosis of liver cancer is in most cases made when curative treatments such as surgery or ablation are not possible. In 2020, after a decade of using only tyrosine kinase inhibitors (TKI), a combination of an immune-check point inhibitor (ICI) and a VEGF antagonist proved superior to a TKI as first line therapy of advanced HCC. In 2022, the addition of an ICI to standard chemotherapy demonstrated an improvement of patient survival in iCCA. Moreover, ICI offer an unprecedented rate of durable responses to HCC and iCCA patients. Nevertheless, still two thirds of patients do not respond to ICI-based combinations, and research efforts are focused on deciphering the mechanisms of immune evasion of these lethal cancers. Reliable predictive and prognostic biomarkers are still lacking, but the molecular phenotyping of the tumor microenvironment is currently providing potential candidates for patient stratification. In this review, we will summarize the current knowledge on the immune biology of the liver, the discovery of cell-intrinsic and immune cell-mediated mechanisms of immune evasion by means of high-resolution single cell data, the main targets of current immunotherapy approaches, and the recent milestones in immunotherapy of HCC and iCCA.

Abbreviations

ACT	adoptive cell therapy
ALD	alcohol-related liver disease
AFP	alpha fetoprotein
CAR	chimeric antigen receptor
DAMP	damage-associated molecular pattern
DC	dendritic cell
GPC3	glypican 3
HBV	hepatitis B virus
HCC	hepatocellular carcinoma
HCV	hepatitis C virus
iCCA	intrahepatic cholangiocarcinoma
ICB	immune checkpoint blockade
ICI	immune checkpoint inhibitor
IFN-γ	interferon gamma
KC	Kuppfer cell
MAFLD	metabolic syndrome-associated fatty liver disease
MDSC	myeloid-derived suppressor cell
MHC	major histocompatibility complex
NK	natural killer cell
NKT	natural killer T cell
OS	overall survival
ORR	overall response rate
PAMP	pathogen-associated molecular pattern
PFS	progression-free survival
TAA	tumor-associated antigen
TAM	tumor-associated macrophage
TGF-β	transforming growth factor beta
TKI	tyrosine kinase inhibitor
TME	tumor micro-environment

1. The immune biology of the liver

The liver is an organ with a blood supply that flows from the gut through the sinusoids, a unique fenestrated vasculature enabling optimal exchange of molecules for metabolic functions, together with generating an immune barrier to microbes and toxins. The liver has a notable capacity to remove gut-derived pathogens, and their derived Pathogen-Associated Molecular Patterns (PAMP), such as lipopolysaccharide (LPS), toxins and Damage-Associated Molecular Patterns (DAMP) from the portal circulation (Lumsden, Henderson, & Kutner, 1988). To cope with this injuring-prone environment the liver is specially armed with several types of innate and adaptive immune cells. But for the same reason, the liver possesses a key immunoregulatory role through its ability to maintain immunotolerance to non-pathological or constant inflammatory stimuli which prevents liver damage and induces systemic tolerance. To enable this immune regulatory function, the liver contains the largest number of resident macrophages in the body, the so called Kupffer cells (KCs). It also retains a high density of natural killer cells (NK cells), natural killer T cells (NKT cells), γδ T cells, and both liver-transiting and resident T lymphocytes (Jenne & Kubes, 2013).

Upon chronic liver damage, such as a viral infection (by hepatitis B virus [HBV] or hepatitis C virus [HCV]), abnormal continuous fat and carbohydrate metabolism disruption (for example in the Metabolic Syndrome-Associated Fatty Liver Disease or MAFLD), to chronic toxic injury (such as in Alcohol-related Liver Disease), or to less common causes of chronic injury, including protein misfolding in the hepatocyte (alpha 1 antitrypsin deficiency) and genetic iron or copper accumulation (in hemochromatosis or Wilson diseases, respectively) this immunological balance is initially deregulated towards an inflammatory environment, clinically detectable through the measurement of circulating hepatic enzymes, such as aminotransferases and/or glutamyl transferases. In these chronic diseases, the low-degree liver damage, if untreated, generates continual cell death and compensatory regeneration with hepatic stellate cell activation, which could lead to advanced liver fibrosis.

During the evolution of these chronic conditions, the relationship between the hepatocyte and the immune environment is maintained by a complex balance between proinflammatory cytokines (IL-2, IL-7, IL-12, IL-15 and IFN-γ) and anti-inflammatory cytokines (IL-10, IL-13 and

TGF-β). Due to the unique immunosuppressive environment and the regenerative potential of the liver, the global functional mass of hepatocytes is kept intact and patients with advanced fibrosis are mostly asymptomatic. In most patients, this chronic low-intensity damage evolves to cirrhosis, which becomes symptomatic after the first decompensation due to portal hypertension, associated with the development of ascites, esophageal variceal bleeding, or encephalopathy.

Primary liver cancers, particularly hepatocellular carcinoma (HCC), but also intrahepatic cholangiocarcinoma (iCCA), typically originate in this inflammatory and fibrotic environment. Inflammation and cell death leads to an intense replicative stress in both hepatocytes and hepatic progenitor cells. For oncogenesis to occur both a somatic mutation in a driver gene and a deficiency on immune surveillance are needed. Reversely, once the tumor clone niche is generated, HCC cells recruit tolerogenic cell types, further fostering evasion from immune mediated cell death.

In this review we summarize the evidence of immune cell dysregulation in HCC to understand the mechanisms of immune evasion and to gain a molecular insight on the mechanism of action of newly established and future immunotherapy-based treatments in HCC. We also detail the current status of immunotherapy for HCC and iCCA.

2. The immunosuppressive environment of hepatocellular carcinoma

Cohort-based histological and flow cytometry studies, bulk tumor RNA sequencing data, and recent single-cell RNA sequencing studies have illustrated the presence and clinical relevance of the immunosuppressive environment in human HCC samples, including regulatory T (Treg) cells, Kuppfer cells (KC) and monocyte-derived tumor-associated macrophages (TAMs), and myeloid-derived suppressor cells (MDSCs) (Zheng et al., 2017). The increase in the amount of these cell types correlates with cytotoxic T lymphocyte (CTL) impairment (Fu et al., 2007) and with the advanced stage (Shen et al., 2010). The following sections will summarize the pathogenic hijacking of these immune cell types by HCC.

2.1 The role of immune and other non-parenchymal cells

CTLs, through the secretion of Th1 type cytokines IFN-γ and IL2, have been associated with antigen-dependent anti-tumor responses. Nearly half of the lymphocytes in the liver express the T cell receptor (TCR), with an enrichment of CD8+ T cells (Parker & Picut, 2005). While CD4+ T cells outnumber CD8$^+$ T cells by approximately two to one in the blood, in the liver this ratio is reversed. CD8$^+$ T cells are considered the primary mediators of cytotoxic anti-tumor effect, but recent findings have highlighted the importance of the specific phenotype of these cells for their cancer-related immune role. CD4+ and CD8+ tissue-resident memory T cells (T_{RM} CD103+ Cells) originate from circulating T cells and, contrary to other circulating T cells, stably reside in an epithelial organ after the recognition of neoantigens (Pallett et al., 2020). In HCC, CD8+ T_{RM} are important for tumor control (Amsen, van Gisbergen, Hombrink, & van Lier, 2018); however, circulating CD8+ T cells targeting neoantigens are only detected in ~15% of patients with HCC and are tolerized by Dendritic Cells (DC) migrating between the tumor and lymph nodes, and by TAMs (Zhang et al., 2019). On the other hand, exhausted CD8+ CXCR6+ T cells accumulate in NASH-derived HCC, promoting resistance to immunotherapy (Pfister et al., 2021). Another exhausted CD8+ T cell cluster expressing high levels of layilin (LAYN) was associated with reduced disease-free survival (Zheng et al., 2017). Two additional high-resolution single cell sequencing works showed that CD8+ *KLRB* + T Cells with innate-like low cytotoxic phenotype T cells are enriched in early HCC with poor prognosis (Sun et al., 2021), whereas CD8+ *XCL1* + T cells are associated with a better prognosis in viral HBV/HCV-related HCC (Song et al., 2020).

The role of CD4+ FOXP3+ T Regulatory Lymphocytes (Treg) is the antigen-specific inhibition of the antitumor immune response (Chen et al., 2003; Wei, Kryczek, & Zou, 2006), to avoid severe autoimmunity (Kim, Rasmussen, & Rudensky, 2007). In HCC, the abundance of Treg correlates with CTL impairment (Fu et al., 2007; Gao et al., 2007) and with advanced HCC stage (Shen et al., 2010). CD4+ Th17 T cells are involved in several autoimmune diseases and chronic inflammatory syndromes. Naive CD4+ T cells preferentially differentiate into the Th17 T cell subset in response to the combined signals of TGF-β and IL-6, commonly secreted in the TME of HCC. IL-23-induced Th17 (Kortylewski et al., 2009) promotes inflammation and angiogenesis, reducing active CD8+ T cell infiltration (Langowski et al., 2006) and antagonizing the tumor-suppressive role of IFN-γ producing CD4+ Th1 cells (Littman & Rudensky, 2010).

B lymphocytes have dual tumor-promoting and antitumoral roles in HCC. Peritumoral CXCR3+ B cells are associated with early HCC recurrence (Liu et al., 2015), while peritumoral PD1^{high} and FcgRII$^{low/-}$ B cells suppress antigen-specific antitumor immunity (Ouyang et al., 2016; Xiao et al., 2016). Immunosuppressive IgA-producing PD1+ plasma cells, inhibit antigen-specific CTL activity and tumor regression (Shalapour et al., 2017). Interestingly, together with T lymphocytes, B lymphocytes can form Tertiary Lymphoid Structures (TLS) that resemble lymphoid organs within inflamed tissues. Although in some cancers TLS have been associated to better prognosis, such as in early stage colorectal and non-small cell lung cancer (Di Caro et al., 2014; Dieu-Nosjean et al., 2008). In inflamed livers TLS promote the formation of microniches for stem cell proliferation and cancer progression and are associated with HCC late recurrence (Finkin et al., 2015). In contrast, intratumoral TLS were associated with lower early recurrence of HCC after surgical resection (Calderaro et al., 2019). Whether this apparent contradiction has to do with the degree of maturation of tumoral TLS, from primary follicles type (FL-I) to secondary follicle type (FL-II), needs to be investigated. On the other hand, plasmacytic B cells can produce antitumor specific antibodies and durable immunity, thus enhancing the response to immunotherapy (DeFalco et al., 2018).

Gamma delta (γδ) T cells are unconventional T cells, defined by expression of T-cell receptors (TCRs) composed of γ and δ chains instead of the conventional ab-containing TCRs. γδ T cells show tissue-specific subsets all of which share the same TCR. The liver contains one of the largest populations of γδ T cells in the body, which comprise 15–25% of the T cells in the liver (Abo, Kawamura, & Watanabe, 2000; Nemeth, Baird, & O'Farrelly, 2009). γδ T cells infiltrate HCCs and other cancers. Recently, the blockade of the immune inhibitory ligand butyrophilin (BTN), belonging to the B7-family of ligands, enabled the coordinated antigen-specific antitumor activity of γδ and ab T and tumor regression in ovarian cancer, constituting a new attractive therapeutic target (Payne et al., 2020). Other T cell subtypes related to HCC promotion are IL-22 producing T cells (Th22), (Chen et al., 2016), IL-21-producing CXCR5$^-$ PD1$^-$ CD69^{high} Follicular Helper T Cells (T$_{FH}$) (Chen et al., 2016) and IL-9 and IL-10 producing T cells (Th9) (Tan, Wang, & Zhao, 2017).

Hepatic Innate Lymphoid Cells (ILC) such as NK cells are lymphoid cells lacking specific antigen receptors which can respond to an array of cell-surface ligands expressed by infected, damaged or transformed cells, and can acquire cytotoxic-like properties and secrete IFN-γ, perforin and granzyme (Notas, Kisseleva, & Brenner, 2009). One-third to one-half of the lymphocytes in the liver are NK cells, more than a three-fold enrichment over that observed in blood (Crispe, 2009; Doherty & O'Farrelly, 2000).

NK are strongly activated when the class I MHC is downregulated in tumor cells, a common mechanism of immune evasion. NK can also induce MHC expression by hepatocytes and hepatic stellate cells (Crispe, 2009). Several studies have shown that the dysfunction or decrease in the abundance of NK cells are associated with higher rate of HCC progression and poor survival (Juengpanich et al., 2019). Despite its immune effector phenotype, their antitumor activity in HCC might be limited by TGF-β-mediated impairment of oxidative phosphorylation (Zecca et al., 2020). Subsets of HCC-infiltrating CD11b−CD27−NK cells have a decreased cytolytic activity and a defective IFN-γ production (Zhang et al., 2017). Macrophages are responsible for NK dysfunction in HCC (Wu et al., 2013). High expression of the inhibitory receptor NKG2A and its ligand HLA-E in HCC cells lead to an exhausted NK CD56dim phenotype and are associated with poor prognosis (Sun et al., 2017). Conversely, NKG2D is an important activating receptor, expressed also in γδ and some T lymphocytes, which promotes cytotoxicity of NK cells against HCC through the interaction with its ligands MHC-I-related chain molecules A and B (MICA/B) in the membrane of cancer cells (Jinushi et al., 2003).

Invariant NK T cells (iNKT) are a distinct population of T cells that express both an invariant αβ TCR, that responds to glycolipid antigens presented in the context of the MHC-I–like molecule CD1d, and several NK-type cell surface molecules. iNKT cells are the only liver-resident lymphocytes to actively patrol the liver vasculature in search of pathogens (Bricard et al., 2009). They can express CD4, CD8 or be double negative. In HCC, the tumor is enriched in CD4+ iNKTs harboring a Treg-phenotype, reduced cytolytic activity, and are capable of antigen-specific CD8+ T cell expansion (Bricard et al., 2009).

KC are CD68+ liver-resident macrophages located in the vasculature, adherent to LSECs and directly exposed to the contents of blood. KCs express several types of scavenger, Toll-like, complement and antibody receptors, which enable KC to phagocyte pathogens and associated molecules. In HCC, KCs seem to have an antitumor role. KC depletion during partial hepatectomy enhances HCC recurrence in mice, where KC-secreted TNF-α plays a key role in tumor rejection (Hastir et al., 2020). Nevertheless, although KCs can activate T cells, continual exposure to PAMPs from the gut, appears to dampen the ability of KCs to activate the adaptive immune response (Huang et al., 2013).

In association with HCC evolution, TAMs originated from blood monocytes become recruited into the tumor microenvironment. In early-stage

HCC, type 1 macrophages (M1-TAM) are recruited to the liver where through pro-inflammatory CXCL19 and CXCL10 and can attract and promote the differentiation of CD4 Th1 and Th17 T cells, as well as NK cells (Biswas & Mantovani, 2010). In this stage, IRF5 induces the expression of IL12, and represses IL10 in TAM (Krausgruber et al., 2011), which should promote immune surveillance. In advanced HCC, type 2 macrophages (M2-TAM) secrete an array of immunosuppressive cytokines such as IL-10, IL-23, IL13, TGF-β, IL-8, CCL17, CCL22 or CCL24, which promote CD4 Th2 differentiation and recruitment (Movahedi et al., 2010). Th2 T Cells, in turn, will promote an M2 phenotype in TAMs. TAMs both increase the expression of HLA-DR and immune checkpoint molecules with inhibitory capacity, such as Programmed Death 1 ligand 1 (PD-L1), which could lead to CTL extinction (Kuang et al., 2009). M2-TAM derived-IL10 promotes the expansion of IL-17-expressing CD4 (T_H17) and CD8+ (Tc17 cells) T cells (Kuang et al., 2010). Th17 and Tc17-secreted IL-17 further induces PD-L1 expression in peritumoral TAMs, which in turn, suppresses CTL function (Zhao et al., 2011). M2-TAMs also support tissue remodeling and angiogenesis through the secretion of VEGFA, VEGFC and EGF (Biswas & Mantovani, 2010; Murdoch, Muthana, Coffelt, & Lewis, 2008). Angiopoietin receptor TIE2-expressing TAMs are increased in the blood and tumor tissue of HCC patients treated with surgical resection or radiofrequency ablation and are associated with higher micro-vessel density (Matsubara et al., 2013). High TAM density is directly correlated with tumor size and advanced stage. Peritumoral macrophage colony-stimulating factors (M-CSF) and macrophage density predicted the patients' death and disease recurrence in resectable HCC (Zhu et al., 2008). In contrast, other studies have found that TAM-associated PD-L1 expression predicts better survival, in contrast to intratumoral PD-L1 that was associated with poor survival (Liu et al., 2018). This latter finding stresses the need of spatial highly multiplexed histological analyses to allow for a better understanding of the immune tumor microenvironment.

MDSCs represent a heterogeneous population of immature myeloid cells not yet committed into macrophages, dendritic cells, or granulocytes (Bronte et al., 2016). Marked by CD11b surface expression, they may be either CD14+, Monocyte-type (M-MDSC) or CD15+ Polymorphonuclear-type (PMN-MDSC). In mouse models, HCC-derived GM-CSF and CXCL1 seem responsible for MDSC recruitment (Kapanadze et al., 2013). The tumor microenvironment inhibits the natural maturation of MDSCs. Instead, MDSC expand and activate as an immature myeloid cell population. MDSCs induce the depletion of essential amino acids L-Arginine (through the enzyme arginase 1 or ARG1) and tryptophan (through the increased expression of indoleamine oxidase1 or IDO1), thus restricting CTL and NK activation and proliferation (Hornyak et al., 2018; Rodriguez, Quiceno, & Ochoa, 2007; Wang et al., 2012). MDSCs also exhibit an increased expression of Nitric Oxide Synthase-2 (NOS2) and secretion of Nitric Oxide (NO), MMP9 and VEGF, which inhibit inflammatory responses and promote blood vessel formation (Shojaei, Zhong, Wu, Yu, & Ferrara, 2008). Finally, MDSCs secrete TGF-β, IL-10, and other immunosuppressive cytokines to promote Treg recruitment and expansion (Condamine & Gabrilovich, 2011; Gabrilovich & Nagaraj, 2009) and to induce tolerogenic Dendritic cell expansion (Li, Harden, Anderson, & Egilmez, 2016).

Dendritic cells (DC) are responsible for capturing cancer antigens and presenting them to naïve T cells in the lymph nodes. Potent proinflammatory CD11c + CD141 + myeloid DCs, present in healthy livers, can stimulate strong T cell responses via IFN-γ and IL-17. Chronically inflamed livers are depleted of these DCs (Kelly et al., 2014). Conversely, a specific subtype of DC expressing CD14 (CD14 + DCs), which is increased in blood of HCC patients, was found to intratumorally express high levels CTLA4 and PD1 and was able to suppress T cell responses through IL10 and IDO (Han et al., 2014).

In early resected HCC, the abundance of intratumoral CD66b + neutrophils was associated with poor prognosis and the ratio neutrophil-to-CD8 + T cell was a good predictor of outcome, a finding confirmed in other malignancies (Ilie et al., 2012; Li et al., 2011). Upon damage, neutrophils are usually capable of platelet recruitment as a rapid initiating mechanism of hemostasis. The aggregation of platelets on adherent neutrophils induces the release of neutrophil extracellular traps (NETs), a potent web-like structure consisting of DNA, histones, and antimicrobial molecules. IL8 produced by cancer cells, MDSCs and fibroblasts, seems to be determinant for neutrophil recruitment to the tumor, the extrusion of NETs and the resistance to immunotherapy (Teijeira et al., 2021). NETs were related to inflammation and HCC development in mouse models (van der Windt et al., 2018).

Cancer Associated Fibroblasts (CAFs), probably derived from HSCs, are the major source of collagen in the HCC stroma. They differ from normal fibroblasts in their ability to secrete high levels of stromal cell-derived factor 1 (SDF-1) and CXCL12 and promote tumor growth and angiogenesis

(Orimo & Weinberg, 2006). Several mechanisms have been linked to the important pro-tumoral role of CAFs in liver cancer (Affo, Yu, & Schwabe, 2017). CAFs can reduce the immune surveillance increasing Treg survival, inhibiting CTL infiltration (Zhao et al., 2011), recruiting MDSCs (Zhao et al., 2014) or promoting the shift of recruited monocytes to a M2-TAM immunosuppressive phenotype (Ji et al., 2015). CAFs are also important producers of TGFb, a potent inhibitor of antitumor immunity that strongly promotes the tolerogenic fate of TAMs, DCs, neutrophils, NK, NKT, Tregs and CD4+ and CD8+ effector T cells (Flavell, Sanjabi, Wrzesinski, & Licona-Limon, 2010).

Liver sinusoidal endothelial cells (LSECs), which conform the highly specialized fenestrated vascular web of the liver, are not mere structural components for inflow transport from arterial and portal circulation, but actively participate in pathogen detection, capture and antigen presentation (Limmer et al., 2000). LSECs comprise ~50% of the nonparenchymal cells in the liver, making these cells more than twice as abundant as either liver-resident macrophages or lymphocytes in the liver (Racanelli & Rehermann, 2006). LSECs express a wide variety of pattern recognition receptors (PRRs), such as Toll-like receptors TLR3, TLR4, TLR7 and TLR9 (Knolle & Limmer, 2003; Wu et al., 2010) and constitutively express major histocompatibility complex (MHC) I and MHC II, costimulatory molecules (CD80 and CD86), and lymphocyte adhesion molecules (Knolle & Limmer, 2003). Recognition of antigens by T cells results in an upregulation of PDL1 in the LSEC (Diehl et al., 2008) and the expansion of antigen-specific CD8+ T cells which do not acquire an effector phenotype, thus being a key element for the induction of central tolerance (Berg et al., 2006; Jenne & Kubes, 2013; Limmer et al., 2000). Constant low-level activation of LSECs by some MAMPs (for example, lipopolysaccharide) and KC-derived IL-10 downregulate major histocompatibility complex and costimulatory molecules and thereby facilitate T cell tolerance (Knolle et al., 1999). In HCC, LSEC can foster an immunosuppressive environment through the secretion of IL10 and the recruitment of Treg, and the secretion of CXCL1, and CXCL2 promoting the recruitment of MDSCs (Wilkinson, Qurashi, & Shetty, 2020). Conversely, LSEC can mediate antitumor surveillance. In mice with modulated commensal bacteria, microbiome-mediated primary-to-secondary bile acid conversion elicited an increased CXCL16 expression in LSEC, leading to a higher cytotoxic NKT activation and tumor growth suppression (Ma et al., 2018).

2.2 The role of tumor cells

Tumor cell-specific genomic alterations can differently shape the tumor microenvironment. Tumor-associated Antigens (TAA) are cancer-specific epitopes originated from newly derepressed oncofetal or cancer-testis antigens, viral peptides (in case of chronic HBV or HCV infection) or neoantigens caused by coding non-synonymous mutations that are expressed, presented through MHC-I and immunogenic. Naturally occurring TAA-specific CD8+ T-cell responses are present as part of the normal T-cell repertoire in patients with HCC and these responses correlate with patient survival (Flecken et al., 2014). The most common derepressed antigens in HCC are the cancer/testis antigen melanoma-associated antigen 3 (MAGEA3), the oncofetal antigen α-fetoprotein (AFP) and glypican 3 (GPC3). Neoantigens can derive from both passenger or driver mutations, and depending on the dominance of the subclone, they could lead to different degrees of exposure, immune tolerance, and T cell exhaustion (Hou, Zhang, Sun, & Karin, 2020). Therefore, it is not clear that the most common driver mutations in HCC (those affecting TERT promoter, TP53, ARID1A, ARID2, CTNNB1, AXIN1 and PIK3CA genes), are good candidates for inducing robust antitumor responses. The tumor mutational burden (TMB), expressed as mutations per megabase (mut/Mb), reflects the amount of non-synonymous mutations in a particular tumor, and is supposed to correlate with the neoantigen load (Rizvi et al., 2015). In highly mutagenic tumors such as melanoma or non-small cell lung cancer, or in microsatellite instability high (MSI-H) colon cancer, high TMB has been linked to response to immunotherapy, suggesting that it is possible to activate exhausted antigen-specific T cell responses in these patients, as well as fostering the development of personalized vaccines or cell therapy directed towards specific tumor neoantigens. Nevertheless, neoantigen editing during tumor evolution, copy-number loss of previously clonal neoantigens, loss of heterozygosity (LOH) in human leukocyte antigens (HLA) or the promoter hypermethylation of neoantigen-bearing coding genes, are just some of the mechanisms of immune evasion enacted by cancer cells under antigen-specific immune pressure (Rosenthal et al., 2019). In patients with HCC, with a reported median TMB of less than 3 mut/Mb, higher TMBs have not been associated with higher levels of immune infiltration (Sia et al., 2017) or response to immunotherapy (Spahn et al., 2020). Importantly, the occurrence of Copy Number Alterations (CNA) and large chromosomal aberrations are associated with LOH in antigen-presentation genes and with immune exclusion in cancer. In line with this, HCC with high levels of chromosomal aberrations displayed immune desertification (Bassaganyas et al., 2020).

Tumor cell-specific signaling cascades can also affect the immune microenvironment of HCC. Around one third of human HCC bear gain-of-function mutations on β-catenin (encoded by CTNNB1 gene) as putative driver gene (Sanchez-Vega et al., 2018). These tumors are usually well differentiated, present a microtrabecular pattern, lack inflammatory infiltrates and are histologically defined by intense nuclear β-catenin staining and strong glutamine synthetase expression (Calderaro et al., 2017). In metastatic melanoma, constitutive activation of β-catenin pathway, by either defined mutations or increased expression in key WNT/CTNNB1 pathway elements results in less infiltration by lymphocytes (Spranger, Bao, & Gajewski, 2015). One of the postulated mechanisms would be the ATF3-mediated repression of the DC-attractant chemokine CCL4 by cancer cells, hampering the recruitment of conventional type 1 DC, known to be a critical factor for T cell and NK infiltration (Spranger et al., 2015). In mice, HCC bearing MYC overexpression and CTNNB1 constitutive activation, endogenously generated in the liver by tail vein hydrodynamic injection, exhibited downregulation of CCL5, recruited less DC1, and were insensitive to immune checkpoint blockade (Ruiz de Galarreta et al., 2019). In human HCC, a downregulation of the NKG2D ligands MIC-A and MIC-B was seen in CTNNB1 activated tumors, which could explain NK recruitment. Interestingly though, this phenotype was related to less aggressiveness, indicating that immune evasion does not necessarily mean worse survival (Cadoux et al., 2021). Glutamine synthesis by cancer cells has been associated with immune depletion and the blockade by glutamine antagonist JHU083 led to enhanced efficacy of immunotherapy and durable responses in syngeneic colon cancer mouse models (Leone et al., 2019). Since glutamine synthesis in human CTNNB1-activating HCC is constitutively active, one could infer that glutamine blockade could be useful to increase immune infiltration in these tumors. However, not all human HCC bearing CTNNB1 mutations are immune excluded. In a recent refinement of transcriptomic classification of HCC, which leverages on bulk RNA sequencing deconvolution to infer immune infiltration, one third of CTNNB1-signature bearing tumors belonged to the so-called inflammatory class, with high expression of IFN-γ signaling, infiltration of CD8 + T cells and M1-TAM, and increased immune cell-attractant chemokines, including

CCL5 (Llovet et al., 2021). Genomic level information from phase III clinical trials with immunotherapy is now needed to firmly determine if and how CTNNB1-activating human HCCs respond differently to immune checkpoint inhibition.

Overexpression of MYC, occurring in a majority of human HCC, led to increased PD-L1 translation transforming KRAS-activated liver cancer from hot to cold tumors in a mouse model. This translational enhancement of PD-L1 was reversed by Eukaryotic Translation Initiation Factor 4E (eIF4E) inhibition (Xu et al., 2019). Around one-third of HCC bear *TP53-null* mutations. These tumors produce high AFP levels and are poorly differentiated and enriched in the macrotrabecullar massive histological subtype, presenting a poor prognosis in the pre-ICB era (Calderaro et al., 2017). Interestingly, in non-small cell lung cancer, TP53 mutations were independently associated with overall survival and longer responses when treated with ICI (Assoun et al., 2019). ARID1A is a core member of the polymorphic BRG/BRM-associated factor chromatin remodeling complex. Mutations in ARID1A gene include around 8–10% of human HCC (Wheeler & Network, 2017). Although ARID1A mutations lead to genomic instability and the generation of unrepaired DNA mismatches and higher TMB-driven immunogenicity, ARID1A aberrations can also limit chromatin accessibility to IFN-responsive genes, impairing IFN gene expression, and response to immunotherapy (Li et al., 2020). TP53-null and ARID1A HCCs are two examples demonstrating that immune exclusion in animal models or histological studies, does not equal resistance to immunotherapy.

Tumor-specific secretome defined by these genomic and functional features can also shape the immune microenvironment. Oncofetal AFP highly expressed by TP53-null HCC can modify the immune infiltrate towards a more tolerogenic one (Pardee, Shi, & Butterfield, 2014; Ritter et al., 2004). HCC bearing VEGF amplification (Horwitz et al., 2014), or IDO overexpression (Li et al., 2018) show increased secretion of VEGFA or Kinurenines, respectively, thus strongly shaping the immune environment. In early stage, resected HCCs, the expression of IDO1 in tumor cells has been associated with higher CD8 + T cell infiltration and patient survival (Li et al., 2018), but in HCC cell lines and animal models, IDO was found to be increased in immune checkpoint-resistant tumors and its inhibition led to higher anti-CTLA4-induced responses (Brown et al., 2018). Equally important, genomic rewiring of HCC cells by driver genes make them increasingly avid for myeloid derived cytokines or growth factors, as exemplified by the TGF-β receptor pathway activation (Chen, Gingold, & Su, 2019), or cMET activation (Wang et al., 2020) cause them to be exquisitely dependent on TGF-β1 and HGF, respectively.

Tumor-specific immune ligandome, defined by the density of ligands related to immune recognition and killing include: the expression and membrane localization of MHC-I, upregulated in HCC cell lines (Huang, Cai, & Wei, 2002), the immune costimulatory molecules CD80 (B7-1), CD86 (B7-2) shown to be downregulated in HCC (Tatsumi et al., 1997), the increased expression of the costimulatory ligand B7-H3 (Sun et al., 2012; Wang et al., 2014), and the increased expression of inhibitory ligands PD-L1 (Huang et al., 2021), or Galectin 9 (Zhou et al., 2018). Whether specific driver gene mutations are associated with a change in expression of these ligands in human HCC is still to be explored.

The complexity of the immune tumor microenvironment speaks against simplistic analysis and conclusions regarding the potential of immunotherapy in liver cancer. Fig. 1 summarizes the most important actors in this play. In the future, a more detailed genomic, transcriptomic, and proteomic characterization of HCC patients treated with immunotherapy in phase III clinical trials will lead to a better understanding of how these driver mutations define the response or resistance, knowing that the genomic profile of a tumor is an evolving landscape affected by immune pressure, Darwinian subclone selection, and the appearance of new immune evasion mechanisms.

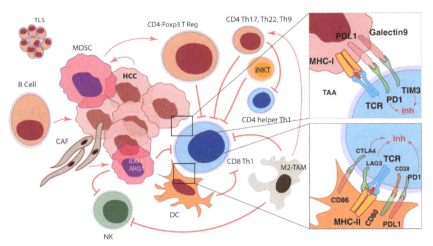

Fig. 1 Mechanisms of immune evasion.

2.3 Etiology of liver cancer and mechanisms of immune evasion

Recent studies have suggested that there may be differences in response to immune checkpoint blockade in patients with diverse cancer etiology. According to some of these studies, patients and animal models of non-alcoholic steatohepatitis (NASH), would not benefit as much from immunotherapies. A hepatocyte-specific Endoplasmic Stress-prone mouse model of NASH-related HCC was used in the preclinical evidence of this phenomenon. Using high resolution single cell mass spectrometry and RNA sequencing, the authors found that livers of NASH mice and humans were enriched in exhausted CD8+ PD1+ T cells and were not responsive to immune checkpoint inhibition, which instead, caused more damage and fibrosis in these already damaged livers. In this model, the depletion of CD8 T cells led to less tumor burden (Pfister et al., 2021). Other mechanisms that could shape NASH livers differently include the activation of dysfunctional immune cells, NKT cells (Wolf et al., 2014), T helper 17 (T_H17) cells (Gomes et al., 2016), and IgA^+ plasma cells (Shalapour et al., 2017) that could impact tumor immune surveillance and favor hepatocarcinogenesis. In the genetically modified mouse model of NASH, the depletion of CD4+ T, induced by fatty acid-dependent oxidative damage, contributes to hepatocarcinogenesis (Ma et al., 2016). In Alcohol-related Liver Disease (ALD), an increased gut permeability leads to translocation of bacterial pathogen-associated molecular patterns (PAMPs) which suppresses KC activation, increases the abundance of M2-TAMs, promotes the recruitment of MDSC infiltration and suppresses CD8+ T cell activation, a mechanism that could dampen HCC immunogenicity.

A meta-analysis using data from the main phase III clinical trials with the anti-PD1 antibodies nivolumab and pembrolizumab and the anti-PDL1 atezolizumab suggested no benefit over sorafenib in patients with non-viral etiology (Pfister et al., 2021). However, trials included in this study employed different therapies and different patient populations, and the meta-analysis showed a high heterogeneity. In fact, recently communicated data from a phase III clinical trial exploring first line anti-PDL1 durvalumab combined with anti-CTLA4 tremelimumab do not support this hypothesis and showed similar benefit in HBV-infected and uninfected patients (Abou-Alfa et al., 2022).

Hepatitis B Virus (HBV) is a DNA virus with hepatic tropism, transmitted through blood contact (transfusions, IV drugs, sexual intercourse, maternal delivery), that causes acute infection and complete surface and core antigen-directed T-cell-mediated clearance in a minority of infected individuals (Knolle & Thimme, 2014). In most patients, CD8+ T cell exhaustion and recruitment of tolerogenic cell types such as granulocytic MDSC, T_{reg} or regulatory B cells (B_{reg}) in the infected livers, mediate disease chronicity (Das et al., 2008; Pallett et al., 2015). The cycle of damage and regeneration, the integration of HBV DNA, and the expression of viral oncoproteins such as HBx, are related to oncogenesis that can arise in non-cirrhotic livers. HBV infection cannot be cured. Antivirals such as lamivudine or entecavir are therapeutic options for patients with high replication rates and HBV-mediated damage or in the preventive setting in immunosuppressed patients. Highly suppressive PD-1^{hi} T_{reg} cells are selectively enriched in HBV-related versus non-viral HCCs and are associated with a poor prognosis, while CD8+ T_{RM} cells are related to good outcomes in patients with HBV-related HCC (Lim et al., 2019). Thus, the liver immune milieu of an HCC-bearing chronically HBV-infected patient trends towards a tolerogenic state and this could hamper the success of immunotherapy.

Hepatitis C Virus (HCV) is another blood-borne transmitted single-stranded RNA which acute infection is cleared in around 20% of infected individuals. This clearance is related to vigorous induction of CD4 and CD8 antigen specific responses and the subsequent narrowing of quasi-species diversity (Farci et al., 2000). In most patients, viral high replication and hypermutation overwhelms the adaptive response and the disease becomes chronic, causing a low-degree inflammatory environment that in the long term promotes neoplastic transformation of injured hepatocytes (Hedegaard et al., 2017). Viral factors inhibit the key sensors of virus infection RIG-I and MDA5 (Cao et al., 2015). High efficacy of direct acting antivirals (DAA) has dramatically impacted the field of hepatology, changing the prevalence of viral etiology in cirrhotic patients and in patients with viral-related HCC. In patients cured from HCV infection, nevertheless, circulating TCF1+CD127+PD1+ HCV-specific exhausted CD8+ T-cell subsets have been found, indicating that once the antigen stimulation has ceased, the immune system is still shaped by the chronic insult (Wieland et al., 2017). In cured patients, the inflammation resolves shortly after the end of treatment (Huang et al., 2021) but it is not clear to what extent the degree of fibrosis regression is, especially in obese (McPhail et al., 2021). Whether the immunotherapy of HCC in HCV-infected versus HCV-cured individuals could have different outcomes is yet unknown, but a probable scenario is the progressive reduction of HCC cases due to this etiology.

The growing debate on the clinical relevance of the etiology in HCC patients has highlighted the need and increased the interest in reviewing the diverse immune context of HCV and HBV infections, NASH and ALD and the study of whether this could account for differences in tumor biology, immune infiltration, and most importantly impact in their response to immunotherapies. Certainly, a difficulty to perform this type of analysis arises from the lack of a good definition of non-viral etiologies of liver cancer in clinical trials and in many observational studies where the immune environment of HCC has been characterized. The animal models of viral-induced HCC are inexistent or poorly relevant to human disease, those of toxic-induced HCC are abnormally mutagenic and those diet-induced need additional toxic or genetical insults to be technically feasible. Another even more important obstacle is the overlap of NASH and ALD and with HCV or HBV infection in real life of humans which should prompt a better reporting of metabolic profile and alcohol consumption in every patient with HCC. But even at the biological level it may be impossible to separate these mechanisms. For example, HCV infection induces liver steatosis and contributes to the development of metabolic dysfunction, and it is precisely the combination of HCV and obesity that increases the risk of developing HCC (Leslie, Geh, Elsharkawy, Mann, & Vacca, 2022). It has been reported that DAA treatment can result in weight gain in a proportion of HCC patients and these metabolic changes could potentially lead to malignant transformation (Leslie et al., 2022).

3. Immunotherapy of HCC

In the previous section, the main features of immune biology of HCC have been delineated. For more than a decade, the multi-tyrosine kinase inhibitor (TKI) sorafenib has been the standard of care for patients with advanced HCC (Llovet et al., 2008). In recent years an array of new TKI such as lenvatinib in first line, and regorafenib, cabozantinib and ramucirumab in second line after sorafenib have been approved. The success of a new way of understanding and treating cancer, through the unleashing of immune evasion, exemplified by the unprecedented responses in advanced melanoma patients, promoted the study of these drugs in other cancers. The immunotherapy aims at reinvigorating exhausted T cell responses, by the activation of costimulatory or the blocking of coinhibitory membrane receptors or their ligands, globally called immune checkpoints. In Fig. 2, the array of immune checkpoints interfacing the tumoral cell,

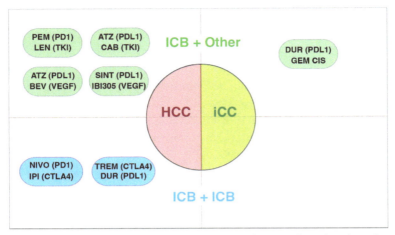

Fig. 2 Immunotherapies with reported or ongoing phase III clinical trials as first line systemic therapy in liver cancer.

the antigen-presenting cell, and the lymphocyte is depicted. Antibodies against the immune checkpoints CTLA4, PD-1 and PD-L1 have reached the clinical stage of development and are responsible for the first successes in liver cancer. Others, such as LAG3, TIM3 or VISTA are currently being tested in phase II trials.

3.1 Combination therapies

In 2022, combination immunotherapy should be considered the preferred option for an HCC patient in need of systemic therapy, including those with vascular invasion or extrahepatic spread, with too much liver burden to be adequately treated by intra-arterial therapies or with liver disease progressing to intra-arterial therapies (Reig et al., 2021). Table 1 summarizes the results of the main phase II and III trials using immune checkpoint inhibitors (ICI) that have been conducted in HCC patients.

The combination of the PD-L1 inhibitor atezolizumab and a relatively high dose (15 mg/kg) of the VEGF-inhibitor bevacizumab resulted in prolonged OS and PFS compared to sorafenib in the IMBRAVE-150 trial with a hazard ratio (HR) of 0.58 (95%CI 0.42–0.79) for overall survival (OS) and 0.59 (95%CI 0.47–0.76) for progression-free survival (PFS) (Finn, Qin, et al., 2020). With more prolonged follow-up, median OS was 19.2 months in patients treated atezolizumab plus bevacizumab (vs. 13.4 with sorafenib) and median PFS was 6.9 months (vs. 4.3 with sorafenib), while ORR was

Table 1 Phase III trials of ICIs in advanced HCC.

Agent (dose)	N	MVI (%)	EHD (%)	AFP>400 ng/mL (%)	ORR (CR) (%)	mPFS (months)	mOS (95% CI) (months)	OS HR (95% CI)	Reference
KEYNOTE-240 (second line after sorafenib)									
Pembrolizumab (200 mg q3w)	278	13	70	46	18 (2)	3	13.9 (11.6–16)	0.71 (0.61–0.96)	Finn et al. (2020), Merle et al. (2021)
Placebo	135	12	69	43	4 (0)	2.8	10.6 (8.3–13.5)		
CheckMate 459 (first line)									
Nivolumab (240 mg q2w)	371	75#	75#	33	15 (4)	3.7	16.4 (13.9–18.4)	0.85 (0.72–1.02)	Yau et al. (2022)
Sorafenib (800 mg qd)	372	70#	70#	38	7 (1)	3.8	14.7 (11.9–17.2)		
IMbrave150 (first line)									
Atezolizumab (1200 mg q3w) + Bevacizumab (15 mg/kg q3w)	336	38	63	38	30 (8)	6.8	19.2	0.66 (0.52–0.85)	Finn et al. (2021), Finn et al. (2020)
Sorafenib (800 mg qd)	165	43	56	37	12 (0)	4.3	13.4 (10.4–NE)		
HIMALAYA (first line)									
STRIDE Tremelimumab (300 mg 1 dose) + Durvalumab (1500 mg q4w)	393	26	53	37	20 (3)	3.8	16.4 (14.2–19.6)	0.78 (0.65–0.92)	Abou-Alfa et al. (2022)
Durvalumab (1500 mg Q4W)	389	24	54	35	17 (2)	3.7	16.6 (14.1–19.1)	0.86 (0.73–1.03)	
Sorafenib (800 mg qd)	389	27	52	32	5 (0)	4.1	13.8 (12.3–16.1)		

Continued

Table 1 Phase III trials of ICIs in advanced HCC.—cont'd

Agent (dose)	N	MVI (%)	EHD (%)	AFP>400 ng/mL (%)	ORR (CR) (%)	mPFS (months)	mOS (95% CI) (months)	OS HR (95% CI)	Reference
ORIENT-32 *(first line)*									
Sintilimab (200 mg q3w) + Bevacizumab biosimilar (15 mg/kg q3w)	380	28	73	43	24 (1)	4.6	NR	0.57 (0.43–0.75)	Ren et al. (2021)
Sorafenib (800 mg qd)	191	26	75	42	8 (0)	2.8	10.8 (8.5–NR)		
COSMIC-312 *(first line)*									
Cabozantinib (40 mg qd) + Atezolizumab (1200 mg q3w)	432	31	54	38	11 (0)	6.8	15.4 (13.7–17.7)	0.90 (0.69–1.18)	Kelley et al. (2022)
Sorafenib (800 mg qd)	217	28	56	30	4 (0)	4.2	15.5 (12.1–NE)		
Cabozantinib (60 mg qd)	188	36#	54	35	6 (0)	4.3	N/A		

N, number of patients; MVI, macrovascular invasion; EHD, extrahepatic disease; AFP, alpha-fetoprotein; mPFS, median progression free survival; mOS, median overall survival; HR, hazard ratio; NR, not reached; #, MVI and EHD were merged.

30% by RECIST 1.1, including 8% complete responses (Cheng et al., 2021). This combination also showed a favorable safety profile and delayed deterioration of health-related quality of life (HRQoL) (11.2 vs. 3.6 months; HR 0.63, 95% CI 0.46–0.85) (Toh et al., 2022). A similar combo of the anti-PD1 sintilimab with a bevacizumab biosimilar (IBI305) was also superior to sorafenib in the ORIENT-32 phase III trial with very comparable HRs of 0.56 (95% CI 0.46–0.70) for OS and 0.56 (95% CI 0.46–0.70) for PFS (Ren et al., 2021a). None of these two trials compared single agent atezolizumab or bevacizumab versus sorafenib. However, in a phase 1b trial which randomized patients to atezolizumab plus bevacizumab or atezolizumab alone, PFS was better with the combination than with single agent atezolizumab (5.6 vs. 3.4 months) despite a similar response rate (20% vs. 17%) (Lee et al., 2020). In an exploratory biomarker analysis of this study using archival or fresh biopsies, TMB was not associated with response to atezolizumab plus bevacizumab or with more prolonged PFS but pre-existing immunity was, as illustrated by a high PD-L1 expression and a simple T effector signature including granzyme B, PRF1, CXCL9 genes (Zhu et al., 2020). High VEGFR2 expression, Treg and myeloid inflammation were also associated with longer PFS in patients treated with the combination compared atezolizumab monotherapy (Table 2).

Very recently, the combination of the PD-L1 inhibitor durvalumab with a single high (300 mg) priming dose of the CTLA4 inhibitor tremelimumab (STRIDE regime) resulted in prolonged OS compared to sorafenib in the HIMALAYA trial with a HR of 0.78 (95%CI 0.65–0.92) (Abou-Alfa et al., 2022). Median OS was 16.4 months in patients treated with durvalumab plus tremelimumab vs.13.8 with sorafenib, while ORR

Table 2 Phase III trials of ICIs in advanced BTC.

Agent (dose)	N	ORR (CR) (%)	mPFS (months)	mOS (95% CI) (months)	OS HR (95% CI)	Reference
TOPAZ-1 (first line)						
GemCis (gemcitabine 1000 mg/m2 + Cisplatin 25 mg/m2 D1, D8 q3w	344	18.7 (0.6)	5.7	11.5 (10.1–11.5)	0.80 (0.66–0.97)	Oh et al. (2020)
GemCis + Durvalumab (1500 mg q3w)	341	26.7 (2.1)	4.2	12.8 (11.1–14.0)		

N, number of patients; mPFS, median progression free survival; mOS, median overall survival; HR, hazard ratio.

was 20.1% by RECIST 1.1 (vs. 5.1%), including 3% complete responses. As it happened in several trials testing single agent ICIs, there was no improvement in PFS. This combination also showed a favorable safety profile with 26% of patients experiencing treatment-related adverse events (TRAE) grade 3 or 4 (vs. 37% with sorafenib) and 8% having TRAE leading to treatment discontinuation (vs. 11% with sorafenib).

There is no question that HCC is sensitive to the CTLA4 blockade. Indeed, the first evidence of the potential benefit of ICIs in HCC was seen in a small trial in patients with concurrent HCC and HCV infection treated with tremelimumab (Sangro et al., 2013) where an ORR of 17% was observed. Interestingly, a reduction in HCV viral load was also detected in some patients indicating the activation of exhausted CD8 HCV-specific responses. In another small trial testing tremelimumab plus subtotal tumor ablation with TACE or RFA, an even higher ORR of 26% was observed. In this study, tremelimumab alone increased the number of activated CD4+ and CD8+ cells in the peripheral compartment and tumor infiltration by CD3+ and CD8+ cells were increased in all patients with paired biopsies before and after tremelimumab and was higher in responders than in non-responders.

When used in combination with PD-(L)1 inhibitors, the dose of CTLA4 is important, as it happens in other tumors. A cohort of the CheckMate-040 trial randomized sorafenib-experienced patients to 3 arms with different doses of the CTLA4 inhibitor ipilimumab and the PD-1 inhibitor nivolumab (Yau et al., 2020). In two of them, lower doses of ipilimumab and higher doses of nivolumab (1 and 3 mg/kg, respectively) were used. Outcomes were better using a higher dose of ipilimumab (3 mg/kg) and a lower dose of nivolumab (1 mg/kg), with an ORR of 32%, median OS of 22.9 months and a remarkable 42% survival rate at 3 years (El-Khoueiry et al., 2021). The HIMALAYA trial has shown that a single high dose of tremelimumab at treatment initiation is able to improve the efficacy of durvalumab to a point where the combination outperforms sorafenib as first-line systemic therapy. Indeed, in a phase 2 study such a priming dose of tremelimumab resulted in the most intense early peripheral burst of proliferating CD8+ cells, a finding associated with response (Kelley et al., 2021). As expected, higher doses of ipilimumab increased the rate of TRAEs and we will have to wait until the results of the CheckMate-9DW trial are presented to learn if this combination prolongs OS compared to sorafenib and shows an acceptable safety profile in the first-line setting.

The COSMIC-312 phase III trial has compared the combination of atezolizumab and cabozantinib vs. sorafenib. In the final analysis (Kelley et al., 2022), the combination resulted in prolonged PFS (median 6.8 vs. 4.2 months, HR 0.63, 95 CI 0.44–0.91, $P=0.0012$) but the same benefit was not shown for OS in the interim analysis (median 15.4 vs. 15.5 months, HR 0.90, 95 CI 0.69–1.18, $P=0.438$). Inhibition of VEGF signaling targeting the ligand or the receptors may enhance T cell stimulation by PD-(L)1 inhibitors through different mechanisms, including normalization of the tumor vasculature and promotion of a more favorable TME by reducing suppressor cells like MDSC and M2-polarized TAMs, and increasing CD8+ T cells (Rahma & Hodi, 2019; Shigeta et al., 2020). In COSMIC-312, the combination of cabozantinib and atezolizumab resulted in a low ORR (11%) compared to the other combinations (15–30%) but we should wait for the final OS analysis and full report of this trial before raising any hypothesis about this somewhat unexpected result.

In animal models, cabozantinib had little effect on tumor infiltration by macrophages and T cells and did not modify PD-L1 expression by tumor cells while the combination with an anti-PD-L1 antibody did not improve the therapeutic efficacy of cabozantinib monotherapy (Shang et al., 2021). The immune effects of cabozantinib have also been studied in a small trial that investigated preoperative cabozantinib and nivolumab in resectable HCC patients (Ho et al., 2021). Cabozantinib alone induced an increase of effector and memory CD4+ and CD8+ T cells in the peripheral compartment. Within the TME (with transcriptional profile analyzed in only four paired biopsy samples), cabozantinib reduced the expression of some immunosuppressive genes like IDO1, TGF-β2, CXCL1 or CX3CR1, but also increased the expression of others like IL8/CXCL8. When nivolumab was added to cabozantinib and patients with major or complete pathological responses were compared to non-responders, the former had higher infiltration by CD4+ and CD8+ T cells and CD20+ B cells, often forming TLS, and an enrichment of IFN-γ producing effector memory CD4+ and CD8+ T cells as well as granzyme B+ effector CD8+ T cells. Besides, responders had more abundant myeloid cells including immunosuppressive M2-macrophage clusters.

Nevertheless, TKIs inhibit a wide spectra of tyrosine kinases and their effect in the TME should not be considered comparable to one another. In a phase 2 trial, the combination of pembrolizumab and lenvatinib doubled (36%) the ORR observed with single agent pembrolizumab and lead to a promising 20-month median OS, at a price of more intense toxicity (TRAE grade 3 or higher in 67% of patients (Finn et al., 2020). The impact of this combination on overall survival as first-line therapy compared to sorafenib is now under investigation in the LEAP-02 trial (Llovet et al., 2019).

3.2 Single agents

Altogether, current scientific evidence shows that, if locally available, patients with advanced HCC should receive atezolizumab plus bevacizumab or durvalumab plus tremelimumab. Contraindications to ICIs include active autoimmune diseases targeting relevant organs and solid organ recipients in whom the risk of provoking a severe flare or rejection, respectively, is high and worrying (Pinter, Scheiner, & Peck-Radosavljevic, 2021). The safety of ICIs has not been studied and therefore ICIs cannot be recommended in patients with HBV infection that is not effectively treated with antiviral agents or in those with HBV and HDV while the use in patients with HIV infection or HBV and HCV coinfection should be considered individually (Sangro, Bruix, Chan, Galle, & Rimassa, 2022).

The results from doublets have certainly curtailed the interest in single agent ICIs. However, there may still be a place for them in selected patients or in health environments where combinations are not available. PD-1 inhibition with nivolumab or pembrolizumab can induce deep, durable responses associated with prolonged survival in around 15% of patients naïve or sorafenib-experienced (El-Khoueiry et al., 2017; Zhu et al., 2018). After sorafenib, a benefit of pembrolizumab over placebo was not proven in the KEYNOTE-224 trial, because of not reaching the predetermined cutoff significance of $P<0.01$ (HR 0.78; 95% CI, 0.61 to 0.99; $P=0.024$), although OS was numerically better with pembrolizumab (median OS 13.9 vs. 10.6 months) (Merle et al., 2021). More recently, an Asian trial with a similar design has shown a significant prolongation of OS vs placebo (HR 0.7; 95% CI 0.63–0.99; $P=0.018$; median 14.6 vs. 13 months) (Qin et al., 2022). In naïve patients, a significant improvement in OS was not observed when nivolumab and sorafenib were compared in the CHECKMATE-459 trial (HR 0.85; 95% CI 0.72–1.02; $P=0.075$), although again OS was numerically better after nivolumab (median OS 16.4 months vs. 14.7 months) and substantially delayed deterioration of HRQoL (Yau et al., 2022). In the HIMALAYA trial, a secondary endpoint was the comparison of OS of durvalumab monotherapy vs. sorafenib with a non-inferiority margin of 1.08 (Abou-Alfa et al., 2022). Non-inferiority was met with a HR of 0.86 (95% CI 0.73–1.03) while median OS was 16.6 vs. 13.8 months for durvalumab and sorafenib, respectively. With a higher ORR (17% vs. 5%) and a better tolerability (TRAE of any grade and grade 3 or 4 occurring in 52% vs. 85% and 13 vs. 37% of patients, respectively), durvalumab seems to be a better option for first-line systemic therapy.

One potential obstacle for efficacy in atezolizumab-containing regimes is the occurrence of anti-drug antibodies (ADA). The prevalence of ADA following treatment with ICB seems to be higher for atezolizumab (54%) than for nivolumab (10%) or pembrolizumab (1.5%) (Enrico, Paci, Chaput, Karamouza, & Besse, 2020). Whether the presence of ADA hampers the efficacy of the drug and impacts on patient's survival is currently under investigation.

Immune checkpoint ligands such as LAG3, TIM3 or VISTA, and others are being investigated as targets in HCC, mainly in multi-cancer phase II trials with no solid data at this moment. Many other new strategies investigated for other cancers have boosted the design of multicancer trials where some HCC patients can be enrolled.

3.3 Biomarkers of response or resistance

Only around one third of the patients treated with current ICI combinations show objective remissions and 20 to 40% have progressive disease as the best response. It is thus crucial to identify upfront those patients that will or will not respond to the ICB. PD(L)1 immunohistochemistry (IHC) has been intensely investigated. In 2015, in a trial testing the efficacy of the anti-PD1 antibody pembrolizumab in patients with non-small cell lung cancer, the PD-L1 positive staining with linearly correlated with overall response rates, which impacted on biomarker-based indication of pembrolizumab in patients with NSCLC (Kang et al., 2017). This was revalidated in several other trials with pembrolizumab. Currently, Head and Neck Squamous Cell, Breast, Cancer and Urothelial treatments with the anti-PD1 pembrolizumab, nivolumab, durvalumab and the anti-PD-L1 atezolizumab are PD-L1 IHC biomarker-based. Unlike in the above-mentioned cancer types, the predictive role of PD-L1 in HCC is not entirely clear. In KEYNOTE-224 trial testing pembrolizumab in the second line the CPS but not the TPS was associated with response (Zhu et al., 2018). In the CHECKMATE-459 trial, the expression of PD-L1 in more than 1% of the area was associated with longer median OS in nivolumab-treated patients (Yau et al., 2022). In the CHECKMATE-040 trial, PD1 and

PDL1 staining were associated with improved OS but only PD1 staining was associated with ORR (Sangro et al., 2020). Median OS was 28.1 (95% CI 18.2–n. a.) vs. 16.6 months (95% CI 14.2–20.2) for patients with tumor PD-L1 $\geq$ 1% vs. <1% ($P=0.03$). Although this could seem promising, complete or partial tumor responses were observed in both PD-L1–positive and PD-L1–negative patients treated with nivolumab monotherapy (Sangro et al., 2020). In the adjuvant setting, the frequency of tumor CD3+ PD1+ expression in post-treatment was higher in responders. No differences were seen in pretreatment tissues. In this study CD8 and CD68 staining pre- or post-treatment were not associated with outcomes (Agdashian et al., 2019). Regarding flow cytometry analyses, baseline frequency of CD4+ PD1+ cells in Peripheral Blood Mononuclear Cells was higher in patients treated with radiofrequency ablation or chemoembolization who responded to tremelimumab adjuvancy (Agdashian et al., 2019). Despite the efforts to find a good predictive marker in HCC, PD-L1 or PD1 IHC show limited positive or negative predictive values and thus have not been implemented for patient stratification in clinical trials or in treatment decisions in clinical practice.

The Tumor Mutational Burden (TMB), a calculation that reflects the number of mutations per megabase of sequenced tumor DNA, has been shown to be associated with both prognosis and predictive power in clinical trials using ICB (Havel, Chowell, & Chan, 2019). Despite this, the fact that HCC has low median TMBs could be the cause of the lack of correlation of Whole Exome Sequencing (WES)-calculated TMB with survival in first line atezolizumab and bevacizumab clinical trial (Zhu et al., 2020). Copy number alterations (CNA), another proxy of high mutational load in cancer cells, which may be related to immune exclusion in HCC (Bassaganyas et al., 2020) has not yet been tested as a predictive biomarker in clinical trials.

Gene expression profiling has been studied mainly in RNA sequencing (RNA-seq) studies of fresh-frozen liver biopsies or surgical pieces. From the first RNA-seq based transcriptomic classifications of HCC, there has always been the evidence of different degrees of inflammatory gene expression signatures in HCC (Rebouissou & Nault, 2020), with certain association of β-Catenin mutation with low immune signature, consistent with findings in melanoma and other tumors. The most recent study in this field identified a subgroup, referred to as an "immune class," that was shown to include 2 distinct clusters with an exhausted immune response, which may help determine why some apparently "inflamed" tumors are poor responders to ICIs

(Sia et al., 2017). A recent study in samples from a first line trial testing nivolumab in patients with advanced HCC, examined these published signatures and its relation to survival and or response to ICB (Sangro et al., 2020). There was a significant or trend association of *T-cell exhaustion (CD274 (PD-L1), CD276, CD8A, LAG3, PDCD1LG2, TIGIT)*, Gajewski 13-gene inflammatory (*CCL2, CCL3, CCL4, CD8A, CXCL10, CXCL9, GZMK, HLA-DMA, HLA-DMB, HLA-DOA, HLA-DOB, ICOS, 1RF1*), 6-gene interferon gamma (*CXCL10, CXCL9, HLA-DRA, IDO1, IFNG, STAT1*), Ribas 10-gene interferon gamma (*CCR5, CXCL10, CXCL11, CXCL9, GZMA, HLA-DRA, IDO1, IFNG, PRF1, STAT1*) and Interferon gamma biology (*CCL5, CD27, CXCL9, CXCR6, IDO1, STAT1*) signatures with response to nivolumab (Sangro et al., 2020). This indicates that the baseline inflammation of the tumor will mark the efficacy of ICB.

Some easier-to-perform laboratory parameters could point to good ICB responders. In nivolumab-treated advanced HCC patients, complete and partial responses to nivolumab were seen in patients with low Neutrophil to Lymphocyte and Platelet to Lymphocyte Ratios (NLR, PLR) (Sangro et al., 2020). In the explorative setting, plasma analysis using ELISA or higher multiplexed techniques have enabled biomarker discovery studies. In a small phase I trial exploring pembrolizumab in the second line, high levels of TGF-β (>200 pg/mL) were associated with poor outcomes. Other proinflammatory and antiinflammatory cytokines (IL1b, IL6, IL8, IL12, IL18, IFN-γ, IL10, CXCL9, CCL4, CCL5) and soluble checkpoint ligands (PDL1 and PDL2) were not associated with prognosis nor were related to objective responses (Feun et al., 2019).

3.4 Beyond immune checkpoint inhibitors

The success of ICIs has set the proof of principle that tumor immune rejection is achievable and can be exploited therapeutically. This has opened the door to testing other forms of immunotherapy in liver cancer. A detailed analysis of these other strategies is beyond the scope of this manuscript, but we will summarize the key strategies that have initiated clinical development.

Adoptive cell therapies (ACT) consist of the use of self-patient lymphocytes which are sensitized and/or expanded in vitro and reinfused to the patient. Several cell types such as lymphokine-activated killer (LAK), cytokine-induced killer (CIK), NK, Tumor Infiltrating Lymphocytes

(TIL) and redirected circulating T cells have been explored. In the case of peripheral blood lymphocytes, two main strategies have been explored, namely the use of transgenic tumor-antigen specific TCR and chimeric antigen receptors (CAR). In ACT, patients are usually treated with a preconditioning cytotoxic regimen to induce lymphodepletion and support in vivo expansion of transferred cells. In a phase III trial in patients with curative HCC resection, the infusion of CIK improved both PFS and OS when compared with patients with no adjuvant therapy (Lee et al., 2015). Feasibility of TIL treatment in HCC was shown in a pilot phase I trial (Jiang et al., 2015). Immunotherapy using ex vivo-expanded allogeneic NK cells in HCC treated with resection was recently shown to be safe in a phase I trial (Kim et al., 2021). Other ongoing trials with NK cells in HCC have been detailed elsewhere (Mantovani, Oliviero, Varchetta, Mele, & Mondelli, 2020). These strategies are nevertheless not regularly used in clinical practice in part due to the lack of in-house cell therapy facilities and the regulatory nuances of ACT.

CAR T cell has shown to be of great success in hematological malignancies (Larson & Maus, 2021), but there are specific challenges, specially the lack of good tumor-specific surface antigens. In HCC, the most explored cancer antigens to be targeted with ACT have been GPC3 and AFP. GPC3 is expressed in the surface of the HCC cell, so GPC3-targeting CAR-T cells has been shown to be efficacious in animal models and are being tested in phase I clinical trials (NCT04121273, NCT02715362, NCT03130712). AFP, instead, is found intracellularly and secreted, so AFP-specific TCR transgenic T lymphocytes are being tested. AFP-specific TCR T cells have obtained some objective tumor remissions reported in an ongoing clinical trial of in HCC (Sangro et al., 2022, NCT03132792). Other strategies such as bispecific GPC3-S-Fabs designed to attract NK cells to GPC3-positive cells, bispecific T-cell Engagers (BiTE) binding GPC3 and CD3, are still in their preclinical phases (Huang et al., 2021). Therapeutic vaccines aim at either de novo priming of T cells against antigens expressed by tumor cells that are unable to spontaneously trigger a response, or enhancing already existing responses, or it can also widen the repertoire and breadth or tumor specific responses (Tagliamonte et al., 2018). In HCC, many trials based on tumor lysates have failed to produce consistent results (Prieto, Melero, & Sangro, 2015). HCC peptide vaccines targeting telomerase, GPC3 or AFP have elicited specific responses in some cases, but none has provided clinically meaningful results (Greten et al., 2010; Sawada et al., 2012). Vaccines targeting true tumor-specific neoantigens identified by a multi-omic approach could be a step forward. An off-the-shelf vaccine including 5 HLA-A*24 and 7 HLA-A*02 as well as 4 HLA-DR restricted peptides identified and selected from human HCC, in combination with a novel RNA-based adjuvant was able to induce immune responses but lacked clinical activity as a single therapy (Buonaguro et al., 2020).

4. Immunotherapy of intrahepatic cholangiocarcinoma

Biliary tract cancer (BTC) includes cholangiocarcinomas and gallbladder cancer. Cholangiocarcinoma (CCA) is a molecularly and clinically heterogeneous entity. The immune environment of cholangiocarcinoma is poorly understood. CCA is an epithelial tumor that arises from the bile ducts and 90% of them are adenocarcinomas (Valle, Kelley, Nervi, Oh, & Zhu, 2021). Anatomically, they are divided into intra- and extrahepatic cholangiocarcinomas. Intrahepatic CCA (iCCA) arises within the hepatic parenchyma whereas extrahepatic (eCCA) arises in the extrahepatic portion of the bile duct. From a clinical perspective they can be divided into resectable versus unresectable disease. Non-resectable disease includes localized or locally advanced disease non amenable to surgical treatment upfront or overt metastatic disease. Due to the low incidence of these entities, randomized clinical trials for non-resectable disease usually include any type of BTC, both iCCA or eCCA. This explains why treatment guidelines for non-resectable or metastatic BTC usually do not differentiate between them, although from a clinical, biological, and molecular perspective they are different entities. There is an obvious need to better understand the best treatment for these entities and guidelines, such as NCCN, describes primary treatment options for patients with unresectable or metastatic BTC as enrolment on a clinical trial, administering systemic treatment or providing best supportive care (Benson et al., 2021).

Standard of care for first line advanced or metastatic ICCA is gemcitabine plus cisplatin based on the ABC-02 trial that showed a median OS of 11.7 months and a mPFS of 8.0 month in the combination arm compared to 8.1 months OS and 5.0 months PFS in the gemcitabine alone arm (Valle et al., 2010). Other accepted combinations based on different phase II trials are gemcitabine plus oxaliplatin (Andre et al., 2008; Jang et al., 2010; Kim et al., 2009), gemcitabine plus capecitabine (Knox et al., 2005; Koeberle et al., 2008; Riechelmann, Townsley, Chin, Pond, & Knox, 2007) gemcitabine plus nab-paclitaxel (Shroff et al., 2019), capecitabine plus oxaliplatin (CAPOX) (Nehls et al., 2008), or the triad including folinic acid, fluorouracil, and oxaliplatin (FOLFOX) (Nehls et al., 2002). In recent years, the interest in better understanding the differences across BTC subtypes has led to the identification of patient populations defined by genomic alterations with improved survival when those alterations are amenable to a therapeutic intervention, enabling the indication of targeted therapies. Some examples include FGFR2 translocations (13–14% of iCCA) targeted with pegatinib or other FGFR kinase inhibitors, IDH1 mutations (10–23% of iCCA) with ivosidenib or similar IDH1 inhibitors, BRAF V600E (3% of iCCA) with BRAF or BRAF plus MEK inhibitors (Goeppert et al., 2014) and NTRK fusions (0.75% of BTC) recently successfully treated (Demols et al., 2020).

4.1 Single agents

Small subgroups of iCC have shown to be sensitive to ICIs. iCCA having mismatch repair deficiency with high microsatellite instability (dMMR/MSI-H) which accounts for around 10% of patients (Ando et al., 2022; Silva et al., 2016), or bearing high TMB defined as 10 or more mutations per megabase, which accounts for 5% of patients (Cao et al., 2020) have responded to monotherapy with the anti-PD1 pembrolizumab. As a result, pembrolizumab has been approved by the US FDA for this subset of patients based on the KEYNOTE-158 phase II trial that tested the efficacy of this drug in non-colorectal MSI-H/dMMR cancers (Marabelle et al., 2020). In iCCA patients ($n=22$) pembrolizumab showed an ORR of 40.9% (95% CI, 20.7%–63.6%) with a median PFS of 4.2 months and a median OS of 24.3 months. iCCA TMB-high patients had 29% of ORR compared to 6% in the non-TMB-high population in the same trial (Marabelle et al., 2020). Similar ORR has been reported with the investigational PD1 antibody dostarlimab in MSI-H/dMMR patients, although only 2 patients with BTC were included in the non-endometrial cohort ($n=106$) (Berton et al., 2021).

In the non-selected BTC population, the role of ICI monotherapy is disappointing. The largest published data is the joint analysis of the KEYNOTE-158 ($n=104$, all comers) and KEYNOTE-028 ($n=24$, tumor PD-L1 expression). The ORR after pembrolizumab in iCCA patients ranged from 6% to 13% with no difference by baseline PD-L1 expression (Piha-Paul et al., 2020). Similar data were seen with Nivolumab in a phase II trial that reported an ORR of 11% (Kim et al., 2020). These were early trials and had several limitations like the use of different PD-L1 expression detection methodologies and the lack of definition by anatomical origin of BTCs (iCCA vs eCCA vs gallbladder cancer). As a positive note, it is interesting to reflect that in the three described trials there were long responders among the few BTC patients that responded. ICI combinations without chemotherapy have not been successful and although data is scarce (Klein et al., 2020; Sahai et al., 2020; Yoon et al., 2021), it does not seem to be the way moving forward at this point, at least with unselected BTC cohorts. Several additional trials are ongoing evaluating anti-PD-1/PD-L1 antibodies as monotherapy or in combination (Hilmi, Vienot, Rousseau, & Neuzillet, 2019; Vogel, Bathon, & Saborowski, 2021).

4.2 Combination therapies

Based on the positive results in other tumor types and the rationale that chemotherapy could reshape the tumor microenvironment and the immune system (Salas-Benito et al., 2021) several trials combining the standard of care with ICIs have been running. Some early data of chemo-immunotherapy combination has already been partially presented in several international meetings (Oh et al., 2020; Sahai et al., 2020) and several trials are ongoing (Manne, Woods, Tsung, & Mittra, 2021; Vogel et al., 2021). Very recently, the TOPAZ-1 trial, testing the anti-DP-L1 antibody durvalumab in combination with gemcitabine plus cisplatin, has become the first phase III positive randomized clinical trial in BTC that may be able to shift the role of immunotherapy in this cancer (Oh et al., 2022). OS was significantly improved with a HR of 0.80 (95% CI 0.66–0.97) and median OS was 12.8 months in the durvalumab plus chemotherapy arm vs 11.5 months in the chemotherapy only group. Based on an initially planned stratification, iCCA showed a larger benefit compared to eCC/GBC. In addition, Asian race and Asia as a region showed higher benefit compared to non-Asian race or the rest of the world. This should be analyzed further as there is data pointing towards genetic differences in iCCA between Eastern and Western patients that may have impacted the overall results of the trial (Cao et al., 2020). Other trials with a similar design using Pembrolizumbab (NCT04003636) or Atezolizumab (NCT04677504) as ICI backbone are also ongoing. Other Immunotherapy approaches in BTC. Based on positive signals shown in HCC, trials with immunotherapy plus TKI are ongoing and some have shared early data with promising responses (Lwin et al., 2020) although it is still unclear the role these combinations may play in the future.

5. Concluding remarks

In only a decade we have witnessed the dawn of immunotherapy of liver cancer leading to a stage where it has become or will soon become part of the standard of care for patients with HCC and iCCA (Fig. 2). While this represents an enormous advance for those patients who will obtain a high-impact benefit, the success comes with important challenges. Financial issues is one of them and not the least important. In coming years, much translational and clinical research is needed to optimize treatment efficacy and patient selection, but the outcome is worth the effort.

Grant support

J.A. has received a grant from the Agencia Estatal de Salud of ICSIII (PI20/01663) related to this topic. M.P.S. reports support by Agencia Estatal de Salud of ICSIII (PI20/01718), B.S. reports support by Agencia Estatal de Salud of ICSIII (PI19/00742).

Conflict of interest

J.A. has no financial or personal disclosures relevant to the contents of this manuscript. M.P.S. reports consultancy fees from Astra Zeneca, Incyte and Ellipses Pharma; research grants (to Institution) from BMS and Roche. B.S. reports consultancy fees from Adaptimmune, Astra Zeneca, Bayer, BMS, Boston Scientific, BTG, Eisai, Eli Lilly, H3 Biomedicine, Incyte, Ipsen, Novartis, Merck, Roche, Sirtex Medical, Terumo; speaker fees from Astra Zeneca, Bayer, BMS, BTG, Eli Lilly, Incyte, Ipsen, Novartis, Merck, Roche, Sirtex Medical, Terumo; research grants (to Institution) from BMS, Roche and Sirtex Medical.

References

Abo, T., Kawamura, T., & Watanabe, H. (2000). Physiological responses of extrathymic T cells in the liver. *Immunological Reviews*, *174*, 135–149. https://doi.org/10.1034/j.1600-0528.2002.017415.x.

Abou-Alfa, G. K., Chan, S. L., Kudo, M., Lau, G., Kelley, R. K., Furuse, J., et al. (2022). Phase 3 randomized, open-label, multicenter study of tremelimumab (T) and durvalumab (D) as first-line therapy in patients (pts) with unresectable hepatocellular carcinoma (uHCC): HIMALAYA. *Journal of Clinical Oncology*, *40*(4_Suppl), 379. https://doi.org/10.1200/JCO.2022.40.4_suppl.379.

Affo, S., Yu, L. X., & Schwabe, R. F. (2017). The role of cancer-associated fibroblasts and fibrosis in liver cancer. *Annual Review of Pathology*, *12*, 153–186. https://doi.org/10.1146/annurev-pathol-052016-100322.

Agdashian, D., ElGindi, M., Xie, C., Sandhu, M., Pratt, D., Kleiner, D. E., et al. (2019). The effect of anti-CTLA4 treatment on peripheral and intra-tumoral T cells in patients with hepatocellular carcinoma. *Cancer Immunology, Immunotherapy*, *68*(4), 599–608. https://doi.org/10.1007/s00262-019-02299-8.

Amsen, D., van Gisbergen, K., Hombrink, P., & van Lier, R. A. W. (2018). Tissue-resident memory T cells at the center of immunity to solid tumors. *Nature Immunology*, *19*(6), 538–546. https://doi.org/10.1038/s41590-018-0114-2.

Ando, Y., Kumamoto, K., Matsukawa, H., Ishikawa, R., Suto, H., Oshima, M., et al. (2022). Low prevalence of biliary tract cancer with defective mismatch repair genes in a Japanese hospital-based population. *Oncology Letters, 23*(1), 4. https://doi.org/10.3892/ol.2021.13122.

Andre, T., Reyes-Vidal, J. M., Fartoux, L., Ross, P., Leslie, M., Rosmorduc, O., et al. (2008). Gemcitabine and oxaliplatin in advanced biliary tract carcinoma: A phase II study. *British Journal of Cancer, 99*(6), 862–867. https://doi.org/10.1038/sj.bjc.6604628.

Assoun, S., Theou-Anton, N., Nguenang, M., Cazes, A., Danel, C., Abbar, B., et al. (2019). Association of TP53 mutations with response and longer survival under immune checkpoint inhibitors in advanced non-small-cell lung cancer. *Lung Cancer, 132*, 65–71. https://doi.org/10.1016/j.lungcan.2019.04.005.

Bassaganyas, L., Pinyol, R., Esteban-Fabro, R., Torrens, L., Torrecilla, S., Willoughby, C. E., et al. (2020). Copy-number alteration burden differentially impacts immune profiles and molecular features of hepatocellular carcinoma. *Clinical Cancer Research, 26*(23), 6350–6361. https://doi.org/10.1158/1078-0432.CCR-20-1497.

Benson, A. B., D'Angelica, M. I., Abbott, D. E., Anaya, D. A., Anders, R., Are, C., et al. (2021). Hepatobiliary cancers, version 2.2021, NCCN clinical practice guidelines in oncology. *Journal of the National Comprehensive Cancer Network, 19*(5), 541–565. https://doi.org/10.6004/jnccn.2021.0022.

Berg, M., Wingender, G., Djandji, D., Hegenbarth, S., Momburg, F., Hammerling, G., et al. (2006). Cross-presentation of antigens from apoptotic tumor cells by liver sinusoidal endothelial cells leads to tumor-specific CD8+ T cell tolerance. *European Journal of Immunology, 36*(11), 2960–2970. https://doi.org/10.1002/eji.200636033.

Berton, D., Banerjee, S. N., Curigliano, G., Cresta, S., Arkenau, H.-T., Abdeddaim, C., et al. (2021). Antitumor activity of dostarlimab in patients with mismatch repair-deficient/microsatellite instability–high tumors: A combined analysis of two cohorts in the GARNET study. *Journal of Clinical Oncology, 39*(15_Suppl), 2564. https://doi.org/10.1200/JCO.2021.39.15_suppl.2564.

Biswas, S. K., & Mantovani, A. (2010). Macrophage plasticity and interaction with lymphocyte subsets: Cancer as a paradigm. *Nature Immunology, 11*(10), 889–896. https://doi.org/10.1038/ni.1937.

Bricard, G., Cesson, V., Devevre, E., Bouzourene, H., Barbey, C., Rufer, N., et al. (2009). Enrichment of human CD4+ V(alpha)24/Vbeta11 invariant NKT cells in intrahepatic malignant tumors. *Journal of Immunology, 182*(8), 5140–5151. https://doi.org/10.4049/jimmunol.0711086.

Bronte, V., Brandau, S., Chen, S. H., Colombo, M. P., Frey, A. B., Greten, T. F., et al. (2016). Recommendations for myeloid-derived suppressor cell nomenclature and characterization standards. *Nature Communications, 7*, 12150. https://doi.org/10.1038/ncomms12150.

Brown, Z. J., Yu, S. J., Heinrich, B., Ma, C., Fu, Q., Sandhu, M., et al. (2018). Indoleamine 2,3-dioxygenase provides adaptive resistance to immune checkpoint inhibitors in hepatocellular carcinoma. *Cancer Immunology, Immunotherapy, 67*(8), 1305–1315. https://doi.org/10.1007/s00262-018-2190-4.

Buonaguro, L., Mayer, A., Loeffler, M., Missel, S., Accolla, R., Ma, Y. T., et al. (2020). Abstract LB-094: Hepavac-101 first-in-man clinical trial of a multi-peptide-based vaccine for hepatocellular carcinoma. *Cancer Research, 80*(16 Suppl), LB-094. https://doi.org/10.1158/1538-7445.Am2020-lb-094.

Cadoux, M., Caruso, S., Pham, S., Gougelet, A., Pophillat, C., Riou, R., et al. (2021). Expression of NKG2D ligands is downregulated by beta-catenin signalling and associates with HCC aggressiveness. *Journal of Hepatology, 74*(6), 1386–1397. https://doi.org/10.1016/j.jhep.2021.01.017.

Calderaro, J., Couchy, G., Imbeaud, S., Amaddeo, G., Letouze, E., Blanc, J. F., et al. (2017). Histological subtypes of hepatocellular carcinoma are related to gene mutations and molecular tumor classification. *Journal of Hepatology*, *67*(4), 727–738. https://doi.org/10.1016/j.jhep.2017.05.014.

Calderaro, J., Petitprez, F., Becht, E., Laurent, A., Hirsch, T. Z., Rousseau, B., et al. (2019). Intra-tumoral tertiary lymphoid structures are associated with a low risk of early recurrence of hepatocellular carcinoma. *Journal of Hepatology*, *70*(1), 58–65. https://doi.org/10.1016/j.jhep.2018.09.003.

Cao, X., Ding, Q., Lu, J., Tao, W., Huang, B., Zhao, Y., et al. (2015). MDA5 plays a critical role in interferon response during hepatitis C virus infection. *Journal of Hepatology*, *62*(4), 771–778. https://doi.org/10.1016/j.jhep.2014.11.007.

Cao, J., Hu, J., Liu, S., Meric-Bernstam, F., Abdel-Wahab, R., Xu, J., et al. (2020). Intrahepatic cholangiocarcinoma: Genomic heterogeneity between eastern and Western patients. *JCO Precision Oncology*, *4*. https://doi.org/10.1200/PO.18.00414.

Chen, J., Gingold, J. A., & Su, X. (2019). Immunomodulatory TGF-beta signaling in hepatocellular carcinoma. *Trends in Molecular Medicine*, *25*(11), 1010–1023. https://doi.org/10.1016/j.molmed.2019.06.007.

Chen, W., Jin, W., Hardegen, N., Lei, K. J., Li, L., Marinos, N., et al. (2003). Conversion of peripheral CD4+CD25- naive T cells to CD4+CD25+ regulatory T cells by TGF-beta induction of transcription factor Foxp3. *The Journal of Experimental Medicine*, *198*(12), 1875–1886. https://doi.org/10.1084/jem.20030152.

Chen, M. M., Xiao, X., Lao, X. M., Wei, Y., Liu, R. X., Zeng, Q. H., et al. (2016). Polarization of tissue-resident TFH-like cells in human hepatoma bridges innate monocyte inflammation and M2b macrophage polarization. *Cancer Discovery*, *6*(10), 1182–1195. https://doi.org/10.1158/2159-8290.CD-16-0329.

Cheng, A. L., Qin, S., Ikeda, M., Galle, P. R., Ducreux, M., Kim, T. Y., et al. (2021). Updated efficacy and safety data from IMbrave150: Atezolizumab plus bevacizumab vs. sorafenib for unresectable hepatocellular carcinoma. *The Journal of Hepatology*. https://doi.org/10.1016/j.jhep.2021.11.030.

Condamine, T., & Gabrilovich, D. I. (2011). Molecular mechanisms regulating myeloid-derived suppressor cell differentiation and function. *Trends in Immunology*, *32*(1), 19–25. https://doi.org/10.1016/j.it.2010.10.002.

Crispe, I. N. (2009). The liver as a lymphoid organ. *Annual Review of Immunology*, *27*, 147–163. https://doi.org/10.1146/annurev.immunol.021908.132629.

Das, A., Hoare, M., Davies, N., Lopes, A. R., Dunn, C., Kennedy, P. T., et al. (2008). Functional skewing of the global CD8 T cell population in chronic hepatitis B virus infection. *The Journal of Experimental Medicine*, *205*(9), 2111–2124. https://doi.org/10.1084/jem.20072076.

DeFalco, J., Harbell, M., Manning-Bog, A., Baia, G., Scholz, A., Millare, B., et al. (2018). Non-progressing cancer patients have persistent B cell responses expressing shared antibody paratopes that target public tumor antigens. *Clinical Immunology*, *187*, 37–45. https://doi.org/10.1016/j.clim.2017.10.002.

Demols, A., Rocq, L., Charry, M., Nève, N. D., Verrellen, A., Ramadhan, A., et al. (2020). NTRK gene fusions in biliary tract cancers. *Journal of Clinical Oncology*, *38*(4_Suppl), 574. https://doi.org/10.1200/JCO.2020.38.4_suppl.574.

Di Caro, G., Bergomas, F., Grizzi, F., Doni, A., Bianchi, P., Malesci, A., et al. (2014). Occurrence of tertiary lymphoid tissue is associated with T-cell infiltration and predicts better prognosis in early-stage colorectal cancers. *Clinical Cancer Research*, *20*(8), 2147–2158. https://doi.org/10.1158/1078-0432.CCR-13-2590.

Diehl, L., Schurich, A., Grochtmann, R., Hegenbarth, S., Chen, L., & Knolle, P. A. (2008). Tolerogenic maturation of liver sinusoidal endothelial cells promotes B7-homolog 1-dependent CD8+ T cell tolerance. *Hepatology*, *47*(1), 296–305. https://doi.org/10.1002/hep.21965.

Dieu-Nosjean, M. C., Antoine, M., Danel, C., Heudes, D., Wislez, M., Poulot, V., et al. (2008). Long-term survival for patients with non-small-cell lung cancer with intratumoral lymphoid structures. *Journal of Clinical Oncology, 26*(27), 4410–4417. https://doi.org/10.1200/JCO.2007.15.0284.

Doherty, D. G., & O'Farrelly, C. (2000). Innate and adaptive lymphoid cells in the human liver. *Immunological Reviews, 174*, 5–20. https://doi.org/10.1034/j.1600-0528.2002.017416.x.

El-Khoueiry, A. B., Sangro, B., Yau, T., Crocenzi, T. S., Kudo, M., Hsu, C., et al. (2017). Nivolumab in patients with advanced hepatocellular carcinoma (CheckMate 040): An open-label, non-comparative, phase 1/2 dose escalation and expansion trial. *Lancet, 389*(10088), 2492–2502. https://doi.org/10.1016/S0140-6736(17)31046-2.

El-Khoueiry, A. B., Yau, T., Kang, Y.-K., Kim, T.-Y., Santoro, A., Sangro, B., et al. (2021). Nivolumab (NIVO) plus ipilimumab (IPI) combination therapy in patients (Pts) with advanced hepatocellular carcinoma (aHCC): Long-term results from CheckMate 040. *Journal of Clinical Oncology, 39*(3_Suppl), 269. https://doi.org/10.1200/JCO.2021.39.3_suppl.269.

Enrico, D., Paci, A., Chaput, N., Karamouza, E., & Besse, B. (2020). Antidrug antibodies against immune checkpoint blockers: Impairment of drug efficacy or indication of immune activation? *Clinical Cancer Research, 26*(4), 787–792. https://doi.org/10.1158/1078-0432.CCR-19-2337.

Farci, P., Shimoda, A., Coiana, A., Diaz, G., Peddis, G., Melpolder, J. C., et al. (2000). The outcome of acute hepatitis C predicted by the evolution of the viral quasispecies. *Science, 288*(5464), 339–344. https://doi.org/10.1126/science.288.5464.339.

Feun, L. G., Li, Y. Y., Wu, C., Wangpaichitr, M., Jones, P. D., Richman, S. P., et al. (2019). Phase 2 study of pembrolizumab and circulating biomarkers to predict anticancer response in advanced, unresectable hepatocellular carcinoma. *Cancer, 125*(20), 3603–3614. https://doi.org/10.1002/cncr.32339.

Finkin, S., Yuan, D., Stein, I., Taniguchi, K., Weber, A., Unger, K., et al. (2015). Ectopic lymphoid structures function as microniches for tumor progenitor cells in hepatocellular carcinoma. *Nature Immunology, 16*(12), 1235–1244. https://doi.org/10.1038/ni.3290.

Finn, R. S., Ikeda, M., Zhu, A. X., Sung, M. W., Baron, A. D., Kudo, M., et al. (2020). Phase Ib study of lenvatinib plus pembrolizumab in patients with unresectable hepatocellular carcinoma. *Journal of Clinical Oncology, 38*(26), 2960–2970. https://doi.org/10.1200/JCO.20.00808.

Finn, R. S., Qin, S., Ikeda, M., Galle, P. R., Ducreux, M., Kim, T. Y., et al. (2020). Atezolizumab plus bevacizumab in unresectable hepatocellular carcinoma. *The New England Journal of Medicine, 382*(20), 1894–1905. https://doi.org/10.1056/NEJMoa1915745.

Finn, R. S., Qin, S., Ikeda, M., Galle, P. R., Ducreux, M., Kim, T.-Y., et al. (2021). IMbrave150: Updated overall survival (OS) data from a global, randomized, open-label phase III study of atezolizumab (atezo) + bevacizumab (bev) versus sorafenib (sor) in patients (pts) with unresectable hepatocellular carcinoma (HCC). *Journal of Clinical Oncology, 39*(3_Suppl), 267. https://doi.org/10.1200/JCO.2021.39.3_suppl.267.

Finn, R. S., Ryoo, B. Y., Merle, P., Kudo, M., Bouattour, M., Lim, H. Y., et al. (2020). Pembrolizumab as second-line therapy in patients with advanced hepatocellular carcinoma in KEYNOTE-240: A randomized, double-blind, phase III trial. *Journal of Clinical Oncology, 38*(3), 193–202. https://doi.org/10.1200/JCO.19.01307.

Flavell, R. A., Sanjabi, S., Wrzesinski, S. H., & Licona-Limon, P. (2010). The polarization of immune cells in the tumor environment by TGFbeta. *Nature Reviews. Immunology, 10*(8), 554–567. https://doi.org/10.1038/nri2808.

Flecken, T., Schmidt, N., Hild, S., Gostick, E., Drognitz, O., Zeiser, R., et al. (2014). Immunodominance and functional alterations of tumor-associated antigen-specific CD8+ T-cell responses in hepatocellular carcinoma. *Hepatology*, *59*(4), 1415–1426. https://doi.org/10.1002/hep.26731.

Fu, J., Xu, D., Liu, Z., Shi, M., Zhao, P., Fu, B., et al. (2007). Increased regulatory T cells correlate with CD8 T-cell impairment and poor survival in hepatocellular carcinoma patients. *Gastroenterology*, *132*(7), 2328–2339. https://doi.org/10.1053/j.gastro.2007.03.102.

Gabrilovich, D. I., & Nagaraj, S. (2009). Myeloid-derived suppressor cells as regulators of the immune system. *Nature Reviews. Immunology*, *9*(3), 162–174. https://doi.org/10.1038/nri2506.

Gao, Q., Qiu, S. J., Fan, J., Zhou, J., Wang, X. Y., Xiao, Y. S., et al. (2007). Intratumoral balance of regulatory and cytotoxic T cells is associated with prognosis of hepatocellular carcinoma after resection. *Journal of Clinical Oncology*, *25*(18), 2586–2593. https://doi.org/10.1200/JCO.2006.09.4565.

Goeppert, B., Frauenschuh, L., Renner, M., Roessler, S., Stenzinger, A., Klauschen, F., et al. (2014). BRAF V600E-specific immunohistochemistry reveals low mutation rates in biliary tract cancer and restriction to intrahepatic cholangiocarcinoma. *Modern Pathology*, *27*(7), 1028–1034. https://doi.org/10.1038/modpathol.2013.206.

Gomes, A. L., Teijeiro, A., Buren, S., Tummala, K. S., Yilmaz, M., Waisman, A., et al. (2016). Metabolic inflammation-associated IL-17A causes non-alcoholic steatohepatitis and hepatocellular carcinoma. *Cancer Cell*, *30*(1), 161–175. https://doi.org/10.1016/j.ccell.2016.05.020.

Greten, T. F., Forner, A., Korangy, F., N'Kontchou, G., Barget, N., Ayuso, C., et al. (2010). A phase II open label trial evaluating safety and efficacy of a telomerase peptide vaccination in patients with advanced hepatocellular carcinoma. *BMC Cancer*, *10*, 209. https://doi.org/10.1186/1471-2407-10-209.

Han, Y., Chen, Z., Yang, Y., Jiang, Z., Gu, Y., Liu, Y., et al. (2014). Human CD14+ CTLA-4+ regulatory dendritic cells suppress T-cell response by cytotoxic T-lymphocyte antigen-4-dependent IL-10 and indoleamine-2,3-dioxygenase production in hepatocellular carcinoma. *Hepatology*, *59*(2), 567–579. https://doi.org/10.1002/hep.26694.

Hastir, J. F., Delbauve, S., Larbanoix, L., Germanova, D., Goyvaerts, C., Allard, J., et al. (2020). Hepatocarcinoma induces a tumor necrosis factor-dependent Kupffer cell death pathway that Favors its proliferation upon partial hepatectomy. *Frontiers in Oncology*, *10*, 547013. https://doi.org/10.3389/fonc.2020.547013.

Havel, J. J., Chowell, D., & Chan, T. A. (2019). The evolving landscape of biomarkers for checkpoint inhibitor immunotherapy. *Nature Reviews. Cancer*, *19*(3), 133–150. https://doi.org/10.1038/s41568-019-0116-x.

Hedegaard, D. L., Tully, D. C., Rowe, I. A., Reynolds, G. M., Bean, D. J., Hu, K., et al. (2017). High resolution sequencing of hepatitis C virus reveals limited intra-hepatic compartmentalization in end-stage liver disease. *Journal of Hepatology*, *66*(1), 28–38. https://doi.org/10.1016/j.jhep.2016.07.048.

Hilmi, M., Vienot, A., Rousseau, B., & Neuzillet, C. (2019). Immune therapy for liver cancers. *Cancers (Basel)*, *12*(1). https://doi.org/10.3390/cancers12010077.

Ho, W. J., Zhu, Q., Durham, J., Popovic, A., Xavier, S., Leatherman, J., et al. (2021). Neoadjuvant cabozantinib and nivolumab converts locally advanced HCC into resectable disease with enhanced antitumor immunity. *Nature Cancer*, *2*(9), 891–903. https://doi.org/10.1038/s43018-021-00234-4.

Hornyak, L., Dobos, N., Koncz, G., Karanyi, Z., Pall, D., Szabo, Z., et al. (2018). The role of indoleamine-2,3-dioxygenase in cancer development, diagnostics, and therapy. *Frontiers in Immunology*, *9*, 151. https://doi.org/10.3389/fimmu.2018.00151.

Horwitz, E., Stein, I., Andreozzi, M., Nemeth, J., Shoham, A., Pappo, O., et al. (2014). Human and mouse VEGFA-amplified hepatocellular carcinomas are highly sensitive to sorafenib treatment. *Cancer Discovery*, 4(6), 730–743. https://doi.org/10.1158/2159-8290.CD-13-0782.

Hou, J., Zhang, H., Sun, B., & Karin, M. (2020). The immunobiology of hepatocellular carcinoma in humans and mice: Basic concepts and therapeutic implications. *Journal of Hepatology*, 72(1), 167–182. https://doi.org/10.1016/j.jhep.2019.08.014.

Huang, J., Cai, M. Y., & Wei, D. P. (2002). HLA class I expression in primary hepatocellular carcinoma. *World Journal of Gastroenterology*, 8(4), 654–657. https://doi.org/10.3748/wjg.v8.i4.654.

Huang, R., Rao, H. Y., Yang, M., Gao, Y. H., Wang, J., Jin, Q., et al. (2021). Histopathology and the predominantly progressive, indeterminate and predominately regressive score in hepatitis C virus patients after direct-acting antivirals therapy. *World Journal of Gastroenterology*, 27(5), 404–415. https://doi.org/10.3748/wjg.v27.i5.404.

Huang, S. L., Wang, Y. M., Wang, Q. Y., Feng, G. G., Wu, F. Q., Yang, L. M., et al. (2021). Mechanisms and clinical trials of hepatocellular carcinoma immunotherapy. *Frontiers in Genetics*, 12, 691391. https://doi.org/10.3389/fgene.2021.691391.

Huang, L. R., Wohlleber, D., Reisinger, F., Jenne, C. N., Cheng, R. L., Abdullah, Z., et al. (2013). Intrahepatic myeloid-cell aggregates enable local proliferation of CD8(+) T cells and successful immunotherapy against chronic viral liver infection. *Nature Immunology*, 14(6), 574–583. https://doi.org/10.1038/ni.2573.

Huang, J., Zhang, L., Chen, J., Wan, D., Zhou, L., Zheng, S., et al. (2021). The landscape of immune cells indicates prognosis and applicability of checkpoint therapy in hepatocellular carcinoma. *Frontiers in Oncology*, 11, 744951. https://doi.org/10.3389/fonc.2021.744951.

Ilie, M., Hofman, V., Ortholan, C., Bonnetaud, C., Coelle, C., Mouroux, J., et al. (2012). Predictive clinical outcome of the intratumoral CD66b-positive neutrophil-to-CD8-positive T-cell ratio in patients with resectable nonsmall cell lung cancer. *Cancer*, 118(6), 1726–1737. https://doi.org/10.1002/cncr.26456.

Jang, J. S., Lim, H. Y., Hwang, I. G., Song, H. S., Yoo, N., Yoon, S., et al. (2010). Gemcitabine and oxaliplatin in patients with unresectable biliary cancer including gall bladder cancer: A Korean Cancer Study Group phase II trial. *Cancer Chemotherapy and Pharmacology*, 65(4), 641–647. https://doi.org/10.1007/s00280-009-1069-7.

Jenne, C. N., & Kubes, P. (2013). Immune surveillance by the liver. *Nature Immunology*, 14(10), 996–1006. https://doi.org/10.1038/ni.2691.

Ji, J., Eggert, T., Budhu, A., Forgues, M., Takai, A., Dang, H., et al. (2015). Hepatic stellate cell and monocyte interaction contributes to poor prognosis in hepatocellular carcinoma. *Hepatology*, 62(2), 481–495. https://doi.org/10.1002/hep.27822.

Jiang, S. S., Tang, Y., Zhang, Y. J., Weng, D. S., Zhou, Z. G., Pan, K., et al. (2015). A phase I clinical trial utilizing autologous tumor-infiltrating lymphocytes in patients with primary hepatocellular carcinoma. *Oncotarget*, 6(38), 41339–41349. https://doi.org/10.18632/oncotarget.5463.

Jinushi, M., Takehara, T., Tatsumi, T., Kanto, T., Groh, V., Spies, T., et al. (2003). Expression and role of MICA and MICB in human hepatocellular carcinomas and their regulation by retinoic acid. *International Journal of Cancer*, 104(3), 354–361. https://doi.org/10.1002/ijc.10966.

Juengpanich, S., Shi, L., Iranmanesh, Y., Chen, J., Cheng, Z., Khoo, A. K., et al. (2019). The role of natural killer cells in hepatocellular carcinoma development and treatment: A narrative review. *Translational Oncology*, 12(8), 1092–1107. https://doi.org/10.1016/j.tranon.2019.04.021.

Kang, S. P., Gergich, K., Lubiniecki, G. M., de Alwis, D. P., Chen, C., Tice, M. A. B., et al. (2017). Pembrolizumab KEYNOTE-001: An adaptive study leading to accelerated approval for two indications and a companion diagnostic. *Annals of Oncology, 28*(6), 1388–1398. https://doi.org/10.1093/annonc/mdx076.

Kapanadze, T., Gamrekelashvili, J., Ma, C., Chan, C., Zhao, F., Hewitt, S., et al. (2013). Regulation of accumulation and function of myeloid derived suppressor cells in different murine models of hepatocellular carcinoma. *Journal of Hepatology, 59*(5), 1007–1013. https://doi.org/10.1016/j.jhep.2013.06.010.

Kelley, R. K., Sangro, B., Harris, W., Ikeda, M., Okusaka, T., Kang, Y.-K., et al. (2021). Safety, efficacy, and pharmacodynamics of tremelimumab plus durvalumab for patients with unresectable hepatocellular carcinoma: Randomized expansion of a phase I/II study. *Journal of Clinical Oncology, 39*(27), 2991–3001. https://doi.org/10.1200/jco.20.03555.

Kelley, R. K., Yau, T., Cheng, A. L., Kaseb, A., Qin, S., Zhu, A. X., et al. (2022). VP10-2021: Cabozantinib (C) plus atezolizumab (A) versus sorafenib (S) as first-line systemic treatment for advanced hepatocellular carcinoma (aHCC): Results from the randomized phase III COSMIC-312 trial. *Annals of Oncology, 33*(1), 114–116. https://doi.org/10.1016/j.annonc.2021.10.008.

Kelly, A., Fahey, R., Fletcher, J. M., Keogh, C., Carroll, A. G., Siddachari, R., et al. (2014). CD141(+) myeloid dendritic cells are enriched in healthy human liver. *Journal of Hepatology, 60*(1), 135–142. https://doi.org/10.1016/j.jhep.2013.08.007.

Kim, J. M., Cho, S. Y., Rhu, J., Jung, M., Her, J. H., Lim, O., et al. (2021). Adjuvant therapy using ex vivo-expanded allogenic natural killer cells in hepatectomy patients with hepatitis B virus related solitary hepatocellular carcinoma: MG4101 study. *Annals of Hepato-Biliary-Pancreatic Surgery, 25*(2), 206–214. https://doi.org/10.14701/ahbps.2021.25.2.206.

Kim, R. D., Chung, V., Alese, O. B., El-Rayes, B. F., Li, D., Al-Toubah, T. E., et al. (2020). A phase 2 multi-institutional study of Nivolumab for patients with advanced refractory biliary tract cancer. *JAMA Oncology, 6*(6), 888–894. https://doi.org/10.1001/jamaoncol.2020.0930.

Kim, H. J., Lee, N. S., Lee, S. C., Bae, S. B., Kim, C. K., Cheon, Y. G., et al. (2009). A phase II study of gemcitabine in combination with oxaliplatin as first-line chemotherapy in patients with inoperable biliary tract cancer. *Cancer Chemotherapy and Pharmacology, 64*(2), 371–377. https://doi.org/10.1007/s00280-008-0883-7.

Kim, J. M., Rasmussen, J. P., & Rudensky, A. Y. (2007). Regulatory T cells prevent catastrophic autoimmunity throughout the lifespan of mice. *Nature Immunology, 8*(2), 191–197. https://doi.org/10.1038/ni1428.

Klein, O., Kee, D., Nagrial, A., Markman, B., Underhill, C., Michael, M., et al. (2020). Evaluation of combination nivolumab and Ipilimumab immunotherapy in patients with advanced biliary tract cancers: Subgroup analysis of a phase 2 nonrandomized clinical trial. *JAMA Oncology, 6*(9), 1405–1409. https://doi.org/10.1001/jamaoncol.2020.2814.

Knolle, P. A., Germann, T., Treichel, U., Uhrig, A., Schmitt, E., Hegenbarth, S., et al. (1999). Endotoxin down-regulates T cell activation by antigen-presenting liver sinusoidal endothelial cells. *Journal of Immunology, 162*(3), 1401–1407. https://www.ncbi.nlm.nih.gov/pubmed/9973395.

Knolle, P. A., & Limmer, A. (2003). Control of immune responses by savenger liver endothelial cells. *Swiss Medical Weekly, 133*(37–38), 501–506. https://doi.org/2003/37/smw-10261.

Knolle, P. A., & Thimme, R. (2014). Hepatic immune regulation and its involvement in viral hepatitis infection. *Gastroenterology, 146*(5), 1193–1207. https://doi.org/10.1053/j.gastro.2013.12.036.

Knox, J. J., Hedley, D., Oza, A., Feld, R., Siu, L. L., Chen, E., et al. (2005). Combining gemcitabine and capecitabine in patients with advanced biliary cancer: A phase II trial. *Journal of Clinical Oncology*, 23(10), 2332–2338. https://doi.org/10.1200/JCO.2005.51.008.

Koeberle, D., Saletti, P., Borner, M., Gerber, D., Dietrich, D., Caspar, C. B., et al. (2008). Patient-reported outcomes of patients with advanced biliary tract cancers receiving gemcitabine plus capecitabine: A multicenter, phase II trial of the Swiss Group for Clinical Cancer Research. *Journal of Clinical Oncology*, 26(22), 3702–3708. https://doi.org/10.1200/JCO.2008.16.5704.

Kortylewski, M., Xin, H., Kujawski, M., Lee, H., Liu, Y., Harris, T., et al. (2009). Regulation of the IL-23 and IL-12 balance by Stat3 signaling in the tumor microenvironment. *Cancer Cell*, 15(2), 114–123. https://doi.org/10.1016/j.ccr.2008.12.018.

Krausgruber, T., Blazek, K., Smallie, T., Alzabin, S., Lockstone, H., Sahgal, N., et al. (2011). IRF5 promotes inflammatory macrophage polarization and TH1-TH17 responses. *Nature Immunology*, 12(3), 231–238. https://doi.org/10.1038/ni.1990.

Kuang, D. M., Peng, C., Zhao, Q., Wu, Y., Zhu, L. Y., Wang, J., et al. (2010). Tumor-activated monocytes promote expansion of IL-17-producing CD8+ T cells in hepatocellular carcinoma patients. *Journal of Immunology*, 185(3), 1544–1549. https://doi.org/10.4049/jimmunol.0904094.

Kuang, D. M., Zhao, Q., Peng, C., Xu, J., Zhang, J. P., Wu, C., et al. (2009). Activated monocytes in peritumoral stroma of hepatocellular carcinoma foster immune privilege and disease progression through PD-L1. *The Journal of Experimental Medicine*, 206(6), 1327–1337. https://doi.org/10.1084/jem.20082173.

Langowski, J. L., Zhang, X., Wu, L., Mattson, J. D., Chen, T., Smith, K., et al. (2006). IL-23 promotes tumor incidence and growth. *Nature*, 442(7101), 461–465. https://doi.org/10.1038/nature04808.

Larson, R. C., & Maus, M. V. (2021). Recent advances and discoveries in the mechanisms and functions of CAR T cells. *Nature Reviews. Cancer*, 21(3), 145–161. https://doi.org/10.1038/s41568-020-00323-z.

Lee, J. H., Lee, J. H., Lim, Y. S., Yeon, J. E., Song, T. J., Yu, S. J., et al. (2015). Adjuvant immunotherapy with autologous cytokine-induced killer cells for hepatocellular carcinoma. *Gastroenterology*, 148(7), 1383–1391 e1386. https://doi.org/10.1053/j.gastro.2015.02.055.

Lee, M. S., Ryoo, B. Y., Hsu, C. H., Numata, K., Stein, S., Verret, W., et al. (2020). Atezolizumab with or without bevacizumab in unresectable hepatocellular carcinoma (GO30140): An open-label, multicentre, phase 1b study. *The Lancet Oncology*, 21(6), 808–820. https://doi.org/10.1016/S1470-2045(20)30156-X.

Leone, R. D., Zhao, L., Englert, J. M., Sun, I. M., Oh, M. H., Sun, I. H., et al. (2019). Glutamine blockade induces divergent metabolic programs to overcome tumor immune evasion. *Science*, 366(6468), 1013–1021. https://doi.org/10.1126/science.aav2588.

Leslie, J., Geh, D., Elsharkawy, A. M., Mann, D. A., & Vacca, M. (2022). Metabolic dysfunction and cancer in HCV: Shared pathways and mutual interactions. *Journal of Hepatology*. https://doi.org/10.1016/j.jhep.2022.01.029.

Li, S., Han, X., Lyu, N., Xie, Q., Deng, H., Mu, L., et al. (2018). Mechanism and prognostic value of indoleamine 2,3-dioxygenase 1 expressed in hepatocellular carcinoma. *Cancer Science*, 109(12), 3726–3736. https://doi.org/10.1111/cas.13811.

Li, Q., Harden, J. L., Anderson, C. D., & Egilmez, N. K. (2016). Tolerogenic phenotype of IFN-gamma-induced IDO+ dendritic cells is maintained via an autocrine IDO-kynurenine/AhR-IDO loop. *Journal of Immunology*, 197(3), 962–970. https://doi.org/10.4049/jimmunol.1502615.

Li, Y. W., Qiu, S. J., Fan, J., Zhou, J., Gao, Q., Xiao, Y. S., et al. (2011). Intratumoral neutrophils: A poor prognostic factor for hepatocellular carcinoma following resection. *Journal of Hepatology*, 54(3), 497–505. https://doi.org/10.1016/j.jhep.2010.07.044.

Li, J., Wang, W., Zhang, Y., Cieslik, M., Guo, J., Tan, M., et al. (2020). Epigenetic driver mutations in ARID1A shape cancer immune phenotype and immunotherapy. *The Journal of Clinical Investigation*, 130(5), 2712–2726. https://doi.org/10.1172/JCI134402.

Lim, C. J., Lee, Y. H., Pan, L., Lai, L., Chua, C., Wasser, M., et al. (2019). Multidimensional analyses reveal distinct immune microenvironment in hepatitis B virus-related hepatocellular carcinoma. *Gut*, 68(5), 916–927. https://doi.org/10.1136/gutjnl-2018-316510.

Limmer, A., Ohl, J., Kurts, C., Ljunggren, H. G., Reiss, Y., Groettrup, M., et al. (2000). Efficient presentation of exogenous antigen by liver endothelial cells to CD8+ T cells results in antigen-specific T-cell tolerance. *Nature Medicine*, 6(12), 1348–1354. https://doi.org/10.1038/82161.

Littman, D. R., & Rudensky, A. Y. (2010). Th17 and regulatory T cells in mediating and restraining inflammation. *Cell*, 140(6), 845–858. https://doi.org/10.1016/j.cell.2010.02.021.

Liu, R. X., Wei, Y., Zeng, Q. H., Chan, K. W., Xiao, X., Zhao, X. Y., et al. (2015). Chemokine (C-X-C motif) receptor 3-positive B cells link interleukin-17 inflammation to protumorigenic macrophage polarization in human hepatocellular carcinoma. *Hepatology*, 62(6), 1779–1790. https://doi.org/10.1002/hep.28020.

Liu, C. Q., Xu, J., Zhou, Z. G., Jin, L. L., Yu, X. J., Xiao, G., et al. (2018). Expression patterns of programmed death ligand 1 correlate with different microenvironments and patient prognosis in hepatocellular carcinoma. *British Journal of Cancer*, 119(1), 80–88. https://doi.org/10.1038/s41416-018-0144-4.

Llovet, J. M., Castet, F., Heikenwalder, M., Maini, M. K., Mazzaferro, V., Pinato, D. J., et al. (2021). Immunotherapies for hepatocellular carcinoma. *Nature Reviews Clinical Oncology*. https://doi.org/10.1038/s41571-021-00573-2.

Llovet, J. M., Kudo, M., Cheng, A.-L., Finn, R. S., Galle, P. R., Kaneko, S., et al. (2019). Lenvatinib (len) plus pembrolizumab (pembro) for the first-line treatment of patients (pts) with advanced hepatocellular carcinoma (HCC): Phase 3 LEAP-002 study. *Journal of Clinical Oncology*, 37(15_Suppl), TPS4152. https://doi.org/10.1200/JCO.2019.37.15_suppl.TPS4152.

Llovet, J. M., Ricci, S., Mazzaferro, V., Hilgard, P., Gane, E., Blanc, J. F., et al. (2008). Sorafenib in advanced hepatocellular carcinoma. *The New England Journal of Medicine*, 359(4), 378–390. https://doi.org/10.1056/NEJMoa0708857.

Lumsden, A. B., Henderson, J. M., & Kutner, M. H. (1988). Endotoxin levels measured by a chromogenic assay in portal, hepatic and peripheral venous blood in patients with cirrhosis. *Hepatology*, 8(2), 232–236. https://doi.org/10.1002/hep.1840080207.

Lwin, Z., Gomez-Roca, C., Saada-Bouzid, E., Yanez, E., Muñoz, F. L., Im, S. A., et al. (2020). LBA41 LEAP-005: Phase II study of lenvatinib (len) plus pembrolizumab (pembro) in patients (pts) with previously treated advanced solid tumors. *Annals of Oncology*, 31, S1170. https://doi.org/10.1016/j.annonc.2020.08.2271.

Ma, C., Han, M., Heinrich, B., Fu, Q., Zhang, Q., Sandhu, M., et al. (2018). Gut microbiome-mediated bile acid metabolism regulates liver cancer via NKT cells. *Science*, 360(6391). https://doi.org/10.1126/science.aan5931.

Ma, C., Kesarwala, A. H., Eggert, T., Medina-Echeverz, J., Kleiner, D. E., Jin, P., et al. (2016). NAFLD causes selective CD4(+) T lymphocyte loss and promotes hepatocarcinogenesis. *Nature*, 531(7593), 253–257. https://doi.org/10.1038/nature16969.

Manne, A., Woods, E., Tsung, A., & Mittra, A. (2021). Biliary tract cancers: Treatment updates and future directions in the era of precision medicine and Immuno-oncology. *Frontiers in Oncology*, 11, 768009. https://doi.org/10.3389/fonc.2021.768009.

Mantovani, S., Oliviero, B., Varchetta, S., Mele, D., & Mondelli, M. U. (2020). Natural killer cell responses in hepatocellular carcinoma: Implications for novel immunotherapeutic approaches. *Cancers (Basel)*, *12*(4). https://doi.org/10.3390/cancers12040926.

Marabelle, A., Fakih, M., Lopez, J., Shah, M., Shapira-Frommer, R., Nakagawa, K., et al. (2020). Association of tumor mutational burden with outcomes in patients with advanced solid tumors treated with pembrolizumab: Prospective biomarker analysis of the multicohort, open-label, phase 2 KEYNOTE-158 study. *The Lancet Oncology*, *21*(10), 1353–1365. https://doi.org/10.1016/S1470-2045(20)30445-9.

Marabelle, A., Le, D. T., Ascierto, P. A., Di Giacomo, A. M., De Jesus-Acosta, A., Delord, J. P., et al. (2020). Efficacy of pembrolizumab in patients with noncolorectal high microsatellite instability/mismatch repair-deficient cancer: Results from the phase II KEYNOTE-158 study. *Journal of Clinical Oncology*, *38*(1), 1–10. https://doi.org/10.1200/JCO.19.02105.

Matsubara, T., Kanto, T., Kuroda, S., Yoshio, S., Higashitani, K., Kakita, N., et al. (2013). TIE2-expressing monocytes as a diagnostic marker for hepatocellular carcinoma correlates with angiogenesis. *Hepatology*, *57*(4), 1416–1425. https://doi.org/10.1002/hep.25965.

McPhail, J., Sims, O. T., Guo, Y., Wooten, D., Herndon, J. S., & Massoud, O. I. (2021). Fibrosis improvement in patients with HCV treated with direct-acting antivirals. *European Journal of Gastroenterology & Hepatology*, *33*(7), 996–1000. https://doi.org/10.1097/MEG.0000000000001821.

Merle, P., Edeline, J., Bouattour, M., Cheng, A.-L., Chan, S. L., Yau, T., et al. (2021). Pembrolizumab (pembro) vs placebo (pbo) in patients (pts) with advanced hepatocellular carcinoma (aHCC) previously treated with sorafenib: Updated data from the randomized, phase III KEYNOTE-240 study. *Journal of Clinical Oncology*, *39*(3_Suppl), 268. https://doi.org/10.1200/JCO.2021.39.3_suppl.268.

Movahedi, K., Laoui, D., Gysemans, C., Baeten, M., Stange, G., Van den Bossche, J., et al. (2010). Different tumor microenvironments contain functionally distinct subsets of macrophages derived from Ly6C(high) monocytes. *Cancer Research*, *70*(14), 5728–5739. https://doi.org/10.1158/0008-5472.CAN-09-4672.

Murdoch, C., Muthana, M., Coffelt, S. B., & Lewis, C. E. (2008). The role of myeloid cells in the promotion of tumor angiogenesis. *Nature Reviews. Cancer*, *8*(8), 618–631. https://doi.org/10.1038/nrc2444.

Nehls, O., Klump, B., Arkenau, H. T., Hass, H. G., Greschniok, A., Gregor, M., et al. (2002). Oxaliplatin, fluorouracil and leucovorin for advanced biliary system adenocarcinomas: A prospective phase II trial. *British Journal of Cancer*, *87*(7), 702–704. https://doi.org/10.1038/sj.bjc.6600543.

Nehls, O., Oettle, H., Hartmann, J. T., Hofheinz, R. D., Hass, H. G., Horger, M. S., et al. (2008). Capecitabine plus oxaliplatin as first-line treatment in patients with advanced biliary system adenocarcinoma: A prospective multicentre phase II trial. *British Journal of Cancer*, *98*(2), 309–315. https://doi.org/10.1038/sj.bjc.6604178.

Nemeth, E., Baird, A. W., & O'Farrelly, C. (2009). Microanatomy of the liver immune system. *Seminars in Immunopathology*, *31*(3), 333–343. https://doi.org/10.1007/s00281-009-0173-4.

Notas, G., Kisseleva, T., & Brenner, D. (2009). NK and NKT cells in liver injury and fibrosis. *Clinical Immunology*, *130*(1), 16–26. https://doi.org/10.1016/j.clim.2008.08.008.

Oh, D.-Y., He, A. R., Qin, S., Chen, L.-T., Okusaka, T., Vogel, A., et al. (2022). A phase 3 randomized, double-blind, placebo-controlled study of durvalumab in combination with gemcitabine plus cisplatin (GemCis) in patients (pts) with advanced biliary tract cancer (BTC): TOPAZ-1. *Journal of Clinical Oncology*, *40*(4_Suppl), 378. https://doi.org/10.1200/JCO.2022.40.4_suppl.378.

Oh, D.-Y., Lee, K.-H., Lee, D.-W., Kim, T. Y., Bang, J.-H., Nam, A.-R., et al. (2020). Phase II study assessing tolerability, efficacy, and biomarkers for durvalumab (D) ± tremelimumab (T) and gemcitabine/cisplatin (GemCis) in chemo-naïve advanced biliary tract cancer (aBTC). *Journal of Clinical Oncology, 38*(15_Suppl), 4520. https://doi.org/10.1200/JCO.2020.38.15_suppl.4520.

Orimo, A., & Weinberg, R. A. (2006). Stromal fibroblasts in cancer: A novel tumor-promoting cell type. *Cell Cycle, 5*(15), 1597–1601. https://doi.org/10.4161/cc.5.15.3112.

Ouyang, F. Z., Wu, R. Q., Wei, Y., Liu, R. X., Yang, D., Xiao, X., et al. (2016). Dendritic cell-elicited B-cell activation fosters immune privilege via IL-10 signals in hepatocellular carcinoma. *Nature Communications, 7*, 13453. https://doi.org/10.1038/ncomms13453.

Pallett, L. J., Burton, A. R., Amin, O. E., Rodriguez-Tajes, S., Patel, A. A., Zakeri, N., et al. (2020). Longevity and replenishment of human liver-resident memory T cells and mononuclear phagocytes. *The Journal of Experimental Medicine, 217*(9). https://doi.org/10.1084/jem.20200050.

Pallett, L. J., Gill, U. S., Quaglia, A., Sinclair, L. V., Jover-Cobos, M., Schurich, A., et al. (2015). Metabolic regulation of hepatitis B immunopathology by myeloid-derived suppressor cells. *Nature Medicine, 21*(6), 591–600. https://doi.org/10.1038/nm.3856.

Pardee, A. D., Shi, J., & Butterfield, L. H. (2014). Tumor-derived alpha-fetoprotein impairs the differentiation and T cell stimulatory activity of human dendritic cells. *Journal of Immunology, 193*(11), 5723–5732. https://doi.org/10.4049/jimmunol.1400725.

Parker, G. A., & Picut, C. A. (2005). Liver immunobiology. *Toxicologic Pathology, 33*(1), 52–62. https://doi.org/10.1080/01926230590522365.

Payne, K. K., Mine, J. A., Biswas, S., Chaurio, R. A., Perales-Puchalt, A., Anadon, C. M., et al. (2020). BTN3A1 governs antitumor responses by coordinating alphabeta and gammadelta T cells. *Science, 369*(6506), 942–949. https://doi.org/10.1126/science.aay2767.

Pfister, D., Nunez, N. G., Pinyol, R., Govaere, O., Pinter, M., Szydlowska, M., et al. (2021). NASH limits anti-tumor surveillance in immunotherapy-treated HCC. *Nature, 592*(7854), 450–456. https://doi.org/10.1038/s41586-021-03362-0.

Piha-Paul, S. A., Oh, D. Y., Ueno, M., Malka, D., Chung, H. C., Nagrial, A., et al. (2020). Efficacy and safety of pembrolizumab for the treatment of advanced biliary cancer: Results from the KEYNOTE-158 and KEYNOTE-028 studies. *International Journal of Cancer, 147*(8), 2190–2198. https://doi.org/10.1002/ijc.33013.

Pinter, M., Scheiner, B., & Peck-Radosavljevic, M. (2021). Immunotherapy for advanced hepatocellular carcinoma: A focus on special subgroups. *Gut, 70*(1), 204–214. https://doi.org/10.1136/gutjnl-2020-321702.

Prieto, J., Melero, I., & Sangro, B. (2015). Immunological landscape and immunotherapy of hepatocellular carcinoma. *Nature Reviews. Gastroenterology & Hepatology, 12*(12), 681–700. https://doi.org/10.1038/nrgastro.2015.173.

Qin, S., Chen, Z., Fang, W., Ren, Z., Xu, R., Ryoo, B.-Y., et al. (2022). Pembrolizumab plus best supportive care versus placebo plus best supportive care as second-line therapy in patients in Asia with advanced hepatocellular carcinoma (HCC): Phase 3 KEYNOTE-394 study. *Journal of Clinical Oncology, 40*(4_Suppl), 383. https://doi.org/10.1200/JCO.2022.40.4_suppl.383.

Racanelli, V., & Rehermann, B. (2006). The liver as an immunological organ. *Hepatology, 43*(2 Suppl. 1), S54–S62. https://doi.org/10.1002/hep.21060.

Rahma, O. E., & Hodi, F. S. (2019). The intersection between tumor angiogenesis and immune suppression. *Clinical Cancer Research, 25*(18), 5449–5457. https://doi.org/10.1158/1078-0432.CCR-18-1543.

Rebouissou, S., & Nault, J. C. (2020). Advances in molecular classification and precision oncology in hepatocellular carcinoma. *Journal of Hepatology, 72*(2), 215–229. https://doi.org/10.1016/j.jhep.2019.08.017.

Reig, M., Forner, A., Rimola, J., Ferrer-Fabrega, J., Burrel, M., Garcia-Criado, A., et al. (2021). BCLC strategy for prognosis prediction and treatment recommendation Barcelona Clinic Liver Cancer (BCLC) staging system. The 2022 update. *The Journal of Hepatology*. https://doi.org/10.1016/j.jhep.2021.11.018.

Ren, Z., Xu, J., Bai, Y., Xu, A., Cang, S., Du, C., et al. (2021). Sintilimab plus a bevacizumab biosimilar (IBI305) versus sorafenib in unresectable hepatocellular carcinoma (ORIENT-32): A randomised, open-label, phase 2–3 study. *The Lancet Oncology*, 22(7), 977–990. https://doi.org/10.1016/S1470-2045(21)00252-7.

Riechelmann, R. P., Townsley, C. A., Chin, S. N., Pond, G. R., & Knox, J. J. (2007). Expanded phase II trial of gemcitabine and capecitabine for advanced biliary cancer. *Cancer*, 110(6), 1307–1312. https://doi.org/10.1002/cncr.22902.

Ritter, M., Ali, M. Y., Grimm, C. F., Weth, R., Mohr, L., Bocher, W. O., et al. (2004). Immunoregulation of dendritic and T cells by alpha-fetoprotein in patients with hepatocellular carcinoma. *Journal of Hepatology*, 41(6), 999–1007. https://doi.org/10.1016/j.jhep.2004.08.013.

Rizvi, N. A., Hellmann, M. D., Snyder, A., Kvistborg, P., Makarov, V., Havel, J. J., et al. (2015). Cancer immunology. Mutational landscape determines sensitivity to PD-1 blockade in non-small cell lung cancer. *Science*, 348(6230), 124–128. https://doi.org/10.1126/science.aaa1348.

Rodriguez, P. C., Quiceno, D. G., & Ochoa, A. C. (2007). L-arginine availability regulates T-lymphocyte cell-cycle progression. *Blood*, 109(4), 1568–1573. https://doi.org/10.1182/blood-2006-06-031856.

Rosenthal, R., Cadieux, E. L., Salgado, R., Bakir, M. A., Moore, D. A., Hiley, C. T., et al. (2019). Neoantigen-directed immune escape in lung cancer evolution. *Nature*, 567(7749), 479–485. https://doi.org/10.1038/s41586-019-1032-7.

Ruiz de Galarreta, M., Bresnahan, E., Molina-Sanchez, P., Lindblad, K. E., Maier, B., Sia, D., et al. (2019). Beta-catenin activation promotes immune escape and resistance to anti-PD-1 therapy in hepatocellular carcinoma. *Cancer Discovery*, 9(8), 1124–1141. https://doi.org/10.1158/2159-8290.CD-19-0074.

Sahai, V., Griffith, K. A., Beg, M. S., Shaib, W. L., Mahalingam, D., Zhen, D. B., et al. (2020). A multicenter randomized phase II study of nivolumab in combination with gemcitabine/cisplatin or ipilimumab as first-line therapy for patients with advanced unresectable biliary tract cancer (BilT-01). *Journal of Clinical Oncology*, 38(15_Suppl), 4582. https://doi.org/10.1200/JCO.2020.38.15_suppl.4582.

Salas-Benito, D., Perez-Gracia, J. L., Ponz-Sarvise, M., Rodriguez-Ruiz, M. E., Martinez-Forero, I., Castanon, E., et al. (2021). Paradigms on immunotherapy combinations with chemotherapy. *Cancer Discovery*, 11(6), 1353–1367. https://doi.org/10.1158/2159-8290.CD-20-1312.

Sanchez-Vega, F., Mina, M., Armenia, J., Chatila, W. K., Luna, A., La, K. C., et al. (2018). Oncogenic Signaling pathways in the Cancer Genome Atlas. *Cell*, 173(2), 321–337, e310. https://doi.org/10.1016/j.cell.2018.03.035.

Sangro, B., Bruix, J., Chan, S. L., Galle, P. R., & Rimassa, L. (2022). Reply to: "An EASL position paper for systemic treatment of hepatocellular carcinoma: Go forward courageously". *Journal of Hepatology*, 76(2), 480–481. https://doi.org/10.1016/j.jhep.2021.10.021.

Sangro, B., Gomez-Martin, C., de la Mata, M., Inarrairaegui, M., Garralda, E., Barrera, P., et al. (2013). A clinical trial of CTLA-4 blockade with tremelimumab in patients with hepatocellular carcinoma and chronic hepatitis C. *Journal of Hepatology*, 59(1), 81–88. https://doi.org/10.1016/j.jhep.2013.02.022.

Sangro, B., Melero, I., Wadhawan, S., Finn, R. S., Abou-Alfa, G. K., Cheng, A. L., et al. (2020). Association of inflammatory biomarkers with clinical outcomes in nivolumab-treated patients with advanced hepatocellular carcinoma. *Journal of Hepatology*, 73(6), 1460–1469. https://doi.org/10.1016/j.jhep.2020.07.026.

Sawada, Y., Yoshikawa, T., Nobuoka, D., Shirakawa, H., Kuronuma, T., Motomura, Y., et al. (2012). Phase I trial of a glypican-3-derived peptide vaccine for advanced hepatocellular carcinoma: Immunologic evidence and potential for improving overall survival. *Clinical Cancer Research*, 18(13), 3686–3696. https://doi.org/10.1158/1078-0432.CCR-11-3044.

Shalapour, S., Lin, X. J., Bastian, I. N., Brain, J., Burt, A. D., Aksenov, A. A., et al. (2017). Inflammation-induced IgA+ cells dismantle anti-liver cancer immunity. *Nature*, 551(7680), 340–345. https://doi.org/10.1038/nature24302.

Shang, R., Song, X., Wang, P., Zhou, Y., Lu, X., Wang, J., et al. (2021). Cabozantinib-based combination therapy for the treatment of hepatocellular carcinoma. *Gut*, 70(9), 1746–1757. https://doi.org/10.1136/gutjnl-2020-320716.

Shen, X., Li, N., Li, H., Zhang, T., Wang, F., & Li, Q. (2010). Increased prevalence of regulatory T cells in the tumor microenvironment and its correlation with TNM stage of hepatocellular carcinoma. *Journal of Cancer Research and Clinical Oncology*, 136(11), 1745–1754. https://doi.org/10.1007/s00432-010-0833-8.

Shigeta, K., Datta, M., Hato, T., Kitahara, S., Chen, I. X., Matsui, A., et al. (2020). Dual programmed death Receptor-1 and vascular endothelial growth factor receptor-2 blockade promotes vascular normalization and enhances antitumor immune responses in hepatocellular carcinoma. *Hepatology*, 71(4), 1247–1261. https://doi.org/10.1002/hep.30889.

Shojaei, F., Zhong, C., Wu, X., Yu, L., & Ferrara, N. (2008). Role of myeloid cells in tumor angiogenesis and growth. *Trends in Cell Biology*, 18(8), 372–378. https://doi.org/10.1016/j.tcb.2008.06.003.

Shroff, R. T., Javle, M. M., Xiao, L., Kaseb, A. O., Varadhachary, G. R., Wolff, R. A., et al. (2019). Gemcitabine, cisplatin, and nab-paclitaxel for the treatment of advanced biliary tract cancers: A phase 2 clinical trial. *JAMA Oncology*, 5(6), 824–830. https://doi.org/10.1001/jamaoncol.2019.0270.

Sia, D., Jiao, Y., Martinez-Quetglas, I., Kuchuk, O., Villacorta-Martin, C., Castro de Moura, M., et al. (2017). Identification of an immune-specific class of hepatocellular carcinoma, based on molecular features. *Gastroenterology*, 153(3), 812–826. https://doi.org/10.1053/j.gastro.2017.06.007.

Silva, V. W., Askan, G., Daniel, T. D., Lowery, M., Klimstra, D. S., Abou-Alfa, G. K., et al. (2016). Biliary carcinomas: Pathology and the role of DNA mismatch repair deficiency. *Clinical Oncology*, 5(5), 62. https://doi.org/10.21037/cco.2016.10.04.

Song, G., Shi, Y., Zhang, M., Goswami, S., Afridi, S., Meng, L., et al. (2020). Global immune characterization of HBV/HCV-related hepatocellular carcinoma identifies macrophage and T-cell subsets associated with disease progression. *Cell Discovery*, 6(1), 90. https://doi.org/10.1038/s41421-020-00214-5.

Spahn, S., Roessler, D., Pompilia, R., Gabernet, G., Gladstone, B. P., Horger, M., et al. (2020). Clinical and genetic tumor characteristics of responding and non-responding patients to PD-1 inhibition in hepatocellular carcinoma. *Cancers (Basel)*, 12(12). https://doi.org/10.3390/cancers12123830.

Spranger, S., Bao, R., & Gajewski, T. F. (2015). Melanoma-intrinsic beta-catenin signalling prevents anti-tumor immunity. *Nature*, 523(7559), 231–235. https://doi.org/10.1038/nature14404.

Sun, T. W., Gao, Q., Qiu, S. J., Zhou, J., Wang, X. Y., Yi, Y., et al. (2012). B7-H3 is expressed in human hepatocellular carcinoma and is associated with tumor aggressiveness and postoperative recurrence. *Cancer Immunology, Immunotherapy*, 61(11), 2171–2182. https://doi.org/10.1007/s00262-012-1278-5.

Sun, Y., Wu, L., Zhong, Y., Zhou, K., Hou, Y., Wang, Z., et al. (2021). Single-cell landscape of the ecosystem in early-relapse hepatocellular carcinoma. *Cell*, 184(2), 404–421, e416. https://doi.org/10.1016/j.cell.2020.11.041.

Sun, C., Xu, J., Huang, Q., Huang, M., Wen, H., Zhang, C., et al. (2017). High NKG2A expression contributes to NK cell exhaustion and predicts a poor prognosis of patients with liver cancer. *Oncoimmunology*, *6*(1), e1264562. https://doi.org/10.1080/2162402X.2016.1264562.

Tagliamonte, M., Petrizzo, A., Mauriello, A., Tornesello, M. L., Buonaguro, F. M., & Buonaguro, L. (2018). Potentiating cancer vaccine efficacy in liver cancer. *Oncoimmunology*, *7*(10), e1488564. https://doi.org/10.1080/2162402X.2018.1488564.

Tan, H., Wang, S., & Zhao, L. (2017). A tumor-promoting role of Th9 cells in hepatocellular carcinoma through CCL20 and STAT3 pathways. *Clinical and Experimental Pharmacology & Physiology*, *44*(2), 213–221. https://doi.org/10.1111/1440-1681.12689.

Tatsumi, T., Takehara, T., Katayama, K., Mochizuki, K., Yamamoto, M., Kanto, T., et al. (1997). Expression of costimulatory molecules B7-1 (CD80) and B7-2 (CD86) on human hepatocellular carcinoma. *Hepatology*, *25*(5), 1108–1114. https://doi.org/10.1002/hep.510250511.

Teijeira, A., Garasa, S., Ochoa, M. C., Villalba, M., Olivera, I., Cirella, A., et al. (2021). IL8, neutrophils, and NETs in a collusion against cancer immunity and immunotherapy. *Clinical Cancer Research*, *27*(9), 2383–2393. https://doi.org/10.1158/1078-0432.CCR-20-1319.

Toh, H. C., Galle, P. R., Zhu, A. X., Nicholas, A., Gaillard, V., Ducreux, M., et al. (2022). IMbrave150: Exploratory efficacy and safety in patients with unresectable hepatocellular carcinoma (HCC) treated with atezolizumab beyond radiological progression until loss of clinical benefit in a global phase III study. *Journal of Clinical Oncology*, *40*(4_Suppl), 470. https://doi.org/10.1200/JCO.2022.40.4_suppl.470.

Valle, J. W., Kelley, R. K., Nervi, B., Oh, D. Y., & Zhu, A. X. (2021). Biliary tract cancer. *Lancet*, *397*(10272), 428–444. https://doi.org/10.1016/S0140-6736(21)00153-7.

Valle, J., Wasan, H., Palmer, D. H., Cunningham, D., Anthoney, A., Maraveyas, A., et al. (2010). Cisplatin plus gemcitabine versus gemcitabine for biliary tract cancer. *The New England Journal of Medicine*, *362*(14), 1273–1281. https://doi.org/10.1056/NEJMoa0908721.

van der Windt, D. J., Sud, V., Zhang, H., Varley, P. R., Goswami, J., Yazdani, H. O., et al. (2018). Neutrophil extracellular traps promote inflammation and development of hepatocellular carcinoma in nonalcoholic steatohepatitis. *Hepatology*, *68*(4), 1347–1360. https://doi.org/10.1002/hep.29914.

Vogel, A., Bathon, M., & Saborowski, A. (2021). Immunotherapies in clinical development for biliary tract cancer. *Expert Opinion on Investigational Drugs*, *30*(4), 351–363. https://doi.org/10.1080/13543784.2021.1868437.

Wang, H., Rao, B., Lou, J., Li, J., Liu, Z., Li, A., et al. (2020). The function of the HGF/c-met Axis in hepatocellular carcinoma. *Frontiers in Cell and Development Biology*, *8*, 55. https://doi.org/10.3389/fcell.2020.00055.

Wang, D., Saga, Y., Mizukami, H., Sato, N., Nonaka, H., Fujiwara, H., et al. (2012). Indoleamine-2,3-dioxygenase, an immunosuppressive enzyme that inhibits natural killer cell function, as a useful target for ovarian cancer therapy. *International Journal of Oncology*, *40*(4), 929–934. https://doi.org/10.3892/ijo.2011.1295.

Wang, F., Wang, G., Liu, T., Yu, G., Zhang, G., & Luan, X. (2014). B7-H3 was highly expressed in human primary hepatocellular carcinoma and promoted tumor progression. *Cancer Investigation*, *32*(6), 262–271. https://doi.org/10.3109/07357907.2014.909826.

Wei, S., Kryczek, I., & Zou, W. (2006). Regulatory T-cell compartmentalization and trafficking. *Blood*, *108*(2), 426–431. https://doi.org/10.1182/blood-2006-01-0177.

Wheeler, D. A., & Network, C. G. A. R. (2017). Comprehensive and integrative genomic characterization of hepatocellular carcinoma. *Cell*, *169*(7), 1327–1341.e1323. https://doi.org/10.1016/j.cell.2017.05.046.

Wieland, D., Kemming, J., Schuch, A., Emmerich, F., Knolle, P., Neumann-Haefelin, C., et al. (2017). TCF1(+) hepatitis C virus-specific CD8(+) T cells are maintained after cessation of chronic antigen stimulation. *Nature Communications*, *8*, 15050. https://doi.org/10.1038/ncomms15050.

Wilkinson, A. L., Qurashi, M., & Shetty, S. (2020). The role of sinusoidal endothelial cells in the axis of inflammation and cancer within the liver. *Frontiers in Physiology*, *11*, 990. https://doi.org/10.3389/fphys.2020.00990.

Wolf, M. J., Adili, A., Piotrowitz, K., Abdullah, Z., Boege, Y., Stemmer, K., et al. (2014). Metabolic activation of intrahepatic CD8+ T cells and NKT cells causes nonalcoholic steatohepatitis and liver cancer via cross-talk with hepatocytes. *Cancer Cell*, *26*(4), 549–564. https://doi.org/10.1016/j.ccell.2014.09.003.

Wu, Y., Kuang, D. M., Pan, W. D., Wan, Y. L., Lao, X. M., Wang, D., et al. (2013). Monocyte/macrophage-elicited natural killer cell dysfunction in hepatocellular carcinoma is mediated by CD48/2B4 interactions. *Hepatology*, *57*(3), 1107–1116. https://doi.org/10.1002/hep.26192.

Wu, J., Meng, Z., Jiang, M., Zhang, E., Trippler, M., Broering, R., et al. (2010). Toll-like receptor-induced innate immune responses in non-parenchymal liver cells are cell type-specific. *Immunology*, *129*(3), 363–374. https://doi.org/10.1111/j.1365-2567.2009.03179.x.

Xiao, X., Lao, X. M., Chen, M. M., Liu, R. X., Wei, Y., Ouyang, F. Z., et al. (2016). PD-1hi identifies a novel regulatory B-cell population in human hepatoma that promotes disease progression. *Cancer Discovery*, *6*(5), 546–559. https://doi.org/10.1158/2159-8290.CD-15-1408.

Xu, Y., Poggio, M., Jin, H. Y., Shi, Z., Forester, C. M., Wang, Y., et al. (2019). Translation control of the immune checkpoint in cancer and its therapeutic targeting. *Nature Medicine*, *25*(2), 301–311. https://doi.org/10.1038/s41591-018-0321-2.

Yau, T., Kang, Y. K., Kim, T. Y., El-Khoueiry, A. B., Santoro, A., Sangro, B., et al. (2020). Efficacy and safety of nivolumab plus Ipilimumab in patients with advanced hepatocellular carcinoma previously treated with Sorafenib: The CheckMate 040 randomized clinical trial. *JAMA Oncology*, *6*(11), e204564. https://doi.org/10.1001/jamaoncol.2020.4564.

Yau, T., Park, J. W., Finn, R. S., Cheng, A. L., Mathurin, P., Edeline, J., et al. (2022). Nivolumab versus sorafenib in advanced hepatocellular carcinoma (CheckMate 459): A randomised, multicentre, open-label, phase 3 trial. *The Lancet Oncology*, *23*(1), 77–90. https://doi.org/10.1016/S1470-2045(21)00604-5.

Yoon, J. G., Kim, M. H., Jang, M., Kim, H., Hwang, H. K., Kang, C. M., et al. (2021). Molecular characterization of biliary tract cancer predicts chemotherapy and programmed death 1/programmed death-ligand 1 blockade responses. *Hepatology*, *74*(4), 1914–1931. https://doi.org/10.1002/hep.31862.

Zecca, A., Barili, V., Canetti, D., Regina, V., Olivani, A., Carone, C., et al. (2020). Energy metabolism and cell motility defect in NK-cells from patients with hepatocellular carcinoma. *Cancer Immunology, Immunotherapy*, *69*(8), 1589–1603. https://doi.org/10.1007/s00262-020-02561-4.

Zhang, Q., He, Y., Luo, N., Patel, S. J., Han, Y., Gao, R., et al. (2019). Landscape and dynamics of single immune cells in hepatocellular carcinoma. *Cell*, *179*(4), 829–845 e820. https://doi.org/10.1016/j.cell.2019.10.003.

Zhang, Q. F., Yin, W. W., Xia, Y., Yi, Y. Y., He, Q. F., Wang, X., et al. (2017). Liver-infiltrating CD11b(−)CD27(−) NK subsets account for NK-cell dysfunction in patients with hepatocellular carcinoma and are associated with tumor progression. *Cellular & Molecular Immunology*, *14*(10), 819–829. https://doi.org/10.1038/cmi.2016.28.

Zhao, Q., Xiao, X., Wu, Y., Wei, Y., Zhu, L. Y., Zhou, J., et al. (2011). Interleukin-17-educated monocytes suppress cytotoxic T-cell function through B7-H1 in hepatocellular carcinoma patients. *European Journal of Immunology, 41*(8), 2314–2322. https://doi.org/10.1002/eji.201041282.

Zhao, W., Zhang, L., Xu, Y., Zhang, Z., Ren, G., Tang, K., et al. (2014). Hepatic stellate cells promote tumor progression by enhancement of immunosuppressive cells in an orthotopic liver tumor mouse model. *Laboratory Investigation, 94*(2), 182–191. https://doi.org/10.1038/labinvest.2013.139.

Zhao, W., Zhang, L., Yin, Z., Su, W., Ren, G., Zhou, C., et al. (2011). Activated hepatic stellate cells promote hepatocellular carcinoma development in immunocompetent mice. *International Journal of Cancer, 129*(11), 2651–2661. https://doi.org/10.1002/ijc.25920.

Zheng, C., Zheng, L., Yoo, J. K., Guo, H., Zhang, Y., Guo, X., et al. (2017). Landscape of infiltrating T cells in liver cancer revealed by single-cell sequencing. *Cell, 169*(7), 1342–1356.e1316. https://doi.org/10.1016/j.cell.2017.05.035.

Zhou, X., Sun, L., Jing, D., Xu, G., Zhang, J., Lin, L., et al. (2018). Galectin-9 expression predicts favorable clinical outcome in solid tumors: A systematic review and meta-analysis. *Frontiers in Physiology, 9*, 452. https://doi.org/10.3389/fphys.2018.00452.

Zhu, A. X., Finn, R. S., Edeline, J., Cattan, S., Ogasawara, S., Palmer, D., et al. (2018). Pembrolizumab in patients with advanced hepatocellular carcinoma previously treated with sorafenib (KEYNOTE-224): A non-randomised, open-label phase 2 trial. *The Lancet Oncology, 19*(7), 940–952. https://doi.org/10.1016/S1470-2045(18)30351-6.

Zhu, A. X., Guan, Y., Abbas, A. R., Koeppen, H., Lu, S., Hsu, C.-H., et al. (2020). Abstract CT044: Genomic correlates of clinical benefits from atezolizumab combined with bevacizumab vs. atezolizumab alone in patients with advanced hepatocellular carcinoma (HCC). *Cancer Research, 80*(16 Suppl), CT044. https://doi.org/10.1158/1538-7445.Am2020-ct044.

Zhu, X. D., Zhang, J. B., Zhuang, P. Y., Zhu, H. G., Zhang, W., Xiong, Y. Q., et al. (2008). High expression of macrophage colony-stimulating factor in peritumoral liver tissue is associated with poor survival after curative resection of hepatocellular carcinoma. *Journal of Clinical Oncology, 26*(16), 2707–2716. https://doi.org/10.1200/JCO.2007.15.6521.

CHAPTER FOURTEEN

Immunotherapy for hepatobiliary cancers: Emerging targets and translational advances

Dan Li, Shaoli Lin, Jessica Hong, and Mitchell Ho*

Laboratory of Molecular Biology, Center for Cancer Research, National Cancer Institute, Bethesda, MD, United States
*Corresponding author: e-mail address: homi@mail.nih.gov

Contents

1. Introduction	417
2. Emerging targets for liver cancer immunotherapy	422
2.1 GPC3 in HCC and MSLN in iCCA	423
2.2 Other tumor targets in HCC and iCCA	423
3. The role of GPC3 in HCC	426
3.1 GPC3 biology and structure	426
3.2 GPC3 expression pattern in human normal tissues and HCC	428
3.3 GPC3-mediated signaling pathways in HCC cells proliferation	429
3.4 GPC3 as a target of HCCs therapy	431
4. The role of mesothelin in iCCA	433
4.1 Mesothelin expression in iCCA	433
4.2 MSLN structure	434
4.3 MSLN promotes tumor progression through diverse pathways	436
4.4 Antibody-based immunotherapies against MSLN	438
5. Conclusion and future perspectives	440
Acknowledgments	440
Conflict of interest statement	440
Grant support	441
References	441

Abstract

Over the past several decades, primary liver cancer (PLC), mostly hepatocellular carcinoma (HCC) and intrahepatic cholangiocarcinoma (iCCA), has become the focus of rising concern mainly due to the increasing rates of incidence and high global mortality. Immunotherapy, as an emerging treatment approach, represents an effective and promising option against PLC. However, the selection of immunotherapeutic targets while considering tumor heterogeneity and immunosuppressive tumor microenvironment is a major challenge. The purpose of this review is to summarize and present the emerging immunotherapeutic targets for HCC and iCCA and to evaluate their translation advances in currently ongoing clinical trials. To better provide a framework for the liver cancer target selection, this chapter will highlight cell surface antigens expressed in both tumor cells and immune cells. Particular focus will be on the development, biology and function of Glypican-3 (GPC3) and Mesothelin (MSLN) in the cancer progress of HCC and iCCA, respectively. By doing so, we will explore the prospects and applications of various immunotherapeutic strategies such as vaccines, monoclonal antibodies, immunotoxins, antibody-drug conjugates (ADCs) and chimeric antigen receptors (CARs) T cells that have been developed targeting GPC3 and MSLN.

Abbreviations

PLC	primary liver cancer
HCC	hepatocellular carcinoma
iCCA	intrahepatic cholangiocarcinoma
GPC3	Glypican-3
MSLN	mesothelin
ADC	antibody-drug conjugate
CAR	chimeric antigen receptor
HBV	hepatitis B virus
HCV	hepatitis C virus
TACE	trans-arterial chemoembolization
TME	tumor microenvironment
ICIs	immune checkpoint inhibitors
TSA	tumor-specific antigen
TAA	tumor-associated antigen
APC	adenomatous polyposis coli
GPI	glycosyl-phosphatidylinositol
TMB	tumor mutation burden
EGFR	epidermal growth factor receptor
AFP	alpha fetal protein
CEA	carcinoembryonic antigen
MHC	major histocompatibility complex
MAGE-A	melanoma-associated gene A
NY-ESO-1	New York esophageal squamous cell carcinoma 1
MRP-3	multidrug resistance-associated protein 3
hTERT	human telomerase reverse transcriptase
GPC1	glypican-1
TIL	tumor-infiltrating lymphocytes
CSC	cancer stem cell
MDSC	myeloid-derived suppressor cell
PD-1	programmed cell death protein 1
CTLA-4	cytotoxic T lymphocyte-associated antigen 4
LAG-3	lymphocyte activation gene 3
TIM-3	T-cell membrane protein 3

AA	amino acids
HS	heparan sulfate
PD-L1	programmed cell death ligand 1
B7-H3	B7 homolog 3
Hhs	hedgehogs
FZD	Frizzled
LRP5/6	lipoprotein receptor-related proteins 5 or 6
DVL	disheveled
LRP1	lipoprotein receptor-related protein-1
PE	pseudomonas exotoxin
ABD	albumin binding domain
PDX	patient-derived xenografts
NK	natural killer
MPF	megakaryocyte potentiating factor
MUC16	mucin 16
MMP-7	matrix metalloproteinase
aPF	activated protal fibroblast
PI3K	phosphoinositide 3-kinase
NSCLC	non-small cell lung cancer
ADCC	antibody-dependent cellular cytotoxicity
TH	target heterogeneity

1. Introduction

According to the Global Cancer Statistics 2020 (GLOBOCAN database) estimates, primary liver cancer (PLC) is the sixth most commonly diagnosed cancer and the third leading cause of cancer-related death worldwide, with approximately 906,000 new cases and an estimated 830,000 deaths in 2020 (Sung et al., 2021). The highest rates of both incidence and mortality of PLC are observed mainly in Asia and Africa, although cases in most western countries including the US are increasing every year (Yang et al., 2019). Among all PLC types, hepatocellular carcinoma (HCC) and intrahepatic cholangiocarcinoma (iCCA) are the most common, accounting for approximately 75–85% and 10–15% of cases, respectively. Most HCC cases occur in people with chronic infection with hepatitis B virus (HBV) or hepatitis C virus (HCV) or liver cirrhosis caused by alcoholism. Key risk factors linked to iCCA include liver flukes, primary sclerosing cholangitis, hepatitis viruses (HBV, HCV), and metabolic conditions (Zucman-Rossi, Villanueva, Nault, & Llovet, 2015).

Despite growing knowledge of PLC tumorigenesis in the past decades (Banales et al., 2020; Ma et al., 2019), the prognosis of both HCC and iCCA continues to be poor. Conventional treatment strategies, including liver resection, ablation, liver transplantation, and selective internal radiotherapy are potentially curative for people with liver cancer at a sufficiently early stage (Vogel et al., 2019). Unfortunately, most HCC patients are diagnosed with intermediate and advanced stage disease, where treatment options are limited and not curative. Phase III trial success for sorafenib in 2007 represents a molecular targeted therapy for advanced HCC in which prognosis is somewhat improved for the patients with late-stage disease (Gauthier & Ho, 2013). However, the clinical benefits of sorafenib monotherapy and its combination with trans-arterial chemoembolization (TACE) remain modest, increasing average survival time of patients by only 3–5 months, mainly due to drug resistance (El-Serag, Marrero, Rudolph, & Reddy, 2008; Gauthier & Ho, 2013). Subsequently, other chemotherapeutic agents, such as lenvatinib (Kudo et al., 2018), regoragenib (Finn et al., 2018), ramucirumab, and cabozantinib (Abou-Alfa et al., 2018) have shown increased therapeutic benefits but the outcomes were below ideal. As a result, none of the current chemotherapy treatments can significantly improve the outcome of this devastating malignant disease (Gauthier & Ho, 2013), and novel targeted strategies are urgently needed.

Immunotherapy has recently emerged as a new promising treatment option for advanced liver cancer patients who are resistant to chemotherapy. Development in this field of research is based on the study of immune cells to specifically recognize tumor cells and reactivate the antitumor immune response to overcome tumor escape in the tumor microenvironment (TME). Multiple immunotherapeutic approaches have been investigated, including immune checkpoint inhibitors (ICIs), adoptive T-cell therapy, and vaccination or virotherapy, which allowed for the identification and validation of predictive tumor antigens (Sangro, Sarobe, Hervas-Stubbs, & Melero, 2021). Tumor-specific antigens (TSAs) and tumor-associated antigens (TAAs) are two major classes of tumor targets for cancer immunotherapies. TSAs includes neoantigens (newly expressed antigens produced from viral protein, normal cellular proteins, or mutated host genes), oncofetal proteins (transiently expressed during embryogenesis and may increase with certain cancers), and testis-associated antigens (normally only expressed in male germ cells in the testis, but aberrantly heterogeneously expressed in cancer cells) (Castellarin, Watanabe, June, Kloss, & Posey, 2018; Finn & Rammensee, 2018; Simpson, Caballero, Jungbluth, Chen, & Old, 2005).

TAAs are overexpressed self-proteins present on tumor cells that are also expressed at lower levels on their normal cell counterparts. Targeting TAAs for liver cancer treatment has been investigated extensively for the field of cancer vaccine research, but with limited success so far (Lu et al., 2021). In comparison, TSAs are ideal targets for cancer immunotherapy because they are exclusively expressed by the cancer cells (Castellarin et al., 2018). We previously summarized the molecular biology of two emerging antigens, glypican-3 (GPC3) and mesothelin (MSLN), and discussed their potential as the therapeutic targets in HCC and iCCA (Ho, 2011). In the past decade, we and others have also shown encouraging results in the development of GPC3 and MSLN as cancer targets of immunotherapeutic therapies. In this context, it is now important to provide an update highlighting more recent developments for these two promising therapeutic targets for PLCs. Furthermore, in addition to the tumor antigens expressed on tumor cells and adenomatous polyposis coli (APCs), there are some cell surface proteins expressed in immune cells that can also be used as therapeutic targets in liver cancer TME. In this review, we will provide an overview of the current emerging targets for liver cancer immunotherapy, with particular emphasis on GPC3 and MSLN. We will describe their respective development, biology, and function in liver cancer progress. We will also summarize the currently available and ongoing clinical trials (Table 1) and discuss the future perspectives and options for liver cancer targeting immunotherapy.

Table 1 Summary of current and ongoing clinical trials for treating liver cancer.
Clinical trials targeting liver cancer

Target	Treatment	Phase	Clinical trial
Hepatocellular carcinoma			
GPC3	CAR T	I	NCT05003895
GPC3	GLYCAR T cells	I	NCT02905188
GPC3	Anti-GPC3 CAR T	I/II	NCT03084380
GPC3	Anti-GPC3 CAR T	I	NCT02395250
GPC3	GPC3 CAR T cells	I	NCT04506983
GPC3	GPC3 CAR T cells	I/II	NCT03130712
GPC3	TAI GPC3 CART cells	I/II	NCT02715362
GPC3	GPC3 and/or TGFβ targeting CAR-T cells	I	NCT03198546

Continued

Table 1 Summary of current and ongoing clinical trials for treating liver cancer.—cont'd
Clinical trials targeting liver cancer

Target	Treatment	Phase	Clinical trial
GPC3	CAR-T cell immunotherapy	I	NCT04121273
GPC3	B010-A i	I	NCT05070156
GPC3	CAR-GPC3 T Cells	I	NCT03884751
GPC3	GLYCAR T cells	I	NCT02905188
GPC3	CT0180 Cells	I	NCT04756648
NY-ESO-1	DEC-205/NY-ESO-1 Fusion Protein CDX-1401	I	NCT01522820
NY-ESO-1	Anti-NY ESO-1 mTCR PBL	II	NCT01967823
CEA	Fowlpox-CEA(6D)/TRICOM vaccine	I	NCT00028496
AFP	Autologous genetically modified AFP^{c332}T cells	I	NCT03132792
AFP	Autologous C-TCR055	I	NCT04368182
AFP	ET1402L1-ARTEMIS™ T cells	I	NCT03888859
VEGFR	Ramucirumab	III	NCT02435433
AFP, GPC3	Peptide cancer vaccine	I	NCT05059821
c-MET	Tepotinib	I/II	NCT02115373
c-MET	Cabozantinib	II	NCT04767906
c-MET	APL-101	I/II	NCT03655613
c-MET	INC280	II	NCT01737827
c-MET/VEGFR	Foretinib	I	NCT00920192
VEGFR, PD-1	Regorafenib/Pembrolizumab	II	NCT04696055
VEGFR	Cabozantinib, Nivolumab	II	NCT05039736
PD-1	Nivolumab	I	NCT04658147
PDL-1	Nanoplexed Poly I:C BO-112/ Pembrolizumab	I	NCT04777708
PD-1	Pembrolizumab	II	NCT02702414
HBV	RECOMBIVAX HB	II	NCT00322361
GM-CSF	JX-594	II	NCT00554372

Table 1 Summary of current and ongoing clinical trials for treating liver cancer.—cont'd
Clinical trials targeting liver cancer

Target	Treatment	Phase	Clinical trial
TIM-3	INCAGN02390	I	NCT03652077
LAG-3	INCAGN02385	I	NCT03538028
CTLA-4, LAG-3	XmAb®22841	I	NCT03849469
CTLA-4	AK104, Lenvatinib	I/II	NCT04444167
PD-1, CTLA-4	XmAb20717	I	NCT03517488
PD-1, CTLA-4	Nivolumab/Ipilimumab	II	NCT03222076
CTLA-4, PDL-1	Tremelimumab/Durvalumab TACE/RFA	II	NCT02821754
CTLA-4, PD-1, VEGFR	Pembrolizumab/Quavonlimab/Lenvatinib	II	NCT04740307
PD-1, CTLA-4	AK104 lenvatinib	II	NCT04728321
CTLA-4	CP 675,206	II	NCT01008358
hTERT	INO-1400, INO-9012, INO-1401	I	NCT02960594
CD147	CD147-CART	I	NCT03993743
Cholangiocarcinoma			
MSLN	Gavocabtagene autoleucel (gavo-cel; TC-210)	I/II	NCT03907852
PDL-1, CD27	CDX-527	I	NCT04440943
MUC-1	MUC-1 CART cell immunotherapy	I/II	NCT03633773
FGFR2	BGJ398	III	NCT03773302
FGFR2	TAS-120	III	NCT04093362
FGFR2	BGJ398 (infigratinib)	II	NCT02150967
FGFR2	RLY-4008	I	NCT04526106
FGFR2	Pemigatinib	II	NCT04256980
FGFR2	Derazantinib	II	NCT03230318
FGFR2	E7090	II	NCT04238715
FGFR2	3D185	II	NCT05039892
FGFR2	TT-00420	II	NCT04919642
PD-1, LAG-3	MGD013	I	NCT03219268

2. Emerging targets for liver cancer immunotherapy

HCCs and some iCCAs may both be derived from the common hepatocyte progenitor, resulting in distinctive tumor phenotypes driven by specific oncogene signatures and TME interactions (Sia, Villanueva, Friedman, & Llovet, 2017). Here, some biomarkers are shared by both HCC and iCCA tumor cells (Chaisaingmongkol et al., 2017), while some are unique, such as GPC3 in HCC (Ho & Kim, 2011) and MSLN in iCCA (Yu et al., 2010). In addition to GPC3 and MSLN, we will broadly categorize antigens by tumor cell targets and immune cell targets based on their expression patterns. Immunotherapeutic targets that are expressed on the membrane of liver cancer cells are summarized in Fig. 1.

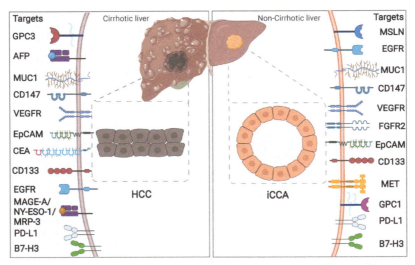

Fig. 1 Cell surface immunotherapeutic targets in liver cancer. The possible surface protein targets for HCC treatment were presented in the left panel while right panel includes possible cell surface protein targets for iCCA treatment. HCC, hepatocellular carcinoma; iCCA, intrahepatic cholangiocarcinoma; GPC3, glypican-3; MSLN, mesothelin; AFP, α-fetoprotein; MUC1, mucin 1; CD147, cluster of differentiation 147; VEGFR, vascular endothelial growth factor receptor; EqCAM, epithelial cellular adhesion molecule; CEA, carcinoembryonic antigen; CD133, cluster of differentiation 133; EGFR, epidermal growth factor receptor; VEGFR, vascular endothelial growth factor receptor; FGFR2, fibroblast growth factor receptor 2; MET, HGF receptor; PD-L1, Programmed cell death ligand 1; B7-H3, B7 homolog 3; Cancer/Testis antigens: MAGE-A, melanoma-associated gene A; NY-ESO-1, New York esophageal squamous cell carcinoma 1; MRP-3, multidrug resistance-associated protein 3.

2.1 GPC3 in HCC and MSLN in iCCA

Glypicans are a family of heparan sulfate proteoglycans that are attached to the cell membrane via a glycosyl-phosphatidylinositol (GPI) anchor (Ho & Kim, 2011). Glypicans include six members, GPC1-GPC6. Homologs of this family, which are expanded throughout the entire Eumetazoa, are highly conserved (Li, Gao, Zhang, & Ho, 2018). As an oncofetal protein, GPC3 is a key member of the glypicans family. Recently, the potential of GPC3 as a HCC target has been evaluated. GPC3 is highly expressed in both HCC tissues and HCC cell lines, but undetectable in normal adult liver, livers with cirrhosis (Fleming et al., 2020; Gao, Kim, et al., 2014; Li et al., 2020), fatty liver or those with hepatic injury (Guo, Zhang, Zheng, & Liu, 2020; Montalbano et al., 2017). Additionally, the expression of GPC3 is undetectable in other normal organs, except for high mRNA levels expressed in the placenta (Li et al., 2020). Furthermore, we and others showed undetectable or very low GPC3 expression in iCCA, indicating that GPC3 is a useful biomarker in differentiating HCC from iCCA (Phung, Gao, Man, Nagata, & Ho, 2012; Zhao et al., 2016). Further proof-of-concept studies using monoclonal antibodies suggested that GPC3 is a suitable target for HCC therapy (Feng et al., 2013; Gao, Kim, et al., 2014; Phung et al., 2012; Zhu et al., 2013).

MSLN is a surface protein that attaches to the cell membrane by a GPI anchor, and it is known to be an attractive oncogenic TAA for cancer immunotherapy including iCCA (Hassan & Ho, 2008; Ho, 2011; Yu et al., 2010). We found that MSLN was highly expressed in iCCA tissues and iCCA cell lines, but not in HCC or normal liver tissues (Yu et al., 2010). Clinically, the aberrant expression of MSLN was associated with the poor overall survival of patients with iCCA (Nomura et al., 2013). However, the precise etiology of high MSLN expression in iCCA remains elusive. MSLN expression is limited in several normal tissues such as mesothelium of peritoneal, pleural, and pericardial cavities (Chang & Pastan, 1994). Therefore, MSLN is a potential target for iCCA therapy.

2.2 Other tumor targets in HCC and iCCA

2.2.1 Tumor cell antigens

Although the neoantigens are highly tumor-specific, the mutated neoepitopes were rarely found on HCC tumor cells, and other tumor types that have low or intermediate tumor mutation burdens (TMBs) (Bassani-Sternberg et al., 2016). Mutated epidermal growth factor receptor

(EGFR) can be a therapeutic neoantigen because it is frequently overexpressed and/or mutated in human cancer, including liver cancer (Wykosky, Fenton, Furnari, & Cavenee, 2011). Besides GPC3, alpha fetal protein (AFP) and carcinoembryonic antigen (CEA) are the most widely used oncofetal protein tumor markers. AFP is an intracellular/secreted glycoprotein that is processed into peptides and presented by class I major histocompatibility complex (MHC) on the tumor cell membrane. Given that AFP is commonly overexpressed in tumors of endodermal origin, including pediatric hepatoblastoma and HCC but rarely in adult normal tissues, chimeric antigen receptor (CAR) T cells targeting AFP peptide-MHC complexes have been utilized to treat HCC (Liu et al., 2017). The plasma-soluble AFP is also used as a diagnostic biomarker in detecting HCC and monitoring therapy progression in clinical status. Approximately 95% of patients with benign liver disease will have AFP concentrations <200 ng/mL, whereas patients with liver cancer will have AFP concentrations greater than 1000 ng/mL. CEA has been well characterized as a diagnostic and prognostic tumor marker in colorectal cancer. It is also highly expressed in rectum, pancreas, gastric, and breast adenocarcinomas (Sarobe, Huarte, Lasarte, & Borras-Cuesta, 2004). Importantly, overexpression of CEA is closely related to liver metastasis, which is the main cause of death of colorectal cancer patients (Lee & Lee, 2017).

It is known that cancer testis-associated antigens are mainly expressed intracellularly, but some antigens with a surface localization via the MHC complex have also been identified, which can be used as potential HCC therapeutic targets, including melanoma-associated gene A (MAGE-A), New York esophageal squamous cell carcinoma 1 (NY-ESO-1), and multidrug resistance- associated protein 3 (MRP-3) (Lu et al., 2021). Human telomerase reverse transcriptase (hTERT) peptides-based vaccines, or hTERT-specific T cells, were correlated with the absence of HCC recurrence, suggesting a potential role of hTERT as a liver cancer target (Greten et al., 2010). Glypican-1 (GPC1) was recently identified as a novel antigen in iCCA for antibody conjugate drug development (Yokota et al., 2021). Moreover, MUC1, CD147, c-Met, VEGFR, and FGFR2 are all cell surface therapeutic targets that were shown to be overexpressed in both HCC and iCCA but low in normal tissues. This phenomenon was correlated with PLC progression and metastasis (Borad, Gores, & Roberts, 2015; Li, Qian, et al., 2021; Miyamoto et al., 2011; Yuan et al., 2005). In addition, the expression of both Programmed cell death ligand 1 (PD-L1) and B7 homolog 3 (B7-H3) have been found at a high proportion on the surface of HCC and iCCA. Their overexpression was associated with tumor-infiltrating lymphocytes (TIL), which are two potential therapeutic TAAs (Cheng et al., 2018; Fontugne et al., 2017; Sun et al., 2012).

Cancer stem cell (CSC) antigens play an essential role in cancer relapse and metastasis, therefore there is an urgent need for novel immunotherapies that target CSC to treat cancer. Currently, several clinical trials are underway, targeting CSC in different types of cancer such as HCC and iCCA (McGrath, Fu, Gu, & Xie, 2020). All CSC makers used as cancer targets that have been mentioned here are examples of cell surface proteins. Some targets are unique for iCCA, whereas some are shared with HCC. For example, EpCAM is a transmembrane glycoprotein involved in cell signaling, migration, proliferation, and differentiation (Sulpice et al., 2014) that has been applied as a prognostic marker for both HCC and iCCA. Patients with cluster of differentiation 133 (CD133)-positive HCC or iCCA experienced poorer overall survival following surgery (Shimada et al., 2010). Other liver cancer stem cell (LCSC) target candidates include CD90, ALDH1, CD44, CD24, CD47, OV-6, CD13, and SOX2 (McGrath et al., 2020).

2.2.2 Immune cell antigens

The liver cancer microenvironment is highly immunosuppressive, consisting of negative regulatory layers. These include regulatory T cells, inhibitory B cells, myeloid-derived suppressor cells (MDSCs), inhibitory soluble factors such as transforming growth factor β, and up-regulation of cell surface co-inhibitory lymphocyte signals such as programmed cell death protein-1 (PD-1) to inhibit T cell attacks (Sangro et al., 2021). Currently, PD-1 and cytotoxic T lymphocyte-associated antigen 4 (CTLA-4) are two major proteins that have been studied in PLC immunotherapy. PD-1 is expressed on activated T cells, B cells, NK cells, Tregs, MDSCs and DCs (Sangro et al., 2021). PD-1 binds to its ligand PD-L1, which results in the suppression of TILs proliferation, migration, and secretion of cytotoxic effectors (Butte, Keir, Phamduy, Sharpe, & Freeman, 2007). CTLA-4 is expressed on the membrane of activated T cells, and functions in binding to its receptors (CD80 and CD86) on antigen-presenting cells, which antagonizes their interaction with CD28 and therefore mediating inhibitory signals to antigen-activated T cells (Chambers, Kuhns, Egen, & Allison, 2001). ICIs against PD-1 and CTLA-4 are currently in clinical trials as monotherapy or combination with other ICIs for PLC patients (Zheng et al., 2021). Additionally, there are other ICI targets in liver cancer including lymphocyte activation gene 3 (LAG-3) and T-cell membrane protein 3 (TIM-3) (Greten & Sangro, 2018).

3. The role of GPC3 in HCC

GPC3 is a promising immunotherapeutic target in HCC. In this section, we will particularly focus on introducing GPC3 development, biology, signaling pathway and application as an emerging target involved in HCC progression.

3.1 GPC3 biology and structure

The GPC3 gene is located on the human X chromosome (Xq26) and encodes the common isoform (isoform 2) of GPC3 protein, comprising a 70-kDa core protein composed of 580 amino acids (AA) and a carboxy terminus with two predicted heparan sulfate (HS) chains (Ho, 2011; Ho & Kim, 2011). There are an additional three isoforms produced by alternative splicing (Ho & Kim, 2011) and another potential isoform is predicted by computational mapping according to the UniPortKB database (UniProtKB/Swiss-Prot P51654). Similar to other glypican members such as GPC1 (Kim, Saunders, Hamaoka, Beachy, & Leahy, 2011; Svensson, Awad, Hakansson, Mani, & Logan, 2012), GPC3 has 14 evolutionarily conserved cysteines that may form intramolecular disulfide bridges connecting the N-terminus and C-terminus (De Cat et al., 2003; Ho & Kim, 2011). The furin cleavage site (RQYR) is predicted in GPC3, which can generate a 40 kDa amino (N)-terminal protein and a 30 kDa membrane-bound carboxyl (C)-terminal protein (Ho & Kim, 2011). We speculated that the N-terminus and C-terminus of GPC3 might remain linked with each other after cleavage, due to the disulfate bonds that are predicted to connect the N-terminus and C-terminus (Ho, 2011; Ho & Kim, 2011), and this is supported by presence of the entire GPC3 protein by Western blotting (Feng et al., 2013).

Among all glypicans, only the GPC1 protein structure has been recently solved in both *Drosophila melanogaster* and humans using crystals of the core protein without the HS domain (Kim et al., 2011; Svensson et al., 2012). We previously generated GPC3 model using the SWISS-MODEL (Li et al., 2019) . Here, we re-predicted GPC3 structure using the entire sequence by AlphaFold, which is a novel machine deep-learning approach that has recently been reported to be able to obtain highly accurate protein structures (Jumper et al., 2021). The GPC3 model showed similar 3D cylindrical-like structures compared with GPC1 because of their highly conserved 14 cystines (Fig. 2A). Similar to GPC1, GPC3 has 14 α-helices and three major

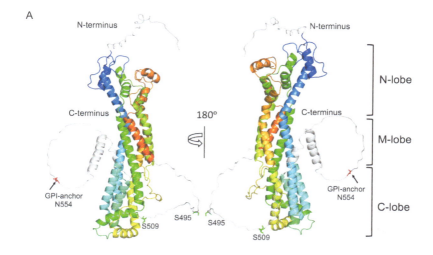
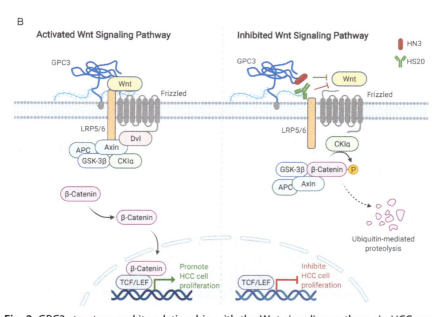

Fig. 2 GPC3 structure and its relationship with the Wnt signaling pathway in HCC progression. (A) An overview of the 3D cylindrical-like structure model (AlphaFold) of GPC3 represented as a schematic diagram in rainbow colors with loops and α-helices (N-terminus in blue and C-terminus in light gray). Also, the predicted are the two HS sites (S495 and S509) represented by green solid sticks, the GPI-anchor site (N554) as red solid sticks, and the different lobes (N-, M-, and C-lobes) in the GPC3 structure. (B). The GPC3/Wnts/FZD complex regulates the Wnt/β-catenin pathway in HCC progression. GPC3 recruits Wnts and interacts with FZD and LRP5/6, leading to dysregulation of β-catenin degradation and subsequently stimulating HCC cell proliferation by β-catenin translocation into the nucleus. HN3 and HS20 antibodies, block the interaction between GPC3 and Wnts on either the GPC3 N-terminal core protein or HS chains, respectively. This potentially inhibits the Wnt/β-catenin pathway during HCC progression.

loops, including two N-terminal loops and one long loop in the C-terminal. The N-, M-, and C-lobes are identified based on their relative spatial positions within the protein (Li et al., 2018). Out of the 14 cystine residues, 12 residues are located in the N-lobe, which is why this lobe is also known as Cys-rich region. Two HS sites (S495 and S509) and a GPI-anchor site (N554) were also predicted in the GPC3 model. Nonetheless, the predicted GPC3 model was not ideal because of low confidence at the cell membrane-proximal region (S477-H580), likely due to the lack of structures in this region from known GPI-anchored proteins in general. Using a combination of approaches including crystallography, small angle X-ray scattering, and chromatographic, Awad and colleagues modeled GPC1 C-terminal region, and they observed that this region is highly flexible and lacks significant secondary or tertiary structure (Awad et al., 2015). Moreover, they considered that the GPC1 protein ostensibly "lies down" in a transverse orientation to the membrane rather an "stand on" on the membrane. We postulate that GPC3 may also appear in a similar pattern as GPC1 because the "lies down" distance is likely sufficient for GPC3 HS or core protein to maximize its interaction with other membrane receptors to form a functional molecular complex.

3.2 GPC3 expression pattern in human normal tissues and HCC

GPC3 is a carcinoembryonic gene that is expressed in human embryos and involved in cell growth, differentiation, and morphogenesis. Hsu and colleagues found that GPC3 (also termed MXR7) might be an HCC marker, because GPC3 mRNA levels were significantly elevated in most HCC tissues compared to normal liver lesions (Hsu, Cheng, & Lai, 1997). They also observed high levels of GPC3 mRNA in placenta, fetal liver, lung, and kidney tissues, whereas there was no expression in adult organs. At the protein level, overexpressed GPC3 was also found in 72% of HCCs. Non-detectable levels GPC3 protein were found either in hepatocytes from benign hepatic lesions of healthy people or patients with hepatitis (Capurro et al., 2003). To characterize and establish GPC3 as a therapeutic target, we and others have made several monoclonal antibodies including YP7, HN3 and GC33 for testing as IgG molecules (Feng et al., 2013; Phung et al., 2012; Takai et al., 2009), antibody-drug conjugates (ADCs) (Fu et al., 2018), immunotoxins (Fleming et al., 2020; Gao et al., 2015; Wang, Gao, Feng, Pastan, & Ho, 2017) and CAR-T cells (Gao, Li, et al., 2014; Li et al., 2020). We isolated a high-affinity anti-GPC3 mAb, namely YP7 (Phung et al., 2012). By IHC staining with YP7 antibody, we showed that 60% of HCC tumor tissues display modest to high levels of GPC3 expression, whereas this protein was undetectable in normal liver and other normal adult tissues (Li et al., 2020). Moreover, by Western blot analysis with a monoclonal antibody targeting GPC3, we also found that YP7 was highly specific for HCC cell lines but not for GPC3 knockout HCC cells. Besides HCC, overexpression of GPC3 has also been detected in melanoma, ovarian clear-cell carcinoma, lung squamous cell carcinoma, yolk sac tumors, neuroblastoma, Wilms' tumors and other tumor types (Ho & Kim, 2011; Li, Spetz, & Ho, 2020). Overall, these evidence suggests that GPC3 expression levels change in a stage- and tissue-specific manner, thereby indicating that they are involved in morphogenesis.

3.3 GPC3-mediated signaling pathways in HCC cells proliferation

As we mentioned above, GPC3 may either promote or inhibit cell growth during tumor progression, which is believed to be related to the different growth factors and signaling pathways induced by GPC3. It has been reported that GPC3 is required for the optimal activity of positively-charged growth factors, including Wnts, Hedgehogs (Hhs), HGFs, FGFs, and bone morphogenetic proteins (Li et al., 2018). Here, three pathways induced by GPC3 will be discussed with a focus on Wnt/β-catenin.

3.3.1 Wnt signaling pathway

Although HCC is highly heterogeneous, the canonical Wnt pathway is one of the most frequent signaling pathways associated with the progression of HCC (Thompson & Monga, 2007). This pathway can be triggered by the interaction of Wnt to two co-receptors: seven-pass transmembrane receptor Frizzled (FZD) and low density lipoprotein receptor-related proteins 5 or 6 (LRP5/6) (He, Semenov, Tamai, & Zeng, 2004). The complex "Wnt-FZD-LRP5/6" recruits intracellular Disheveled (DVL) and Axin, and inhibits β-catenin phosphorylation, which subsequently migrates into the nucleus in order to promote cell proliferation and survival. Multiple studies have reported that overexpression of GPC3 stimulates HCC growth *via* the canonical Wnt/β-catenin pathway (Capurro, Xiang, Lobe, & Filmus, 2005; Gao & Ho, 2011; Gao, Kim, et al., 2014; Gao et al., 2015; Gao, Xu, Liu, & Ho, 2016; Kolluri & Ho, 2019; Li et al., 2019) (Fig. 2B). We previously showed that both the HS chains and the core protein of GPC3 participate in Wnt binding and activation. The HS chains on

GPC3 bind to Wnts, which encourages the GPC3/Wnts/FZD complex formation by promoting the interaction between Wnt and FZD (Gao, Kim, et al., 2014; Gao et al., 2016). Specifically, we identified a human monoclonal antibody, HS20, by phage display (Kim & Ho, 2018), which targets the HS chains of GPC3 and blocks GPC3 and Wnt3a interaction. We found that the HS20 antibody inhibited HCC cell proliferation by blocking Wnt/β-catenin signaling (Gao, Kim, et al., 2014; Gao et al., 2016). Using a panel of different lengths of synthetic HS oligosaccharides, we further demonstrated that the binding of HS20 and Wnt to a unique heparan sulfate motif with sulfation occurs at both the C2 position (2-O-sulfation) and C6 position (6-O-sulfation) (Gao et al., 2016). We also isolated a single domain antibody, HN3 (Feng et al., 2013), which binds to a novel conformational epitope in the GPC3 core protein and is capable of blocking Wnt recognition (Gao et al., 2015). In subsequent work using computational modeling, we predicted a Wnt-binding groove in the GPC3 core protein, and found that mutation of F41 residue in this groove significantly reduced Wnt binding, which blocked the Wnt/β-catenin pathway *in vitro* and inhibited HCC tumor growth in nude mice (Li et al., 2019). Based on these observations, we proposed GPC3 to be a new Wnt co-receptor: overexpression of GPC3 binds Wnts through the Wnt-binding groove containing the F41 residue in the N-lobe. Then, FZD is locally concentrated to form the GPC3/Wnts/FZD complex in which GPC3 may serve as a bridge *via* HS chains to stabilize the Wnt and FZD interaction in order to modulate downstream β-catenin signaling (Kolluri & Ho, 2019).

3.3.2 Yap and Hh signaling pathways

Aberrant activation of Yap has been observed in 50% of HCC and its overexpression causes HCC in mouse models, suggesting that Yap is a potential target for liver cancer therapy (Johnson & Halder, 2014; Liu, Wang, & Yang, 2020). In our research on HN3 and HS20 antibody activity, we for the first time suggested that Yap is the downstream oncogenic gene that is regulated by GPC3 in HCC tumorigenesis. In several HN3-treated HCC cell lines, we found up-regulated expression of inactive phosphorylated Yap, decreased total levels of Yap, and decreased level of cyclin D1 that acts as the target gene of Yap. Moreover, GPC3-knockdown in HCC cell lines showed that reduction of the Yap pathway and constitutive activation of Yap signaling could reverse the inhibitory effect of GPC3 knockdown and HN3 inhibition (Feng et al., 2013). We also observed that the

HN3-based immunotoxin HN3-PE38 inhibited Yap signaling in GPC3$^+$ HCC tumor cells in the presence of Wnt3a (Gao et al., 2015). Furthermore, we identified FAT1, an upstream receptor for the Hippo/Yap pathway in *Drosophila*, as a novel cell surface partner for GPC3 in humans and found that both gene have similar patterns of expression for metastasis in HCC (Meng et al., 2021). In another recent study, FAT1 has also been suggested to act as the cell surface receptor of Yap signaling in cancer (Martin et al., 2018).

The Hedgehog (Hh) pathway plays a pivotal role in embryonic development. This pathway is essential for liver regeneration, promotes liver fibrosis, and is significantly activated in HCCs. GPC3 was shown to be a potent inhibitor of the Hh pathway (Capurro et al., 2008) and can bind to Shh and Ihh at a high affinity, which prevents Patched (Hh receptor) from Hh binding (Capurro, Li, & Filmus, 2009). Capurro and colleagues found that the Hh-induced endocytosis of the GPC3/Hh complex was induced by low-density-lipoprotein receptor-related protein-1 (LRP1), which is essential for the Hh-inhibitory activity of GPC3 (Capurro, Shi, & Filmus, 2012).

3.4 GPC3 as a target of HCCs therapy

GPC3 has been studied as a new therapeutic target for HCC immunotherapy due to its specific overexpression in most HCCs and its correlation with poor prognosis as described above. Various therapeutic strategies have been developed based on GPC3 as a target including vaccines, monoclonal antibodies, immunotoxins, ADCs and CARs.

3.4.1 Antibody-based strategies

The most studied monoclonal antibodies targeting GPC3 are GC33 (Nakano et al., 2009), YP7 (Phung et al., 2012), and HN3 (Feng et al., 2013). All three antibodies showed potent anti-tumor activities in mice bearing HCC xenografts through ADCC (YP7 and GC33), CDC (YP7) and/or novel direct inhibition (HN3) mechanisms. However, GC33 showed a modest antitumor effect in some HCC patients in phase I clinical trials (Zhu et al., 2013). Unfortunately, GC33 treatment failed to show a significant clinical benefit in a randomized phase II trial, suggesting that antibody treatment alone may not be sufficient to treat liver cancer. Ishiguro and colleagues developed a bispecific antibody (ERY974) that targets GPC3 and the T cell antigen CD3 in order to specifically redirect T cells to GPC3-postive tumor cells (Ishiguro et al., 2017). More recently, we constructed the bispecific antibodies with CD3-targeting antibody OKT3 (scFv) paired with humanized YP7 (hYP7; scFv) or HN3 (VH only) and found that hYP7-OKT3 was potent in suppressing HCC tumor growth in mice (Chen et al., 2022). The anti-tumor efficacy of the hYP7-OKT3 bispecific antibody can be further improved by combination with Irinotecan.

ADCs and immunotoxins are two potent antibody-based therapeutic strategies used in cancer treatment. We postulate that GPC3 is a potential target of antibody-cytotoxin conjugates against liver cancer. To validate our hypothesis, we constructed anti-GPC3 ADC using hYP7 antibody with two DNA damaging agents (Duocarmycin SA and pyrrolobenzodiazepine) as the payloads, respectively. The GPC3 ADC (hYP7-DC) significantly killed a panel of GPC3-positive HCC cell lines, but not GPC3-negative cells. Moreover, hYP7-DC showed a synergetic effect with the approved drug Gemcitabine when treating HCCs *in vitro* and *in vivo* (Ying et al., 2019). In addition, we engineered an anti-GPC3 immunotoxin by infusing the HN3 antibody with a truncated from of the Pseudomonas exotoxin (PE38) (Gao, Kim, et al., 2014). To reduce immunogenicity, the HN3 was fused to a deimmunized toxin fragment (HN3-mPE24) after removing the B-cell epitopes. The HN3-PE24 immunotoxin caused liver tumor regressions and prolonged mouse survival with no obvious side effects (Wang et al., 2017). To further reduce immunogenicity and improve serum half-life of GPC3 immunotoxins, we engineered a novel HN3-ABD-T20 by removing T-cell epitopes and adding a streptococcal albumin binding domain (ABD) (Fleming et al., 2020). We found that the HN3-ABD-T20 extended serum half-life 45-fold longer than HN3-T20, and it mediated tumor regression at low dose at 1 mg/kg. Future clinical trials and studies will evaluate the efficacy and therapeutic potential of GPC3-targeted antibody therapeutics, including ADCs, immunotoxins and bispecific antibodies for treating HCC.

3.4.2 Cell-based strategies
T-cell based strategies targeting GPC3 have been investigated in preclinical and clinical studies. The development of anti-GPC3 CAR-T was first reported by Gao, Kim, et al. (2014). By using the scFv sequence of GC33, they engineered both first generation (GPC3-Z) with CD3ζ domain only and third generation CAR-T (GPC3-28BBZ) with CD28, 4-1BB, and CD3ζ domains. Encouragingly, GPC3-28BBZ CAR-T cells could

efficiently kill GPC3-positive HCC cells (HepG2, Huh-7, Hep3B, and PLC/PRF/5) and suppressed the growth of established low GPC3 expression HCC xenografts. Moreover, GPC3-targeted CAR-T cells could suppress the tumor growth in patient-derived xenografts (PDX) of HCC (Jiang et al., 2016). Clinically, the phase I trial (ClinicalTrials.gov: NCT02395250) of GPC3 CAR-T has been completed in 13 Chinese patients, and the CAR-T showed tolerance and partial antitumor activity in these HCC patients (Shi et al., 2020). A series of studies were later carried out to examine armed GPC3 CAR-T aiming to improve GPC3 CAR-T efficiency. For example, Batra et al. engineered GPC3 CAR with IL15 and IL21 co-expression in order to enhance CAR T expansion and persistence (Batra et al., 2020). Wu et al. analyzed a combination therapy of GPC3 CAR-T with sorafenib and found that the synergistic effects of this combination in HCC mouse models (Wu et al., 2019).

Our laboratory has generated GPC3 CAR-T using humanized YP7 (hYP7) and HN3 targeting the proximal epitope and distal epitope of GPC3, respectively. We demonstrated that hYP7 antibody derived CAR-T had potent and persistent anti-tumor activity in orthotopic HCC mouse models and that was accomplished by using functional genomics sequencing and single cell-based T cells analysis (Li et al., 2020). A clinical trial using our hYP7 CAR-T cells for treatment in patients with advanced HCC is ongoing at the National Institute of Health (ClinicalTrials.gov Identifier: NCT05003895). Furthermore, we recently engineered PD-L1-targeted CAR-T using a novel shark V_{NAR} single domain antibody targeting PD-L1 and we found that the combination of hYP7 CAR-T and PD-L1 CAR-T could synergically promote regression of HCCs in mice (Li, English, et al., 2021). Additionally, GPC3-specific CAR engineered natural killer (NK) cells showed potent anti-tumor activity in multiple HCC xenograft types with both low and high GPC3 expression levels (Yu et al., 2018).

4. The role of mesothelin in iCCA
4.1 Mesothelin expression in iCCA

MSLN is a cell surface antigen found in ovarian cancer and mesothelioma initially (Chang & Pastan, 1996; Chang, Pastan, & Willingham, 1992) and was later shown to be present on a variety of tumor types (Hassan & Ho, 2008), such as lung cancer (Ho et al., 2007), pancreatic cancer (Hassan & Ho, 2008; Ho et al., 2005), and triple-negative breast cancer

(Parinyanitikul et al., 2013). In liver cancer, mesothelin is a tumor-associated antigen elevated in about 30% of the surgically resected iCCA specimens (Ho, 2011; Yu et al., 2010). In an early study, we did immunohistochemistry staining to test expression levels of MSLN protein in 87 patient liver tumor specimens. We found that 33% of iCCA specimens had MSLN expression. We further examined six iCCA cell lines and found that three of them express high levels of MSLN with the dominant form being the 40 kDa mature form as shown on Western blot (Yu et al., 2010). To compare the protein levels of the clinical iCCA specimens with the iCCA cell line model, Western blotting was carried out to quantify the expression of MSLN. We found that 22% of iCCA specimens express a high level of MSLN comparable to the iCCA tumor cell lines. The expression of MSLN in normal tissue was confined to mesothelial cells (Hassan & Ho, 2008; Ho, 2011), thus making it a potential target for tumor therapy. More recently, Manzanares et al. confirmed the expression of the 40-kDa MSLN in both the orthotopic liver tumors and in iCCA cells derived from a 3D culture (Manzanares, Campbell, Maldonado, & Sirica, 2018). In another study, MSLN expression was also detected in 8 out of 25 (32%) iCCA patients and five of them were stage IV (Nomura et al., 2013). High expression levels of MSLN are also correlated with the poor prognosis of other cancers, such as ovarian cancer (Ho et al., 2005), lung adenocarcinoma (Ho et al., 2007), colorectal cancer (Shiraishi et al., 2019), and acute myeloid leukemia (Kaeding et al., 2021). Overall, MSLN expression is elevated in approximately 30% of iCCA as detected by anti-MSLN antibodies, indicating that MSLN is a potential target for antibody-based therapies.

4.2 MSLN structure

Human MSLN is a 71 kDa protein expressed as a glycophosphatidylinositol-linked cell surface glycoprotein with four predicted N-linked glycosylation sites (Hassan & Ho, 2008). The N-terminal 31 kDa portion is released with the treatment of furin, known as the megakaryocyte potentiating factor (MPF) (Kojima et al., 1995) (Fig. 3A and B). There are multiple predicted protease cleavage sites including a trypsin cleavage site at Arg 286 and a furin cleavage site at Arg 295 found in a region (residues 286–296) between the MPF and the cell surface MSLN (Hassan & Ho, 2008). In a rat cholangiocarcinoma model, the 40 kDa form is found positively correlated with the malignant progression of tumors, indicating that it can be used as a predicator of cholangiocarcinoma (Manzanares et al., 2018).

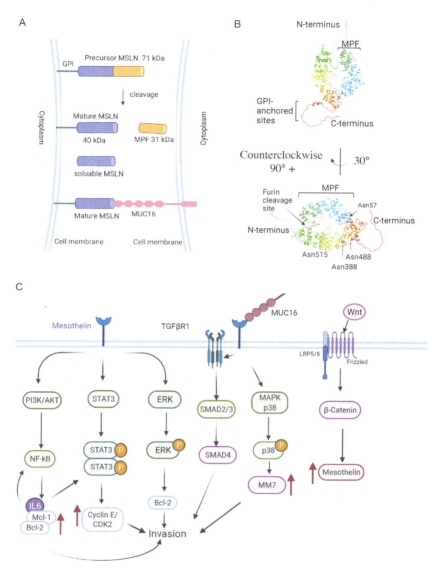

Fig. 3 Mesothelin protein structure, biology, and regulated signaling pathways involved in iCCA progression. (A) Presented is a schematic of mesothelin. Two forms of MSLN are produced by cells. One is the GPI-anchored membrane MSLN (71 kDa), which will further yield a 31 kDa megakaryocyte potentiating factor (MPF) that circulates in the blood, and the second is a mature membrane-bound protein (40 kDa), which is responsible for the binding of an interaction partner (such as MUC16) and leading to intracellular signaling transduction. (B) The presented is a structure of MSLN. The structure of MSLN is extracted from AlphaFold (Uniprot Q13421) and presented from two angles. The MPF located near the N terminus and mature MSLN located near the GPI anchored area. The furin cleavage site is labeled as red, and the predicted glycosylation residues are denoted as black. (C). MSLN is involved in signaling cascades. MSLN activates multiple signaling pathways, such as NF-kB, PI3K/AKT, STAT3, TGF-β, and MAPK/ERK to enhance pro-survival activity of cancer tissue. In addition, Wnt signaling activation upregulates MSLN expression.

In addition to the membrane-bound MSLN, Ho et al. found that the protein could be shed from the tumor cell surface (Ho et al., 2006) (Fig. 3A). Using an affinity column, we purified shed MSLN from tumor culture with the molecular weight of 40 kDa. The N-terminal sequence (EVEK) of purified shed MSLN was determined by Edman degradation, and peptides corresponding to multiple regions of MSLN were determined using mass spectrometry. We confirmed that the shed soluble MSLN was the extracellular domain of membrane-bound MSLN. Using anti-cross-reacting determinant antibodies, we found that MSLN shedding did not require phosphatidylinositol-specific phospholipase C phospholipolysis. More recently, the MMP/ADAM/BACE family members were found to be responsible for the shedding of MSLN (Liu, Chan, Tai, Andresson, & Pastan, 2020). The knockdown of the family members, ADAM10, MMP15, BACE1, and TACE, decreased the shedding of MSLN, thus indicating that multiple mechanisms are involved in MSLN shedding process (Zhang, Chertov, Zhang, Hassan, & Pastan, 2011).

4.3 MSLN promotes tumor progression through diverse pathways

MSLN can regulate cell survival through engagement with its binding partner Mucin 16 (MUC16, also known as CA125) (Fig. 3A and C). Mucin 16 is a well-known binding partner of MSLN in tumor progression (Kaneko et al., 2009; Rump et al., 2004). We identified that the binding region (residues 296–359) of MUC16, IAB, consists of 64 amino acids at the N-terminus of cell surface MSLN for MUC16 (Kaneko et al., 2009). This binding relies heavily on the N-linked oligosaccharides of MUC16, since the removal of N-linked oligosaccharides abrogates the binding of MUC16 and MSLN (Gubbels et al., 2006). The engagement of MSLN and MUC16 induces the phosphorylation of p38, facilitating the expression of matrix metalloproteinase (MMP-7). The activation of MMP-7 further enhanced cancer cell migration and invasion (Chen, Hung, Wang, Paul, & Konstantopoulos, 2013). Liver specimens from patients with cholestatic liver fibrosis were shown to have elevated levels of MSLN positive activated portal fibroblasts (aPFs). These aPFs can mediate the hepatic fibrosis in which MSLN/MUC16 signaling regulated TGF-β signaling, thereby leading to the nucleus translocation of SMAD2/3/4 and the expression of the target genes to promote tumor progression (Koyama et al., 2017). Thus, the molecular interaction of MSLN and MUC16 is a major finding related to the MSLN function. Inhibition of this interaction has been proposed as an attractive strategy for cancer treatment (Xiang et al., 2011).

Wnt signaling is a canonical pro-tumor pathway induced by engagement of Wnt molecules and membrane receptors, such as Frizzled, LRP5/6 and GPC3 (Ho, 2011; Kolluri & Ho, 2019). In mouse mammary epithelial cells, Prieve and Moon found that overexpressed Wnt-1 can upregulate the expression of MSLN, thus indicating that MSLN is a downstream factor of Wnt signaling involved in tumor progression (Prieve & Moon, 2003) (Fig. 3C).

MSLN has been indicated to promote tumor cell progression and cancer invasion through various mechanisms (Tang, Qian, & Ho, 2013). First, it downregulates the apoptotic signaling, as high MSLN expression confers tumor cells more resistance to cell apoptosis. In an ovarian tumor model, MSLN was demonstrated to accelerate tumor burden and peritoneal dissemination in nude mice. Mechanistically, MSLN enhances cell adherence, anoikis (a type of apoptosis induced by detachment from the substratum), cell survival, and anchorage-independent cell growth, which results in advanced invasion of cancer cells (Coelho et al., 2020). Similar phenomena have been noted in breast cancer, and the high levels of MSLN induces anchorage-independent growth, which triggers suppression of the proapoptotic gene, Bim. Suppression of Bim leads to the loss of sensitivity to anoikis *via* stimulation of the ERK pathway (Uehara, Matsuoka, & Tsubura, 2008). Overexpression of MSLN also upregulates the expression of anti-apoptotic Bcl-2 family members, such as Bcl-2 and Mcp-1, through activation of the phosphoinositide 3-kinase (PI3K)/AKT and MARP/ERK pathways (Chang et al., 2009). On the other hand, NF-κB signaling activated by MSLN, leads to the elevated cyclin A for cell cycle progression, decrease of caspase activity and enhanced expression of anti-apoptotic members, such as Mcl-1, BAX, and p-BAD. The series of signaling changes confers the cells antagonization of TNF-a induced apoptosis (Bharadwaj, Marin-Muller, Li, Chen, & Yao, 2011a). In addition, MSLN enhances pro-survival signaling. In pancreatic cancer cells, exogenous MSLN expression induces the excessive activation of STAT3 and increased cyclin E and CDK2 complex formation. This activation promotes the G1 to S phase transition during cell cycle progression (Bharadwaj, Li, Chen, & Yao, 2008). Moreover, IL6 secretion is found positively related to MSLN expression through NF-κB signaling. Elevated IL6 in turn serves as a growth factor *via* initiating an auto/paracrine signaling pathway by engagement with soluble IL6 receptors to increase cell survival and proliferation (Bharadwaj et al., 2011a, 2011b).

Overall, MSLN promotes tumor progression through diverse pathways. Further study into these available mechanisms can help to expand knowledge of the MSLN molecular function and aid in the design of MSLN-targeted anti-tumor therapy.

4.4 Antibody-based immunotherapies against MSLN

Currently, a large number of antibody-based therapies against MSLN have been designed and applied to clinical trials (Hassan & Ho, 2008; Pastan & Hassan, 2014). For instance, SS1 is a murine antibody raised against the MSLN membrane-distal region (N-terminal region I, 296–390aa), and based on the study using this antibody, multiple modifications such as immunotoxin and scFv engineering have been conducted for clinical trials. SS1 based immunotoxin, SS1P, showed significant growth inhibition of multiple iCCA and non–small cell lung cancer (NSCLC, MSLN positive) cell lines, indicating that MSLN-directed antibody therapy can be a powerful tool to restrain tumor cell growth (Ho et al., 2007; Yu et al., 2010). However, due to the short half-life of immunotoxin and the neutralizing antibody produced during the treatment, most therapies using SS1P have not entered or passed the phase III clinical trials. In our lab, we developed HN1 scFv against MSLN from a human scFv phage display library. The scFv recognizes a conformation-sensitive epitope since it only binds to full-length MSLN but not any of its fragments. The scFv was further engineered into human IgG and shows potent antibody-dependent cellular cytotoxicity (ADCC) activity on cancer cells as well as significant blockage of the interaction of MSLN-MUC16 (Ho, Feng, Fisher, Rader, & Pastan, 2011). This suggests that HN1 has great potential for MSLN-directed cancer treatment. We also isolated high affinity rabbit monoclonal antibodies against MSLN, YP223, YP218, and YP3 by immunizing the rabbit with a rabbit Fc (rFc)-human MSLN fusion protein, and the antibodies showed high specificity in tumor tissues. Among the antibodies, YP158 binds to membrane-distal MSLN region I (residues 296–390), YP223 and YP187 bind to MSLN region II (residues 391–486), and YP3 binds to a conformation epitope that requires the full-length of MSLN while YP218 targets the membrane-proximal region (residues 487–581) (Zhang et al., 2015). YP218 has been humanized by protein engineering and the YP218-based immunotoxin shows the most potent cell killing ability in MSLN positive tumors (Zhang & Ho, 2017; Zhang et al., 2015). Another human IgG1 antibody, HN125, was generated by fusion of hFc with the MUC16-binding fragment (296–359aa) of MSLN. This antibody was demonstrated to block the heterotypic cell adhesion efficiently on OVCAR3 cell model through disruption of MUC16 and MSLN interaction. Also, it shows significant ADCC activity on OVCAR3 cells with no MUC16-negative tumor cell death, indicating the specificity of the anti-tumor activity (Xiang et al., 2011). Collectively, these available antibodies provide a solid platform for MSLN-directed cancer therapy.

To overcome the limitations of antibody-based therapy and enhance the persistence of the anti-MSLN effects, CAR-T therapy has been introduced for targeting MSLN. CAR-T cells can be generated by transduction of lentiviral systems that contain MSLN-targeted CAR, or the CAR mRNA can be transiently electroporated into T cells for expression (Y. Zhao et al., 2010). The generated MSLN-directed CAR-T cells can be administered systemically or locally and show great potency for tumor treatment *in vivo* (Castelletti, Yeo, van Zandwijk, & Rasko, 2021). Anti-MSLN CAR-T cells have shown effectiveness against ovarian, colorectal, and breast cancers *in vivo* (Zhang et al., 2021). It was demonstrated that humanized YP218 antibody (hYP218)-based CAR-T cells exhibit excellent antitumor potency in gastric cancer and ovarian cancer models that express MSLN. The cell toxicity is higher than 60% when using an Effector:Target ratio of 1:1. Different from SS1, which recognizes the membrane-distal region (Region I) of MSLN, hYP218 binds to the membrane proximal region (Region III). Compared to SS1-based CAR-T cells, hYP218-based CAR-T cells induce expression of higher levels of cytokines (IL2, IFN-γ, and TNF-α) and show much stronger cell toxicity against the MSLN-positive tumor models both *in vitro* and *in vivo*. Notably, the hYP218 CAR-T cells not only targeted ovarian tumors at early stages but also significantly inhibited large tumors *in vivo* (Zhang et al., 2020). Recent studies also showed that disruption of PD-1 facilitates MSLN-targeted CAR-T treatment (Cherkassky et al., 2016), thereby further aiding in the efficacy of CAR-T therapy. Several related therapies have entered clinical trials (Lv & Li, 2019). This progress has provided a promising direction for MSLN-directed tumor treatment.

Collectively, MSLN is a therapeutic target in cholangiocarcinoma and is closely correlated with disease progression. MSLN-directed therapies have a great potential for clinical treatment of cholangiocarcinoma.

5. Conclusion and future perspectives

In summary, the field of liver cancer immunotherapy has never been as inspiring as it is now. There are, however, some major challenges including tumor heterogeneity, immunosuppressive TME and target selection. Cancer gradually becomes more heterogeneous throughout disease progression or target heterogeneity (TH), likely resulting in non-uniform distribution of genetic or epigenetic alterations between cancer cells within the same tumor nodule (intratumor heterogeneity) or between different tumor sites within the same patient (intertumor heterogeneity) (Losic et al., 2020). HCC is characterized by high heterogeneity, and this heterogeneity has emerged as the most common cause of "treatment resistance" to current therapeutic agents. Using IHC technologies, we and others have demonstrated a significant intratumor heterogeneity in the expression behavior of both GPC3 and MSLN in liver cancer tissues (Anatelli, Chuang, Yang, & Wang, 2008; Ho, 2011; Li et al., 2020; Moentenich et al., 2020; Yu et al., 2010), In solid tumors such as PLCs, the CAR T cells must overcome the TME to reach the tumor, recognize TSA or TAA and infiltrate tumor, perform their effector function under pressure of negative Treg, and then differentiate and persist as memory T cells that provide long-term protection. Multiple TME factors may hinder the effect of CAR T treatment. Thus, a better understanding of the tumor/target heterogenicity and suppressive TME will shape the design of new agents and combinatorial therapies including antibody and cell-based therapies, checkpoint blockade, oncolytic viruses, and radiotherapy for treating liver cancer.

Acknowledgments

We thank NIH Fellows Editorial Board for editorial assistance. Cartoons in Figs. 1, 2B, 3A and C were created with BioRender.com. The content of this publication does not necessarily reflect the views or policies of the Department of Health and Human Services, nor does mention of trade names, commercial products, or organizations imply endorsement by the U.S. Government.

Conflict of interest statement

The authors declare no competing interests.

Grant support
This work was supported by the Intramural Research Program of NIH, Center for Cancer Research (CCR), National Cancer Institute (NCI) (Z01 BC010891 and ZIA BC010891 to M.H.).

References
Abou-Alfa, G. K., Meyer, T., Cheng, A. L., El-Khoueiry, A. B., Rimassa, L., Ryoo, B. Y., et al. (2018). Cabozantinib in patients with advanced and progressing hepatocellular carcinoma. *New England Journal of Medicine, 379*(1), 54–63. https://doi.org/10.1056/NEJMoa1717002.

Anatelli, F., Chuang, S. T., Yang, X. J., & Wang, H. L. (2008). Value of glypican 3 immunostaining in the diagnosis of hepatocellular carcinoma on needle biopsy. *American Journal of Clinical Pathology, 130*(2), 219–223. https://doi.org/10.1309/WMB5PX57Y4P8QCTY.

Awad, W., Adamczyk, B., Ornros, J., Karlsson, N. G., Mani, K., & Logan, D. T. (2015). Structural aspects of N-glycosylations and the C-terminal region in human Glypican-1. *Journal of Biological Chemistry, 290*(38), 22991–23008. https://doi.org/10.1074/jbc.M115.660878.

Banales, J. M., Marin, J. J. G., Lamarca, A., Rodrigues, P. M., Khan, S. A., Roberts, L. R., et al. (2020). Cholangiocarcinoma 2020: The next horizon in mechanisms and management. *Nature Reviews. Gastroenterology & Hepatology, 17*(9), 557–588. https://doi.org/10.1038/s41575-020-0310-z.

Bassani-Sternberg, M., Braunlein, E., Klar, R., Engleitner, T., Sinitcyn, P., Audehm, S., et al. (2016). Direct identification of clinically relevant neoepitopes presented on native human melanoma tissue by mass spectrometry. *Nature Communications, 7*, 13404. https://doi.org/10.1038/ncomms13404.

Batra, S. A., Rathi, P., Guo, L., Courtney, A. N., Fleurence, J., Balzeau, J., et al. (2020). Glypican-3-specific CAR T cells coexpressing IL15 and IL21 have superior expansion and antitumor activity against hepatocellular carcinoma. *Cancer Immunology Research, 8*(3), 309–320. https://doi.org/10.1158/2326-6066.CIR-19-0293.

Bharadwaj, U., Li, M., Chen, C., & Yao, Q. (2008). Mesothelin-induced pancreatic cancer cell proliferation involves alteration of cyclin E via activation of signal transducer and activator of transcription protein 3. *Molecular Cancer Research, 6*(11), 1755–1765. https://doi.org/10.1158/1541-7786.mcr-08-0095.

Bharadwaj, U., Marin-Muller, C., Li, M., Chen, C., & Yao, Q. (2011a). Mesothelin confers pancreatic cancer cell resistance to TNF-alpha-induced apoptosis through Akt/PI3K/NF-kappaB activation and IL-6/Mcl-1 overexpression. *Molecular Cancer, 10*, 106. https://doi.org/10.1186/1476-4598-10-106.

Bharadwaj, U., Marin-Muller, C., Li, M., Chen, C., & Yao, Q. (2011b). Mesothelin overexpression promotes autocrine IL-6/sIL-6R trans-signaling to stimulate pancreatic cancer cell proliferation. *Carcinogenesis, 32*(7), 1013–1024. https://doi.org/10.1093/carcin/bgr075.

Borad, M. J., Gores, G. J., & Roberts, L. R. (2015). Fibroblast growth factor receptor 2 fusions as a target for treating cholangiocarcinoma. *Current Opinion in Gastroenterology, 31*(3), 264–268. https://doi.org/10.1097/MOG.0000000000000171.

Butte, M. J., Keir, M. E., Phamduy, T. B., Sharpe, A. H., & Freeman, G. J. (2007). Programmed death-1 ligand 1 interacts specifically with the B7-1 costimulatory molecule to inhibit T cell responses. *Immunity, 27*(1), 111–122. https://doi.org/10.1016/j.immuni.2007.05.016.

Capurro, M. I., Li, F., & Filmus, J. (2009). Overgrowth of a mouse model of Simpson-Golabi-Behmel syndrome is partly mediated by Indian hedgehog. *EMBO Reports*, *10*(8), 901–907. https://doi.org/10.1038/embor.2009.98.

Capurro, M. I., Shi, W., & Filmus, J. (2012). LRP1 mediates hedgehog-induced endocytosis of the GPC3-hedgehog complex. *Journal of Cell Science*, *125*(Pt 14), 3380–3389. https://doi.org/10.1242/jcs.098889.

Capurro, M., Wanless, I. R., Sherman, M., Deboer, G., Shi, W., Miyoshi, E., et al. (2003). Glypican-3: A novel serum and histochemical marker for hepatocellular carcinoma. *Gastroenterology*, *125*(1), 89–97. https://doi.org/10.1016/s0016-5085(03)00689-9.

Capurro, M. I., Xiang, Y. Y., Lobe, C., & Filmus, J. (2005). Glypican-3 promotes the growth of hepatocellular carcinoma by stimulating canonical Wnt signaling. *Cancer Research*, *65*(14), 6245–6254. https://doi.org/10.1158/0008-5472.CAN-04-4244.

Capurro, M. I., Xu, P., Shi, W., Li, F., Jia, A., & Filmus, J. (2008). Glypican-3 inhibits hedgehog signaling during development by competing with patched for hedgehog binding. *Developmental Cell*, *14*(5), 700–711. https://doi.org/10.1016/j.devcel.2008.03.006.

Castellarin, M., Watanabe, K., June, C. H., Kloss, C. C., & Posey, A. D., Jr. (2018). Driving cars to the clinic for solid tumors. *Gene Therapy*, *25*(3), 165–175. https://doi.org/10.1038/s41434-018-0007-x.

Castelletti, L., Yeo, D., van Zandwijk, N., & Rasko, J. E. J. (2021). Anti-mesothelin CAR T cell therapy for malignant mesothelioma. *Biomarker Research*, *9*(1), 11. https://doi.org/10.1186/s40364-021-00264-1.

Chaisaingmongkol, J., Budhu, A., Dang, H., Rabibhadana, S., Pupacdi, B., Kwon, S. M., et al. (2017). Common molecular subtypes among Asian hepatocellular carcinoma and cholangiocarcinoma. *Cancer Cell*, *32*(1), 57–70 e53. https://doi.org/10.1016/j.ccell.2017.05.009.

Chambers, C. A., Kuhns, M. S., Egen, J. G., & Allison, J. P. (2001). CTLA-4-mediated inhibition in regulation of T cell responses: Mechanisms and manipulation in tumor immunotherapy. *Annual Review of Immunology*, *19*, 565–594. https://doi.org/10.1146/annurev.immunol.19.1.565.

Chang, M. C., Chen, C. A., Hsieh, C. Y., Lee, C. N., Su, Y. N., Hu, Y. H., et al. (2009). Mesothelin inhibits paclitaxel-induced apoptosis through the PI3K pathway. *Biochemical Journal*, *424*(3), 449–458. https://doi.org/10.1042/BJ20082196.

Chang, K., & Pastan, I. (1994). Molecular cloning and expression of a cDNA encoding a protein detected by the K1 antibody from an ovarian carcinoma (OVCAR-3) cell line. *International Journal of Cancer*, *57*(1), 90–97. https://doi.org/10.1002/ijc.2910570117.

Chang, K., & Pastan, I. (1996). Molecular cloning of mesothelin, a differentiation antigen present on mesothelium, mesotheliomas, and ovarian cancers. *Proceedings of the National Academy of Sciences of the United States of America*, *93*(1), 136–140. https://doi.org/10.1073/pnas.93.1.136.

Chang, K., Pastan, I., & Willingham, M. C. (1992). Isolation and characterization of a monoclonal antibody, K1, reactive with ovarian cancers and normal mesothelium. *International Journal of Cancer*, *50*(3), 373–381. https://doi.org/10.1002/ijc.2910500308.

Chen, S. H., Hung, W. C., Wang, P., Paul, C., & Konstantopoulos, K. (2013). Mesothelin binding to CA125/MUC16 promotes pancreatic cancer cell motility and invasion via MMP-7 activation. *Scientific Reports*, *3*, 1870. https://doi.org/10.1038/srep01870.

Chen, X., Chen, Y., Liang, R., Xiang, L., Li, J., Zhu, Y., et al. (2022). Combination therapy of hepatocellular carcinoma by GPC3-targeted bispecific antibody and irinotecan is potent in suppressing tumor growth in mice. *Molecular Cancer Therapeutics*, *21*(1), 149–158. https://doi.org/10.1158/1535-7163.Mct-20-1025.

Cheng, R., Chen, Y., Zhou, H., Wang, B., Du, Q., & Chen, Y. (2018). B7-H3 expression and its correlation with clinicopathologic features, angiogenesis, and prognosis in intrahepatic cholangiocarcinoma. *Acta Pathologica, Microbiologica, et Immunologica Scandinavica, 126*(5), 396–402. https://doi.org/10.1111/apm.12837.

Cherkassky, L., Morello, A., Villena-Vargas, J., Feng, Y., Dimitrov, D. S., Jones, D. R., et al. (2016). Human CAR T cells with cell-intrinsic PD-1 checkpoint blockade resist tumor-mediated inhibition. *Journal of Clinical Investigation, 126*(8), 3130–3144. https://doi.org/10.1172/JCI83092.

Coelho, R., Ricardo, S., Amaral, A. L., Huang, Y. L., Nunes, M., Neves, J. P., et al. (2020). Regulation of invasion and peritoneal dissemination of ovarian cancer by mesothelin manipulation. *Oncogene, 9*(6), 61. https://doi.org/10.1038/s41389-020-00246-2.

De Cat, B., Muyldermans, S. Y., Coomans, C., Degeest, G., Vanderschueren, B., Creemers, J., et al. (2003). Processing by proprotein convertases is required for glypican-3 modulation of cell survival, Wnt signaling, and gastrulation movements. *Journal of Cell Biology, 163*(3), 625–635. https://doi.org/10.1083/jcb.200302152.

El-Serag, H. B., Marrero, J. A., Rudolph, L., & Reddy, K. R. (2008). Diagnosis and treatment of hepatocellular carcinoma. *Gastroenterology, 134*(6), 1752–1763. https://doi.org/10.1053/j.gastro.2008.02.090.

Feng, M., Gao, W., Wang, R., Chen, W., Man, Y.-G., Figg, W. D., et al. (2013). Therapeutically targeting glypican-3 via a conformation-specific single-domain antibody in hepatocellular carcinoma. *Proceedings of the National Academy of Sciences, 110*(12), E1083–E1091. https://doi.org/10.1073/pnas.1217868110.

Finn, R. S., Merle, P., Granito, A., Huang, Y. H., Bodoky, G., Pracht, M., et al. (2018). Outcomes of sequential treatment with sorafenib followed by regorafenib for HCC: Additional analyses from the phase III RESORCE trial. *Journal of Hepatology, 69*(2), 353–358. https://doi.org/10.1016/j.jhep.2018.04.010.

Finn, O. J., & Rammensee, H. G. (2018). Is it possible to develop Cancer vaccines to neoantigens, what are the major challenges, and how can these be overcome? Neoantigens: Nothing new in spite of the name. *Cold Spring Harbor Perspectives in Biology, 10*(11). https://doi.org/10.1101/cshperspect.a028829.

Fleming, B. D., Urban, D. J., Hall, M., Longerich, T., Greten, T., Pastan, I., et al. (2020). The engineered anti-GPC3 immunotoxin, HN3-ABD-T20, produces regression in mouse liver cancer xenografts via prolonged serum retention. *Hepatology, 71*(5), 1696–1711. https://doi.org/10.1002/hep.30949.

Fontugne, J., Augustin, J., Pujals, A., Compagnon, P., Rousseau, B., Luciani, A., et al. (2017). PD-L1 expression in perihilar and intrahepatic cholangiocarcinoma. *Oncotarget, 8*(15), 24644–24651. https://doi.org/10.18632/oncotarget.15602.

Fu, Y., Urban, D. J., Nani, R. R., Zhang, Y.-F., Li, N., Fu, H., et al. (2018). Glypican-3-specific antibody drug conjugates targeting hepatocellular carcinoma. *Hepatology*. https://doi.org/10.1002/hep.30326.

Gao, W., & Ho, M. (2011). The role of glypican-3 in regulating Wnt in hepatocellular carcinomas. *Cancer Reports, 1*(1), 14–19. Retrieved from http://www.ncbi.nlm.nih.gov/pubmed/22563565.

Gao, W., Kim, H., Feng, M., Phung, Y., Xavier, C. P., Rubin, J. S., et al. (2014). Inactivation of Wnt signaling by a human antibody that recognizes the heparan sulfate chains of glypican-3 for liver cancer therapy. *Hepatology, 60*(2), 576–587. https://doi.org/10.1002/hep.26996.

Gao, H., Li, K., Tu, H., Pan, X., Jiang, H., Shi, B., et al. (2014). Development of T cells redirected to glypican-3 for the treatment of hepatocellular carcinoma. *Clinical Cancer Research, 20*(24), 6418–6428. https://doi.org/10.1158/1078-0432.CCR-14-1170.

Gao, W., Tang, Z., Zhang, Y.-F., Feng, M., Qian, M., Dimitrov, D. S., et al. (2015). Immunotoxin targeting glypican-3 regresses liver cancer via dual inhibition of Wnt signalling and protein synthesis. *Nature Communications, 6*, 6536. https://doi.org/10.1038/ncomms7536.

Gao, W., Xu, Y., Liu, J., & Ho, M. (2016). Epitope mapping by a Wnt-blocking antibody: Evidence of the Wnt binding domain in heparan sulfate. *Scientific Reports, 6*, 26245. https://doi.org/10.1038/srep26245.

Gauthier, A., & Ho, M. (2013). Role of sorafenib in the treatment of advanced hepatocellular carcinoma: An update. *Hepatology Research, 43*(2), 147–154. https://doi.org/10.1111/j.1872-034X.2012.01113.x.

Greten, T. F., Forner, A., Korangy, F., N'Kontchou, G., Barget, N., Ayuso, C., et al. (2010). A phase II open label trial evaluating safety and efficacy of a telomerase peptide vaccination in patients with advanced hepatocellular carcinoma. *BioMed Central Cancer, 10*, 209. https://doi.org/10.1186/1471-2407-10-209.

Greten, T. F., & Sangro, B. (2018). Targets for immunotherapy of liver cancer. *Journal of Hepatology, 68*, 157–166. https://doi.org/10.1016/j.jhep.2017.09.007.

Gubbels, J. A. A., Belisle, J., Onda, M., Rancourt, C., Migneault, M., Ho, M., et al. (2006). Mesothelin-MUC16 binding is a high affinity, N-glycan dependent interaction that facilitates peritoneal metastasis of ovarian tumors. *Molecular Cancer, 5*, 50. https://doi.org/10.1186/1476-4598-5-50. Artn 50.

Guo, M., Zhang, H., Zheng, J., & Liu, Y. (2020). Glypican-3: A new target for diagnosis and treatment of hepatocellular carcinoma. *Journal of Cancer, 11*(8), 2008–2021. https://doi.org/10.7150/jca.39972.

Hassan, R., & Ho, M. (2008). Mesothelin targeted cancer immunotherapy. *European Journal of Cancer, 44*(1), 46–53. https://doi.org/10.1016/j.ejca.2007.08.028.

He, X., Semenov, M., Tamai, K., & Zeng, X. (2004). LDL receptor-related proteins 5 and 6 in Wnt/beta-catenin signaling: Arrows point the way. *Development, 131*(8), 1663–1677. https://doi.org/10.1242/dev.01117.

Ho, M. (2011). Advances in liver cancer antibody therapies: A focus on glypican-3 and mesothelin. *BioDrugs, 25*(5), 275–284. https://doi.org/10.2165/11595360-000000000-00000.

Ho, M., Bera, T. K., Willingham, M. C., Onda, M., Hassan, R., FitzGerald, D., et al. (2007). Mesothelin expression in human lung cancer. *Clinical Cancer Research, 13*(5), 1571–1575. https://doi.org/10.1158/1078-0432.CCR-06-2161.

Ho, M., Feng, M., Fisher, R. J., Rader, C., & Pastan, I. (2011). A novel high-affinity human monoclonal antibody to mesothelin. *International Journal of Cancer, 128*(9), 2020–2030. https://doi.org/10.1002/ijc.25557.

Ho, M., Hassan, R., Zhang, J., Wang, Q. C., Onda, M., Bera, T., et al. (2005). Humoral immune response to mesothelin in mesothelioma and ovarian cancer patients. *Clinical Cancer Research, 11*(10), 3814–3820. https://doi.org/10.1158/1078-0432.Ccr-04-2304.

Ho, M., & Kim, H. (2011). Glypican-3: A new target for cancer immunotherapy. *European Journal of Cancer, 47*(3), 333–338. https://doi.org/10.1016/j.ejca.2010.10.024.

Ho, M., Onda, M., Wang, Q. C., Hassan, R., Pastan, I., & Lively, M. O. (2006). Mesothelin is shed from tumor cells. *Cancer Epidemiology Biomarkers & Prevention, 15*(9), 1751. https://doi.org/10.1158/1055-9965.Epi-06-0479.

Hsu, H. C., Cheng, W., & Lai, P. L. (1997). Cloning and expression of a developmentally regulated transcript MXR7 in hepatocellular carcinoma: Biological significance and temporospatial distribution. *Cancer Research, 57*(22), 5179–5184. Retrieved from https://www.ncbi.nlm.nih.gov/pubmed/9371521.

Ishiguro, T., Sano, Y., Komatsu, S. I., Kamata-Sakurai, M., Kaneko, A., Kinoshita, Y., et al. (2017). An anti-glypican 3/CD3 bispecific T cell-redirecting antibody for treatment of solid tumors. *Science Translational Medicine, 9*(410), 4291. https://doi.org/10.1126/scitranslmed.aal4291.

Jiang, Z., Jiang, X., Chen, S., Lai, Y., Wei, X., Li, B., et al. (2016). Anti-GPC3-CAR T cells suppress the growth of tumor cells in patient-derived xenografts of hepatocellular carcinoma. *Frontiers in Immunology*, 7, 690. https://doi.org/10.3389/fimmu.2016.00690.

Johnson, R., & Halder, G. (2014). The two faces of hippo: Targeting the hippo pathway for regenerative medicine and cancer treatment. *Nature Reviews. Drug Discovery*, 13(1), 63–79. https://doi.org/10.1038/nrd4161.

Jumper, J., Evans, R., Pritzel, A., Green, T., Figurnov, M., Ronneberger, O., et al. (2021). Highly accurate protein structure prediction with AlphaFold. *Nature*, 596(7873), 583–589. https://doi.org/10.1038/s41586-021-03819-2.

Kaeding, A. J., Barwe, S. P., Gopalakrishnapillai, A., Ries, R. E., Alonzo, T. A., Gerbing, R. B., et al. (2021). Mesothelin is a novel cell surface disease marker and potential therapeutic target in acute myeloid leukemia. *Blood Advances*, 5(9), 2350–2361. https://doi.org/10.1182/bloodadvances.2021004424.

Kaneko, O., Gong, L., Zhang, J., Hansen, J. K., Hassan, R., Lee, B., et al. (2009). A binding domain on mesothelin for CA125/MUC16. *Journal of Biological Chemistry*, 284(6), 3739–3749. https://doi.org/10.1074/jbc.M806776200.

Kim, H., & Ho, M. (2018). Isolation of antibodies to heparan sulfate on glypicans by phage display. *Current Protocols in Protein Science*, 94(1). https://doi.org/10.1002/cpps.66, e66.

Kim, M. S., Saunders, A. M., Hamaoka, B. Y., Beachy, P. A., & Leahy, D. J. (2011). Structure of the protein core of the glypican dally-like and localization of a region important for hedgehog signaling. *Proceedings of the National Academy of Sciences of the United States of America*, 108(32), 13112–13117. https://doi.org/10.1073/pnas.1109877108.

Kojima, T., Oh-eda, M., Hattori, K., Taniguchi, Y., Tamura, M., Ochi, N., et al. (1995). Molecular cloning and expression of megakaryocyte potentiating factor cDNA. *Journal of Biological Chemistry*, 270(37), 21984–21990. https://doi.org/10.1074/jbc.270.37.21984.

Kolluri, A., & Ho, M. (2019). The role of Glypican-3 in regulating Wnt, YAP, and hedgehog in liver cancer. *Frontiers in Oncology*, 9, 708. https://doi.org/10.3389/fonc.2019.00708.

Koyama, Y., Wang, P., Liang, S., Iwaisako, K., Liu, X., Xu, J., et al. (2017). Mesothelin/mucin 16 signaling in activated portal fibroblasts regulates cholestatic liver fibrosis. *Journal of Clinical Investigation*, 127(4), 1254–1270. https://doi.org/10.1172/JCI88845.

Kudo, M., Finn, R. S., Qin, S., Han, K. H., Ikeda, K., Piscaglia, F., et al. (2018). Lenvatinib versus sorafenib in first-line treatment of patients with unresectable hepatocellular carcinoma: A randomised phase 3 non-inferiority trial. *Lancet*, 391(10126), 1163–1173. https://doi.org/10.1016/S0140-6736(18)30207-1.

Lee, J. H., & Lee, S. W. (2017). The roles of carcinoembryonic antigen in liver metastasis and therapeutic approaches. *Gastroenterology Research and Practice*, 2017, 7521987. https://doi.org/10.1155/2017/7521987.

Li, D., English, H., Hong, J., Liang, T., Merlino, G., Day, C.-P., et al. (2021). A novel PD-L1-targeted shark VNAR single domain-based CAR-T strategy for treating breast cancer and liver cancer. *bioRxiv*. https://doi.org/10.1101/2021.07.20.453144. 2021.2007.2020.453144.

Li, N., Gao, W., Zhang, Y.-F., & Ho, M. (2018). Glypicans as cancer therapeutic targets. *Trends in Cancer*, 4(11), 741–754. https://doi.org/10.1016/j.trecan.2018.09.004.

Li, D., Li, N., Zhang, Y.-F., Fu, H., Feng, M., Schneider, D., et al. (2020). Persistent polyfunctional chimeric antigen receptor T cells that target Glypican 3 eliminate orthotopic hepatocellular carcinomas in mice. *Gastroenterology*, 158(8), 2250–2265. https://doi.org/10.1053/j.gastro.2020.02.011.

Li, K., Qian, S., Huang, M., Chen, M., Peng, L., Liu, J., et al. (2021). Development of GPC3 and EGFR-dual-targeting chimeric antigen receptor-T cells for adoptive T cell therapy. *American Journal of Translational Research*, 13(1), 156–167. Retrieved from https://www.ncbi.nlm.nih.gov/pubmed/33527015.

Li, N., Spetz, M. R., & Ho, M. (2020). The role of glypicans in cancer progression and therapy. *Journal of Histochemistry & Cytochemistry, 68*(12), 841–862. https://doi.org/10.1369/0022155420933709.

Li, N., Wei, L., Liu, X., Bai, H., Ye, Y., Li, D., et al. (2019). A frizzled-like cysteine rich domain in glypican-3 mediates Wnt binding and regulates hepatocellular carcinoma tumor growth in mice. *Hepatology, 70*(4), 1231–1245. https://doi.org/10.1002/hep.30646.

Liu, X., Chan, A., Tai, C. H., Andresson, T., & Pastan, I. (2020). Multiple proteases are involved in mesothelin shedding by cancer cells. *Communications Biology, 3*(1), 728. https://doi.org/10.1038/s42003-020-01464-5.

Liu, Y., Wang, X., & Yang, Y. (2020). Hepatic hippo signaling inhibits development of hepatocellular carcinoma. *Clinical and Molecular Hepatology, 26*(4), 742–750. https://doi.org/10.3350/cmh.2020.0178.

Liu, H., Xu, Y., Xiang, J., Long, L., Green, S., Yang, Z., et al. (2017). Targeting alpha-fetoprotein (AFP)-MHC complex with CAR T-cell therapy for liver cancer. *Clinical Cancer Research, 23*(2), 478–488. https://doi.org/10.1158/1078-0432.CCR-16-1203.

Losic, B., Craig, A. J., Villacorta-Martin, C., Martins-Filho, S. N., Akers, N., Chen, X., et al. (2020). Intratumoral heterogeneity and clonal evolution in liver cancer. *Nature Communications, 11*(1), 291. https://doi.org/10.1038/s41467-019-14050-z.

Lu, L., Jiang, J., Zhan, M., Zhang, H., Wang, Q. T., Sun, S. N., et al. (2021). Targeting tumor-associated antigens in hepatocellular carcinoma for immunotherapy: Past pitfalls and future strategies. *Hepatology, 73*(2), 821–832. https://doi.org/10.1002/hep.31502.

Lv, J., & Li, P. (2019). Mesothelin as a biomarker for targeted therapy. *Biomarker Research, 7*(1), 18. https://doi.org/10.1186/s40364-019-0169-8.

Ma, L., Hernandez, M. O., Zhao, Y., Mehta, M., Tran, B., Kelly, M., et al. (2019). Tumor cell biodiversity drives microenvironmental reprogramming in liver Cancer. *Cancer Cell, 36*(4). https://doi.org/10.1016/j.ccell.2019.08.007. 418–430.e416.

Manzanares, M. A., Campbell, D. J. W., Maldonado, G. T., & Sirica, A. E. (2018). Overexpression of periostin and distinct mesothelin forms predict malignant progression in a rat cholangiocarcinoma model. *Hepatology Communications, 2*(2), 155–172. https://doi.org/10.1002/hep4.1131.

Martin, D., Degese, M. S., Vitale-Cross, L., Iglesias-Bartolome, R., Valera, J. L. C., Wang, Z., et al. (2018). Assembly and activation of the hippo signalome by FAT1 tumor suppressor. *Nature Communications, 9*(1), 2372. https://doi.org/10.1038/s41467-018-04590-1.

McGrath, N. A., Fu, J., Gu, S. Z., & Xie, C. (2020). Targeting cancer stem cells in cholangiocarcinoma (review). *International Journal of Oncology, 57*(2), 397–408. https://doi.org/10.3892/ijo.2020.5074.

Meng, P., Zhang, Y. F., Zhang, W., Chen, X., Xu, T., Hu, S., et al. (2021). Identification of the atypical cadherin FAT1 as a novel glypican-3 interacting protein in liver cancer cells. *Scientific Reports, 11*(1), 40. https://doi.org/10.1038/s41598-020-79524-3.

Miyamoto, M., Ojima, H., Iwasaki, M., Shimizu, H., Kokubu, A., Hiraoka, N., et al. (2011). Prognostic significance of overexpression of c-met oncoprotein in cholangiocarcinoma. *British Journal of Cancer, 105*(1), 131–138. https://doi.org/10.1038/bjc.2011.199.

Moentenich, V., Comut, E., Gebauer, F., Tuchscherer, A., Bruns, C., Schroeder, W., et al. (2020). Mesothelin expression in esophageal adenocarcinoma and squamous cell carcinoma and its possible impact on future treatment strategies. *Therapeutic Advances in Medical Oncology, 12*. https://doi.org/10.1177/1758835920917571, 1758835920917571.

Montalbano, M., Georgiadis, J., Masterson, A. L., McGuire, J. T., Prajapati, J., Shirafkan, A., et al. (2017). Biology and function of glypican-3 as a candidate for early cancerous transformation of hepatocytes in hepatocellular carcinoma (review). *Oncology Reports, 37*(3), 1291–1300. https://doi.org/10.3892/or.2017.5387.

Nakano, K., Orita, T., Nezu, J., Yoshino, T., Ohizumi, I., Sugimoto, M., et al. (2009). Anti-glypican 3 antibodies cause ADCC against human hepatocellular carcinoma cells. *Biochemical and Biophysical Research Communications, 378*(2), 279–284. https://doi.org/10.1016/j.bbrc.2008.11.033.

Nomura, R., Fujii, H., Abe, M., Sugo, H., Ishizaki, Y., Kawasaki, S., et al. (2013). Mesothelin expression is a prognostic factor in cholangiocellular carcinoma. *International Surgery, 98*(2), 164–169. https://doi.org/10.9738/Intsurg-D-13-00001.1.

Parinyanitikul, N., Blumenschein, G. R., Wu, Y., Lei, X., Chavez-Macgregor, M., Smart, M., et al. (2013). Mesothelin expression and survival outcomes in triple receptor negative breast cancer. *Clinical Breast Cancer, 13*(5), 378–384. https://doi.org/10.1016/j.clbc.2013.05.001.

Pastan, I., & Hassan, R. (2014). Discovery of mesothelin and exploiting it as a target for immunotherapy. *Cancer Research, 74*(11), 2907–2912. https://doi.org/10.1158/0008-5472.CAN-14-0337.

Phung, Y., Gao, W., Man, Y.-G., Nagata, S., & Ho, M. (2012). High-affinity monoclonal antibodies to cell surface tumor antigen glypican-3 generated through a combination of peptide immunization and flow cytometry screening. *MAbs, 4*(5), 592–599. https://doi.org/10.4161/mabs.20933.

Prieve, M. G., & Moon, R. T. (2003). Stromelysin-1 and mesothelin are differentially regulated by Wnt-5a and Wnt-1 in C57mg mouse mammary epithelial cells. *BMC Developmental Biology, 3*, 2.

Rump, A., Morikawa, Y., Tanaka, M., Minami, S., Umesaki, N., Takeuchi, M., et al. (2004). Binding of ovarian cancer antigen CA125/MUC16 to mesothelin mediates cell adhesion. *Journal of Biological Chemistry, 279*(10), 9190–9198. https://doi.org/10.1074/jbc.M312372200.

Sangro, B., Sarobe, P., Hervas-Stubbs, S., & Melero, I. (2021). Advances in immunotherapy for hepatocellular carcinoma. *Nature Reviews. Gastroenterology & Hepatology, 18*(8), 525–543. https://doi.org/10.1038/s41575-021-00438-0.

Sarobe, P., Huarte, E., Lasarte, J. J., & Borras-Cuesta, F. (2004). Carcinoembryonic antigen as a target to induce anti-tumor immune responses. *Current Cancer Drug Targets, 4*(5), 443–454. https://doi.org/10.2174/1568009043332916.

Shi, D., Shi, Y., Kaseb, A. O., Qi, X., Zhang, Y., Chi, J., et al. (2020). Chimeric antigen receptor-Glypican-3 T-cell therapy for advanced hepatocellular carcinoma: Results of phase I trials. *Clinical Cancer Research, 26*(15), 3979–3989. https://doi.org/10.1158/1078-0432.CCR-19-3259.

Shimada, M., Sugimoto, K., Iwahashi, S., Utsunomiya, T., Morine, Y., Imura, S., et al. (2010). CD133 expression is a potential prognostic indicator in intrahepatic cholangiocarcinoma. *Journal of Gastroenterology, 45*(8), 896–902. https://doi.org/10.1007/s00535-010-0235-3.

Shiraishi, T., Shinto, E., Mochizuki, S., Tsuda, H., Kajiwara, Y., Okamoto, K., et al. (2019). Mesothelin expression has prognostic value in stage IotaIota/IotaIotaIota colorectal cancer. *Virchows Archiv, 474*(3), 297–307. https://doi.org/10.1007/s00428-018-02514-4.

Sia, D., Villanueva, A., Friedman, S. L., & Llovet, J. M. (2017). Liver cancer cell of origin, molecular class, and effects on patient prognosis. *Gastroenterology, 152*(4), 745–761. https://doi.org/10.1053/j.gastro.2016.11.048.

Simpson, A. J., Caballero, O. L., Jungbluth, A., Chen, Y. T., & Old, L. J. (2005). Cancer/testis antigens, gametogenesis and cancer. *Nature Reviews. Cancer*, *5*(8), 615–625. https://doi.org/10.1038/nrc1669.

Sulpice, L., Rayar, M., Turlin, B., Boucher, E., Bellaud, P., Desille, M., et al. (2014). Epithelial cell adhesion molecule is a prognosis marker for intrahepatic cholangiocarcinoma. *Journal of Surgical Research*, *192*(1), 117–123. https://doi.org/10.1016/j.jss.2014.05.017.

Sun, T. W., Gao, Q., Qiu, S. J., Zhou, J., Wang, X. Y., Yi, Y., et al. (2012). B7-H3 is expressed in human hepatocellular carcinoma and is associated with tumor aggressiveness and postoperative recurrence. *Cancer Immunology, Immunotherapy*, *61*(11), 2171–2182. https://doi.org/10.1007/s00262-012-1278-5.

Sung, H., Ferlay, J., Siegel, R. L., Laversanne, M., Soerjomataram, I., Jemal, A., et al. (2021). Global cancer statistics 2020: GLOBOCAN estimates of incidence and mortality worldwide for 36 cancers in 185 countries. *A Cancer Journal for Clinicians*, *71*(3), 209–249. https://doi.org/10.3322/caac.21660.

Svensson, G., Awad, W., Hakansson, M., Mani, K., & Logan, D. T. (2012). Crystal structure of N-glycosylated human glypican-1 core protein: Structure of two loops evolutionarily conserved in vertebrate glypican-1. *Journal of Biological Chemistry*, *287*(17), 14040–14051. https://doi.org/10.1074/jbc.M111.322487.

Takai, H., Kato, A., Kinoshita, Y., Ishiguro, T., Takai, Y., Ohtani, Y., et al. (2009). Histopathological analyses of the antitumor activity of anti-glypican-3 antibody (GC33) in human liver cancer xenograft models: The contribution of macrophages. *Cancer Biology & Therapy*, *8*(10), 930–938.

Tang, Z., Qian, M., & Ho, M. (2013). The role of mesothelin in tumor progression and targeted therapy. *Anti-Cancer Agents in Medicinal Chemistry*, *13*(2), 276–280. https://doi.org/10.2174/1871520611313020014.

Thompson, M. D., & Monga, S. P. (2007). WNT/beta-catenin signaling in liver health and disease. *Hepatology*, *45*(5), 1298–1305. https://doi.org/10.1002/hep.21651.

Uehara, N., Matsuoka, Y., & Tsubura, A. (2008). Mesothelin promotes anchorage-independent growth and prevents anoikis via extracellular signal-regulated kinase signaling pathway in human breast cancer cells. *Molecular Cancer Research*, *6*(2), 186–193. https://doi.org/10.1158/1541-7786.MCR-07-0254.

Vogel, A., Cervantes, A., Chau, I., Daniele, B., Llovet, J. M., Meyer, T., et al. (2019). Hepatocellular carcinoma: ESMO clinical practice guidelines for diagnosis, treatment and follow-up. *Annals of Oncology*, *30*(5), 871–873. https://doi.org/10.1093/annonc/mdy510.

Wang, C., Gao, W., Feng, M., Pastan, I., & Ho, M. (2017). Construction of an immunotoxin, HN3-mPE24, targeting glypican-3 for liver cancer therapy. *Oncotarget*, *8*(20), 32450–32460. https://doi.org/10.18632/oncotarget.10592.

Wu, X., Luo, H., Shi, B., Di, S., Sun, R., Su, J., et al. (2019). Combined antitumor effects of sorafenib and GPC3-CAR T cells in mouse models of hepatocellular carcinoma. *Molecular Therapy*, *27*(8), 1483–1494. https://doi.org/10.1016/j.ymthe.2019.04.020.

Wykosky, J., Fenton, T., Furnari, F., & Cavenee, W. K. (2011). Therapeutic targeting of epidermal growth factor receptor in human cancer: Successes and limitations. *Chinese Journal of Cancer*, *30*(1), 5–12. https://doi.org/10.5732/cjc.010.10542.

Xiang, X., Feng, M., Felder, M., Connor, J. P., Man, Y. G., Patankar, M. S., et al. (2011). HN125: A novel immunoadhesin targeting MUC16 with potential for cancer therapy. *Journal of Cancer*, *2*, 280–291.

Yang, J. D., Hainaut, P., Gores, G. J., Amadou, A., Plymoth, A., & Roberts, L. R. (2019). A global view of hepatocellular carcinoma: Trends, risk, prevention and management. *Nature Reviews Gastroenterology & Hepatology*, *16*(10), 589–604. https://doi.org/10.1038/s41575-019-0186-y.

Ying, F., Daniel, J. U., Roger, R. N., Yi-Fan, Z., Nan, L, Haiying, Fu, et al. (2019). Glypican-3-specific antibody drug conjugates targeting hepatocellular carcinoma. *Hepatology*, 70(2), 563–576. https://doi.org/10.1002/hep.30326.

Yokota, K., Serada, S., Tsujii, S., Toya, K., Takahashi, T., Matsunaga, T., et al. (2021). Anti-Glypican-1 antibody-drug conjugate as potential therapy against tumor cells and tumor vasculature for Glypican-1-positive cholangiocarcinoma. *Molecular Cancer Therapeutics*, 20(9), 1713–1722. https://doi.org/10.1158/1535-7163.MCT-21-0015.

Yu, L., Feng, M., Kim, H., Phung, Y., Kleiner, D. E., Gores, G. J., et al. (2010). Mesothelin as a potential therapeutic target in human cholangiocarcinoma. *Journal of Cancer*, 1, 141–149. https://doi.org/10.7150/jca.1.141.

Yu, M., Luo, H., Fan, M., Wu, X., Shi, B., Di, S., et al. (2018). Development of GPC3-specific chimeric antigen receptor-engineered natural killer cells for the treatment of hepatocellular carcinoma. *Molecular Therapy*, 26(2), 366–378. https://doi.org/10.1016/j.ymthe.2017.12.012.

Yuan, S. F., Li, K. Z., Wang, L., Dou, K. F., Yan, Z., Han, W., et al. (2005). Expression of MUC1 and its significance in hepatocellular and cholangiocarcinoma tissue. *World Journal of Gastroenterology*, 11(30), 4661–4666. https://doi.org/10.3748/wjg.v11.i30.4661.

Zhang, Y., Chertov, O., Zhang, J., Hassan, R., & Pastan, I. (2011). Cytotoxic activity of immunotoxin SS1P is modulated by TACE-dependent mesothelin shedding. *Cancer Research*, 71(17), 5915–5922. https://doi.org/10.1158/0008-5472.CAN-11-0466.

Zhang, Y.-F., & Ho, M. (2017). Humanization of rabbit monoclonal antibodies via grafting combined Kabat/IMGT/Paratome complementarity-determining regions: Rationale and examples. *MAbs*, 9(3), 419–429. https://doi.org/10.1080/19420862.2017.1289302.

Zhang, Z., Jiang, D., Yang, H., He, Z., Liu, X., Qin, W., et al. (2020). Correction: Modified CAR T cells targeting membrane-proximal epitope of mesothelin enhances the antitumor function against large solid tumor. *Cell Death & Disease*, 11(4), 235. https://doi.org/10.1038/s41419-020-2450-z.

Zhang, Q., Liu, G., Liu, J., Yang, M., Fu, J., Liu, G., et al. (2021). The antitumor capacity of mesothelin-CAR-T cells in targeting solid tumors in mice. *Molecular Therapy-Oncolytics*, 20, 556–568. https://doi.org/10.1016/j.omto.2021.02.013.

Zhang, Y. F., Phung, Y., Gao, W., Kawa, S., Hassan, R., Pastan, I., et al. (2015). New high affinity monoclonal antibodies recognize non-overlapping epitopes on mesothelin for monitoring and treating mesothelioma. *Scientific Reports*, 5, 9928. https://doi.org/10.1038/srep09928.

Zhao, Y., Moon, E., Carpenito, C., Paulos, C. M., Liu, X., Brennan, A. L., et al. (2010). Multiple injections of electroporated autologous T cells expressing a chimeric antigen receptor mediate regression of human disseminated tumor. *Cancer Research*, 70(22), 9053–9061. https://doi.org/10.1158/0008-5472.CAN-10-2880.

Zhao, Q., Yu, W. L., Lu, X. Y., Dong, H., Gu, Y. J., Sheng, X., et al. (2016). Combined hepatocellular and cholangiocarcinoma originating from the same clone: A pathomolecular evidence-based study. *Chinese Journal of Cancer*, 35(1), 82. https://doi.org/10.1186/s40880-016-0146-7.

Zheng, Y., Li, Y., Feng, J., Li, J., Ji, J., Wu, L., et al. (2021). Cellular based immunotherapy for primary liver cancer. *Journal of Experimental & Clinical Cancer Research*, 40(1), 250. https://doi.org/10.1186/s13046-021-02030-5.

Zhu, A. X., Gold, P. J., El-Khoueiry, A. B., Abrams, T. A., Morikawa, H., Ohishi, N., et al. (2013). First-in-man phase I study of GC33, a novel recombinant humanized antibody against glypican-3, in patients with advanced hepatocellular carcinoma. *Clinical Cancer Research*, 19(4), 920–928. https://doi.org/10.1158/1078-0432.CCR-12-2616.

Zucman-Rossi, J., Villanueva, A., Nault, J. C., & Llovet, J. M. (2015). Genetic landscape and biomarkers of hepatocellular carcinoma. *Gastroenterology*, 149(5). https://doi.org/10.1053/j.gastro.2015.05.061. 1226–1239.e1224.